INORGANIC MEMBRANES
FOR SEPARATION AND REACTION

Membrane Science and Technology Series

Membrane Science and Technology Series, 3

INORGANIC MEMBRANES
FOR SEPARATION AND REACTION

H.P. HSIEH

Alcoa Technical Center, 100 Technical Drive, Alcoa Center, PA 15069, USA

1996

ELSEVIER

Amsterdam – **Lausanne** – **New York** – **Oxford** – **Shannon** – **Tokyo**

ELSEVIER SCIENCE B.V.
Sara Burgerhartstraat 25
P.O. Box 211, 1000 AE Amsterdam, The Netherlands

ISBN 0-444-81677-1

This book is printed on acid-free paper.

Printed in The Netherlands

This book is dedicated to my grandfather who spared nothing for my education.

PREFACE

Books on science and technology of organic polymer membranes are abundant. But there has been only one book available today that exclusively addresses inorganic membranes and the book is an edited volume of individually contributed chapters. In addition, the potentially important subject of inorganic membrane reactors was only briefly covered. While an edited book provides useful information of the referenced technology, it generally lacks continuity in the treatment of the subject from chapter to chapter or from article to article. A fully integrated reference book with a more consistent and up-to-date perspective, better organized contents and easier cross referencing seems to be particularly opportune for the emerging field of inorganic membranes.

This book is, therefore, written to provide in one place the essential data and background materials on various aspects of inorganic membranes with a major coverage on inorganic membrane reactors. It is intended for the following technologists so they do not need to gather scattered information from the current and past literature: industrial as well as institutional researchers, application scientists and engineers with an interest in inorganic membranes as separators and/or reactors and students pursuing advanced separation and reaction studies.

After a brief introduction to some unique advantages of inorganic membranes in Chapter 1, an overview of historical developments of these special types of membranes is presented in Chapter 2. This is followed by discussions on an array of preparation methods (Chapter 3) that lead to various microstructural and separation properties of both dense and porous inorganic membranes. Different manufacturer- and end user-oriented test schemes and their measured application-generic characteristics and properties of the membranes are then described in Chapter 4. Commercial inorganic membranes and their general application considerations are introduced in Chapter 5. Traditional separation and unconventional uses of inorganic membranes treating liquid-phase and gas-phase mixtures are detailed in subsequent chapters (Chapter 6 and Chapter 7, respectively). Examples of production as well as test cases are used to illustrate the niche applications using the unique characteristics of inorganic membranes.

The challenging field of membrane reactor has emerged as a promising technology when inorganic membranes begin to rise to the horizon of commercial reality. Close to half of the book provides a state-of-the-art review on the science and technology of this conceptually new unit operation. High-temperature catalytic reaction applications where separation of reaction components is either beneficial or necessary are most amenable to inorganic membrane reactors. Application examples involving various reactions are discussed at some length in Chapter 8 with demonstrated enhancement in conversion or selectivity. The catalytic and material aspects of the inorganic membrane reactor technology are treated in detail in Chapter 9. Reaction engineering analysis through mathematical modeling is presented in Chapter 10 followed by discussions on the effects of various transport, reaction and design parameters in Chapter 11. Finally, those critical issues that need to be resolved before the technology becomes both technically feasible and economically viable are highlighted with a brief discussion of the future trends in Chapter 12. For the readers' convenience, a summary section is included at the end of each chapter.

Emphasis is placed on the key concepts, available data and their practical significance or implications. With the exception of Chapter 10 which is entirely devoted to mathematical modeling of inorganic membrane reactors, only essential mathematical equations or formulas to illustrate important points are presented.

The extensive literature data included in the book highlight research results published up to mid-1995. As the field of inorganic membranes as separators and reactors is evolving, various branches of the technology are all advancing at a fast pace. It is then understandable that a broad survey like this book cannot possibly cover every topic in equal depth. I fully recognize that unintentional omissions of some relevant studies are inevitable in view of the rapidly expanding volume of the literature.

I would like to express my deep appreciation to Dr. Jim F. Roth for his getting me interested in the writing of this book and Dr. Paul K. T. Liu for his generous supply of many useful references and countless discussions on several aspects of inorganic membranes including the joy and frustration of running a small membrane technology company. I am indebted to Ms. Marilyn A. Sharknis who has never failed to demonstrate her great efficiency and patience during the preparation of the book. Ms. Feirong Li's help in arranging to have many figures reproduced and Ms. Jackie Blackburn's assistance in getting permissions to use figures and tables are appreciated. Finally, this book would have been indefinitely delayed without the support of my wife, Meishiang, who constantly excused me from numerous social gatherings.

My special thanks also go to several staff members of Elsevier Science: Mr. Robert L. Goodman (retired), Mr. Adri van der Avoird, Mrs. Sabine Plantevin and Dr. Huub Manten-Werker for their patience and advice.

CONTENTS

CHAPTER 1

MEMBRANES AND MEMBRANE PROCESSES

In its earlier stages of development, membrane science and technology focused mostly on naturally occurring membranes despite evidences of studies of synthetic membranes in the mid-nineteenth century. This is not surprising in light of the fact that all life forms use natural membranes for separation of nutrients, selective protection from toxins, photosynthesis, etc. In the early to middle twentieth century, it began to evolve from a fairly narrow scientific discipline with limited practical applications to a broader field with diverging applications that support a unique industry which started in the late 1960s and early 1970s. Higher energy costs in recent decades have made the membrane processes even more economically competitive with conventional separation technologies such as distillation, crystallization, absorption, adsorption, solvent extraction, or cryogenics. An excellent overview on the historic growth of membrane technology has been published previously by one of the membrane pioneers, H. K. Lonsdale (1982).

1.1 MEMBRANE PROCESSES

A simplified working definition of a membrane can be conveniently stated as a semipermeable active or passive barrier which, under a certain driving force, permits preferential passage of one or more selected species or components (molecules, particles or polymers) of a gaseous and/or liquid mixture or solution (Figure 1.1).

The primary species rejected by the membrane is called retentate(s) or sometimes just "solute" while those species passing through the membranes is usually termed permeate(s) or sometimes "solvent." The driving force can exist in the form of pressure, concentration, or voltage difference across the membrane. Depending on the driving force and the physical sizes of the separated species, membrane processes are classified accordingly: microfiltration (MF), ultrafiltration (UF), reverse osmosis (RO), dialysis, electrodialysis (ED) and gas separation (Table 1.1). Their current status of the technology is summarized in Table 1.2 [Baker et al., 1991]. Any major and minor technical problems associated with each membrane process that need to be solved for further process improvement and applications development are listed along with the technical hurdles that are essentially overcome.

Membrane processes can be operated in two major modes according to the direction of the feed stream relative to the orientation of the membrane surface: dead-end filtration and crossflow filtration (Figure 1.1). The majority of the membrane separation applications use the concept of crossflow where the feed flows parallel to and past the membrane surface while the permeate penetrates through the membrane overall in a direction normal to the membrane. The shear force exerted by the flowing feed stream on the membrane surface help remove any stagnant and accumulated rejected species that may reduce permeation rate and increase the retentate concentration in the permeate. Predominant in the conventional filtration processes, dead-end filtration is used in membrane separation only in a few cases such as laboratory batch separation. In this mode, the flows of the feed stream and the permeate are both perpendicular to the membrane surface.

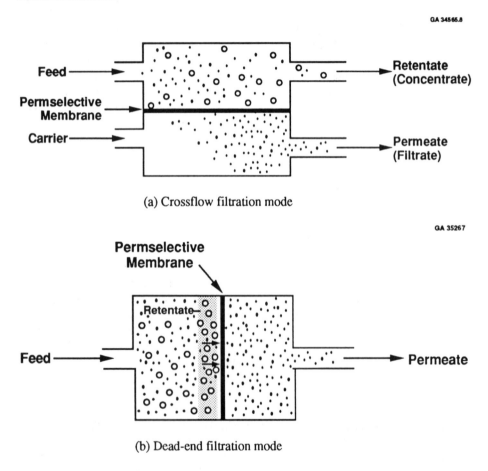

Figure 1.1 Schematic diagram of a membrane process: (a) crossflow filtration mode and (b) dead-end filtration mode

Two of the most important parameters that describe the separation performance of a membrane are its permselectivity and permeability. Permeability is typically used to provide an indication of the capacity of a membrane for processing the permeate; a high permeability means a high throughput. A high throughput is useless, however, unless

TABLE 1.1
Membrane separation processes

Process	Driving force	Typical permeate	Typical retentate
Microfiltration	Pressure difference, 10 psi	Water and dissolved species	Suspended materials
Ultrafiltration	Pressure difference, 10-100 psi	Water and salts	Biologicals, colloids and macromolecules
Gas separation	Pressure difference, 1-100 atm	Gases and vapors	Membrane-impermeable gases and vapors
Reverse osmosis	Pressure difference, 100-800 psi	Water	Virtually all suspended and dissolved materials
Dialysis	Concentration difference	Ions and low molecular-weight organics	Dissolved and suspended materials with molecular weight >1,000
Electrodialysis	Voltage	Ions	All non-ionic and macromolecular species

(Adapted from Lonsdale [1982])

another important membrane property, permselectivity, also exceeds an economically acceptable level. On the other hand, a membrane with a high permselectivity but a low flux or permeability may require such a large membrane surface area that it becomes economically unattractive. Simply put, permselectivity is the ability of the membrane to separate the permeate from the retentate. For MF, UF and RO, it is usually conveniently expressed in terms of a rejection or retention coefficient:

$$R = 1 - C_p/C_r \qquad (1\text{-}1)$$

where C_p and C_r represent the concentrations of the rejected species in the permeate and retentate, respectively. The rejection coefficient essentially gives a percentage of the rejected species that "leaks" through the membrane. Since the coefficient refers to retention by the membrane of a given species in certain defined experimental conditions, caution should be exercised in its extensive use especially beyond the test conditions.

As a general rule, the membrane technology is a competitive separation method for small to medium volumetric flowrate applications and for either primary separation or when the purity level required is in the 95 to 99% range.

TABLE 1.2
Current status of membrane technology

Process	Problems Major	Problems Minor	Problems Mostly solved	Comments
Microfiltration	Reliability (fouling)	Cost	Selectivity	Better fouling control could improve membrane lifetime significantly
Ultrafiltration	Reliability (fouling)	Cost	Selectivity	Fouling remains the principal operational problem of ultrafiltration. Current fouling control techniques are a substantial portion of process costs
Reverse osmosis	Reliability	Selectivity	Cost	Incremental improvements in membrane and process design will gradually reduce costs
Electrodialysis	Fouling Temperature stability	Cost	Selectivity Reliability	Process reliability and selectivity are adequate for current uses. Improvements could lead to cost reduction, especially in newer applications
Gas separation	Selectivity Flux	Cost	Reliability	Membrane selectivity is the principal problem in many gas separation systems. Higher permeation rates would help to reduce costs
Pervaporation	Selectivity Reliability	Cost	-	Membrane selectivities must be improved and systems developed that can reliably operate with organic solvent feeds before major new applications are commercialized
Coupled and facilitated transport	Reliability (membrane stability)	-	-	Membrane stability is an unsolved problem. It must be solved before this process can be considered for commercial applications

(Adapted from Baker et al. [1991])

1.2 POLYMERIC MEMBRANES

While the market size of the membrane industry worldwide varies from one estimate to another, it is generally agreed that the industry grew to be approximately $1 billion annually in 1986 and was expected to exceed $4 billion annually in the mid-1990s. The breakdown of four major market components according to various sources can be summarized as given in Table 1.3. The industry so far is dominated by polymeric membranes which after many years of research and development and marketing by several giant chemical companies particularly in the U.S. start to enjoy niche applications ranging from desalination of sea and brackish waters, food and beverage processing, gas separations, hemodialysis to controlled release.

TABLE 1.3
Worldwide membrane market size

Process	Major Applications	1986 Sales	1996 Sales (Forecast)
Microfiltration	Biotechnology, chemicals, electronics, environmental control, food & beverage, pharmaceutical	$550 million	$1.5 billion
Ultrafiltration	Biotechnology, chemicals, environmental control, food & beverage	$350 million	$1.1 billion
Reverse osmosis	Desalinating sea water, electronics, food & beverage	$120 million	$530 million
Gas separation	Environmental control	$20 million	$1.5 billion
Total		$1 billion	$4.6 billion

1.2.1 Membrane Materials and Preparation

The first widespread use of polymeric membranes for separation applications dates back to the 1960-70s when cellulose acetate was cast for desalination of sea and brackish waters. Since then many new polymeric membranes came to the market for applications extended to ultrafiltration, microfiltration, dialysis, electrodialysis and gas separations. So far ultrafiltration has been used in more diverse applications than any other membrane processes. The choice of membrane materials is dictated by the application environments, the separation mechanisms by which they operate and economic considerations. Table 1.4 lists some of the common organic polymeric materials for various membrane processes. They include, in addition to cellulose acetate, polyamides, polyimides, polysulfones, nylons, polyvinyl chloride, polycarbonate and fluorocarbon polymers.

TABLE 1.4
Commonly used membrane materials and their properties

Material	Application(s)	Approximate maximum working temperature (°C)	pH range
Cellulose acetates	RO, UF, MF	50	3-7
Aromatic polyamides	RO, UF	60-80	3-11
Fluorocarbon polymers	RO, UF, MF	130-150	1-14
Polyimides	RO, UF	40	2-8
Polysulfone	UF, MF	80-100	1-13
Nylons	UF, MF	150-180	
Polycarbonate	UF, MF	60-70	
Polyvinyl chloride		120-140	
PVDF	UF	130-150	1-13
Polyphosphazene		175-200	
Alumina (gamma)	UF	300	5-8
Alumina (alpha)	MF	>900	0-14
Glass	RO, UF	700	1-9
Zirconia	UF, MF	400	1-14
Zirconia (hydrous)	DM (RO, UF)	80-90	4-11
Silver	MF	370	1-14
Stainless steel (316)	MF	>400	4-11

The preparation methods of organic polymeric membranes depend on the structural characteristics of the membranes suitable for specific applications. For example, to prepare dense symmetrical membranes where the structures are more or less homogeneous throughout, solution casting and melt forming have been used. To produce microporous symmetrical membranes, common techniques such as irradiation (will be discussed in Chapter 3), stretching or template leaching can be employed. For asymmetric membranes, a large array of methods can be applied: phase inversion (or called solution-precipitation), interfacial polymerization, solution casting, plasma polymerization and reactive surface treatment. The phase inversion process, used in making the earliest commercial microporous membranes (cellulosic polymers), remains one of the most widely used techniques for fabricating a wide variety of commercial membranes today. The phase inversion process is applicable to any polymers that can be dissolved in a solvent and precipitated in a continuous phase by a miscible nonsolvent. It is capable of making polymeric membranes with a wide range of the pore size by varying polymer content, temperature, composition of the precipitation medium and the type of solvent.

It is not the intent of this book to get into any details of organic polymeric membranes. The readers, therefore, are referred to some recently published books in this field for the synthesis, characteristics and applications of various organic membranes [Belfort, 1984; Lloyd, 1985; Sourirajan and Matsuura, 1985; Baker, 1991].

1.2.2 Membrane Elements and Modules

Polymeric membrane elements and modules which consist of elements come in different shapes. The shape strongly determines the "packing density" of the element or module which is indicative of the available membrane filtration area per unit volume of the element or module; the packing density, in turn, can affect the capital and operating costs of the membranes. The packing density is often balanced by other factors such as ease and cost of maintenance and replacement, energy requirements, etc. Most of the polymeric membranes are fabricated into the following forms: tube, tubes-in-shell, plate-and-frame, hollow-fiber, and spiral-wound.

As a technology borrowed from the filtration industry, the plate-and-frame design is comprised of a series of planar composite membranes sandwiched between spacers that act as rigid porous supports and flow channels. The most common usage of this configuration is the electrodialysis cells. The tube-in-shell configuration resembles that in a shell-and-tube heat exchanger. Each tube can be cleaned, plugged off or replaced independently of other tubes. This can be an advantage but also a disadvantage particularly when hundreds or thousands of tubes are packed in a shell. The spiral-wound configuration can be viewed as a plate-and-frame system on flexible porous supports that has been wrapped around a central porous tube several times. The porous supports also provide the flow path for the permeate. The feed flows axially into the feed channels. Retentate continues through these channels to the exit end. The permeate flows spirally to the central tube where it is collected. The available membrane filtration area per unit volume of the module is about one and a half times that for a plate-and-frame design. This very compact configuration, however, makes it difficult for mechanical cleaning and is not suitable for particulate feed streams. Hollow fiber has the maximum packing density, even higher than the spiral-wound design. A hollow fiber module may consist of thousands of hair-like hollow membrane fibers, usually 50 to 100 μm in diameter, assembled into a bundle and contained in a vessel. Hollow fiber membranes have been used extensively in reverse osmosis. This configuration is also susceptible to fouling of particles or macromolecules.

1.3. GENERAL COMPARISONS OF ORGANIC AND INORGANIC MEMBRANES

The advantages of inorganic membranes have been recognized for a long time. In fact, studies of the use of some inorganic membranes such as platinum and porous glass were evident even in the last century. Thermal and pH resistance characteristics of inorganic and organic membranes are compared in Table 1.4 where approximate ranges of operable temperature and pH are given. Although inorganic polymers have not been used commercially yet, polyphosphazene which is under development is included in the table. It can be seen that thermal stabilities of organic polymers, inorganic polymers and inorganic materials as membrane materials can be conveniently classified as <100-150°C, 100-350°C and >350°C, respectively. It should be stressed that the use of pH stability is only a crude indication of the chemical stability of the membrane material. An example is silver which can withstand strong bases and certain strong acids. Although silver is resistant to strong hydrochloric or hydrofluoric acid, it is subject to the attack by nitric and sulfuric acids and cyanide solutions. Therefore, the wide operable pH range of silver can not be construed that it is resistant to all acids.

The operable temperature limits of inorganic membranes are obviously much higher than those of organic polymeric membranes. The majority of organic membranes begin to deteriorate structurally around 100°C. Thermal stability of membranes is becoming not only a technical problem but also an economic issue. In gas separation applications, for example, if the membrane can withstand the required high process temperatures, the need to ramp down the temperature to maintain the physical integrity of an organic membrane and to ramp up the temperature again after separation can be eliminated.

Inorganic membranes generally can withstand organic solvents, chlorine and other chemicals better than organic membranes. This also permits the use of more effective and yet corrosive cleaning procedures and chemicals. Many organic membranes are susceptible to microbial attack during applications. This is not the case with inorganic types, particularly ceramic membranes. In addition, inorganic membranes in general do not suffer from the mechanical instability of many organic membranes where the porous support structure can undergo compaction under high pressures and cause decrease in permeability.

It is obvious that in a high temperature or harsh chemical environment, inorganic membranes could become the only recourse to many challenging separation applications.

Another very important operating characteristics of inorganic membranes that is not shown in Table 1.4 has to do with the phenomena of fouling and concentration polarization. Concentration polarization is the accumulation of the solutes, molecules or particles retained or rejected by the membrane near its surface. It is deleterious to the purity of the product and the decline of the permeate flux. Fouling is generally believed to occur when the adsorption of the rejected component(s) on the membrane surface is strong enough to cause deposition. How to maintain a clean membrane surface so that the membrane can be continuously used without much interruption has been a key operational issue with many membranes. First of all, some inorganic membranes such as microporous alumina membranes and surface treated porous glass membranes are more fouling resistant due to their low protein adsorption. Secondly, many inorganic membranes are less susceptible to biological and microbial degradation. And finally with some inorganic membranes, for example porous ceramic and metallic membranes, it is possible to apply short bursts of permeate streams in the reverse direction through the membrane to dislodge some clogged pores of the membrane. This is referred to as backflush. This way the maintenance cycle of the membrane system can be prolonged. More details of this aspect of operation will be discussed in Chapter 5.

Because of the unique characteristics of inorganic membranes mentioned above, the search for inorganic membranes of practical significance has been continuing for several decades. With the advent of ceramic membranes with superior stabilities coming to the separation markets, the potentials for inorganic membranes as separators and/or reactors are being explored at an accelerated rate never witnessed before.

1.4 TYPES OF INORGANIC MEMBRANE STRUCTURES

To facilitate discussions on the preparation methods, characteristics and applications of inorganic membranes in the following chapters, some terminologies related to the types of membranes according to the combined structures of the separating and support layers, if applicable, will be introduced.

Membranes can be divided into two categories according to their structural characteristics which can have significant impacts on their performance as separators and/or reactors (membrane reactors or membrane catalysts): dense and porous membranes. Dense membranes are free of discrete, well-defined pores or voids. The difference between the two types can be conveniently detected by the presence of any pore structure under electron microscopy. The effectiveness of a dense membrane strongly depends on its material, the species to be separated and their interactions with the membrane.

The microstructure of a porous membrane can vary according to the schematic in Figure 1.2. The shape of the pores is strongly dictated by the method of preparation which will be reviewed in Chapter 3. Those membranes that show essentially straight pores across the membrane thickness are referred to as straight pore or nearly straight pore membranes. The majority of porous membranes, however, have interconnected pores with tortuous paths and are called tortuous pore membranes.

When the separating layer and the bulk support designed for mechanical strength are indistinguishable and show an integral, homogeneous structure and composition in the direction of the membrane thickness, it is called a symmetric or isotropic membrane. Since the flow rate through a membrane is inversely proportional to the membrane thickness, it is very desirable to make the homogeneous membrane layer as thin as possible. However, very thin stand-alone membranes typically do not exhibit mechanical integrity to withstand the usual handling procedures and processing pressure gradients found in many separation applications. A practical solution to the dilemma has been the concept of an asymmetric or composite membrane where the thin, separating membrane layer and the open-cell mechanical support structure are distinctly different. In this "anisotropic" arrangement, separation of the species in the feed stream and ideally the majority of the flow resistance (or pressure drop) also takes place primarily in the thin membrane layer. The underlying support should be mechanically strong and porous enough that it does not contribute to the flow resistance of the membrane element to any significant extent.

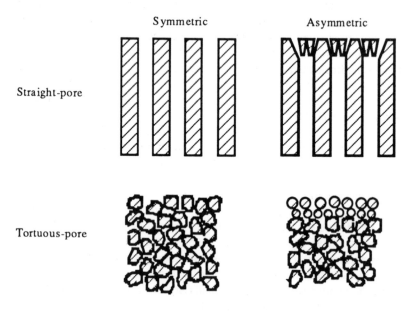

Figure 1.2 Cross-sectional schematic diagrams of various types of porous membranes

If a membrane has a graded pore structure but is made in one processing step, frequently from the same material across its thickness, it is called an asymmetric membrane. If, on the other hand, the membrane has two or more distinctively different layers made at different steps, the resulting structure is called a composite membrane. Almost invariably in the case of a composite membrane, a predominantly thick layer provides the necessary mechanical strength to other layers and the flow paths for the permeate and is called the support layer or bulk support. Composite membranes have the advantage that the separating layer and the support layer(s) can be tailored made with different materials. Permselective and permeation properties of the membrane material are critically important while the material for the support layer(s) is chosen for mechanical strength and other consideration such as chemical inertness. The composite membranes can have more than two layers in which the separating layer is superimposed on more than one support layer. In this case the intermediate layer serves a major purpose of regulating the pressure drops across the membrane-support composite by preventing any appreciable penetration of the very fine constituent particles into the pores of the underlying support layer(s). Typically the intermediate support layer(s) is also thin.

The aforementioned membranes are made prior to end use. There is a different type of inorganic membranes that are formed in-situ while in the application environment. They are called dynamic membranes which have been studied a great deal especially in the 1960s and 1970s. The general concept is to filter a dispersion containing suspended inorganic or polymeric colloids through a microporous support to form a layer of the colloids on the surface of the support. This layer becomes the active separating layer (membrane). Over time this permselective layer is eroded or dissolves and must be replenished as they are washed away in the retentate. Commonly used materials for the support are porous stainless steel, carbon or ceramics. A frequently used dynamic membrane material is hydrous metal oxides such as zirconium hydroxide although some organic colloids such as polyvinyl methyl ether or acrylic acid copolymers have been used as well. Dynamic membranes have been investigated for reverse osmosis applications such as desalination of brackish water, but found to be difficult to provide consistent performance and the added cost of the consumables makes the process unattractive economically. Today only limited applications for recovering polyvinyl alcohol in the field of textile dyeing is commercially practiced.

There is another type of membrane called liquid membrane where a liquid complexing or carrier agent supported or immobilized in a rigid solid porous structure function as the separating transport medium (membrane). The liquid carrier agent completely occupies the pores of the support matrix and reacts with the permeating component on the feed side. The complex formed diffuses across the membrane/support structure and then releases the permeant on the product side and at the same time recovers the carrier agent which diffuses back to the feed side. Thus permselectivity is accomplished through the combination of complexing reactions and diffusion. This is often referred to as facilitated transport which can be used for gas separation or coupled transport which can separate metal compounds through ion transport. An example of the former is some molten salts supported in porous ceramic substrate that are selective toward oxygen and of the latter is some liquid ion exchange reagent for selectively transporting copper ions. In this configuration, the composite of the liquid membrane and its support can be considered to be a special case of dense membranes. Despite their potentials for very high selectivities, liquid membranes suffer from physical instability of the membrane in the support and chemical instability of the carrier agent. Consequently, liquid membranes have not seen any significant commercial applications and is not likely to be a major commercial force in the separation industry in the next decade [Baker et al., 1991]. The liquid membrane process will not be treated in this book, but the use of inorganic membranes as the carriers for liquid membranes will be briefly discussed in Chapter 7.

1.5 INORGANIC MEMBRANES ACTIVELY PARTICIPATING IN CHEMICAL REACTIONS

Among the frontier developments in the field of inorganic membranes, a general area is particularly promising. It involves the use of a membrane as an active participant in a chemical transformation. This process integration stems from the concept of a membrane reactor where the membrane not only plays the role as a separator but also as part of a reactor. With their better thermal stability than organic membranes, inorganic membranes logically become potential attractive candidates for the use in many industrially important chemical reactions that are frequently operated at high temperatures and, in some cases, under harsh chemical environments.

For the above reason, a significant portion of this book will be devoted to the various issues centering around the utilization of inorganic membranes as both a separator and a reactor. Chapters 8 through 11 will deal with those subjects.

1.6 SUMMARY

While organic membranes have enjoyed over two decades of commercial success, inorganic membranes are just making their inroads to a growing number of commercial applications. This new market development is fueled by the availability of consistent quality ceramic membranes introduced in recent years. Compared to their organic counterparts, inorganic membranes made of metals, ceramics and inorganic polymers typically exhibit stabilities at high temperatures and extreme pH conditions.

Their historical developments and various membrane preparation methods will be discussed in Chapters 2 and 3, respectively. Chapter 4 reviews the general separation and non-separation properties of the membranes and the methods by which they are measured. Chapter 5 presents commercial membrane elements and modules and their application features which are followed by discussions of liquid-phase separation applications in Chapter 6. Many of those applications are commercially practiced. Potential gas separation and other applications (such as sensors and supports for liquid membranes) will be discussed in Chapter 7.

Those higher thermal and chemical stabilities not only make inorganic membranes very suitable for separation applications, but also for reaction enhancement. The duel use of an inorganic membrane as a separator as well as reactor offers great promises which along with potential hurdles to overcome will be treated in Chapters 8 through 11. The concept and related material, catalytic and engineering issues are addressed.

Market and economic aspects as well as major technical hurdles to be resolved prior to widespread usage of inorganic membranes are finally summarized in Chapter 12.

REFERENCES

Baker, R.W., E.L. Cussler, W. Eykamp, W.J. Koros, R.L. Riley and H. Strathmann, 1991, Membrane Separation Systems - Recent Developments and Future Directions (Noyes Data Corp., New Jersey, USA).

Belfort, G. (Ed.), 1984, Synthetic Membrane Processes (Academic Press, Orlando, USA).

Lonsdale, H.K., 1982, J. Membrane Sci. **10**, 81.

Lloyd, D.R. (Ed.), 1985, Materials Science of Synthetic Membranes (ACS Symposium Ser. 269, American Chemical Society, Washington, D.C., USA).

Sourirajan, S., and T. Matsuura (Eds.), 1985, Reverse Osmosis and Ultrafiltration (ACS Symposium Ser. 281, American Chemical Society, Washington, D.C., USA).

CHAPTER 2

HISTORICAL DEVELOPMENT AND COMMERCIALIZATION OF INORGANIC MEMBRANES

As stated in the previous chapter, for the convenience of discussions throughout this book, inorganic membranes will be classified into two broad categories: dense and porous membranes. The difference between the two categories is not so much a matter of the absolute size of the "pores" in the membrane structures, but rather more a distinction of the generally accepted transport mechanisms for the permeate. For dense membranes, the general solution-diffusion type mechanisms, similar to that prevailing in reverse osmosis membranes, are believed to be responsible for the transport and separation of species. With detectable pores throughout their structures, the porous membranes operate in various pore-related transport mechanisms which will be discussed in Chapter 4.

2.1 DENSE INORGANIC MEMBRANES

Historically there are two major types of dense inorganic membranes that have been studied and developed extensively. They are metal membranes and solid electrolyte membranes.

2.1.1 Metal Membranes

The dense metal membranes are dominated by palladium and its alloys. Palladium has long been recognized to possess the characteristics of a membrane. It is also probably the first well-studied and documented inorganic membrane. Thomas Graham (1866) first observed the unusually high hydrogen-absorption ability of metallic palladium. It is permeable only to hydrogen. Since then, the study of this metal hydride system has been rather extensive, particularly in Russia and, to a lesser degree, in Japan for the past three decades.

The first commercially available dense inorganic membrane system appears to be the palladium-silver (77%/23%) alloy membrane tubes (Hunter, 1956; Hunter, 1960) made by Johnson Matthey for hydrogen purification in the early 1960s (Shu et al., 1991). The process is still in use today in small to medium plants with an individual unit production capacity of up to 56 m^3/h. Recently, Johnson Matthey developed a palladium-based hydrogen generator using methane-water mixture as the feed and a catalyst located within the alloy tube. The reported production capacity reaches 25 m^3/h of hydrogen. Johnson and Matthey continue to hold several key patents in the manufacturing of thin Pd-based membranes. A large plant using these membranes was installed by Union

Carbide in the 1960s (McBride and McKinley, 1965) to separate hydrogen from refinery off-gas. It utilized 25 μm thick Pd-Ag membrane foils (McBride et al., 1967).

V. M. Gryaznov and his co-workers (e.g. [Gryaznov, 1986]) have extensively explored the permselective properties of palladium and its alloys as dense membranes and membrane reactors. While their studies will be discussed in later chapters, it suffices to say that the palladium-based membranes have reached the verge of a commercialization potential for the process industry.

Besides Pd and its alloys, other dense metal materials are also possible candidates for separation of fluid components. Notable examples are tantalum, vanadium and niobium which have high selectivities for hydrogen. Dense silver is known to be permselective to oxygen gas.

2.1.2 Solid Electrolyte Membranes

Solid electrolytes are another class of dense materials that are selective to certain ionic species. They are impervious to gases or liquids but allow one or more kinds of ions to pass through their lattices under some driving forces which can be an applied voltage difference or a chemical potential gradient of the migrating ions. Their unique ion conductivities also make it possible to use them for measuring concentrations of the migrating species, even at high temperatures.

Since Kiukkola and Wagner (1957) found that doped zirconia and thoria, due to their defect ionic crystals, selectively conduct oxygen ions and Ure (1957) similarly demonstrated that calcium fluoride, in pure or doped form, exhibits the fluorine ion conductivity, the search for and understanding of potential solid electrolytes have been extensive. Some other emerging materials for practical applications are certain fluorides, and silver ion conductors. Common solid electrolytes so far basically fall under one of three types: simple or complex oxides (e.g. β-aluminas), simple or complex halides (e.g. $RbAg_4I_5$) and oxide solid solutions (e.g. ZrO_2-Y_2O_3, ZrO_2-CaO and ThO_2-Y_2O_3). Developments of applications over the years have focused on oxygen meters for gas and liquid streams, oxygen pumps, fuel cells, high-energy storage batteries, power generation, etc.

2.2 POROUS INORGANIC MEMBRANES

In contrast to dense inorganic membranes, the rate of advances toward industrial-scale applications of porous inorganic membranes has been rapid in recent years. In the early periods of this century, microporous porcelain and sintered metals have been tested for microfiltration applications and, in the 1940s, microporous Vycor-type glass membranes became available. Then in the mid-1960s porous silver membranes were commercialized. These membranes, however, have not seen large scale applications in the separation industry to become a significant player. Their limited usage can often be attributed to some product quality issues such as inconsistency in the pore size distributions.

2.2.1 Gaseous Diffusion Applications

Much of the impetus for the awakened interest and utilization of inorganic membranes recently came from a history of about forty or fifty years of some large scale successes of porous ceramic membranes for gaseous diffusion to enrich uranium in the military weapons and nuclear power reactor applications. In the gaseous diffusion literature, the porous membranes are referred to as the porous barriers. For nuclear power generation, uranium enrichment can account for approximately 10% of the operating costs [Charpin and Rigny, 1989].

Gaseous diffusion is the first large scale, and probably so far still the single largest application of porous inorganic membranes. And these technical successes led to refinements of manufacturing technologies and development of commercial applications. The technology and applications in the U.S. in the 1940s from the World War II Manhattan Project are still classified, although there have been some attempts to declassify them in recent years. The underlying concern is that once declassified the technology of making the secret membranes may enable some nations to enrich uranium for making nuclear weapons to cause nuclear proliferation. In contrast, the series of developments and applications in France in the 1950s essentially have led to the developing market of ceramic membranes today.

Natural uranium contains close to 0.72% by weight of the only fissionable isotope ^{235}U and more than 99% non-fissionable ^{238}U, and a trace quantity of ^{234}U. For uranium to be useful as the fuel to the nuclear reactor , the ^{235}U level needs to reach 1 to 5% (more often 3 to 5%) while most of the nuclear weapons and submarines require a concentration of at least 90% ^{235}U. The separation of those uranium isotopes, very similar in properties, can not be effected by chemical means.

Among the proven uranium isotope separation processes are distillation, monotherm chemical exchange, gaseous diffusion and gas centrifuge, with their separation efficiencies increasingly ranked in the order listed. Their efficiencies, however, are all typically low [Benedict et al., 1981]. The most successful and extensively employed physical process so far is the gaseous diffusion method. The method is based on the notion that, in a gaseous isotopic mixture at low pressures, the lighter isotopes have slightly higher velocities than the heavier ones through the membrane pores. Currently approximately 90% of uranium enrichment is achieved by gaseous diffusion with the remaining by ultracentrifugation. The membrane or barrier is a critical element in the gaseous diffusion process.

Since uranium itself is not a gas, separation of the isotopes by gaseous diffusion is carried out by using a sufficiently volatile and stable gaseous compound such as uranium hexafluoride, UF_6. Uranium hexafluoride is a solid at the ambient temperature but has a relatively low sublimation temperature and vaporizes readily. UF_6 can be obtained from uranium trioxide by successive treatments with hydrogen, anhydrous fluoride and fluorine gas. Uranium trioxide, in turn, is prepared from uranium ores and ore concentrates. The membrane pore diameters are smaller than the mean free path of the gas mixture. Thus the predominant mechanism in gaseous diffusion with membranes is not bulk flow of the gases but rather Knudsen diffusion where two gases of different molecular weights diffuse through the membrane pores at rates inversely proportional to the square root of their molecular weights. The degree of separation of isotopes can be expressed in terms of the separation factor to be defined and discussed in Chapter 7. The molecular weights of $^{235}UF_6$ and $^{238}UF_6$ are nearly equal and the theoretical separation factor between the two in a single pass through the membrane can be calculated to be the square root of the molecular weight ratio of $^{238}UF_6$ to $^{235}UF_6$ and is equal to only 1.00429. In contrast, the actual separation factor using alumina membranes is only 1.0030 [Isomura, et al., 1969]. In practice, up to thousands of membrane tubes are arranged in a counter-current cascade configuration to achieve the required degree of separation: for example, over 1,200 stages required for 3% ^{235}U and over 4,000 stages for 97% ^{235}U even with gas recirculation.

Starting in the 1940s, the developments in France involved in the making of composite membranes where a thin selective layer (to improve permeability) through which separation takes place is formed on a much thicker microporous support (to ensure mechanical strength of the selective layer) as schematically shown in Figure 1.2. SFEC (a subsidiary of CEA, the French Atomic Energy Commission) was responsible for developing and preparing the selective layer while the major parties fabricating the microporous supports included Desmarquest and LeCarbone Lorraine (both as subsidiaries of Pechiney), CGEC (as a subsidiary of CE, then as Carefree, CST which was a subsidiary of Alice and most recently as a subsidiary of US Filters) and later Euroceral in which Norton had an equal partnership.

While information on the materials tested and actually used in gaseous diffusion plants is sketchy at best, there is indication that the materials developed in the French technology include sintered and anodic alumina, gold-silver alloys, nickel and Teflon [Frejacques et al., 1958], porcelain [Albert, 1958] and possibly zirconia [Molbert, 1982]. These porous barriers (membranes) were prepared by a number of processes: etching with a strong acid such as nitric acid (e.g. gold-silver alloys), pressing or extrusion of finely powdered metals or their oxides (e.g. nickel and alumina) and anodic oxidation (e.g. alumina). The preference of oxide materials such as alumina and zirconia over metals probably stems from their better resistance toward the corrosive uranium hexafluoride. In the presence of water and many organic or inorganic compounds, uranium hexafluoride breaks down quickly and generates the very corrosive hydrofluoric acid and various harmful oxyfluorides such as UO_2F_2. This not only destroys many membrane materials but also causes gradual clogging of the membrane pores. One of the effective means of improving the corrosion resistance of the membranes toward uranium hexafluoride is to pretreat the membranes with fluorine or chlorine trifluoride (ClF_3). Typical membranes used for the above gaseous diffusion applications have a pore diameter in the 0.006 to 0.04 μm range and are operated at a temperature of 60 to 200°C (more often about 65°C to prevent condensation of UF_6) and under a pressure of 0.4 to 3.3 bars [Charpin and Rigny, 1989]. The support for the membrane layer has a pore diameter of 2 to 10 μm.

Concurrently in the U.S., membrane development efforts were sponsored and initiated by the Atomic Energy Commission for gaseous diffusion. Large scale plants were built at Oak Ridge (Tennessee) and subsequently at Paducah (Kentucky) and Portsmouth (Ohio) in the early 1950s with a total capacity of approximately 150 t/d processed. The confidential membranes are believed to be made of nickel and high-nickel alloys but other membrane materials possibly have also been considered.

In addition to the U.S. and France, other countries such as the Soviet Union, China and England were also involved in using presumably inorganic membranes for its gaseous diffusion operations although little has been documented. Ceramic membranes were also made by the anodic oxide process (to be discussed later in Chapter 3) in Sweden for military and nuclear applications.

The interest of using fine-pore thin-film ceramic or metal membranes for isotope separation (e.g. uranium) is still apparent even after years of production practice [Miszenti and Nannetti, 1975; Sumitomo Electric Industry, 1981]. Isotopes other than uranium, such as those of Ar or Ne [Isomura, et al., 1969; Fain and Brown, 1974], can also be separated by gaseous diffusion. The membrane materials having been successfully tested for these specific applications include alumina, glass and gold.

2.2.2 Porous Glass and Metal Membranes

Porous glass has been available for transport studies for several decades, but its use as a membrane material commercially started only in the early 1980s. Its uses have been mostly in the biotechnology area. Porous silver membranes were available in the mid-1960s in the form of tubes and later in the shape of a disk. The use of silver membranes, however, has been limited. Porous stainless steel membranes have been used for years as high-quality filters or supports for dynamic membranes.

2.3 EARLY COMMERCIALIZATION OF INORGANIC MEMBRANES

Practically all the major industrial producers of inorganic membranes in the 1980s were involved in various aspects of the gaseous diffusion operations discussed above.

It is estimated that a total of more than 100 million porous membrane tubes have been made using the French technology. The process has been found to be very reliable and the membranes produced for uranium isotope separation typically lasted a very long time. In some cases, the membranes were still functional even after 20 years of service life. In addition, gaseous diffusion was about to face stiff competition from new enrichment technologies such as gas centrifuging and atomic vapor laser isotope separation. These factors triggered the closing of many membrane production plants more than ten years ago and the membrane technology developed for gaseous diffusion needed to find new commercial applications.

In the late 1970s, CEA started to realize that the technology developed for the uranium isotope separation had great potential for use in liquid phase microfiltration and ultrafiltration. The selective layer of the membrane for uranium enrichment has a pore diameter in the ultrafiltration range while its underlying support falls in the microfiltration spectrum. And so a new generation of ceramic membranes emerged [Charpin and Rigny, 1989].

Some of those developments at Oak Ridge were believed to spin off in some form at Union Carbide and some aspects of the efforts led to the commercialization of dynamically formed membranes primarily for ultrafiltration and hyperfiltration (reverse osmosis) applications. In these dynamic membranes, a mixture of zirconium hydroxide and polyacrylic acid deposited on a porous support which provides the necessary mechanical strength. The support is mostly made of porous carbon although porous ceramic and stainless steel are also used. These non-sintered membranes, in great contrast to most of the membranes discussed in this book, are formed in situ and require periodic regeneration with new zirconium hydroxide and polyacrylic acid.

The dynamic membranes originally developed by Union Carbide are protected by three core patents: U.S. 3977967, 4078112, and 4412921 (Trulson and Litz, 1976; Bibeau, 1978; and Leung and Cacciola, 1983) and their foreign equivalents. Those patents cover a broad range of metal oxides such as zirconia, gamma alumina, magnesia-alumina spinel, tantalum oxide and silica as the membrane materials and carbon, alumina, aluminosilicates, sintered metals, fiberglass or paper as the potential porous support materials. However, their marketed product, trade named Ucarsep® membranes, focused on dynamic membranes of hydrous zirconium oxide on porous carbon support.

These membranes were designed for the following proven applications: concentration and separation of emulsified oil from the bulk aqueous phase of metal working or metal washing waste streams which may contain valuable detergents, removal of dirt and oil in electropaint operations, separation of polyvinyl alcohol from textile sizing solutions, concentration of the black liquor from paper mill liquids containing paper pulp, etc. Union Carbide later sold the worldwide licensing right to SFEC (Societe de Fabrication d'Elements Catalytiques in France) excluding those applications of treating process streams in textile and metalworking industries. The licensing right to the latter application was transferred to Gaston County Dyeing Machine Company (USA). In the textile application, the principal use is in the recovery of polyvinyl alcohol as a sizing agent.

The above early commercial developments of inorganic membranes, although slow at the beginning, have stimulated sufficient market interest to entice more companies to enter the field with new types of membranes. These activities and various features of currently available commercial inorganic membranes will be highlighted in Chapter 5.

2.4 SUMMARY

Although several dense metallic and solid electrolyte membranes have been produced and characterized for a long time, only palladium-based membranes have seen some commercial activities. The use of Pd and Pd alloy membranes is still limited due to their high cost and low permeabilities. Porous inorganic membranes have been used extensively for uranium isotope separation for both military and nuclear power plant applications. That technology led to the commercialization of porous ceramic membranes in the late 1970s and early 1980s. The market entry of those consistent-quality ceramic membranes, largely for microfiltration and to a lesser degree for ultrafiltration, has created significant interest in inorganic membranes as a whole due to their thermal and chemical stabilities.

REFERENCES

Albert, H., 1958, Design of a gaseous diffusion uranium hexafluoride isotopic concentration experimental line, in: Proc. 2nd U.N. Int. Conf. on Peaceful Uses of Atomic Energy **4** (Production of Nuclear Materials and Isotopes), 412.

Benedict, M., T.H. Pigford and H.W. Levi, 1981, Nuclear Chemical Engineering, McGraw-Hill, New York.

Bibeau, A.A., 1978, U.S. Patent 4,078,112.

Charpin, J., and P. Rigny, 1989, Inorganic membranes for separative techniques: from uranium isotope separation to non-nuclear fields, in: Proc. 1st Int. Conf. on Inorganic Membranes, Montpellier, France, 1989 (Trans Tech Publ., Zürich) p. 1.

Fain, D.E., and W.K. Brown, 1974, U.S. Atomic Energy Commission Report "Neon isotope separation by gaseous diffusion transport in the transition flow regime with regular geometries."

Frejacques, C., O. Bilous, J. Dizmier, D. Massignon and P. Plurien, 1958, Principal results obtained in France in studies of the separation of the uranium isotopes by gaseous diffusion, in: Proc. 2nd U.N. Int. Conf. on Peaceful Uses of Atomic Energy **4** (Production of Nuclear Materials and Isotopes), 418.

Graham, T., 1866, Phil. Trans. Roy. Soc. (London) **156**, 399.

Gryaznov, V.M., 1986, Plat. Met. Rev. **30**, 68.

Hunter, J.B., 1956, U.S. Patent 2,773,561.

Hunter, J.B., 1960, Plat, Met. Rev. **4**, 130.

Isomura, S., T. Watanabe, R. Nakane and S. Kikuchi, 1969, Nippon Genshiryoku Gakkaishi **11**, 417.

Kiukkola, K., and C. Wagner, 1957, J. Electrochem. Soc. **104**, 379.

Leung, P.S., and A.R. Cacciola, 1983, U.S. Patent 4,412,921.

McBride, R.B., and D.L. McKinley, 1965, Chem. Eng. Prog. **61**, 81.

McBride, R.B., R.T. Nelson and R.S. Hovey, 1967, U.S. Patent 3,336,730.

Miszenti, G.S., and C.A. Nannetti, 1975, U.S. Patent 3,874,899.

Molbert, M., 1982, AIChE Symp. Ser. **78** (221), 10.

Shu, J., B.P.A. Grandjean, A. van Neste and S. Kaliaguine, 1991, Can. J. Chem. Eng. **69**, 1036.

Sumitomo Electric Ind., 1981, Japanese Patent 81,008,643.

Trulson, O.C., and L.M. Litz, 1976, U.S. Patent 3,977,967.

Ure, R.W., 1957, J. Chem. Phys. **26**, 1363.

CHAPTER 3

MATERIALS AND PREPARATION OF INORGANIC MEMBRANES

As will be shown in Chapter 4, inorganic membranes vary greatly in their microscale and macroscale characteristics. The variations come as a result of the precursor materials and the ways by which the membrane elements are prepared and the systems are fabricated. While many of these synthesis methods are still in varying stages of development, the rapid growth of the inorganic membrane market pulls some of these methods, many of them patented, into commercial practice. This is particularly true with porous membranes. Therefore, a significant portion of the discussions to follow will focus on the known techniques for making those commercial membranes.

For convenience of discussion, the preparation and fabrication methods will be grouped into three categories of inorganic membranes according to the difference in their morphology: dense, tortuous-pore and nearly-straight-pore. As the inorganic membrane technology exists today, the preferred microstructure of membrane elements dictates the use of certain types of preparation and fabrication techniques.

3.1 DENSE MEMBRANES

Some metals and oxides (many as solid electrolytes) have been well-studied and documented as dense (or nonporous) membrane materials. Among them palladium and its alloys stand out as a result of their unique resistance to surface oxidation. Since palladium was first found to absorb a large amount of hydrogen in 1866 [Graham, 1866], this metal hydride system and its alloys have been investigated extensively. In addition to palladium and its alloys, other metals and oxides in nonporous forms are known to be selectively permeable to some gases. Dense refractory metals (e.g., Nb, V, Ta, Ti, Zr) or their alloys have been found to be selectively permeable to hydrogen. For example, due to its permeability to hydrogen, dense zirconium has been investigated for the removal of hydrogen and deuterium from fusion blanket fluids [Hsu and Buxbaum, 1987]. It is well known that dense silver is highly permselective to oxygen.

Nonporous silica glass is also known to be highly permselective to hydrogen at high temperatures [Altemose, 1961]. In addition, a class of materials called solid electrolytes has been found to be selectively permeable to certain gases. They consist of either crystalline solids such as AgI, β-Al_2O_3, Bi_2O_3, CeO_2, Li_3N, $SrCeO_3$, stabilized ThO_2 and, probably best known, stabilized ZrO_2 or non-crystalline glasses (e.g., B_2O_3-Li_2O-LiI, $LiNbO_3$, etc.). Defect solid state has been a major scientific study in the past few decades. Solid electrolytes represent an important area of defect solid state.

Most of these solid electrolytes are either oxygen or hydrogen conductors but, there are also solid electrolyte membranes that conduct other ions such as silver, fluoride, sodium, nitrogen, carbon, and sulfur. While solid electrolyte membranes in general have been studied for a wide range of applications, the following applications have received the most attention: gas sensors, fuel cells, production of hydrogen by electrolysis of water at high temperatures, solid state rechargeable batteries and oxygen and hydrogen pumps. Although organic solid electrolyte membranes are available, the inorganic ones have the advantages of being able to withstand high temperatures because generally greater ion transport is possible at higher temperatures and having better fouling and microbial resistances. These membranes have been generally fabricated into two major shapes: disk and tube. In all cases studied, solid electrolytes in polycrystalline form are used.

3.1.1 Dense Metal Membranes

The manufacture of dense metal membranes or thin films can be effected by a number of processes: casting/rolling, vapor deposition by physical and chemical means, electroplating (or electroforming) and electroless plating. By far, casting in combination with rolling is the predominant preparation and fabrication technique. It is noted that many of these processes have been demonstrated with palladium and its alloys because of their low oxidation propensity. Preparation of dense metal membranes is summarized in some detail as follows.

<u>**Casting/rolling**</u>. This method has long been employed on a production scale to manufacture metallic plates and sheets. It involves a number of steps. First, the membrane precursor material with a desired composition is melted at a high temperature. Next, it is cast in ingot form which then goes through a high temperature homogenization step before hot and cold forging or pressing. The material finally receives repeated sequences of alternate cold rolling and anneals to reach the required thickness. Pd alloy tubes with a thickness of 50 μm and an outside diameter of 0.6 mm have also been made commercially in Russia by seamless tube drawing [Gryaznov, 1992]. As the metal foil becomes thinner, the control of the metal impurity at various processing steps proves to be more critical in affecting the performance of the finished membranes.

Traces of certain specific elements (such as C, S, Si, Cl and O) and inclusion of foreign particles or gases may connect both sides of the membrane (with a thickness of 10 μm or less) and thus render it unsuitable for separation purposes. Aluminum foils have been made down to a thickness of 10 μm and special fabrication methods can be used to produce palladium (or its alloys) foils with a thickness under 1 μm [Shu et al., 1991]. Commercially available Pd alloy foils, however, are in the 10 to 100 μm range. Cold rolling often generates lattice dislocation and it can enhance hydrogen solubility in palladium and some of its alloys due to the accumulation of excess hydrogen around the dislocation.

Vapor deposition. Both physical and chemical vapor deposition methods can be used to prepare dense inorganic membranes. In either process, vaporization of the membrane material to be deposited is effected by physical means (such as thermal evaporation and sputtering) or chemical reactions.

In physical vapor deposition (or called thermal evaporation), a solid material is first vaporized by heating at a sufficiently high temperature followed by depositing a thin film onto a cooler substrate by condensation of the vapor. Physical vapor deposition is usually performed under high vacuum (approximately 10^{-5} torr). Conventional resistive heating of the metal to be deposited can be used as the evaporation energy source, but the issue of potential contamination of the metal with the crucible or lining materials can be serious. Other evaporation schemes such as electron gun, laser beam or flash evaporation do not require direct contact of the metal with the crucible and are also more energy efficient. Thin films of Pt, Ag, Au and Al have been deposited on porous supports by physical vapor deposition [Ilias and Govind, 1989]. Due to the different partial pressures and evaporation rates of the metal components in an alloy melt, physical vapor deposition processes suffer from the shortcoming that direct deposition from solid alloys can not be well controlled. Instead, dense alloy membranes may be produced by simultaneous or sequential evaporation from separate sources of pure metals with the resulting compositions determined by the relative evaporation rates of the constituent metals.

Another kind of physical vapor deposition technique that can handle alloy deposition in a more controllable fashion is sputtering. The rates of evaporation for various metals using this method are similar. It does not involve thermal evaporation. Instead the method entails the use of rapid ion bombardment with an inert gas such as argon to dislodge atoms of a selected solid material and then deposit them onto a nearby target substrate. Its rate of deposition is in general lower than those of thermal evaporation processes. For comparison, the deposition rates for the resistive heating and the sputtering methods are approximately > 1 μm/min and < 0.1 μm/min, respectively. Films of Pd and Pd-Ag alloys as thin as 30 to 60 nm have been prepared by several researchers according to this technique [Shu et al., 1991]. Thin dense metal membranes on porous supports have been widely studied in recent years to impart a high permselectivity and a relatively high flux at the same time. Two parameters are found to be most critical to the synthesis of gas-tight and mechanically strong composite membranes by sputter deposition: the surface roughness of the support and the deposition temperature [Jayaraman et al., 1995]. Fine-pore supports result in better quality membranes than coarse-pore supports. The optimal deposition temperature appears to be about 400°C for coating an alumina support with thin palladium membranes (less than 500 nm thick).

In the chemical vapor deposition (CVD) process, heat is supplied through resistive heating, infrared heating, laser beam or plasma to effect a gas-phase chemical reaction involving a metal complex. The metal produced from the reaction deposits by nucleation and growth on the hot substrate which is placed in the CVD reactor. Effective reactants should be quite volatile and for this reason metal carbonyls, hydrides, halides or organometallic compounds are commonly used.

Palladium films on the order of 10 to 300 nm thick have been made by CVD at low temperatures in the 27-197°C range. Nickel film has been formed to seal the pores of a mixed oxide support using a nickel organometallic complex, $Ni(hfac)_2$ where hfac is hafnium acetate with hydrogen as the reducing agent [Carolan et al., 1994]. The organometallic complex and the reducing agent enter the support from the opposite sides and react within the pores of the support to form plugs as a dense inorganic membrane.

Electroplating. Basically in electroplating, a substrate is coated with a metal or its alloy in a plating bath where the substrate is the cathode and the temperature is maintained constant. Membranes from a few microns to a few millimeters thick can be deposited by carefully controlling the plating time, temperature, current density and the bath composition. Dense membranes made of palladium and its various alloys such as Pd-Cu have been prepared. Porous palladium-based membranes have also been made by deposition on porous support materials such as glass, ceramics, etc.

While this method is capable of making thick and ductile metal membranes, some technical problems may be encountered such as hydrogen pumping which requires annealing the deposited membranes under vacuum to ensure their ductility and difficulty in controlling the thickness over a wide range of alloy composition.

Electroless plating. This process involves autocatalyzed decomposition or reduction of a few selected metastable metallic salt complexes on substrate surfaces. The reaction should be carefully controlled to avoid potential decomposition of the metal film thus formed and to control the film thickness. The induction period of the reaction can be appreciable and an effective way of reducing the period is to preseed the substrate in advance with nuclei of the metal to be deposited in an activation solution [Uemiya et al., 1990].

The technique has the advantages of providing hardness and uniform deposits on complex shapes. Purity of the membranes thus prepared, however, may be compromised. Thin palladium-based membranes have been made by using amine complexes $Pd(NH_3)_4X_2$ where X can be NO_2, Cl or Br in the presence of a reducing agent such as hydrazine or sodium hypophosphite [Uemiya et al., 1988; Athavale and Totlani, 1989]. Composite membranes of palladium on microporous glass or stainless steel support have also been prepared in the same fashion with a thickness of approximately 10 to 15 μm [Uemiya et al., 1988; Shu et al., 1993]. Silver has also been deposited in porous stainless steel supports by electroless plating [Shu et al., 1993].

In selected cases two different metals can be codeposited onto a substrate by electroless plating. Shu et al. [1993] have attempted simultaneous deposition of palladium and silver on porous stainless steel supports using an electroless bath containing EDTA (ethylene diamine tetraacetic acid) as a complexing agent. Silver is preferentially deposited, but, after effective activation of the support by palladium predeposition, both Pd and Ag can be successfully codeposited in separate phases. The heterogeneity of the composite membrane thus obtained can be substantially eliminated by annealing in hydrogen atmosphere.

3.1.2 Dense Solid Electrolyte and Oxide Membranes

Dense solid electrolyte and non-electrolyte oxide membranes can be prepared by conventional ceramic or powder metallurgy methods such as pressing, extrusion, and slip casting. These techniques are often preceded by powder preparation, mixing, and calcination and followed by sintering. In the mixing step, the major component particles such as zirconia or thoria are usually wet blended with a desired amount of particles of the dopant which typically comes in the chemical form of an oxide, carbonate or sulfate. The blend is then dried and often calcined to convert the salts to oxides. Sometimes grinding and pelletizing are combined with de-airing and repeated calcination to ensure homogeneous oxides before fabrication.

Small-size starting particles with an appropriate size distribution that lead to a maximum packing density and final fired density are essential in the sintering step to form dense membranes. The required sintering temperatures are lower because of the active state of the fine particulate materials used. For example, when particles of an average size of 5 nm made by the alkoxide approach are used to make stabilized zirconia, 1450°C instead of the usual 2000°C is all that is necessary to produce fully dense material [Mazdiyasni et al., 1967]. The amount of additives such as the stabilizers affects densification of the final solid electrolyte membranes as well. Generally there is an optimum amount of stabilizer for maximum densification. Excess addition actually can lead to lower densification.

Other methods such as glass fabrication techniques and different deposition methods are also used.

Extrusion/sintering. A number of metal oxide powders can be mixed with certain organic additives to form a paste called slip for fabrication of dense ceramic membranes of a tubular or rod form by plastic extrusion. The slip generally has sufficient plasticity for ease of processing or is added with plasticizers such as glycerol or rubber solution in an organic solvent to be formed into various shapes and yet is strong enough to maintain the physical integrity of the molded shape at its green state (i.e., before firing or calcining/sintering). The organic additives usually consist of solvents, dispersants, binders, plasticizers, viscosity modifiers, etc. The composition of the additives required depends on the characteristics of the ceramic powders, such as size distribution and morphology, and the forming conditions.

The plastic mass of the slip is then forced through an extrusion die at high pressure to produce tubes or multichannel, honeycomb structures. The extruded shapes are typically heat treated slowly in the lower temperature range to remove gases from the decomposition of the organic additives. The heating rate is then escalated to the sintering range.

For example, Balachandran et al. [1993] describes the above procedures for making ceramic membrane tubes from perovskite-type oxides. They prepared ceramic powders of La-Sr-Fe-Co-O perovskites by solid-state reaction of the constituent carbonates and nitrates. The powders are mixed and milled in methanol with zirconia media for 15 hours, then dried and calcined in air at 850°C for 16 hours followed by further grinding. The resultant powders are then made into a slip by mixing with organic additives which facilitate shaping and maintaining green strength of the slip. The slip is extruded into tubes of an outside diameter of 6.5 mm and up to 0.3 m long at an extrusion die pressure of approximately 20 MPa. The tubes are then heated in air slowly at 5°C/h around 150-400°C range and then faster at 60°C/h to reach and stay at a high sintering temperature of 1200°C for 5 to 10 hours.

The final composition and phase structure of the dense solid electrolyte membrane can be strongly influenced by the oxygen partial pressure employed during the sintering step. Sintering temperature or rate can depend on a number of factors. Slow densification of the membrane may be attributed to a low cation diffusion rate. For the ZrO_2-CaO solid electrolyte, the maximum densification occurs at the lowest CaO content required for stabilization [Subbarao, 1980]. The sintering temperature should be high enough but not too high that the ion permselective properties are lost. The material can undergo substantial volume expansion as a result of phase transition, thus potentially leading to destruction of the physical integrity.

Pressing/sintering. Membranes in the form of disks and plates can be prepared by the conventional ceramic processing techniques such as cold pressing (from 2,000 to 100,000 psi) and isostatic pressing which applies uniformly distributed hydrostatic pressure from all directions and achieves a higher and more uniform green density. Sometimes partially sintered compact is subject to the isostatic pressing. Typically some solvent such as water and some binders (e.g., polyvinyl alcohol, zirconium oxychloride or wax) and lubricants (e.g., carbowax) are used [Subbarao, 1980]. The organic binders are burnt out at a temperature above approximately 300°C.

For pressing as well as extrusion, the solid electrolyte precursor particles (e.g., zirconia) are often mixed or reacted with an inorganic cementing substance. It is preferred that such adhesive materials also have ion permselective properties as the precursor particles. Phosphates of zirconium, titanium and zinc are examples of such cements although other materials such as calcium aluminate and calcium aluminosilicates are candidates as well [Arrance et al., 1969]. For these cementing materials to be effective, the metal oxides must be only partially hydrated so that they are reactive with the bonding compounds.

The subsequent sintering step for the pressed parts is similar to that for the extrusion method and the considerations for achieving a maximum density are therefore for the most part applicable here as well.

Slip casting/sintering. Tubes can be slip cast in plaster molds using the so called slip prepared from the stabilized powders and a binder such as polyvinyl alcohol which is widely used in ceramic processing to provide adequate green strength. The pH of the slip is usually controlled between 3 and 4. The slip is often degassed to remove any entrapped air before fabrication. The rheological behavior of the slip depends on the type and dosage of the binders and sometimes viscosity modifiers are also added to control the rheology for ease of processing.

Mixed ion and electronic conducting ceramic membranes (e.g., yttria-stabilized zirconia doped with titania or ceria) can be slip cast into a tubular form from the pastes containing the constituent oxides in an appropriate proportion and other ingredients and the cast tubes are then subject to sintering at 1,200 to 1,500°C to render them gas impervious [Hazbun, 1988].

Chemical vapor deposition. Physical vapor deposition techniques in principle can be used to prepare dense non-metal membranes, but the control of the deposition thickness, low deposition rate (e.g., in sputtering) and problem of adhesion of the deposited material make them less practical.

In contrast, chemical vapor deposition techniques have been successfully demonstrated to produce dense layers of oxide membranes, particularly those silica-based membranes. CVD is often carried out at high temperatures. This process has the advantage of eliminating the need for drying and calcination required in liquid-phase methods for preparing oxide membranes. In addition, membranes produced by CVD are usually denser than those prepared by liquid-phase processes. The process offers the promise of being able to fine tune the composition, growth and uniformity of the deposited films. Variations of the chemical vapor deposition process have yielded membranes with different morphologies. It is probably difficult to make a mechanically strong self-supporting dense oxide membrane by the vapor deposition technique alone. When deposited on a porous support, a dense oxide membrane is formed first by partially plugging the pores of the support with the deposit before deposition grows into an integral layer covering the surface of the support.

There are two major different approaches to deposit oxide layers using vapors of the appropriate chemical reactants. The difference lies in the manner and the location in which the reactants are brought into contact with respect to the support surface and pores.

In the first method, the support is exposed in a reaction chamber to two reactants that are introduced on the same side of the support surface. One reactant is the metal source and the other is an oxidizing or reducing agent with both diluted in an inert carrier gas such as nitrogen. The reactive agent can be oxygen, air, water, ozone or other oxidizing sources or hydrogen or carbon monoxide as the reducing agent. Hydrolysis of chlorosilane compounds of silicon, titanium, aluminum and boron at 600 to 800°C leads to the formation of silica membranes on porous Vycor glass [Tsapatsis et al., 1991]. A kinetic model of dense membrane formation by CVD has been proposed for silica and alumina membranes [Tsapatsis and Gavalas, 1992]. Megiris and Glezer [1992b] also used such a chemical vapor deposition process to produce uniformly thin amorphous silica films, consisting of clusters of tightly packed grains, on porous α- and γ–alumina supports under a low pressure of 0.02 bar. The mean pore diameters of the supports used are 0.2 μm and 4 nm, respectively. The deposition rate increases with the temperature employed. Triisopropylsilane oxidized at 750°C results in a deposition rate of 50 nm/min which has been found to be rather insensitive to the ratio of the two reactants used. The interface between the support and the membrane formed is rather well-defined and the membrane has completely plugged the pores in the support structure. The choice of triisopropylsilane over silane was made because triisopropylsilane is non-explosive, non-toxic and easier to handle. In addition, silane oxidation at a lower temperature (e.g., 450°C) yields thermally less stable silica membranes. Similarly, Ha et al. [1993] prepared dense silica membranes by pyrolysis of tetraethylorthosilicate.

The second approach is more novel and complex. Its reaction configuration has been referred to as the opposing-reactants geometry [Isenberg, 1981; Gavalas et al., 1989; Megiris and Glezer, 1992a] or chemical vapor infiltration [Carolan et al., 1994]. In this approach, one or more vaporized organometallic compounds come into contact with and permeate into one surface of a porous support. A gaseous reactive agent (an oxidizing or reducing agent) enters the other surface of the support. The two reactants diffuse opposite to each other and react within a narrow front inside the pores of the support to deposit a thin film of metal oxide (or carbonate) or just metal. Depending on the ratio of the reactant vapor concentrations, the deposition location can be varied from the surface to the inside of the substrate [Cao et al., 1993]. When deposition takes place inside the support, pore narrowing first occurs and further sufficient deposition may lead to pore closure. Once a pore is plugged, the normal chemical vapor deposition process stops. Dense silica membranes have been prepared this way.

If, however, the deposit material is oxygen permeable such as yttria-stabilized zirconia, another type of deposition, electrochemical vapor deposition, starts [Lin et al., 1989]. In the electrochemical vapor deposition process which was originally developed for making metal oxide layers for solid oxide fuel cells, the oxygen source (e.g., oxygen or water vapor) permeates through the solid oxide electrolyte plug and reacts with the organometallic vapor to continue deposition. Given sufficient time, the plug can advance to the pore mouth and the deposition can even progress further to form a film on the surface of the substrate that is in contact with the organometallic compound(s). Thus, the membrane obtained essentially does not have through-pores. A schematic of the development of pore modification/narrowing, pore closure and film growth on the support surface is depicted in Figure 3.1 [Lin et al., 1989]. In stage 1, the walls of the support pores are coated with a very thin deposit layer which modify the surface chemistry of the pores. In stage 2, the pores are narrowed by further deposit. When a sufficient amount of deposit occurs in a pore, plugging can take place as in stage 3 when the conventional chemical vapor deposition stops. The oxygen from the gas reactant can permeate through such plugs in the form of an oxygen ion and continue the chemical reaction, thus the film grows and can eventually form a gas-tight layer on the surface of the support (stage 4).

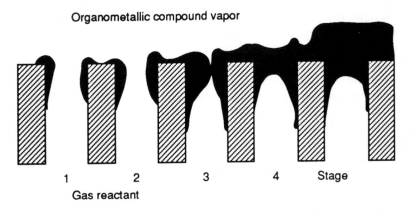

Figure 3.1. Schematic of the various stages of pore narrowing and closure by the chemical vapor deposition/electrochemical vapor deposition process [Adapted from Lin et al., 1989]

Deposition from having both reactants introduced to the same side of the support yields thinner and more permeable dense membranes than that from the opposing-reactant geometry.

Since the thickness of the deposit layer directly affects the permeability of the resulting membrane, it is highly desirable to produce as thin membranes as practically possible while retaining their physical integrity and maintaining uniform thickness along the membrane length. To ensure uniform membrane thickness by CVD, Kim and Gavalas [1995] proposed a novel reaction scheme called "alternating reactant vapor deposition" which has been known in electronic devices manufacturing as atomic layer epitaxy. Take the example of forming silica membranes from chlorosilanes and water vapor. When flowing both reactants simultaneously, particles are formed in the gas phase as well as on the surfaces of the support or its pores. The particles formed in the gas phase also deposit on the support, thus causing non-uniform thickness along the membrane length. A proven effective means of producing more uniform, thinner and denser membranes with higher permeabilities and permselectivities is to flow the two reactants alternately instead of simultaneously. The alternating deployment of the reactants for deposition is repeated multiple times to arrive at the desired thickness.

To further increase the membrane permeability, the idea of placing temporary carbon plugs in the support pores prior to CVD to limit the penetration of the metal-compound (e.g., chlorosilane) and hence reduce the thickness of the final membrane was proposed by Jiang et al. [1995]. The carbon barriers or plugs are introduced into the pores of the support by vapor deposition polymerization of furfuryl alcohol, catalyzed by p-toluene sulfonic acid. After polymerization and crosslinking, the polymer is carbonized by slow heating to a temperature of 600°C. CVD of silica is then carried out also at 600°C by alternating flows of chlorosilane and water vapor. The carbon plugs are removed by oxidation after silica deposition is completed. The use of temporary carbon barriers increases the permeabilities of the final membranes by 2-4 times.

It has been found that a higher deposition temperature causes a wider deposition zone. The above deposition process has been analyzed through a theoretical model that reveals the importance of the vapor and gas diffusivities, bulk phase reactant concentrations and the pore size of the support [Lin and Burggraaf, 1992]. Practically speaking, in order for the pores of the support to be plugged, their pore diameters should not be too large, preferably significantly smaller than 1 μm.

In addition to those solid electrolyte and non-electrolyte oxide membranes mentioned above, other membrane materials have also been investigated. Yttria-stabilized zirconia membranes can be prepared from $ZrCl_4$ and YCl_3 [Lin et al., 1989], silica/carbon membranes from triisopropylsilane[Megiris and Glezer, 1992a], oxygen-conducting mixed-oxides membranes containing La, Ba, Co and Fe from their organometallic complexes, $La(thd)_3$, $(Ba)_4(thd)_8$, $Co(thd)_2$ and $Fe(thd)_3$ and a $BaO/BaCO_3$ mixture membrane from $(Ba)_4(thd)_8$ where (thd) represents (2,2,6,6-tetramethyl-3,5-heptanedionate) [Carolan et al., 1994].

Similarly, impervious yttria-stabilized zirconia membranes doped with titania have been prepared by the electrochemical vapor deposition method [Hazbun, 1988]. Zirconium, yttrium and titanium chlorides in vapor form react with oxygen on the heated surface of a porous support tube in a reaction chamber at 1,100 to 1,300°C under controlled conditions. Membranes with a thickness of 2 to 60 μm have been made this way. The dopant, titania, is added to increase electron flow of the resultant membrane and can be tailored to achieve the desired balance between ionic and electronic conductivity. Brinkman and Burggraaf [1995] also used electrochemical vapor deposition to grow thin, dense layers of zirconia/yttria/terbia membranes on porous ceramic supports. Depending on the deposition temperature, the growth of the membrane layer is limited by the bulk electrochemical transport or pore diffusion.

Other methods. Suspension of fused β-alumina powder in an organic solvent such as amyl alcohol is electrically charged by adsorption of positive charges from dissociation of a dissolved organic acid (e.g., trichloracetic acid). Homogeneous deposition by this electrophoresis method has been used to prepare solid electrolytes in tubular form [Powers, 1975; Kennedy and Foissy, 1977].

3.1.3 Dense Inorganic Polymer Membranes

Structural deformation and decomposition of organic membranes upon exposure to high temperatures or harsh chemical environments motivate the development of not only ceramic and metal membranes but also inorganic or organometallic polymeric membranes. New families of inorganic or organometallic polymers with the backbone molecules other than carbon have received increasing attention for many potential applications in recent years. One such a promising application is membrane separation and reaction. Although they can not withstand the very high temperatures that ceramic or metal membranes do, these inorganic membranes fill the needs of "medium" temperature range applications. Due to the wide variety of the organic side groups possible in their structures, these polymers can be tailored to different properties. Perhaps the most publicized family of inorganic polymers as a membrane material is polyphosphazenes.

Polyphosphazenes. Phosphazene polymers belong to the class of the amorphous rubbery polymers in contrast to the glassy polymers which are presently used in many gas separation applications. Generally rubbery polymer membranes exhibit higher permeabilities but lower permselectivities compared to glassy polymer membranes. Polyphosphazene membranes, however, show two unique properties. They have very good permselectivity in favor of acid gases (e.g., carbon dioxide and hydrogen sulfide) against non-acid gases, particularly hydrocarbons (e.g., methane). And they are thermally and chemically more stable than organic membranes because of their backbone molecules. Polyphosphazenes consist of alternating phosphorus-nitrogen single and double bonds in the polymer backbone with two organic substituents attached to the phosphorus atoms. Their basic building block can be simplified as follows:

$$\left[\begin{array}{c} YR \\ | \\ -P = N- \\ | \\ Y'R' \end{array} \right]_n$$

where Y and Y' can be the same or different and selected from oxygen, nitrogen or sulfur and R and R' can be the same or different and selected from substituted alkyl and aryl and n is an integer of 10 or more (to about 70,000). Typically polyphosphazenes with $-OR$ side groups on the phosphorous are more thermally and chemically stable, for example, to hydrolysis, than those with $-NR_2$ groups [Kraus and Murphy, 1987]. It is noted that polyphosphazenes can exist not only in a linear polymer form as might be implied in the above formulae but also in a cyclolinear or cyclomatrix form.

These side groups can be easily modified with various organic groups by nucleophilic substitution and exchange reactions. Depending on the polymer backbone structure and the side groups, polyphosphazenes can exhibit a multitude of "design" physical and chemical properties, thus leading to varying separation properties. Phosphazene which is usually made as a cyclic trimer can form three types of polymers with the (-PN-) skeletons: linear polyphosphazene by ring cleavage polymerization of a cyclic trimer, and cyclolinear and cyclomatrix polyphosphazenes which are synthesized by reacting the cyclic trimer with difunctional monomers.

Many of the fabrication methods developed for organic polymers often can be used to process inorganic polymers. Polyphosphazenes can be cast into films by the widely used method for preparing organic polymers: phase inversion from solution. For example, dense and homogeneous poly[bis(trifluoroethoxy)phosphazene], poly[bis(phenoxy)phosphazene], copoly[(phenoxy, p-ethyl phenoxy)phosphazene], poly[cumyl(phenoxy)phosphazene], and poly[cumylphosphazene] membranes have been made by first preparing the polymers and then dissolving the polymers in a suitable solvent such as ethyl acetate, tetrahydrofuran methanol, acetone or dimethylacetone for casting [Kraus and Murphy, 1987; McCaffrey et al., 1987; Gaeta et al., 1989; Peterson, et al., 1993]. The polymer solution, in the range of 1 to 5% concentration, is then cast by evaporation followed by coagulation (with water) using the solution casting technique. Transport and thus separation properties of polyphosphazene membranes greatly depend on the microstructure and morphology of the membranes which in turn are a strong function of the casting conditions. Two such critical factors are the choice of the solvent system [Peterson et al., 1993] and the rate of solvent evaporation [McCaffrey and Cummings, 1988]. The polymer concentration can also be a critical parameter. Too low a concentration produces a membrane with too much porosity that is mechanically weak while too high a concentration may result in unstable solutions. The coagulation temperature affects the membrane permeability and permselectivity in the opposite directions: the permeability increases while the permselectivity decreases with increasing coagulation temperature [Gaeta et al., 1989].

A characteristic of most rubbery polymers such as polyphosphazenes is their mechanical weakness. For this reason, polyphosphazene membranes need an underlying layer of mechanical support which, for permeation consideration, should be porous. Multiple coatings of polyphosphazenes a few microns thick on the porous support may be required. Polyphosphazene membranes so obtained are expected to be operable reliably for an extended period of time at a temperature of up to about 175 to 200°C. Cross-linking the unsaturated polyphosphazenes can further enhance their thermal and chemical stabilities toward more aggressive application environments involving higher temperatures, solvating compounds or swelling impurity components. This, for example, can be accomplished either by treating polyphosphazenes with disodium salt of highly fluorinated alkyl diols in solution near ambient temperature or in solid state at a higher temperature [Kraus and Murphy, 1987].

Spray forming of polyphosphazene membranes. Like many organic polymer membranes, polyphosphazene membranes can be fabricated by conventional methods such as evaporative knife casting as briefly discussed earlier and in Chapter 1. Recently, Idaho National Engineering Laboratory has developed a new and faster technique for preparing polyphosphazene membranes, called controlled aspiration process. It is based on the concept of spray forming in which atomized droplets of a solution (e.g., 7 wt. %) of linear poly[bis(phenoxy)phosphazene] (with a molecular weight of 100,000 to 750,000 daltons) in tetrahydrofuran are deposited onto a support to form a polymer film. The atomized droplets are generated from a converging/diverging nozzle having a throat at which the polymer is introduced and an exit from which the nebulized droplets leave entrained in a carrier gas such as argon, as schematically shown in Figure 3.2 [McHugh and Key, 1994]. The solvent molecules are shed from the atomized droplets during their flight and the remainder evaporate at the substrate.

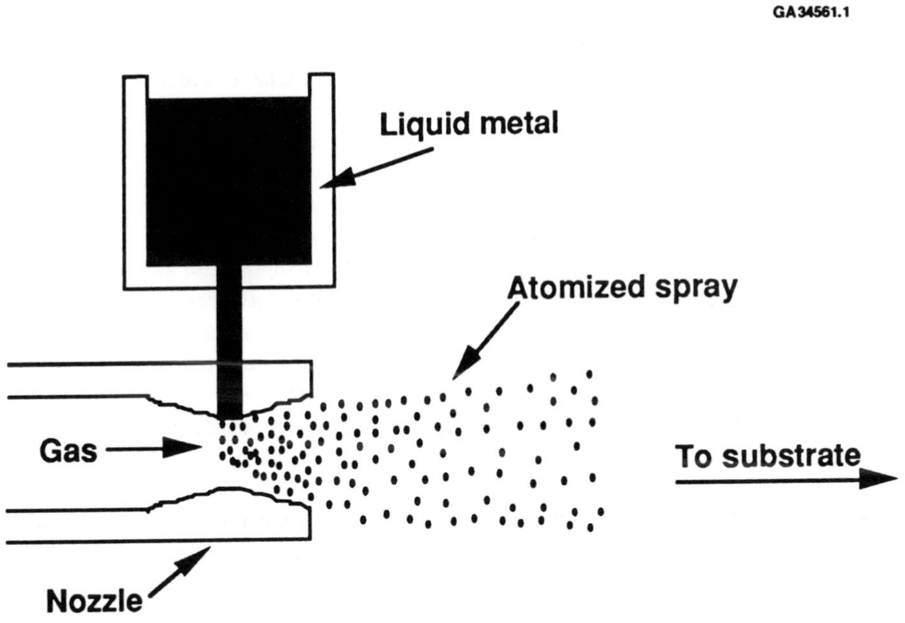

GA34561.1

Figure 3.2 Schematic diagram of spray forming of inorganic polymer membrane [McHugh and Key, 1994]

The resulting membranes exhibit very good separation properties. For example, the separation factors for the gaseous mixtures of 10%SO$_2$/90%N$_2$ and 10%H$_2$S/90%CH$_4$

by the spray formed polyphosphazene membranes at 80 and 130°C are orders of magnitude higher than those exhibited by knife cast polyphosphazene membranes [McHugh et al., 1993]. At 130°C, the separation factors are 303 and 344, respectively. The spray formed membranes also show high performance in separating the liquid mixture of halogenated hydrocarbons and alcohols. Other advantages of spray formed membranes over the conventional knife or spin casting method are the fabrication speed (drying in seconds versus hours) and the ability to produce near-net shape membrane elements.

Polysilazanes. Another organometallic polymer which has the potential for being an inorganic membrane material is polysilazane. There is indication that this family of Si- and N-based inorganic polymers has been studied as a membrane material [Johnson and Lamoreaux, 1989] but no detail information is available.

It should be noted that the aforementioned polymers and many other inorganic or organometallic polymers containing non-carbon backbones can also be pre-ceramic polymers. Upon pyrolysis they can form corresponding ceramics. Broadly speaking, inorganic polymers include ceramics, silica, silicates, metals and ionic crystals. Many of them possess high temperature stabilities and excellent elastomeric properties at low temperatures for ease of fabrication. This potentially opens a large window of possibility for various kinds of inorganic membranes. The field of inorganic polymers and its potentials for inorganic membranes are beyond the scope of this book.

3.2 TORTUOUS-PORE MEMBRANES

The methods of preparing inorganic membranes with tortuous pores vary enormously. Some use rigid dense solids as the templates for creating porous structures while many others involve the deposition of one or more layers of smaller pores on a pre-manufactured microporous support with larger pores. Since ceramic membranes have been studied, produced and commercialized more extensively than any other inorganic membrane materials, more references will be made to the ceramic systems.

Most of the remaining discussions in this chapter will be grouped according to the preparation and fabrication methods. Certain types of materials are more amenable to specific manufacturing methods. They will be referenced in the discussions. However, separate treatments will be made on those emerging materials, for example, sol-gel and ultramicroporous materials.

3.2.1 Pressing, Casting and Extrusion

To make membranes or typically membrane supports possessing relatively large pores (i.e., diameters larger than 1 μm), traditional polymer or ceramic shape forming processes involving compaction of precursor particles can be employed. The most noted examples are pressing and extrusion.

Pressing. Pregraded powders can be pressed or consolidated, in a dry or damped state, under relatively high pressures into porous aggregates which are then heated to convert into metallic or ceramic membranes or supports having relatively large pores in disc or other forms. The use of high pressures not only ensures the strength of the "green" body but also reduces shrinkage as a result of subsequent heat treatment. The precursor particles typically require a combination of some preparation steps such as milling (dry or wet), screening, de-ironing, and blending to obtain a desired particle size distribution. Organic or inorganic binders (or both) are added to hold the constituent particles together during drying and sintering, particularly for naturally non-plastic materials including alumina and other oxides. Some examples of organic binders are polyvinyl alcohol and waxes.

It is not uncommon to observe cracks or pinholes in the porous membranes made by pressing. Cracks can be caused by a number of factors: air entrapment, die wear and uneven pressure distribution. One way to combat the problem of uneven pressures is to use the method of isostatic pressing where the pre-ceramic sample is placed in a rubber sheath and pressure is applied through a fluid. Porous membranes have been made this way with many materials including alumina, zirconia, titania, silica and potassium titanate. Potassium titanate can be made into an alkali resistant filter or battery separator [Smatko, 1973]. Isostatic pressing of powders finer than 10 μm has been applied to fabricate stainless steel tubular membranes with extremely high mechanical strength and a mean pore diameter of 3 μm and a porosity of 42% [Boudier and Alary, 1989]. It has been claimed [Sumitomo Electric Ind., 1973] that finer-pore nickel or alumina membrane tubes can be made by pressing superfine constituent powders (0.05-0.2 μm in diameter) between two cylindrical bodies (made of Ni) followed by heating the material and bodies at elevated temperatures. After cooling, the inner body is then removed from the sintered membrane material by dissolution and the outer body is subjected to a corrosive treatment to form many pores through the membrane thickness. The resulting membrane pore diameters are in the 1-10 nm range.

One of the latest techniques to produce porous materials, including membranes, is hot isostatic processing (HIP) applied to powder compacts or pre-sintered bodies. HIP is normally used to fully densify materials. However, if the powder compacts or pre-sintered bodies are sintered directly at high temperatures under high gas pressures, the densification is delayed and open pores are prevented from closing. The hot isostatic processed porous materials have a higher open porosity, higher mechanical strength, narrower pore size distribution and higher fluid permeability than conventional sintered porous materials. The enhanced properties of the porous materials by HIP is probably due to increased surface diffusion caused by the high gas pressure used. The narrower pore size distribution is the result of the well-grown necks between particles and disappearance of fine pores and the improved permeability is attributed to interconnected more or less round-shaped pores. Titania membranes made by the HIP have been reported and porous membranes made of superconducting oxides, such as $YBa_2Cu_3O_7$, can separate paramagnetic from diamagnetic gases such as oxygen from argon [Ishizaki and Nanko, 1995].

In attempts to prepare inorganic membranes for gaseous diffusion applications, Mårtensson et al. [1958] produced reinforced sintered membranes by roll pressing or hydraulic pressing of metal and/or ceramic powders into the mesh of a wire gauze, followed by sintering. The powders are prepared by the sputtering of metal electrodes in an electric arc in various gas atmospheres. For example, powders comprised of aluminum nitride and aluminum have been produced by sputtering aluminum in a nitrogen atmosphere. Predominantly alumina powders can be obtained by sputtering aluminum in an oxygen-nitrogen atmosphere (10 to 20% oxygen). Nickel powders have also been made by sputtering nickel in nitrogen. The wire gauze is deposited with the powders from the electric arc sputtering step. After deposition, the powder is forced into the mesh by passing the gauze through a rolling mill or by applying a hydraulic press. The pressed gauze containing the powders is then sintered at a temperature of up to 650°C for strength consideration.

Tape casting. Tape casting is a well established practice for making substrates in microelectronics. The same type of process can be employed to form either a support for a membrane or a homogeneous membrane layer by itself. Tape casting has the advantage of being able to cast a large piece of flat support or membrane but suffers from the potential drawback of not being able to control the contact time between the deposition slurry and the membrane support. This technique has been known to fabricate films on the order of 0.01 to 1 mm thick.

A slip containing properly sized particles, dispersants, binders and plasticizers is often ball milled to enhance dispersion of the particles. This is followed by deaeration. The slip is then spread onto a substrate (usually a plastic sheet) on a moving blade caster using a knife edge called doctor blade. The "green" membrane thickness is controlled by adjusting the height of the doctor blades. The binder (e.g., polyvinyl butyral, methyl cellulose, or acrylic [Winnick, 1990]) in the slip helps to hold the tape together when dried, the plasticizer (e.g., polyethylene glycol or polyethylene oxide) is designed to make the dried tape flexible for ease of removal from the substrate and the dispersant facilitates the coating of the binder on the constituent solid particles. Drying rate can be critical to the success of making a crack-free membrane by tape casting. On one hand, the drying should be fast enough to prevent the constituent particles from settling and consequently segregating. On the other hand, if drying is too fast, cracking occurs.

After air drying, the film is separated from the substrate and stamped to the desired shape and dimensions. This is followed by burning out the organic matter at a temperature of, say, 400-500°C and then sintering at a much higher temperature. A solid content approaching 70% and an excess of the binders (e.g., 20% by weight) to increase the membrane porosity can be used. The heat treating cycle required to make the final membranes consists of multiple stages with different ramping rates.

In the case of tape casting a membrane on a porous support, the membrane thickness and the quality of the support surface appear to be important issues. It has been found that alumina, but not titania or zirconia, can be tape cast into crack-free membranes on porous supports and any defect on the support surface (e.g., a protrusion or cavity) can yield a deposit layer that is either bare or cracked [Simon et al., 1991]. Alumina membranes have been tape cast on top of tape cast alumina supports. Borosilicate glass, but not silica, membranes have also been tape cast [Winnick, 1990].

Extrusion. Extrusion is a well established process for manufacturing plastic shapes and has also been applied to ceramic parts such as bricks and tubes. The process has been applied to making dense membranes in tubular shape as described earlier in this chapter. It has also been used to fabricate commercial ceramic and metallic supports or membranes with tubular and monolithic honeycomb forms. Stainless steel powders smaller than 10 μm can be extruded to fabricate thin, porous, tubular stainless steel membranes with a highly uniform porous structures [Boudier and Alary, 1989]. For porous ceramic membranes of high quality, the procedures appear to be more involved. High-consistency, plastic mass of pastes are prepared from several types of powders of well controlled particle sizes by screening and undergo several steps of mixing, deaeration and deagglomeration before molding into shapes. During the wet mixing or milling stage, ball mills or various paste mixers can be used. It is often necessary to remove some impurities (e.g., iron) from the pastes in order to maintain high quality or meet regulations governing certain types of applications such as food and beverages. To enhance formability and prevent cracks from occurring in the final membrane, frequently added to the paste are organic binders such as polyvinyl alcohol, polyvinyl acetal, water soluble cellulose ether, polyglycols, etc. and plasticizers such as glycerol or rubber solution in an organic solvent. Sometimes inorganic binders are also added. Care, however, should be exercised in the use of binders and lubricants as they may give rise to such problems as material slumping and air entrapment. Entrapped air can cause blisters on the surface of the extruded parts which in turn can make uniform and pinhole-free deposition of a membrane difficult. These organic additives should be burned off during subsequent calcination or sintering steps.

The constituent particles, various binders and other additives used for making porous supports or membranes have been shown to affect the plastic properties of the extrusion paste. The plastic properties, in turn, have significant impacts on the final support and membrane characteristics [Shkrabina et al., 1995a and 1995b].

The paste is then forced through a die at a given speed which is controlled by the piston pressure. Some details of this process have been given in the literature (e.g., Schleinitz [1981]). Sometimes painstaking care and seemingly repetitive procedures are taken to ensure that the membranes so produced are crack- and pinhole-free, dimensionally consistent and devoid of surface roughness. It is possible to simultaneously extrude a double-layer structure where the two layers are made of particles of different sizes [Charpin and Rigny, 1989]. This way the fabrication can be carried out in one operation instead of two. The "green" body thus formed has sufficient strength to hold its shape before it is transferred to the drying step.

The precursor powder compositions can be critical in determining the final microporous structures in pressed or extruded porous ceramic bodies. One very important, but not very well documented, factor is the size distribution of the precursor particles. Coarse particles generally result in pastes that have less plasticity and "green" strength but are easier to dewater and the final heat treated products are subject to less shrinkage. The particle size distribution determines how compactly particles pack, thus affecting the porosity and the pore size distribution. For example, Clark et al. [1988] demonstrated this through a series of designed experiments of extrusion (of supports) and a simple particle packing theory by Furnas [1931]. They showed how the mean and the largest pore sizes of a ceramic support and a membrane-support asymmetric composite vary with the sizes and the percentages of the precursor or constituent particles used. A ceramic membrane shown in Figure 3.3 has a mean pore diameter of approximately 10-15 μm. It is noted that the constituent particles are about 20 to 60 μm in diameter. Many ceramic membrane manufacturers have employed this type of large-pore membranes as supports for finer-pore membranes.

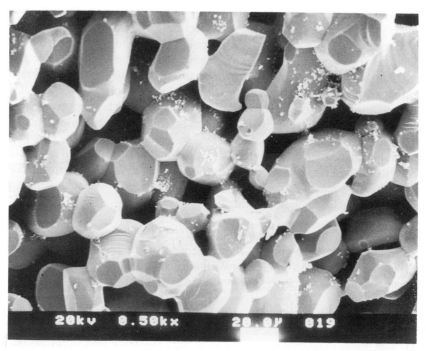

Figure 3.3 Scanning electron micrograph of an alumina membrane with 10-15 μm pores

Another just as important factor in affecting the final pore size distribution of a ceramic membrane is the degree of aggregation of precursor particles in the "slip." The degree of aggregation, in turn, is largely determined by the pH of the slip. Terpstra and Elferink [1991] have shown that the pH of the otherwise identical slips has significant effects on the pore size (Figure 3.4) and consequently the water permeability (Figure 3.5) of the resultant alpha–alumina membranes. For example, the maxima of the pore size and the water permeability curves approximately correspond to the isoelectric point of alpha-alumina at 8.5.

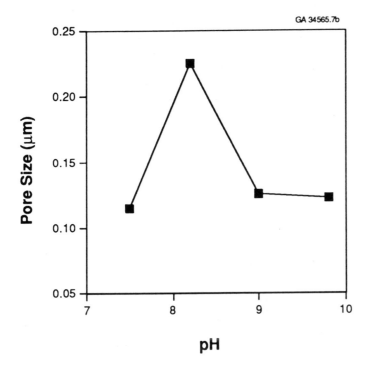

Figure 3.4. Effect of slip pH on mean pore diameter of alpha-alumina membrane [Terpstra and Elferink, 1991]

Many commercial ceramic membranes nowadays come in the form of a monolith consisting of multiple, straight channels parallel to the axis of the cylindrical structure (Figure 3.6). The surfaces of the open channels are deposited with permselective membranes and possibly one or more intermediate support layers. The porous support of these multi-channel structures are produced by extrusion of ceramic pastes described above with a channel diameter of a few millimeters. Their lengths are somewhat limited by the size of the furnaces used to dry, calcine and sinter them and also by such practical considerations as the total compact weights to be supported during heat treatment and the risk of distortion in the middle section. It should be noted that this type of honeycomb structure in principle can be made with a template material that can be burnt out. For example, more or less straight, uniform-sized channels can be created by using carbon filaments aligned in a matrix with a ceramic compact (e.g., alumina or zirconia) and, upon firing, channels of controlled diameters are formed [Kato and Mizumoto, 1986]. The diameter of the channels generated by this method can be smaller than one micron.

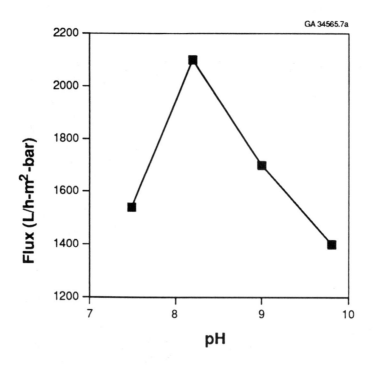

Figure 3.5. Effect of slip pH on water flux of alpha-alumina membrane [Terpstra and Elferink, 1991]

The porous mass of the monolithic structure has a low pressure drop. If further reduction of the pressure drop in the bulk support is desired, there are a few possibilities [Atomic Energy Commission (France), 1971]. Additional grooves can be machined or swaged with dowels which will produce channels, or a network of additional channels perpendicular to the honeycomb feed channels is provided by using materials such as carbon that can be burned off.

The "green" macroporous compact is next subject to carefully controlled drying and subsequent calcination and sintering conditions. The drying and pre-sintering treatments are typically very time-consuming and require days instead of just hours. The thermal treating conditions (e.g. heating and cooling schedules and procedures) and the required facilities and tools (e.g. associated handling materials such as the type of ovens and furnaces and the materials to hold the green materials during thermal treatments) are often guarded as proprietary information [Charpin and Rigny, 1989]. The temperature-time schedule not only affects the pore size of the final membranes but also can determine the final phase compositions. Programmable rates of heating and cooling can avoid the costly hairline cracks in the product. But the sensitivity of heat treatment has to be balanced with productivity to search for the window of robust and yet not complicated heating conditions.

Figure 3.6 Cross-sectional view of a multi-channel monolithic membrane element [Courtesy of U.S. Filter]

The organic binders are burnt off during the early stages of heat treatment. Calcination or pre-sintering is performed at a moderate temperature depending on the material (e.g., at a temperature of 400°C or higher in air to calcine alumina [Atomic Energy Commission (France), 1971]). The sintering process is performed at a temperature below but approaching the melting point of the metal alloy or ceramics. For sintering metals such as stainless steel, the atmosphere is reducing in nature and serves to remove surface oxide films. At the sintering conditions, particles are bonded to their neighboring particles via the process of solid state diffusion and form an interconnected porous membrane structure. The sintering temperature required varies with the materials involved. For α-alumina, sintering starts at approximately 1,100°C and completes around 1,400°C. For many refractory oxide mixtures that contain alumina as the primary component, their sintering temperatures for making porous bodies are very similar at 1,400 to 1,420°C [Siemens Aktiengesellschaft, 1975]. For titania, the sintering temperature lies in the range of 1,400 to 1,500°C. In practice, the operating sintering temperature may be even higher at 1,500 - 1,800°C [Jain and Nadkarni, 1990]. For other materials such as silicon carbide, sintering takes place at an even higher temperature, up to about 2,300°C [Ibiden Co., 1986]. This type of information on production practice is obviously either sketchy or lacking in the open literature.

After sintering, the extruded and heat treated membrane is applied with end-sealing compounds on both ends of the membrane element to prevent remixing of the permeate and retentate during application. Furthermore, gaskets or other joining schemes are used to connect the membrane to processing vessels or piping to form a membrane module.

A wide range of relatively large-pore membranes (approximately 1 to 100 μm) can be made by this method through careful control of the powder particle size and the extent of aggregation. Equally attractive is the wide range of materials that are amenable to this technique. Not only metal oxides (alumina, magnesia, titania, zirconia and quartz), spinel ($MgO \cdot Al_2O_3$), mullite ($3Al_2O_3 \cdot 2SiO_2$), cordierite ($2MgO \cdot 2Al_2O_3 \cdot 5SiO_2$), zirconium silicate, aluminum phosphates, graphites, silicon carbide, etc. have been used, but also metals such as stainless steel [Guizard et al., 1989]. Due to the typical broad size distribution of these large pores, the permselectivity of these membranes is usually not very high. These self-supporting porous materials have also been widely used as the underlying primary structural supports for permselective membranes with finer pore sizes for microfiltration, ultrafiltration, gas separation, pervaporation and catalytic membrane reactor applications.

When used as a support structure, these porous materials made with extrusion or compaction can affect the quality of the deposited membranes. For example, if the surface of such a porous support exhibits cracks or "peaks" and "valleys," it is very difficult, if not impossible, to form a uniform and smooth layer of membrane on top of it. Therefore, the desirable characteristics of a membrane support can be summarized as follows. It should be easy to deposit one or more layers of fine particles on top of it, should provide the necessary mechanical strength, and should impose relatively low flow resistance.

3.2.2 Preparation of Porous Hollow Fiber Membranes by Extrusion

Hollow fiber represents perhaps the highest packing density geometry of membrane element. Organic membranes in the form of hollow fibers have been quite successful in the marketplace. Although more challenging, inorganic hollow fiber membranes are a logical extension of their organic counterparts. So are the basic processes for fabricating them. Because of their small diameters and great potentials, their preparation methods are treated separately here.

Based on similar techniques for making organic hollow fibers, Dobo [1980] has provided details for making porous inorganic hollow fiber membranes with an outer diameter of 100 to 550 μm and a wall thickness of 50 to 200 μm. Although the inorganic material can be Ni or its alloys, nickel oxide or its mixture with iron oxide, alumina, β-alumina, glass, mullite, silica, cermets etc., details have been provided to make porous nickel oxide membranes. Basically, the process involves the following major steps:

(1) Prepare an organic fiber-forming polymer solution containing nickel oxide. It is desirable to disperse nickel oxide in a solvent before adding the polymer. The solvent can be dimethylacetamide, dimethylformamide or dimethyl sulfoxide. The choices of the polymer are very broad and can be dictated by the cost issue. Polyacrylonitrile and polymers of acrylonitrile with one or more other monomers polymerizable are desirable. The ratio of the inorganic material to polymer, by weight, is preferably 4.5 to 10. Extrusion and fiber-forming considerations suggest the use of the polymer at a concentration of 15 to 30%, by weight, of its solution in the solvent. Small particles of nickel oxide (5 μm or less in diameter) produce hollow fibers with fewer cracks and other surface defects. Surfactants such as sorbitan monopalmitate can help wet nickel oxide by the solvent and plasticizers such as N,N-dimethyl lauramide improve the plasticity of the mixture for easier extrusion and fiber forming.

(2) Extrude the aforementioned mixture through a hollow fiber spinneret. Degassing is recommended to reduce any void formation in the final membrane. The mixture is often maintained under a substantially inert (e.g., nitrogen) atmosphere to prevent coagulation of the polymer and fire hazard of the solvent prior to extrusion. A pressure of 1 to 5 atm is used.

(3) Form a polymeric precursor hollow fiber. The fiber spinning techniques practiced in the synthetic fibers industry can be utilized. Depending on the wet or dry spinning process, the spinneret may be in or outside the coagulating bath which, maintained at 2 to 25°C, coagulates the fibers after extrusion. The wet process is preferred. Ethylene glycol, methanol or water can be used as the coagulating agent, especially ethylene glycol. Frequently the extrusion and fiber-forming speeds are set in the 0.1 to 1.7 m/s range. Drying of the fibers occurs at 15-35°C and 40-60 % relative humidity.

(4) Remove the organic polymer in an ammonia gas atmosphere and sinter the remaining in a furnace. Upon heating, the volatile components escape and the nickel oxide fiber surface becomes uniformly porous to form a nickel oxide membrane. Significant shrinkage can be expected.

3.2.3 Dip and Spin Coating

Porous membranes are composed of constituent particles or polymers. A dispersion of such particles with a controlled size distribution or polymer in a solvent (in most cases, water and/or alcohol) is called a slip. The slip can be deposited onto a rigid microporous support by the so-called dip coating (or sometimes referred to as wash coating or slip casting) or spin coating process to form a thin layer of well-ordered solids which, upon proper drying and calcining, is converted to a porous membrane. The deposition process is facilitated by capillary forces.

Porous membranes with high separation efficiency require narrow pore size distributions. This, in turn, calls for precursor particles with narrow size distributions which can be attained by such classification methods as sedimentation or sieving. It is, therefore, essential to have a de-flocculated or de-agglomerated slip to make a pinhole-free membrane layer. Dispersants (e.g., Darvan C as described by Gillot [1987]) and peptizing agents (e.g., nitric, hydrochloric or perchloric acid) are used and, when necessary, additional energy such as ultrasonification [Berardo et al., 1986] or prolonged grinding or milling [Gillot, 1987] can be applied as well. Typically other chemical agents such as organic and/or inorganic binders and viscosity modifiers are added to enhance pumping and deposition characteristics of the slip. Thermally reactive inorganic binders have been suggested to reduce the firing temperature necessary to achieve strong bonding between membrane constituent particles and between these particles and the underlying support. The inorganic binder particles react at a temperature substantially lower than that of the membrane constituent particles. These inorganic binders (such as certain phases of alumina, zirconia, titania, silica, silicon carbide or silicon nitride) are added in the form of fine reactive ceramic particles less than 1 μm [Goldsmith, 1989]. Organic binders help the particles (including any inorganic binder particles) adhere to each other and minimize cracking during drying and subsequent heat treatment steps. The viscosity modifiers (such as polyethylene glycol or an acetal resin) control the fluidity of the slip to ensure a uniform and crack-free deposit layer. To ensure no air pockets exist in the slip which can have adverse effects on making pinhole-free membranes, the slip may require a deaeration step prior to deposition on the support.

In dip or spin coating, some mechanical means (such as pumping and/or spinning) is provided to bring the slip in direct contact with a dry support uniformly over its surface. For example, the inner cavity of a porous tube is filled with the slip by dipping or pumping for a few seconds to a few minutes before being emptied. The resulting capillary pressure difference causes the dispersion medium (in most cases water or water-alcohol mixtures) to be drawn into the pores of the support, leaving behind at the pore entrance a concentrated and well packed layer of particles on the contact surface of the support. Often the side of the porous support that is not to be deposited with the slip is kept at a slightly lower pressure to facilitate the deposition process by more quickly removing the dispersing liquid of the slip in a filtration mode. However, to prevent pinholes from occurring, the capillary suction of the pores should be minimum [Terpstra et al., 1988]. To avoid penetration of the particles into the support pores, the sizes of the particles or their agglomerates should not be too small. The concentration of the particles should not be too low so that run-off of the concentrated particle layer will not occur.

For microfiltration applications, pore diameters of ceramic membranes in the range of 0.1 to 10 μm are typical. These membranes can be prepared by dip or spin coating of particle slips. The size of the particles and the pH of the slips determine the pore size distribution of the membrane. Surfactants may be needed to reduce aggregation of particles. Although the exact recipes used by the ceramic membrane manufacturers to prepare microfiltration membranes are guarded trade secrets, indications are that superfine and narrow size distribution particles in the range of 0.5 to a few microns can be made into a slip or stable dispersion of a high solid concentration. For example, slips containing about 35% solids can be prepared as a precursor to ceramic microfiltration membranes by carefully controlling the pH of the dispersion [Terpstra and Elferink, 1991].

Prior to spin coating, both ends of a pre-fabricated porous support, either in the form of a tube or a monolithic honeycomb element, are filled with the slip and temporarily sealed with gaskets. Slips of Ni, Al and Al_2O_3 have been spin coated at a rotational speed of up to about 1,500 rpm to form membrane layers [Sumitomo Electric Ind., 1981; NGK Insulators, 1986]. The development of a coating layer is expected to be faster with the spin coating process than the dip coating process due to the centrifugal force.

Two parameters emerge as the critical factors during the dip or spin coating process in affecting the thickness of the membrane: viscosity of the slip and the coating speed or time. The membrane thickness, generally on the order of 20 to 100 μm, depends on how compactly the precursor particles pack which, in turn, is determined by the slip viscosity or pH. For a microfiltration membrane, Figure 3.7 depicts how the coating speed and the pH (and consequently viscosity) of the slip can determine the deposition or coating thickness of the membrane layer [Terpstra and Elferink, 1991]. The thickness is approximately proportional to the square root of the dipping or coating speed. Similarly, the deposition thickness of an alumina membrane has been shown to vary linearly with the square root of the dipping time [Leenaars and Burggraaf, 1985]. This is illustrated in Figure 3.8. The slope of the linear relationship depends on the slip concentration which determines the viscosity.

The dip coating process can be illustrated with a pressure drop profile across the membrane/support structure from the slip to the other side of the support not in direct contact with the slip. This can be schematically shown in Figure 3.9 to represent a certain stage (time) of the coating process. A gel layer of concentrated particles with a thickness L_g is formed on top of the support. A portion of the support close to the slip becomes saturated with the dispersion liquid (e.g., water). The wetted part of the support occupies a thickness of L_s and the rest of the support is assumed to be dry. The menisci in the boundary zone generate the capillary pressure drop which drives the coating process. The capillary pressure drop can be divided into the contribution due to the wetted support and that due to the gel layer and depends on the surface tension of the liquid in the pores of the support, contact angle between the liquid and the pore surface of the support, and the pore radius of the support. Based on this reasoning, Leenaars and Burgraaf [1985] have developed a model describing the gel layer (coating) thickness as a function of the coating time as follows.

Dip coating is analogous to a slip casting process for making ceramic parts. The membrane deposition behavior by slip casting can be described by a theory of colloidal filtration for incompressible cakes [Aksay and Schilling, 1984] and compressible cakes [Tiller and Tsai, 1986]. The theory predicts that the thickness of the consolidated layer, L, is given by

$$L = (C_g t/\mu)^{1/2} \tag{3-1}$$

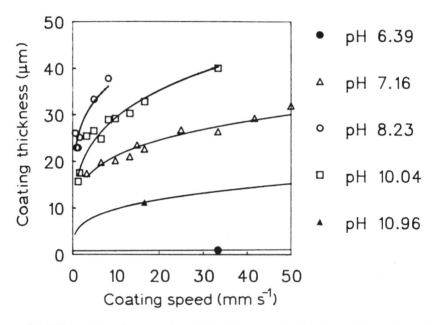

Figure 3.7 Effects of coating speed and slip pH on coating thickness of membrane layer [Terpstra and Elferink, 1991]

where t is the coating time, μ the slip viscosity and the constant C_g depends on the properties of the support and the gel layer including porosity, permeability and capillary pressure of the support and the volume fraction of solids in both the slip and the gel layer. Thus, Eq. (3-1) describes the behavior shown in Figure 3.8 except the non-zero intercept in the figure which is attributed to a thin layer of sol adhering to the gel layer when the support is removed from the sol [Leenaars and Burggraaf, 1985].

Alternatively, the coating thickness can be related to the coating or withdrawal speed by theories of dip coating:

$$L = C_g{'} (U\mu/\rho)^{1/2} \tag{3-2}$$

where U is the coating speed, ρ the slip density and the constant $C_g{}'$ depends on the density and surface tension of the slip. According to Eq. (3-2), the dependency of the coating thickness on coating speed also shows a square root relationship. Strawbridge and James [1986] found that experimental data of sol-gel derived silica coatings prepared by dipping follow Eq. (3-2) satisfactorily within the range of experimental capability.

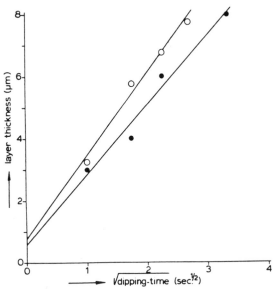

Figure 3.8 Deposition thickness of an alumina membrane as a function of dip coating time [Leenaars and Burggraaf, 1985]

While dip coating and spin coating are used most often for making microporous ceramic membranes, the filtration technique for forming dynamic membranes can also be used [Okubo et al., 1991]. The filtration method is particularly suitable when the support does not have enough pore volume to supply sufficient slip particles at the support surface to form a layer of membrane.

The deposit layer, after drying, is subject to a carefully programmed time-temperature calcination profile depending on the material. Calcination produces a membrane adhered strongly to the porous support. This step is very critical in determining if a crack-free membrane is made and, if so, what the final porosity and pore size are. During calcination combustible materials introduced as the processing aids in previous steps are burned off. The calcination temperatures for common metal oxides such as alumina, zirconia, silica and titania are usually in the range of 300 to 800°C, depending on the desired membrane properties.

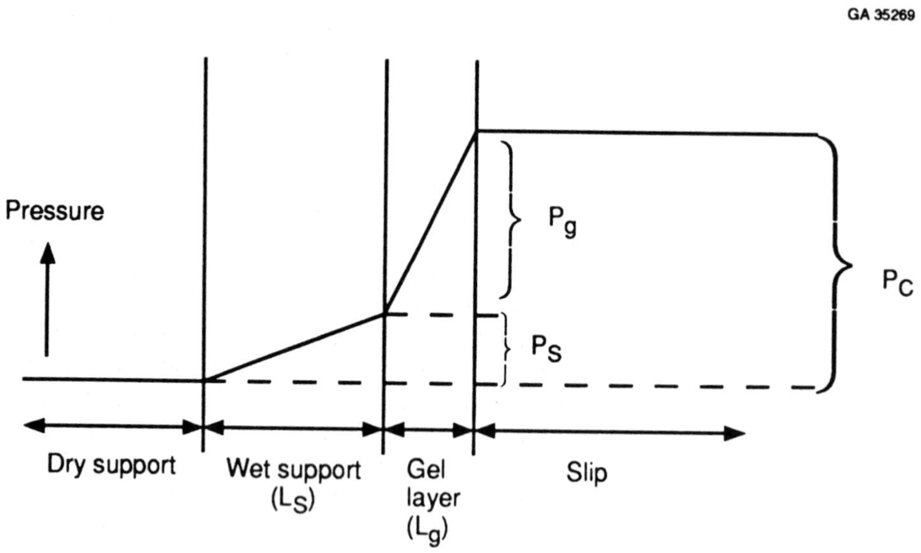

GA 35269

Figure 3.9 Pressure drop profile across the membrane/support structure during dip coating process [Leenaars and Burggraaf, 1985]

For making mechanically strong porous ceramic supports with larger pores, the calcined materials may continue to be heat treated to their sintering temperatures. It has been claimed that, between the calcination and sintering steps, a cold compaction step to the calcined membrane-support composite can help make the final membrane pore sizes finer and more narrowly distributed [Mitszenti and Nannetti, 1975]. Compaction is best carried out by subjecting the parts to an isostatic compression, for example, using deformable unvulcanized rubber.

Considerations should be given to the calcination or sintering temperature when the membrane material differs from the support material. For example, porous cordierite having a melting temperature of 1,450°C should not be used as a support for alpha-alumina membrane particles larger than 1 μm which require a sintering temperature exceeding 1,500°C [Goldsmith, 1989]. Similarly, it is essential that the thermal expansion coefficients of the membrane and the support materials are closely matched to avoid peeling, crazing or cracking during the fabrication (heat treatment) as well as the application stages. For example, the same two materials, alpha-alumina and cordierite, have distinctively different thermal expansion coefficients at $7-8 \times 10^{-6}$ and $1.0-1.2 \times 10^{-6}$°C^{-1}, respectively, and therefore will cause the aforementioned problem when the two materials are used in the same membrane composite.

Another form of spin coating is to use a molten material, rather than a slip or slurry, and spray onto the cooling surface of a metal roller. The material sprayed is then solidified by injecting liquefied nitrogen or carbon dioxide onto both sides of the membrane formed. This combination of spin coating and cryogenic solidification has been applied to such materials as barium titanate ($BaTiO_3$), metal nitrides or borides or ceramic alloys [Kuratomi, 1988].

Both dip and spin coating can be applied multiple times to produce multi-layered composite membranes with drying, calcining and possibly sintering steps as part of a repeating cycle. Intermediate layer(s) of support serves two major purposes. One is to control or avoid penetration of the particles from the membrane or finer-pore support layer into the support structure underneath the intermediate layer. Penetration of particles into the pores of the underlying, larger-pore layer causes hydraulic resistance of the membrane composite to increase. By eliminating any infiltration of particles into the underlying support layer, the membrane permeability can be increased by as much as 70%. One way to prepare a non-infiltrated membrane is to impregnate the support with a polymer solution of methyl cellulose and, upon the drying of the polymer film, to deposit a layer of the membrane on the support by dip or spin coating. The polymer film prevents the slip particles from penetrating into the support pores. The polymer film is burned off during the calcination step. Titania membranes on alumina supports have been prepared by this approach with an increase of the membrane permeation rate by 30-70% [Elmaleh et al., 1994].

The other reason for having intermediate layer(s) is to provide a better quality surface for the membrane layer to deposit. A membrane free of pin holes demands a smooth surface of the underlying support layer. Gillot [1987] described a criterion on the roughness of a support layer for making a good quality membrane on top of it. He argued that the roughness level of the support layer should be less than 10% of the mean diameter of its constituent particles and used the methodology to prepare three-layered membrane-support composites.

The dip and spin coating methods are the most practiced techniques as indicated in the open literature. Many materials have been made into porous membranes by this route including the more commonly referenced materials such as metal oxides, silicon carbide, silicon nitride, silicon and aluminum oxynitrides and glasses.

To make economically viable ultrafiltration or gas separation composite or asymmetric inorganic membranes, the separating layer should be as thin as possible to retain high fluxes and its pores should be small enough and the associated pore size distribution needs to be narrow to impart the required permselectivity. For ceramic membranes of this type, the pore size required are generally smaller than 100 nm. An emerging family of techniques for preparing these microporous membranes is the sol-gel process. Since the sol-gel process plays a very important role in making many of the commercial ceramic membranes today and potentially new membranes in the future, the materials science and engineering aspects of this process will be treated in more details in a separate section that follows.

3.2.4 Sol-gel Processes

Nanoscale materials are drawing increasing attention in recent years due to their potentials in various high technology applications. Among the processes that have been studied extensively, the sol-gel process seems to be by far the most suitable for making thin porous films with macropores (for microfiltration), mesopores (for ultrafiltration) or nanopores (for nanofiltration), high-purity and uniform constituent particles in an industrial production environment. It also offers the versatility of introducing catalytic activity to the membranes through different means: preparation of inherently catalytic membranes such as LaOCl, dispersing a catalyst into the ceramic matrix the same time when the membrane is made, or impregnation of a formed membrane with catalyst from gas or solution. Since there have been many recent developments and commercialization in this field, it deserves a separate and more comprehensive treatment here.

Basically sol-gel is a general process which converts a colloidal or polymeric solution, called sol, to a gelatinous substance called gel. It involves the hydrolysis and condensation reactions of alkoxides or salts dissolved in water or organic solvents. In most of the sol-gel processes for preparing microporous membranes, a stable sol is first prepared as an organometallic oxide precursor, followed by the addition, if necessary, of some viscosity modifiers and binders as discussed earlier in Section 3.2.3. The thickened sol is then deposited as a layer on a porous support as a result of the capillary forces induced by dip or spin coating. This is followed by gelation of the layer, upon drying, to form a gel which is the precursor to a ceramic membrane prior to controlled calcination and/or sintering. The general steps involved in a sol-gel process for making ceramic membranes are shown schematically in Figure 3.10. In the majority of the sol-gel processes, controlled hydrolysis of an organometallic compound (such as metal alkoxide) or salt leads to metal hydroxide particles, monomers or clusters, depending on the materials and conditions used as will be discussed in detail as follows. It has been suggested that not only oxides, but also nitrides, oxinitrides, sulfides and carbides can be prepared by the sol-gel process [Dislich, 1986].

<u>Sol</u>. A sol is basically a stable dispersion of nanometer-sized particles and most of the voluminous research that has been conducted in this field is related to the formation of metal oxide sols. Metal oxide sols can be prepared by the following various techniques as summarized by Livage [1986]:

(1) polycondensation of metal salts to form, for example, Al_2O_3 and V_2O_5;

(2) redox reactions of aqueous solutions of metal salts with inorganic or organic reagents to form, for example, MnO_2; and

(3) hydrolysis and condensation or polymerization of organometallic compounds to form, for example, TiO_2, Al_2O_3 and SiO_2.

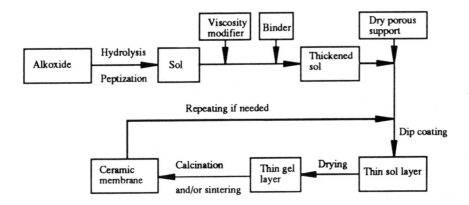

Figure 3.10 Simplified process flow diagram for making fine-pore membranes by sol-gel process

The most versatile for making metal oxide sols is the last technique: controlled hydrolysis, and simultaneously condensation or polymerization reaction, of an organometallic compound (such as metal alkoxide). This process is well known for making high-purity and ultrafine particles that can be converted to ceramic films and monoliths at sintering temperatures substantially lower than what conventional processes require. It also allows for an intimate mixing of different precursors at a molecular level capable of yielding homogeneous multicomponent ceramics. The sol-gel processes can be grouped into two major routes according to the microstructural nature of the sol (Figure 3.11) [Burggraaf et al., 1989]. A major difference in their practices is the amount of water involved in the hydrolysis step and the resulting hydrolysis rate relative to the polycondensation rate.

In the first route where typically a metal salt or a hydrated oxide is added to a non-stoichiometrically excess amount of water, a "particulate" sol is precipitated which consists of gelatinous hydroxide particles. The hydrolysis rate is fast. The primary colloidal particles so obtained are usually in the range of 5 to 15 nm.

The other route often involves hydrolysis in an organic medium such as alcohol, with a small amount of water, which leads to the formation of soluble intermediate species which then in turn undergoes condensation of oxygen bridges to form inorganic polymers or "polymeric" sol. Besides controlling water addition, the hydrolysis rate is intentionally kept lower than that by the particulate sol route by selecting a sol precursor that hydrolyzes relatively slow such as silica. In contrast to particulate sol, the polymeric sol gels more readily and the resulting gels usually contain lower solid concentrations. The sols often go through aging treatments before gelling.

GA 35270.3

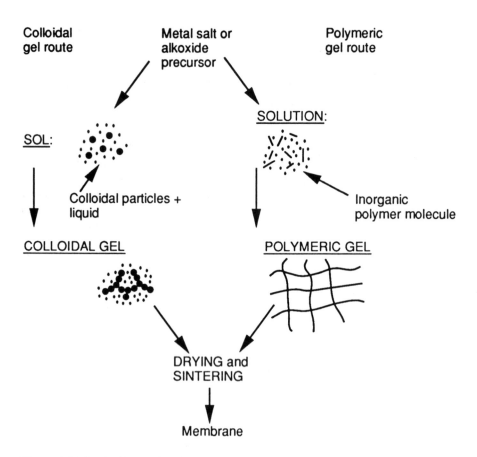

Figure 3.11 Particulate and polymeric routes of sol-gel process [Burggraaf et al., 1989]

The hydrolysis can be acid- or base-catalyzed. In the case of silica, acid-catalyzed hydrolysis results in sols with weakly-branched polymers that lead to final products having smaller pores while base-catalyzed hydrolysis produces particulate sols and final porous bodies showing larger pores [Brinker et al., 1982; Shafer et al., 1987].

Alkoxysilane or alkyl- or phenyl-bridged silsesquioxane can undergo sequential hydrolysis (a condensation reaction) to form polymeric sols [Brinker et al., 1994]. It is essential to optimize the sols for obtaining the desired final membrane of fine pores by adjusting the catalyst concentration and the aging time. The catalyst concentration determines the silica condensation rate which, in turn, affects the interpenetration of clusters and the collapse of polymer network during the final drying stage. Minimizing condensation rate is desirable for making fine pore membranes. The aging time controls the size and fractal nature of the polymer which dictates interpenetration of polymer clusters during deposition.

The fine colloids obtained through the colloidal sol route need to be stabilized (peptized) to avoid significant aggregation of primary particles in the liquid dispersion medium with an electrolyte such as acid. Peptization requires addition of some threshold amount of the acid and heat for a sufficient period of time [Yoldas, 1975b]. The peptizing agent presumably yields electric charges of the same sign on the particle surfaces and these charges cause the particles to repel each other, thus maintaining a stable dispersion (sol).

The membrane pore diameter appears to depend on the acid concentration (Table 3.1). In the case of alumina membranes, as the ratio of acid to alkoxide increases, the pore diameter of the final membrane decreases [Leenaars et al., 1984]. It should be mentioned that the mean pore diameter may not always decrease as the acid to sol concentration ratio increases. In fact, Yoldas [1975b] demonstrated a maximum pore diameter at a value of the acid to sol ratio near 0.05-0.06. Below that ratio, the pore diameter actually increases slightly as the ratio increases.

TABLE 3.1
Effect of acid concentration on diameters of precusor particles and membrane pores

H^+ / Al (mole ratio)	Particle diameter[1] (nm)	Pore diameter[2] (nm)
0.05	3.7	4.1
0.07	4.7	3.8
0.09	3.7	3.7
0.11	3.9	3.5

Note: (1) Based on x-ray line-broadening measurements.
    (2) Calculated on the assumption of cylinder-shaped pores; membrane samples heat treated at 500°C.
[Leenaars et al., 1984]

It has been argued that the acid concentration affects the size of the sol particles which, in turn, determines the pore diameter of the final membrane. Shown in Table 3.2 is the diameter of the titania sol particles as determined from quasi-elastic light scattering measurements. The particle size in this case systematically varies with the acid to titanium molar ratio. A minimum particle size exists at an optimum ratio. The optimum acid concentration corresponds to the condition of a stable sol. At higher acid concentrations, aggregates occur leading to a higher particle size. The trend relating the size of sol particles to the acid concentration as given in Table 3.2 may not hold for all cases. For example, Table 3.1 indicates no consistent relationship between the acid concentration and the sol particle size. More definitive studies with conclusive evidences are needed to resolve the discrepancy.

It appears that the particular acid used must have a non-complexing anion with the metal and have sufficient strength to produce the required charges for repulsive forces among particles at low concentrations. Nitric, hydrochloric, perchloric, acetic, chloroacetic and formic acids meet those requirements. Ordered aggregation of primary particles may occur [Livage, 1986], but careful shrinkage of such aggregates can lead to membranes with well controlled pore size and orientation. Additional measures such as ultrasonification may be imposed to further control the aggregation. Thus, the stability of the sol depends on the pH, sol concentration, temperature, hydrolysis ratio (e.g., water over alkoxide) and also, to a greater extent, the nature and the amount of an electrolyte as a peptizing agent. It has also been found that when the acid is added during the hydrolysis step together with a cationic exchange resin, the resultant pore size is smaller and pore size distribution is narrower [de Albani and Arciprete, 1992].

TABLE 3.2
Effect of acid concentration on particle size of a titania sol

H^+ / Ti (mole ratio)	Particle size[1] (nm)
0.2	87
0.4	82
0.7	96
1.0	139

Note: (1) Average effective diameter measured by light scattering.
        (2) H_2O/Ti (mole ratio) = 200; TiO_2 (wt. %) = 2%
[Adapted from Anderson et al., 1988]

Other additives are also frequently used to improve the processing characteristics of the sol and ultimately the properties of the final membranes. For example, to prevent crack formation in subsequent process steps of drying and heat treatments, some binders and/or plasticizers are added to the sol as discussed in the previous section. Polyvinyl alcohol has been found to serve the purpose very well.

Membrane pores are formed as the voids generated by the packing of primary particles. As in the case of larger particles used for making a large-pore membrane or support by pressing or extrusion, particle packing in a "particulate" sol plays an important role in affecting the resulting pore size of the final membrane. In fact, it has been suggested that the narrow pore size distribution of ultrafiltration membranes made of boehmite or transition alumina membranes is the result of having card-pack stacking structure of plate-shaped crystallites. The orderly microporous structure of these crystallites, due to the capillary compaction and repulsion forces between them, yields slit-shaped pores in the membranes [Leenaars et al., 1984].

Layer deposition. The sol is then brought into contact with a dry porous support by means of dip or spin coating discussed in Section 3.2.3. Capillary forces drive the dispersion medium through the support, leaving behind a layer of concentrated sol on the surface of the support. There is only a narrow window of acceptable thickness of the layer deposited in order to prevent cracks from occurring during drying and further heat treatment steps.

Gel. As the solvent in the sol evaporates upon drying and the concentration of particles reaches a threshold level or as the surface charges of the sol particles are changed, the colloidal suspension is transformed to a semi-solid material having an interlinked network structure of particles or agglomerates called gel. Partlow and Yoldas [1981] have demonstrated that the pH and the nature of the electrolyte can substantially influence the gelling point and volume. When the zeta potential of the sol before gelling is near the isoelectric point, flocculation is likely to occur with the resulting membrane having either defects or a broad pore size distribution. The behavior of gelling depends on the route by which the sol is prepared. In the particulate sol route, gelling can be detected by the sharp increase in the sol viscosity. In contrast, the viscosity increase is more gradual in the case of polymeric sols.

During gelling and subsequent drying stages, shrinkage and the effect of capillary forces may cause cracks in the porous gels as a result of stresses between the support and the membrane. This is a critical step in the preparation of membranes as the capillary forces involved can easily cause "hairlines" or microcracking. Generally the drying process consists of a constant drying rate period followed by a falling rate period [Strumillo and Kudra, 1987]. During constant rate drying, the gel layer will shrink and, being restrained from shrinking in a direction parallel to the support, it is subject to stress development. During the second drying period, capillary forces present in the pores due to the fluid menisci also contribute to stress in the gel layer. Once the stress exceeds the elastic strength of the gel, or if the relaxation time is insufficient, cracking will occur [Voncken et al., 1992]. Thus, stress control during membrane formation is critical to making membrane of high physical integrity. This consideration is particularly valid when drying supported sol-gel membranes with a thickness far exceeding approximately 10 μm.

The magnitude of the stress has been studied for a boehmite gel layer on a porous α-alumina support. Voncken et. al [1992] have found that among various stress measurement methods the cantilever principle is most suitable for studying porous thin films like gels. Using a laser displacement meter to detect the deflection of gel and support layers, they found that the tensile stress exerted on the drying membrane (gel) layer by the porous support increases sharply in the first 20 to 30 minutes of drying at 25 to 40° C and under various relative humidity conditions. Maximum stress in the range of 150 to 200 MPa for a gel thickness of 2 to 3 μm has been reached without cracking. However, a stress much exceeding a few hundred MPa is very likely to cause crack formation. Stressed gel layers can be relaxed by increasing relative humidity. It is, therefore, imperative to apply proper drying rate and humidity condition to avoid unnecessary cracks, especially in the initial period of drying. Various aspects related to drying of gels in general have been reviewed by Scherer [1990]. Dwivedi [1986] has provided a detailed experimental study on the drying behavior of alumina gels.

Calcination/sintering. The membrane pore size is a strong function of the primary crystal size which in turn depends on the operating parameters in the sol and gel steps as well as in the calcination/sintering step. As the temperature increases, the constituent crystallites (primary particles) of the membrane grow and so do the membrane pores. The change in the crystallite size is particularly profound around the phase transition temperatures. Table 3.3 highlights some pore diameter-temperature relationships for alumina, titania and zirconia membranes [Leenaars et al., 1984; Anderson et al., 1993]. As the calcination/sintering temperature increases, the pore diameter of the resulting membrane is enlarged. As the temperature becomes exceedingly high, sintering or crystallization can induce grain growth, decrease the porosity and increase the pore diameter.

TABLE 3.3
Effect of heat treating temperature on mean pore diameters of alumina, titania and zirconia membranes

Heat treating temp. (°C)	Alumina[1]		Titania[2]	Zirconia[2]
	Slit-shaped	Cylinder-shaped		
200	2.5	3.7	1.3	1.4
300			1.4	1.4
350			3.0	3.5
400	2.7	4.0	3.3	3.9
500	3.2	4.9		
600	3.5	5.5		
700	3.7	5.9		
800	4.8	7.9		
900	5.4	8.9		
1,000	39	78		

Note: (1) Heat treated at temperature for 34 hours [Leenaars et al., 1984]
      (2) Heat treated at temperature for 0.5 hour [Anderson et al., 1993]

Cracks due to significant shrinkage can occur during drying and calcination of the gel. Some suggestions have been made to reduce or eliminate that shrinkage by adding to the sol some calcined materials. Calcined alumina (at 25 to 75% by weight) has been mixed with alumina sol to eliminate the cracking problem [Mitsubishi Heavy Ind., 1984c]. The trade-off, however, is the resulting larger pore diameter.

Sol-gel derived membrane materials. Most sol-gel processes operate in aqueous media which lead to the formation of oxide materials. Many oxide materials have been made into microporous membranes via the sol-gel route. Better known cases are alumina, zirconia, titania, silica and porous glass (e.g., Zr containing glass [Agency of Ind. Sci. Tech. (Japan), 1986]). The majority of them are derived from the associated alkoxides. The literature data in this field is voluminous in recent years as the drive to make fine-pore membranes becomes obvious. Summarized in Table 3.4 are the typical ranges of conditions for preparing the more commonly used materials: alumina, zirconia, silica and titania. For more detailed information on the sol-gel process as applied to these four oxide materials for making ceramic membranes, readers are referred to the following selected and many other references in this growing field: alumina membranes [Yoldas, 1975a and 1975b; Leenaars et al., 1984; Leenaars and Burggraaf, 1985; Burggraaf et al., 1989; Okubo et al., 1991; Dimitrijewits de Albani and Arciprete, 1992], silica membranes [Klein and Gallagher, 1988; Langlet et al., 1992; Brinker et al., 1993; Anderson and Chu, 1993; Hyun and Kang, 1994], titania membranes [Larbot et al., 1987; Anderson et al., 1988; Burggraaf et al., 1989; Larbot et al., 1989; Xu and Anderson, 1991 and 1994] and zirconia membranes [Guizard et al., 1986b; Larbot et al., 1989; Xu and Anderson, 1991 and 1994].

Alumina is much more adapted to the particulate sol route while silica and titania are more flexible in that they can lead to both particulate and polymeric gels as the membrane precursors. This is particularly true with the silica system. Hydrolysis and condensation reactions in silica systems are known to be slower than most other alkoxides. Their reactions, therefore, are easier to control. The fact that it is much easier to prepare alumina sol-gel membranes than their titania equivalents may suggest that the plate-shaped particles like alumina can be more readily processed to form defect-free membranes than spherical particles such as titania [van Praag et al., 1989].

In addition to the aforementioned common oxide materials, other sol-gel derived membrane systems have also been studied. Ceria membranes with a mean pore diameter approaching 2 nm have been obtained by the sol-gel route [Burggraaf et al., 1989]. Hackley and Anderson [1992] have investigated two variations of the sol-gel process for making ferric oxide ceramic membranes. In one approach, the starting material of asymmetric goethite (α-FeOOH) leads to the formation of hematite (α-Fe$_2$O$_3$) membranes with a mean pore diameter of about 10 nm. In another scheme, microporous ferric oxide membranes are derived from the hydrolytic ferric oxide polymer [Fe(O,OH,H$_2$O)$_6$]$_n$ and exhibit a mean pore diameter of less than 3.4 nm. The polymer-based membranes appear to suffer from particle coarsening at high temperatures. Mullite membranes ($3Al_2O_3 \cdot 2SiO_2$) with a pore diameter of 3.4 nm and aluminosilicate membranes [$(Al_2O_3)_x(SiO_2)_y$] with a pore diameter of 4.4 nm have also been prepared by the sol-gel process [Sheng et al., 1992]. Furthermore, imogolite, another aluminosilicate with an effective composition of $(HO)_3Al_2O_3SiOH$, is being investigated as a potential inorganic membrane material [Huling et al., 1992]. It comprises nanoscale one-dimensional pore channels with a diameter of 0.8-1.2 nm.

TABLE 3.4
Examples of common metal oxide membranes prepared by sol-gel process

Material	Alumina	Silica	Titania	Zirconia
Alkoxide(s)	Al tri-sec-butoxide	Tetra-ethyl orthosilicate	Ti tetra-isopropoxide; Ti tert-amyloxide	Zr tert-amyloxide
H_2O/alkoxide (molar ratio)	6-200	6-82	2-300	1-2
Hydrolysis temp. (°C)	<80-90	25-80	25-80	25
Peptizing agent (pH)	HNO_3, HCl, $HClO_4$, CH_3CO_2H (1.3-3.2)	HNO_3 (3)	HNO_3 (2-6.8)	HNO_3 (2)
Calcination temperature (°C)	390-800	500-900	350-500	400
Pore dia. of final membrane (nm)	2.2-5.8	1.6-2	1.4-4.0	1.4
Reference(s)	Anderson et al.. [1988]	Anderson & Chu [1993]	Anderson et al.. [1988]	Xu and Anderson [1994]

The smallest pore diameters of ceramic membranes that can be obtained consistently by the straight sol-gel process without any modification has been in the range of 2.5 to 3.0 nm. This minimum pore size obtainable is determined by the smallest size of the primary sol particles that can be crystallized and the consideration of the effect of subsequent calcination on the pore size. Stable, non-aggregated nuclei finer than 5 nm are very difficult to generate. To produce membranes via the sol-gel route that are smaller than 2.5-3.0 nm requires further modifications by such techniques as adsorption or precipitation in the pores as will be discussed in Section 3.4.1.

Mixed metal oxide membranes. Mixed metal oxide membranes are very interesting because potentially they have enhanced properties over those properties of single oxide membranes. The improved properties include thermal stability, chemical resistance, separation efficiency, catalytic property and electrical conductivity.

First, the thermal stabilities of the membranes are improved by the addition of other oxide(s). Specifically, the mixed oxide membranes will retain their pore sizes and porosities when heated to a temperature at least about 100°C higher than the temperatures which would degrade the pore size and porosity of single oxide membranes. TiO_2-ZrO_2 membranes with varying compositions have been proven to be the case. Moreover, other mixed oxide membranes containing Ti-Nb and Ti-V [Anderson et al., 1993] and Al-Ba and Al-La [Chai et al., 1994] are additional examples.

Another major reason for studying mixed metal oxide membranes from double metal alkoxides is the potential for preparing zeolite-like membranes which can exhibit not only separation but also catalytic properties. It has been suggested that combinations of silica and alumina in a membrane could impart properties similar to those of natural and synthetic zeolites [Anderson and Chu, 1993]. Membranes with a pore diameter of 10 to 20 nm and consisting of combinations of titania, alumina and silica have been demonstrated by using a mixture of a meta-titanic acid sol, an alumina sol and silicic acid fine particles followed by calcining at a temperature of 500 to 900°C [Mitsubishi Heavy Ind., 1984d].

Ultrafiltration permselectivity of a ceramic membrane can be enhanced by applying an electric field to the surface of the electrically conductive membrane [Guizard et al., 1986a]. The electric field is created using two electrodes with the conductive membrane being the anode or cathode. Such conductive membranes can be prepared with the introduction of a proper amount of RuO_2 to TiO_2.

Normally when one of the two performance indicators of a porous ceramic membrane for gas separation (i.e., separation factor and permeability) is high, the other is low. It is, therefore, necessary to make a compromise that offers the most economic benefit. Often it is desirable to slightly sacrifice the separation factor for a substantial increase in the permeation flux. This has been found to be feasible with a 5% doping of silica in an alumina membrane [Galán et al., 1992].

In principle, mixed oxide membranes can be prepared by mixing of the individual sols or co-hydrolysis or co-polymerization of respective alkoxides. Mixing of different sols is often used to prepare mixed oxide membranes. For example, a sol made from hydrolyzing $Si(OC_2H_5)_4$ is mixed with another sol from the hydrolysis of $Zr(OC_3H_7)_4$. The resulting mixed sols can then be used to prepare water- and alkali-resistant porous glass membranes.

Some selective double alkoxides have been synthesized and can be utilized to prepare mixed metal oxide membranes. Preparation and stabilities of double or multiple metal alkoxides are sometimes challenging. Aluminum alkoxides are particularly interesting as they are reactive with other metal alkoxides [Bradley et al., 1978]. Some alkoxides undergo hydrolysis more rapidly than other alkoxides; therefore, the ratio of the different alkoxides that can be used may be limited because of undesirable reaction paths. An example of this is the concurrent hydrolysis of $Si(OC_2H_5)_4$ and $Zr(OC_3H_7)_4$ to prepare water- and alkali-resistant porous glass membranes. The latter hydrolyzes faster than the former. When the molar ratio of $ZrO_2:SiO_2$ in the alkoxide mixture exceeds 0.5, $Zr(OC_3H_7)_4$ can react with a peptizing agent, say, HCl, to form powdery white precipitate $ZrOCl_2$ which is not desirable [Yazawa et al., 1991]. Another example is magnesium aluminum spinel $(MgAl_2O_4)$ membranes for ultrafiltration applications. They have been successfully made by hydrolyzing the double alkoxides, $Mg[Al(OR)_4]_2$ [Pflanz et al., 1992]. Finally, mixed titanium and zirconium oxide membranes with an approximate composition of $Zr_{0.1}Ti_{0.9}O_2$ and other more general double oxides in the form of $Zr_xTi_yO_z$ have also been prepared [Anderson and Xu, 1991; Anderson et al., 1993]. The $Zr_xTi_yO_z$ composition encompasses both titanium-rich or zirconium-rich systems. They can be prepared as follows. First the corresponding metal alkoxides are converted to metal alcohol solutions (e.g., titanium or zirconium tert-amyl alcohol) by an alcohol exchange method. The metal alcohols are then mixed and dissolved in a suitable alcohol (e.g., tert-amyl alcohol). A solution of water in the same alcohol is slowly added to the mixed metal alcohol solution under vigorous stirring. The resulting sol has a molar ratio of Ti to Zr equal to 9 and water to total metal alkoxides equal to 2. After coating onto a support, the sol is converted into a gel and subsequently a mixed oxide membrane.

3.2.5 Phase Separation and Leaching

Commercial glass membranes with homogeneous porous structures have been available for some time. They are known to be made by a combination of heat treatment and chemical leaching. Conventional and modified glass ingredients such as silica sand, boron sand, and other chemicals such as sodium compounds and/or alumina are blended and made into melts at very high temperatures (1,000 to 1,500°C). The melts are then molded into desired shapes (e.g. plate and tube). The silica content in the melt usually lies between 20 and 75%. A phase diagram of the SiO_2-B_2O_3-Na_2O system shown in Figure 3.12 [Schnabel and Vaulont, 1978] indicates incomplete mixing. Immiscibility of such an ideal system occurs at a temperature between 500 and 750°C. Phase separation refers to the formation of two immiscible, mutually penetrating phases upon heat treatment. Thus the degree of phase separation depends on the heat treatment temperature and time and the glass composition. Phase separation is effected by at least two mechanisms: the classical nucleation and growth mechanism and the spinodal decomposition mechanism in which small fluctuations in composition grow rapidly when the glasses are in the spinodal region of a miscibility gap [Hammel et al., 1987a]. Generally speaking, a lower heat treating temperature requires a longer time and vice versa.

The soluble phase which consists primarily of Na_2O and B_2O_3 can be leached out by a leaching agent. If larger pores are desired, additional leaching steps can be taken.

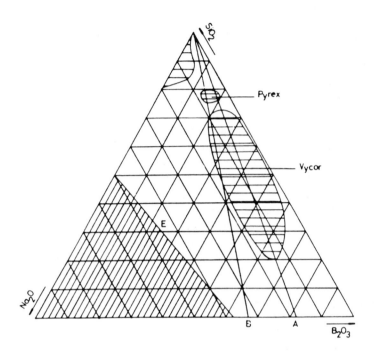

Figure 3.12 Phase diagram of the SiO_2-B_2O_3-Na_2O system [Schnabel and Vaulont, 1978]

Symmetric porous membranes. Various glass compositions containing different minor ingredients, if phase-separable, are then generally heat-treated at a temperature between 150 and 800°C for a period of several minutes to several days, resulting in the separation of two distinct phases: a boron-rich phase containing, for example, sodium borate and boric acid and a silica-rich phase with a small amount of boric oxide. Longer treatment times afford lower temperatures and vice versa. Phase separation is usually accompanied by the pronounced opaqueness of the glass. The boron-rich phase can be removed from the glass structure with a leaching agent such as water, acidic or alkaline solution, leaving behind an insoluble silica-rich microporous skeleton which consists of a matrix of interconnected, tortuous pores. The pore diameters generated can vary over a wide range from a few to over a few hundred nm. Heat treatment at the higher end of the temperature range (e.g. 700 to 800°C) produces smaller pore diameters on the order of less than 0.5 nm. The leaching agent can be hydrochloric, sulfuric, nitric or oxalic acid and applied in the 80 - 100°C range for several minutes to a few hours.

The pore size of the membrane can also be enlarged by partially dissolving the silica structure through the use of an alkali. Membranes with a pore diameter of 2 to 300 nm can be prepared this way. However, such a porous glass, due to the presence of the rather active silanol groups [Schnabel and Vaulont, 1978], can only be used in acidic or neutral pH environments. To enhance chemical resistance of the glass membranes thus obtained, chemical surface treatments may be necessary as will be discussed in Section 3.4.1 and Section 4.6. The presence of the silanol groups also provides opportunities to "design' porous glass membranes for specific applications. One such an application area is blood purification and protein separation which call for hydrophobicity that can be offered by some organic functional groups which readily react with the silanol groups.

Porous metal membranes similarly can be prepared by this technique of heat treatment followed by leaching with a strong acid, base or hydrogen peroxide [Croopnick and Scruggs, 1986].

Hollow fiber porous membranes. Hollow fiber porous inorganic membranes can be made by using dense hollow glass fibers as the template. Hollow glass fiber technology is well established. Dense hollow glass fibers can be produced by first melting the glass composition at a high temperature (e.g., between 1,100 and 1,650°C for sodium borosilicate glass) for a few hours and then mechanically attenuating the hollow glass fibers at a speed of 500 to nearly 50,000 ft/min through the orifices of a bushing which is connected to the furnace and has tubes aligned and connected to a supply of gas at a super-atmospheric pressure for making continuous or intermittent fibers. The fibers thus obtained are then air cooled to ambient temperature.

Beaver [1986] revealed that hollow fiber, porous, noncrystalline silica-rich membranes can be formed by acid leaching preceded with or without heat treatment to separate the glass composition into two groups: extractable (primarily boric oxide or anhydride, calcia and magnesia) and non-extractable (mostly silica and sometimes zirconia, titania or hafnia) components. The porosity of the resulting hollow fiber porous glass membranes is controlled by the ratio of the amounts of extractable and non-extractable components and the mean pore diameter is affected by whether heat treatment is provided or not. When heat treatment is applied, higher temperatures generate larger pore diameters.

Similarly, Hammel et al. [1987a; 1987b] have produced hollow fibers having a filament diameter of 1 to 200 μm, a wall thickness of 1 to 30 μm, a mean pore diameter of 0.5 to 5 nm and boron and/or alkali metal glass compositions (with at least 20 % leachable) by the phase separation and leaching technique. Heat treatment is performed at a temperature of 300 to less than 480°C for a few minutes to a day for fully separable compositions, and 400 to 600°C for hindered phase-separable compositions. The extractable content (boron oxide, alkali metal oxide, CaO, MgO, Al_2O_3, etc.) lies between 20 and 60% by weight. Too much extractables will decrease the softening point of the glass fibers which can lead to fibers sticking together during any heat treatment. The non-extractables include silica and any refractory glass modifiers such as ZrO_2, TiO_2 or HfO_2. These modifiers improve the alkaline stability and porosity control of the final hollow fiber membranes by decreasing porosity without adding silica. The amount of siliceous material should be between 30 to 80% of the total glass composition.

Hammel et al. [1987a; 1987b] have also found that: (1) a faster attenuation speed helps providing an open glass structure in the fiber to facilitate extraction and (2) rapid cooling of the fibers assists in producing open network structure which enables extractables to be leached at a faster rate. The leaching agents include water or acids except hydrofluoric or phosphoric acids which can also attack the siliceous structure. Commonly used is nitric acid. The acid strength ranges from 2 to 3 N. Leaching can be carried out with agitation at a temperature ranging from about 50 to 60°C for a few minutes to a few hours. Maintaining the acid concentration at a low pH or the water at near neutral pH in the leaching bath favors the leaching kinetics. It is preferred not to remove any sizing compound from the glass because the compound may actually protect the glass during the leaching step. It has been indicated that symmetric porous membranes so obtained can be further treated to yield asymmetric pore structures by heating to condense some pores or by selective leaching with acidic or alkaline materials but no details are given.

Composite porous membranes. Porous glass membranes have also been made onto a porous ceramic or metal support by the following methods. In the first method, the surface of the porous support is covered with a layer of phase-separable glass particles. These precursor particles are then transformed into a dense glassy membrane on the porous support by applying heat. Subsequently the glassy membranes are subject to phase separation and acid leaching as discussed above to generate porous glass membranes [Abe and Fujita, 1987]. Alternatively, porous ceramic supports are coated with a metal alkoxide solution such as silicon tetraethoxide and boric acid in a mixture of ethanol, hydrochloric acid and water. The coating is then dried, heated and then acid leached to produce porous glass membranes with a pore diameter of 4 to 64 nm [Ohya et al., 1986].

3.2.6 Polymerization

The same type of polymerization and membrane casting methods discussed in Section 3.1.3 for dense inorganic polymer membranes are also applicable to making porous membranes by adjusting the casting parameters. Porous membranes can be cast from the polymer-solvent systems that are doped with pore-forming salts instead of single solvent systems [Peterson et al., 1993]. However, currently there is no evidence of commercial or intensified developmental activities on inorganic polymer membranes.

3.2.7 Pyrolysis or Sintering of Organic Polymers

In the process of phase separation by heat treatment followed by chemical leaching as reviewed in Section 3.2.5, the final membrane geometry (e.g., plate or tube) is determined by that of the precursor material (such as glass or metals) which serves as a template. Similarly in the pyrolysis processes involving organic polymers, the precursor materials act as the templates for the membranes to be prepared. It has been shown on a bench scale that porous inorganic membranes can be made by controlled pyrolysis of organic polymers of a given shape under an inert atmosphere. In some cases, the first pyrolysis step is followed by sintering in air. A few polymers have been tested as the precursor materials.

Carbons produced by pyrolysis of a wide variety of polymers and natural coals in an inert atmosphere are usually highly porous. Commercial carbon membranes have been made by the pyrolysis of certain thermosetting polymers (such as polyacrylonitrile) in, say, the tubular form. Pyrolysis is performed under a controlled inert atmosphere. Porous carbon supports and one or more layers of carbon membranes on top of them are made without complete graphitization which can result in a physically weak and hydrophobic material [Fleming, 1988]. The support has a fibrous structure. A second layer of a different polymer (e.g., phenolic resin in an organic solution) is deposited and then carefully pyrolyzed to the desired pore size down to about 4 nm. Deposition can be achieved by adsorption (for a thin layer) of appropriate monomer or oligomer, followed by in-situ polymerization. Depending on the type and degree of pyrolysis, membrane surface properties (hydrophilicity versus hydrophobicity), pore size and morphology can be adjusted. The pyrolysis temperature is higher than those used for preparing activated carbons. A great advantage of this pyrolysis process is that end sealing required of single membranes to form a module can be provided by pyrolysis as well. The thickness of the membranes prepared in this fashion ranges from 0.1 to 1 µm.

To obtain carbon membranes with molecular sieving properties, pore diameters in the range of a few angstroms are required. The associated pyrolysis procedures and post treatments are more involved. This subject will be treated later under section 3.2.10 Molecular Sieving Membranes.

Silica membranes have been successfully prepared by carefully pyrolyzing silicone rubber (polydimethyl siloxane) [Lee and Khang, 1987]. Silicone rubber hollow fibers are subject to a two-step process. In the first step, inert gas is used to pyrolyze the polymer below 800°C for more than 3 hours to remove hydrocarbons and low-boiling-point silicon compounds from the polymer chains. It is believed that the material after the first step mainly consists of a partially cross-linked Si-O chain structure. The second step is performed in air below 950°C for 3 to 15 hours. During this oxidation step the membrane structure becomes completely cross-linked to form a refractory silica material. Optionally the membrane can be re-pyrolyzed using an inert gas as in the first step. The pore diameters of the resulting silica membranes, although not specified, are most likely in the range of 5 to 10 nm as the gas flow has been found to follow the Knudsen diffusion mechanism under a pressure gradient. Thus, the final membrane microstructure, and consequently the gas permeability, can be controlled by the temperature and the duration of the pyrolysis and sintering steps.

A variation of the above methods of pyrolyzing straight polymers into inorganic membranes is to introduce inorganic compounds in the precursor materials when casting polymers. When a sufficient amount of inorganic material(s) is present, the precursor material can be sintered in air without going through a pyrolysis step as in the case of using straight polymers as the precursors. Alumina and alumina-silica membranes have been prepared in this manner [Lee and Kim, 1991]. First a mixed dispersion of alumina (or alumina-silica) powder and polysulfone in a solvent such as dimethyl acetamide is prepared and a precursor hollow fiber having a radially anisotropic structure (i.e., a finger with sponge structure) is made by wet spinning. The precursor material is then sintered above 1,100°C in air. As the sintering temperature increases, both the porosity and the gas permeability decrease. The resulting pore diameter falls in the range of about 1 μm. The required alumina or alumina-silica content in the oxide-polysulfone mixture for making anisotropic hollow fiber membranes is at least 50% by weight.

3.2.8 Thin-film Deposition

The thin-film deposition technology has been studied extensively recently and encompasses conventional physical and chemical deposition techniques such as vapor deposition in several variations, sputtering, ion plating, and metal plating. Although its original objective was to make essentially nonporous layers for uses in electronic manufacturing, thermal protection and wear and corrosion resistance, the technology can be applied to produce porous membranes by adapting appropriate conditions.

Thin film deposition for producing dense membranes has been presented in Sections 3.1.1 and 3.1.2. The processes can also be used to prepare porous membranes by adjusting the operating conditions. For example, transition metals and their alloys can be deposited on a porous ceramic, glass, or stainless steel support by the thin-film deposition process to produce porous metal membranes with small pore sizes [Teijin, 1984].

There have been many attempts to prepare ultramicroporous ceramic membranes suitable for gas separation and membrane reactors with high separation efficiency. Chemical vapor deposition (CVD) is commonly adopted to make predominantly silica-modified membranes. Silica membranes were deposited on alumina supports by CVD using tetraethyl orthosilicate as a precursor and the deposition time, temperature, pressure and carrier gas flow rate determine the membrane pore size and layer thickness [Gallaher et al., 1994; Lin et al., 1994]. The resulting pore diameter can be made smaller than 0.5 nm. Okubo and Inoue [1989] used tetraethoxysilane as the precursor agent in CVD to modify porous glass membranes. The pore size of zeolite mordenite was reduced by CVD modification using silicon alkoxides [Niwa et al., 1984]. These are just a few examples.

3.2.9 Other Methods

There are other methods which are either some substantial variations of the above techniques or new processes that are not well studied.

In those processes where metal oxide particles dispersed in a liquid medium are molded (pressing, tape casting, extrusion, and dip or spin coating), only a small amount of organic polymers is used as binders or viscosity modifiers. A variation of this practice has been claimed to improve porosity control and reduce cracking of the membrane formed [Jain and Nadkarni, 1990]. The variation lies in the use of a significant quantity of a polymer such as polyacrylonitrile and its copolymers and terpolymers in the suspension or slip and the subsequent step of carbonization before sintering of metal oxide particles. The polymer selected should have high carbon yield upon carbonization which takes place in a non-oxidizing atmosphere (e.g., nitrogen or argon) at 500 to 750°C. The membrane porosity can be adjusted by varying the ratio of the metal oxide to the carbon and the particles size of the oxide. Porosity can be controlled especially when nitrogen or other gas (e.g., ammonia or amine) is used. The membrane thickness, however, is limited to within 5 mm to avoid the potential problem of having a membrane with a dense core and a porous surface. Some of the metal oxides that can be applied are alumina, zirconia, titania, yttria and tungsten oxide.

Ultra fine ceramic particles produced from a chemical vapor deposition reactor are charged between a creeping discharge electrode and another electrode for particle migration. The particles form a layer on the surface of a porous support located between the two electrodes by the electrostatic force. The porous layer thus formed is then subject to sintering. The creeping discharge electrode comprises an induction electrode, an alumina layer and a linear electrode, and the particle migration electrode is situated in the opposite direction of the linear electrode. Alumina, zirconia, silica, titania and silicon nitride membranes having pores on the order of a few nanometers have been made by this process [Masuda, 1989].

Having a large available membrane area in a unit volume (i.e., a high "packing density") is highly desirable in membrane separation applications. A high packing density geometry of organic membranes has been the so-called spiral wound module briefly mentioned in Chapter 1. Only a few attempts have been made to fabricate inorganic membranes in a similar geometry. An example is a columnar porous ceramic body prepared by impregnating a planar synthetic resin foam with a ceramic slurry and rolling the impregnated foam material into a column followed by its drying and sintering [Kyocera Corp., 1985]. The membranes so produced have a low void volume in the center and the void volume increases spirally outward (Figure 3.13). The membranes have relatively large pores and thus can be used as a filter for trapping carbon particles in exhaust gas. Another interesting inorganic membrane shape having a high packing density is a three-dimensional criss-crossing network of porous hollow ceramic filament [Baker et al., 1984]. The network is made first by coating a continuous combustible core (e.g., a cotton yarn) with a slip consisting of such refractory oxides as alumina, mullite, silica, cordierite, magnesia, spinel, forsterite, enstatite or their mixtures. The coated filament is then made into a three-dimensional network by a criss-cross pattern extending in the length-width dimensions. This is followed by firing the filament to burn off the core and form a rigid porous spirally wound hollow fiber inorganic membrane.

A new technique for making precision particle or macromolecular microfilter is based on the technology used to fabricate microelectronics. The method uses a two-step self-aligning process involving lateral boron doping by diffusion and silica undercut etching [Stemme and Kittilsland, 1988]. The structure consists of two patterned silicon layers displaced laterally relative to each other and separated by an exact and constant distance with silica spacers. The top layer comprises 1-μm-thick silica deposited on 1 μm polysilicon and the bottom layer is single-crystal silicon. The permselectivity of the structure (membrane) is determined by the thickness of the silica spacers separating the two layers. Particles larger than this precision distance will be rejected. Such microporous structures reportedly have a separation size of 50 to 200 nm.

GA 35288.3

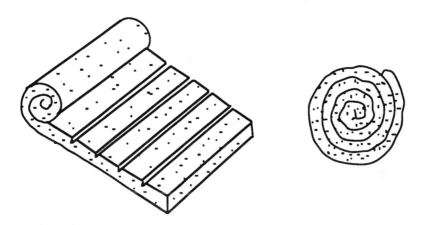

Figure 3.13 Spiral-wound ceramic membrane [Kyocera Corp., 1985]

Some metals, ceramics and glasses are melted in a hydrogen-containing atmosphere under a preset pressure to provide the desired hydrogen concentration. The cooling rate and direction and the hydrogen pressure at which the melts solidify can be adjusted to control the resulting orientation, size and morphology of the pores. Applying the gas-eutectic transformation, a wide range of porous structures (including honeycombed, irregular and finely spaced) have been observed under different solidification rates. A number of materials such as nickel, copper, bronze, magnesium, aluminum, stainless steel, cobalt, chromium and other metals and alumina-magnesia and glass have been made into porous membranes having different microstructures by this solid-gas eutectic solidification technique [Walukas, 1992].

3.2.10 Molecular Sieving Membranes

There are a class of materials, although made by various methods, that deserve a separate discussion here due to their great promises for making very fine-pore and, in some cases, inherently catalytic membranes. They are molecular sieves which separate molecules based on whether the molecules can penetrate the regularly spaced channels of molecular dimensions (pore diameter in the 0.2 to 1 nm range) in the molecular sieve crystals. This "sieving" mechanism on the molecular level is the most distinct characteristics of these materials.

There are a number of materials that exhibit molecular sieving properties. The best known among them are zeolites. Others include some porous glass, fine-pore silicas, montmorillonite and molecular sieve carbons.

Carbon molecular sieve membranes. Molecular sieve carbons can be produced by controlled pyrolysis of selected polymers as mentioned in 3.2.7 Pyrolysis. Carbon molecular sieves with a mean pore diameter from 0.25 to 1 nm are known to have high separation selectivities for molecules differing by as little as 0.02 nm in critical dimensions. Besides the separation properties, these amorphous materials with more or less regular pore structures may also provide catalytic properties. Carbon molecular sieve membranes in sheet and hollow fiber (with a fiber outer diameter of 5 μm to 1 mm) forms can be derived from cellulose and its derivatives, certain acrylics, peach-tar mesophase or certain thermosetting polymers such as phenolic resins and oxidized polyacrylonitrile by pyrolysis in an inert atmosphere [Koresh and Soffer, 1983; Soffer et al., 1987; Murphy, 1988].

An important requirement of the choice of materials is that the material does not soften or liquefy during any stage of the pyrolytic carbonization process. And the pyrolysis temperature should be between the decomposition temperature of the carbonaceous material and the graphitization temperature (about 3,000°C). Although in principle pyrolysis temperature can be between 250 and 2,500°C, pyrolysis is preferably conducted in the 500 - 800°C range. It is essential that an even, flawless thermosetting polymer membrane is used as the precursor in order to obtain a membrane free of pores larger than molecular dimension.

Formation of these "ultramicropores" depends greatly on the chemistry of pyrolysis and is probably the result of the small gaseous molecules channeling their way out of the solid matrix during pyrolysis [Koresh and Soffer, 1983]. There is evidence that these ultramicropores consist of relatively wide openings with narrow constrictions which are responsible for the stereoselectivity of gas molecules [Koresh and Soffer, 1987].

Apparently degassing carbon molecular sieves below 650 to 750°C removes some surface oxygen functional groups as carbon oxides, thus enlarging the pore size. On the other hand, heat treatment at higher temperatures tends to anneal the carbon and cause pore closure [Koresh and Soffer, 1977]. At this annealing stage, the higher the final pyrolysis temperature, the smaller the pores formed. The pores thus obtained can be enlarged by activation through mild and well controlled oxidation steps (oxidative burnoff) [Walker et al., 1966] or narrowed by high temperature sintering [Lamond et al., 1965]. Besides oxygen or air, other oxidizing agents include steam, carbon dioxide, nitrogen oxide, chlorine oxides and solutions of nitric acid and its mixture with sulfuric acid, chromic acid and peroxides. Whether activation is conducted in a stationary (equal pressurization on both sides of the membrane) or dynamic (with a pressure difference across the membrane thickness) mode, the resultant pore diameter can be quite different.

Therefore, by fine-tuning the pyrolysis temperature and conditions and subsequent activation or sintering temperature and duration, the desired pore sizes and associated permeability and permselectivity can be achieved. In addition, by introducing non-volatile organic compounds such as ethene, propane, ethylene, benzene or other hydrocarbons into the pores of the carbon membrane precursor, it is possible to further increase the permselectivity of the final membrane [Soffer et al., 1987]. Theoretically, the membrane pore diameter can be adjusted to the range of 0.28 to 0.52 nm, potentially suitable for separating many gaseous mixtures. While the carbon molecular sieve membranes are thermally stable and mechanically stronger than organic polymeric membranes, they are brittle and require extra care during handling.

Carbon molecular sieve membranes have been prepared on porous supports by controlled pyrolysis. For example, Chen and Yang [1994] prepared carbon molecular sieve membranes on porous graphite supports by coating a layer of polyfurfuryl alcohol followed by controlled pyrolysis with a final temperature of 500°C. The procedure can be repeated to deposit a desired thickness of the carbon membrane. The choice of a graphite support is partially based on the consideration of the compatibility in thermal expansion between the carbon and the support.

In addition to the pyrolysis method, carbon molecular sieve membranes can also be prepared by chemical vapor deposition and plasma deposition on a porous support [Soffer et al., 1987]. Chemical vapor deposition of various organic gases has been performed at 900°C over a porous graphite tube. Plasma deposition has been conducted using a mixture of ethane, ethylene, propane, and benzene in argon gases to pass through an electric discharge zone where porous graphite supports are coated with the resulting membrane layer under reduced pressures. Due to their pore sizes, carbon molecular sieve membranes have potential uses in such industrially important applications as separation of nitrogen from oxygen, helium from natural gas, and individual components from the $H_2/CO_2/CO/CH_4$ gas mixtures resulting from coal gasification.

Zeolite membranes. Zeolites represent a class of highly ordered, porous and crystalline silica-containing materials which typically have rather uniform and very small pore diameters. They can have a wide variety of compositions with some of the silicon atoms replaced by bivalent ions such as Be, or by trivalent ions such as Al, B, Ga or Fe, or by tetravalent ions such as Ti or Ge, or by a combination of the aforementioned ions. Hydrated aluminosilicates with the simplified empirical formula of $M_{2/n} \cdot Al_2O_3 \cdot SiO_2 \cdot yH_2O$ are common. Upon dehydration, zeolite crystals develop a very uniform pore structure of channels in the diameter range of 0.3 to 1 nm capable of "sieving" gas molecules according to their molecular dimensions. Having high surface-to-volume ratios, zeolites are capable of discriminating molecules that are different in size less than 0.1 nm. The attained pore size depends on the type of zeolite, the cation present and the nature of treatments such as calcination or leaching [Satterfield, 1980].

When made into a membrane, their uniform pore sizes potentially will permit only certain molecules to pass through the pores. The challenge, however, lies in the sealing of the intercrystal voids. Zeolites are usually prepared in the sodium form and the sodium can be replaced by many different cations or by a hydrogen ion. Naturally occurring zeolites are called mordenite while synthetic zeolites having different chemical forms are typically designated by a letter or groups of letters such as A, X, Y, ZSM, etc. With many possible structural forms, zeolites vary in their separation and catalytic properties.

Zeolite membranes capable of molecularly sieving a number of gases have been prepared on temporary non-porous substrates [Haag and Tsikoyiannis, 1991]. The membrane thus obtained can be separated from the substrate in a number of ways depending on the nature of the substrate. This type of membranes consists of a continuous array of densely packed and intertwined crystals having a size of 10 to 100 μm. The zeolite synthesis solution can be prepared under hydrothermal condition, for example, by slowly adding colloidal silica solution to a solution of tetrapropylammonium bromide in sodium hydroxide with continuous stirring. The non-porous substrate is then dipped into the synthesis solution for an extended period of time (up to about 200 hours) at a temperature of 110 to 200°C to form a layer of membrane of 0.1 to 400 μm thick. The resulting membrane before or after being removed from the substrate is then calcined in nitrogen at 560°C and in air at 600°C. When the non-porous substrate is a metal (silver, nickel or stainless steel), the zeolite membrane can be readily removed from the metal with a solvent such as acetone at room temperature or by dissolving the metal with an acid such as hydrochloric or nitric acid. The zeolite membranes thus obtained are inherently catalytic for a variety of cracking and dehydrogenation reactions and other hydrocarbon conversion processes. Their catalytic and separation properties can be adjusted by introducing other compounds via ion exchange (e.g., with K, Rb, Cs or Ag) or deposition techniques (e.g., deposition of coke or pyridine, silica, phosphorus compounds or metal salts).

Zeolite membranes have also been made atop of porous supports. To make a zeolite membrane on a porous support, a sol or gel layer of metal oxides (such as alumina or gallium oxide and silica or germanium oxide) is first deposited on the support. This is followed by a hydrothermal treatment at a relatively high temperature (e.g., ≥ 100°C) to convert the gel layer to the desired cage-shaped zeolite film. Zeolite membranes have be made by reacting sodium silicate with caustic soda directly on the surface of a porous sintered alumina to form a gel layer which is then treated hydrothermally [Matsushita Electric Ind., 1985].

Suzuki [1987] has described in detail a large array of methods for preparing an ultra thin zeolite membrane on one surface of a porous metal (e.g., etched Raney nickel), Vycor glass or ceramic support layer having ten to less than a hundred nm thickness. The zeolites thus obtained comprise a 6-, 8-, 10- or 12-member oxygen ring window and X-ray Fourier analysis indicates that their pore diameters range from 0.22 to 1.2 nm. The gel layer can be prepared and deposited on the support by one of the following methods:

(1) reacting sodium aluminate (or sodium gallate) and sodium silicate (or sodium germanate) with sodium hydroxide (or potassium hydroxide) solution, followed by depositing the reaction products, a sol, on the support;

(2) mixing sodium silicate (or germanium silicate) and aluminum sulfate (or gallium sulfate) and depositing the resulting oxide on the support by vacuum deposition, cathode sputtering or molecular epitaxy, followed by reaction with sodium (or potassium) hydroxide solution to form a gel layer;

(3) depositing an ultra thin layer of an alloy composed of aluminum (or gallium) and silicon (or germanium) by plating, flame spraying, vacuum deposition or cathode sputtering, followed by chemically oxidizing (with nitric acid) or electrochemically anodizing (with chromic acid) the alloy layer and then reacting the oxidized alloy with sodium hydroxide (or potassium hydroxide) solution;

(4) electroless plating of an aluminum (or gallium) salt and a silicon (or germanium) salt via the oxidation-reduction reaction to form a layer of the metals on the support, followed by the oxidation and the sodium (or potassium) hydroxide reaction steps in the above method (3).

The gel thus obtained is then hydrothermally treated and can be ion exchanged by conventional methods. The preferred pore diameter of the support is 10 nm and may be obtained alternatively by using a dense support followed by chemical after-treatment to the membrane-support composite. A wide variety of zeolites including NaA-, CaA-, HA-, H-, NaY-, CaY-, REY-, CaX-types all have been produced and they have shown very efficient separation of hydrocarbons.

Jia et al. [1993] have prepared thin, dense pure silicalite zeolite membranes on porous ceramic supports by an in-situ synthesis method. A sol consisting of silica, sodium hydroxide, tetrapropylammonium bromide and water is prepared with thorough mixing. A ceramic support is immersed in the sol which is then heated and maintained at 180°C for a few hours. The membrane thus obtained is calcined at a temperature range of 400 to 450°C for 8 hours. It has been confirmed by X-ray diffraction and scanning electron microscopy that individual crystallites have intergrown in three dimensions into a fused polycrystal zeolite layer. Permeation tests, however, provide evidences that the membrane layer contains both zeolite channels and larger non-zeolite channels (defects). The size of the defects is estimated to be less than 2 nm. This seems to support the general belief that it is a great challenge to produce zeolite membranes free of intercrystal voids.

Similarly, Sano et al. [1994] added colloidal silica to a stirred solution of tetrapropylammonium bromide and sodium hydroxide to synthesize a hydrogel on a stainless steel or alumina support with a mean pore diameter of 0.5 to 2 μm. The composite membrane is then dried and heat treated at 500°C for 20 hours to remove the organic amine occluded in the zeolite framework. The silicalite membranes thus obtained are claimed to be free of cracks and pores between grains, thus making the membranes suitable for more demanding applications such as separation of ethanol/water mixtures where the compound molecules are both small. The step of calcination is critical for synthesizing membranes with a high permselectivity.

Aluminophosphate molecular sieve membranes. In addition to zeolites, Haag & Tsikoyiannis [1991] have also briefly described another type of molecular sieve membranes consisting of $AlPO_4$ units whose aluminum or phosphorous constituent may be substituted by other elements such as silicon or metals. These membranes are made from aluminophosphates, silico-aluminophosphates, metalo-aluminophosphates or metalo-aluminophosphosilicates. Like zeolites, these materials have ordered pore structures that can discriminate molecules based on their molecular dimensions. Their separation and catalytic properties can also be tailored with similar techniques employed for zeolites. The procedures for calcining the membranes or separating them from non-porous substrates are essentially the same as those described earlier for zeolites.

Partially pyrolyzed polysilastyrene membranes. Molecular sieve membranes can also be made from porous solids other than carbon and zeolitic materials. One such potential candidate is the family of precursors of silicon carbide. Among the possible precursors, polysilastyrene (phenylmethylsilane-dimethylsilane copolymer) is commercially available and soluble in many common solvents such as toluene and tetrahydrofuran and can be crosslinked by UV radiation. Polysilastyrene is comprised of long chains of silicon atoms:

Polysilastyrene in solution has the advantage of being amenable to fiber drawing or sheet casting operations. When pyrolyzed in an inert atmosphere, its side-group carbon atoms insert themselves into the polysilane chain and forms a polymer with alternating silicon and carbon atoms. At higher temperatures, hydrogen and hydrocarbons are given off. And above 1,300°C, the pyrolytic reaction yields β–SiC. Grosgogeat et al. [1991] have prepared the polysilastyrene membrane by coating the inside surfaces of porous Vycor glass tubes with a toluene solution of polysilastyrene. The composite membranes were then crosslinked by a UV light for two weeks, dried in a vacuum oven at 60°C for a day and then partially pyrolyzed at temperatures up to 735°C in a nitrogen atmosphere. The membranes, however, cracked when the pyrolysis temperature exceeded 470°C. The membranes produced showed molecular sieving effects with good separation factors. For more details on the preparation of polysiloxane membranes, readers are referred to Hwang et al. [1989].

Molecular sieving silica membranes. Other porous inorganic membranes having molecular sieving capabilities have also been prepared. Raman and Brinker [1995] applied an organic "template" approach to making relatively high-flux, high-permselectivity silica membranes. In their novel technique, fugitive organic ligands are used as micropore templates. The organic ligands are embedded in a dense inorganic matrix (e.g., via co-polymerization) and removed by pyrolysis to create a continuous network of ultramicropores. In principle, the volume fraction of the organic ligands controls membrane porosity and permeate flux while the ligand size and shape determines the permselectivity.

The technical challenges are that: (1) the ligands must be uniformly distributed in the inorganic matrix; (2) the inorganic matrix is dense and "gas-tight"; and (3) the removal of the template does not result in collapse of the matrix. As a demonstration of the technique, Raman and Brinker [1995] co-polymerized tetraethoxysilane (TEOS) and methyltriethoxysilane (MTES) on a microporous support and heat treated the hybrid organic-inorganic polymers to densify the inorganic matrix and pyrolyze the methyl ligands. The final membranes exhibited a very high CO_2 permeance with a moderate CO_2/CH_4 separation factor (~12). The separation factor can be further improved (to ~72) without appreciable reduction of the CO_2 permeance by subsequent derivatization of the pore surfaces with monomeric TEOS.

3.3 NEARLY STRAIGHT-PORE MEMBRANES

Membranes having nearly straight pores have long been attractive to many researchers as a model system for fundamental transport studies because of their simple and well defined geometry. Many bench studies have been conducted to synthesize inorganic membranes with approximately straight pores extending from the high pressure (feed) to the low pressure (permeate) sides. These membranes have remained as laboratory curiosities until recently when anodic alumina membranes became commercialized. The commercial applications, however, are still limited to the laboratory scale.

Two well known processes have been used to prepare inorganic membranes with nearly straight pores: anodic oxidation and track etching. The former method has been practiced commercially while the latter remains as a laboratory investigation.

3.3.1 Anodic Oxidation

It has long been observed that when aluminum oxidizes on its surface, the resulting structure of the oxide is rather unique in having pores of a columnar shape. Researchers have turned this unique phenomenon into synthesis of a desirable and controlled membrane pore morphology.

A thin aluminum foil or sheet of a few hundred micron thick first undergoes a pretreatment of cleaning, degreasing (e.g., by washing) and polishing (e.g., by electrolytic polishing) to render a smooth surface to avoid pinholes in the final membrane. If one side of the treated aluminum is oxidized anodically (using aluminum as the anode) at a moderate voltage (e.g., 10 to 160 volts) in an acid electrolyte (such as phosphoric, sulfuric, oxalic or chromic acid or their mixtures) at a low to moderate temperature (e.g., 0 to 60°C), a porous oxide film showing a uniform array of hexagonally close-packed cells is formed in the direction perpendicular to the macroscopic surface of the metal (Figure 3.14). Each cell contains a cylindrical pore with a nearly circular shape. This porous film, however, does not penetrate the entire thickness of the original folia or sheet and is attached to the unaffected aluminum by an essentially nonporous "skin" layer of alumina known as the barrier layer. To convert such a regularly shaped porous film into a membrane with pores connecting both sides of the film, the aluminum substrate needs to be detached from the porous film and the barrier layer has to be removed or modified to generate pores.

The aluminum substrate can be dissolved in hydrochloric acid or cupric chloride or bromide solution or, more preferably, separated by etching the barrier layer at an ambient temperature briefly in a concentrated acid such as phosphoric acid. Hydrogen evolved from the etching reaction helps lift the porous film from the aluminum [Rigby et al., 1990]. One effective way of reducing the thickness of the barrier layer and causing it to be porous, and thus making it easier to separate the porous membrane layer from the unaffected aluminum substrate, is to lower the anodizing voltage in a series of small steps after the growth of the porous film is completed. This method causes uniform thinning of the barrier layer by nucleation of small pores. A large voltage reduction in one step can create only localized and non-uniform thinning.

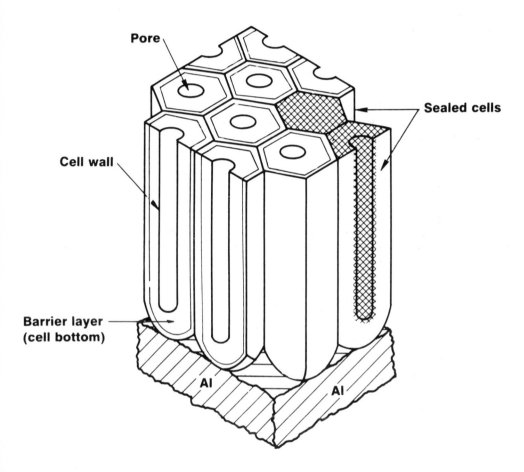

Figure 3.14. Porous oxide film showing a uniform array of hexagonally close-packed cells by anodic oxidation [Wefers and Misra, 1987]

After etching, rinsing and drying at 60°C, the anodic membrane is lifted from the aluminum. The porous membrane thus formed is amorphous and has an asymmetric pore structure: larger pores (0.1 to 0.2 µm in diameter) extending through the bulk of its thickness interconnected with an array of smaller pores which appear as a "skin" at the surface originally attached to the aluminum substrate [Rigby et al., 1987; Furneaux et al., 1989]. Apparently, the bulk of the pores are generated during anodizing while the small pores in the "skin" layer are created during the controlled voltage reduction stage. When this layer of small pores (down to about 0.01 to 0.02 μm) is not desirable, it can be removed by a second etching possibly with a different acid than the electrolyte acid or the first etchant [Toyo Soda Mfg., 1985]; thus, a symmetric membrane is produced. It has been evident that the chemical etching process actually enlarges the pores.

The resulting pore diameter and pore density strongly depends on the electrolyte used [Diggle, et al., 1969] and the anodizing voltage. For example, phosphoric acid is preferred over other electrolytes because it makes the pore nucleation process faster at a given voltage and it inhibits hydration of the alumina which may result in loss of control of the pore size [Rigby et al., 1990]. However, to make membranes with very small pores, sulfuric acid is actually preferred. The pore size increases directly with the voltage. The pore density is inversely proportional to the square root of the anodizing voltage [Rai and Ruckenstein, 1975]. The membrane thickness is determined by the current density and anodizing time. Alumina membranes with very narrow pore size distributions in sheet and tubular forms have been made by this process. Pore diameters in the 10 to 250 nm range, pore densities between 10^{12} and $10^{15}/m^2$ and membrane thickness up to 100 μm have been obtained [Furneaux et al., 1989].

A novel, but not commercially practiced, method of making symmetric membranes is to press two aluminum sheets (or foils) together in a plastic holder and anodize from only one side. The nonporous barrier layer is displaced into the second aluminum sheet and the first alumina sheet becomes a membrane containing cylindrical pores extending across its entire thickness [Rai and Ruckenstein, 1975].

The membranes produced by anodic oxidation often require further hydrothermal treatments using water or a base at temperatures in the 35 to 80°C range or calcination to a high temperature (up to 1,000°C) to significantly improve their stabilities upon long exposure to water [Mitrovic and Knezic, 1979]. The amorphous alumina membranes as produced by the straight anodic oxidation process typically are fragile and susceptible to acid and base attack. Polycrystalline alumina membranes with improved mechanical strength and acid and base resistances can be prepared by carefully calcining the above amorphous alumina membranes at a temperature greater than 850°C [Mardilovich et al., 1995].

Other metals such as zirconium, in principle, can also be synthesized in the same way to form oxide membranes with essentially straight pores.

3.3.2 Electroless Metal Deposition Using Anodic Alumina as a Template

A novel method of preparing metal membranes with nearly straight pores has been described by Masuda et al. [1993]. Basically it involves a two-step molding process (Figure 3.15). A porous anodized aluminum structure without the aluminum substrate removed is used as a template and the pores of the anodic alumina is filled with a monomer such as methylmethacrylate (MMA) and an initiator (e.g., benzoyl peroxide) for polymerization to form polymethylmethacrylate (PMMA). The polymer thus becomes a negative of the anodic structure. The polymer is separated from the alumina and aluminum by dissolving the latter two in a sodium hydroxide solution. Metal is then deposited into the polymer "mold" by electroless plating. Finally, a metal membrane is formed by selectively dissolving the polymer (PMMA) with acetone.

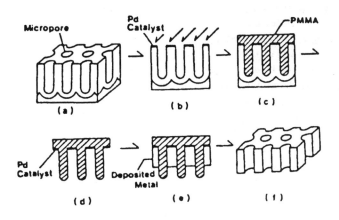

Figure 3.15. A two-step molding process for fabricating a microporous metal membrane by using polymethylmethacrylate (PMMA) as an intermediate template [Masuda et al., 1993]

The resulting metal membrane has a structure similar to the bulk pores of the anodic alumina but without the dense "skin" layer and has straight through-pores. Nickel and platinum membranes having pore diameters in the 15 to 200 nm range can be produced this way. One of the critical steps of this method is depositing palladium catalyst on the surface, but not inside the pores, of the anodic alumina. The catalyst is used to facilitate the deposition of metal from the bottom to the surface of the cylindrical pores.

3.3.3 Track Etching

Another approach of making inorganic membranes containing very uniform, straight, and parallel pores is to use a radioactive radiation source to fire highly charged particles (fission fragments) through a thin layer of material. The radiation creates tracks of permanent change by breaking chemical bonds. The tracks so produced are highly sensitive to an etchant (e.g., concentrated hydrofluoric acid). After the tracks are dissolved away in an etching bath, the material becomes a membrane with essentially straight capillary pores of uniform shape (typically cylindrical) and size. The pore diameter can be controlled by etching time and the pore density (i.e., the number of pores per unit area) is controlled by the dosage of the charged particles. Membranes made by this method have very narrow pore size distributions and their pore size range can be varied widely; however, they suffer from low porosity due to a relatively small percentage of the surface areas that is affected to avoid overlap of the pores and the pore length (or depth) is limited due to the limited energy of the charged particles.

This method has been employed to commercialize polycarbonate membranes with extremely narrow pore size distributions and a wide range of pore size. The pore length (depth) is limited due to the energy constraints of the charged particles. No inorganic membranes are commercially produced this way but mica membranes with pore diameters of 6 nm to 1.2 mm have been prepared in the laboratory [Quinn et al., 1972; Riedel and Spohr, 1980]. Membranes prepared by this technique are good candidates for fundamental transport studies due to their very uniform pore shape.

Metal membranes can also be made similarly. Aluminum foil can be irradiated with an oxygen ion beam, ignited and the residual aluminum is removed by treating it with aqueous sodium hydroxide. Aluminum is converted to alumina [Tokyo Shibaura Electric Co., 1974]. Thus an alumina membrane is produced.

High-speed charging particles such as He^+, O^+, H^+ or F^+ particles, protons or heavy protons are irradiated to penetrate through an amorphous metal or alloy (such as Fe-Cr alloy) film. By treating the film in an etchant such as hydrochloric acid at 40°C, the affected tracks are removed, thus forming essentially straight-pore membranes. The resulting membranes show a pore diameter of 0.1 to 0.3 μm [Agency of Ind. Sci. Technol., 1986].

3.3.4 Other Methods

Other methods have also been adopted to produce inorganic membranes with essentially straight pores. Witte [1988] subjected a metal foil such as nickel to a two-step photolithographic procedure. The pore density of the final membrane exceeds 150,000/cm^2 and the pore sizes formed fall under the microfiltration and ultrafiltration ranges. Although both flat and cylindrical shapes can be handled, the flat shape is preferred.

3.4 SURFACE-MODIFIED MEMBRANES

The processes discussed so far produce various types of inorganic membranes in one production process prior to applications and the membrane structures are fixed to the supports in the case of composite membranes. There are, however, other special types of inorganic membranes that are prepared either by a second process to modify the separation properties of the membranes or by an in-situ formation process during the application. Both types will be treated briefly in this section.

3.4.1 Surface Coatings or Pore Modifications Prior to Applications

Membranes prepared by the majority of the established methods have limitations in their pore sizes. To make membranes with finer pore diameters suitable for more demanding separation and membrane reactor applications, a widely practiced technique is to modify the pores or the surface of an existing membrane structure which has already been made. This encompasses a variety of techniques. Some of them are based on gas or vapor phase reactions. Others modifications occur in liquid phase. Some progress having pore diameters in the molecular sieving range has been made.

<u>Gas/vapor phase modifications</u>. Many inorganic membrane materials display functional groups that have chemical affinity to selected chemical agents. A well known example is a gamma-alumina membrane which has hydroxyl groups on the surfaces of the alumina crystallites. These hydroxyl groups present on the pore walls and the macroscopic surface of the membrane can act as the reactive sites for modifications of the pore structure with a chemical agent such as the diversified family of silane compounds (chloro- or alkoxy-silanes).

Miller and Koros [1990] applied this concept to modifying gamma-alumina membranes. The pore sizes of these membranes are modified by using the trifunctional silane, tridecafluro-1,1,2,2-tetrahydrooctyl-1-trichlorosilane (abbreviated as TDFS), which has three reactive chlorine atoms. The reaction between the silane and the available hydroxyl groups on the membrane surface leads to the formation of Al-O-Si bonds and the release of HCl. The release of HCl in combination with the pressure measurement can be used to monitor the reaction.

Schematically the reaction on the membrane or pore surface can be depicted in Figure 3.16. The reaction occurs at two of the three reactive chlorine sites. It has been found that the treatment duration (and therefore the degree of modification) can have significant effects on the permeability and permselectivity of the resulting membrane. Performance data on the separation of toluene from a high molecular weight (about 660) lube oil indicates an increase in permselectivity when the membrane is modified with TDFS for a short time (one day) but not for a long time (two weeks). Highly modified membranes with closely spaced silane groups as a result of long reaction can actually have deleterious effects on selectivity of toluene/lube oil separation. There is also evidence of preferential oil adsorption on the membrane surface which is highly packed with silanes. It is noted that assumed in the structure displayed in Figure 3.16 are the retained plate shape of boehmite crystallites and the "card-stack" configuration of the crystallites. As a result the pore structure is slit-shaped.

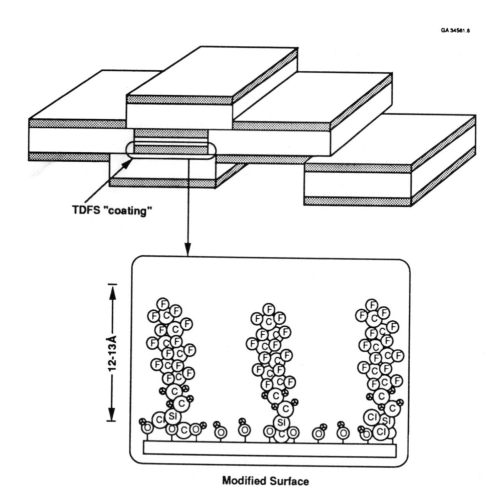

GA 34561.6

TDFS "coating"

12-13Å

Modified Surface

Figure 3.16 Schematic diagram of a gamma-alumina membrane surface modified by tridecafluro-1,1,2,2-tetrahydrooctyl-1-trichlorosilane (TDFS) [Miller and Koros, 1990]

Other oxide, glass and metal membranes have also been modified with silanes. Modifications with alpha-alumina, silica, tin oxide, ruthenium oxide and gold all have been studied for various applications. Alpha-alumina membranes, which also exhibit some, although few, hydroxyl groups, have been modified with ethoxy and chlorosilanes for bioreactor applications [Shimizu et al., 1987]. Interfacial polymerization of silanes has been investigated on anodic alumina membranes [Sugawara et al., 1989]. They all show reduction of the membrane pore sizes as a result of chemical modifications. Similarly, when tetraethoxysilane is introduced into the pores of porous glass membranes and decomposed on the pore walls upon exposure to oxygen at 200°C, pores near the reaction side become smaller. This yields an asymmetric membrane structure.

which is believed to be responsible for doubling the separation efficiency of the He-O2 gaseous mixture [Okubo and Inoue, 1989].

With many possible tail groups having differing chemical and steric properties, silanes can potentially tailor the desired permeability and selectivity of an inorganic membrane. The surface of a porous glass membrane contains the silanol groups with a typical surface density of about $4.7/nm^2$ [Schnabel et al., 1989], making the surface very active. These groups can be chemically modified with many other functional groups such as silanes. Different functional groups of silanes can impart, for example, different adsorption properties of the porous glass membranes.

A more complex and novel type of vapor deposition is the opposing-reactants chemical vapor deposition. Chemical vapor deposition in an opposing reactant geometry has been applied to reduce the pore size of a membrane as explained in Section 3.1.2. Care, however, needs to be exercised to control the deposition process so that pore closure does not occur. To ensure that the deposition is confined to the inside of the membrane pores, the concentration of the organometallic compound should be higher and/or the concentration of the gas stream (oxygen) should be lower. Cao et al. [1993] have reduced the mean pore diameters of prefabricated yttria-stabilized zirconia membranes from 9 to 6 nm in one case and from 80 to 20 nm in the other. $ZrCl_4$ and YCl_3 are used to react with air/water at 800°C to form yttria-stabilized zirconia layers inside the membrane pores. The rate of pore reduction has been found to be larger for those membranes with larger pore sizes [Lin and Burggraaf, 1992].

Liquid phase modifications. Alternatively a porous membrane can be reduced in pore size by a liquid deposition process where the membrane is dipped into a solution or sol to form deposits inside the membrane pores. For example, a silicon nitride tube with a mean pore diameter of 0.35 μm is first immersed in a solution of aluminum alcoholate (aluminum isopropylate or 2-butylate) or chelate (aluminum tris(ethyl acetoacetate) or ethyl acetoacetate aluminum diisopropylate) in an organic solvent (hexane, cyclohexane, benzene, isopropanol, etc.). It is then treated with saturated water vapor to hydrolyze the alcoholate or chelate to form boehmite inside the pores, thus changing the pore diameter to as small as 20 nm [Mitsubishi Heavy Ind., 1984a and 1984b]. Upon calcining at 800°C, boehmite transforms into transition-alumina.

Porous glass membranes can be converted to inorganic ion-exchange membranes by modifications with silane coupling reagents. For example, when reacting with a 3% 3-aminopropyltrimethoxysilane or [3-(2-aminoethylamino)propyl]triethoxysilane solution at room temperature, the glass membranes show anion-selective properties because the amino groups are protonated [Chiang et al., 1989]. When these modified membranes are further modified with succinic anhydride in pyridine also at room temperature, the membranes become cation selective due to the dissociated carboxyl groups from the succinic anhydride.

Trocha and Koros [1994] applied a diffusion-controlled caulking procedure with colloidal silica to plug large pores or defects in ceramic membranes. An important feature of this procedure is the proper selection of the colloid particle size to eliminate or reduce large, less selective pores while minimizing deposition in the small desirable pores. The technique has been successfully proven with anodic alumina membranes (having the majority of the pores about 200 nm in diameter) using 10-20 nm silica colloids.

Modifications with metal membranes. The above modification techniques mostly deposit non-metal materials onto the membrane pores or surfaces. For enhanced separation or catalytic effects, metal layers are sometimes added to porous supports.

Many composite inorganic membranes have been prepared where a layer of dense separating membrane is deposited onto a porous support. The deposited layer is typically thin to avoid significant reduction in the permeation rate.

To make a dense metal membrane on a porous metal support, diffusion welding of a thin metal foil in the range of tens of microns thick can be used. Chemical plating has also been used to deposit a dense Pd membrane on a porous Ag disk [Govind and Atnoor, 1991]. If the dense membrane to be deposited onto a porous support is a multi-component alloy, a two-step process involving liquid spreading followed by sputtering has been developed [Gryaznov, 1992]. For example, liquid indium is first distributed over a porous stainless steel (or magnesia) plate and, after hardening, Pd and Ru are introduced by magnetron sputtering to produce a 2 μm thick dense membrane.

To make dense Pd membrane on a porous glass having a mean pore diameter of 0.3 μm, Uemiya et al. [1988] apply electroless plating using the solution of a Pd salt, $[Pd(NH_3)_4]Cl_2$. The hydrogen permeation rate of the composite membrane is ten times that of a single-layered Pd membrane. At the same time, the composite membrane is 100% selective to hydrogen when processing hydrogen-nitrogen and hydrogen-carbon monoxide gaseous mixtures. The composite membrane also has the advantage of having high mechanical strength imparted by the ceramic support. Therefore, the Pd layer in the composite membrane can be only 13 μm thick while a single-layered Pd membrane normally requires a thickness of about 150 μm for the strength consideration.

Dense palladium membranes have been prepared on porous anodic alumina employing a liquid phase deposition process [Konno et al., 1988]. The alumina is first sputtered with a mixture of Pd and Pt (at a 20/80 weight ratio) to improve the resistance of the anodic alumina toward water. The sputtering treatment has also been found to promote the reaction involved in the subsequent deposition of palladium using the aqueous $[Pd(NH_3)_4]Cl_2$ solution at a concentration of 2 g/L. The solution is filtered through the alumina under an applied pressure difference for two hours at 25°C. It is hypothesized that the promotion of palladium deposition by sputtering is caused by autocatalytic effect as in electroless plating. The resulting composite membranes have much better hydrogen/nitrogen separation performance than the untreated alumina membranes. The performance improves with increasing temperature, which is different from the typical behavior of organic polymeric membranes.

Furthermore, vacuum evaporation has been used to deposit a dense membrane layer of Pd or Pd-Pt alloys on a porous alumina support [Saito, 1988].

As discussed earlier, many composite porous membranes have one or more intermediate layers to avoid substantial penetration of fine particles from the selective layer into the pores of the bulk support matrix for maintaining adequate membrane permeability and sometimes to enhance the adhesion between the membrane and the bulk support. The same considerations should also apply when forming dense membranes on porous supports. This is particularly true for expensive dense membrane materials like palladium and its alloys. In these cases, organic polymeric materials are sometimes used and some of them like polyarilyde can withstand a temperature of up to 350°C in air and possess a high hydrogen selectivity [Gryaznov, 1992].

3.4.2 Membranes Formed In-situ during Applications

There is another type of membrane that is conceptually different from the membranes prepared according to the above methods. It is called dynamic membranes. They are formed, during application, on microporous carriers or supports by deposition of the colloidal particles or solute components that are present in the feed solution. This in-situ formation characteristic makes it possible to tailor them for specific applications in ultrafiltration and reverse osmosis (hyperfiltration).

Although the pioneering research on dynamic membranes at the Oak Ridge National Laboratory mostly dealt with retention of electrolytes, these membranes are also known to be selective toward uncharged molecules or particles. Generally, formation of a dynamic membrane consists of the following steps: (1) conditioning of the porous support; (2) deposition of a colloid to form an ultrafilter on the support; (3) deposition of some organic polymers to form a hyperfilter on the colloid layer; and (4) treatment on the dynamic membrane to enhance separation properties and membrane stability. Several inorganic materials have been tested for dynamic membranes. They include oxides of zirconium, aluminum, iron, silicon, uranium, rhodium and bentonite clays. Among those materials hydrous zirconium oxide was found to perform best as an ultrafilter and has been extensively studied. Porous carbon, stainless steel, and alumina have been employed as the porous supports. Many polymers as organic polyelectrolytes have been examined and the most commonly used is poly(acrylic acid). Currently, the hydrous zirconium oxide ultrafilters and the hydrous zirconium oxide-poly(acrylic acid) hyperfilter are available commercially.

This subject has been extensively studied and is beyond the scope of this book. Comprehensive reviews of dynamic membranes can be found elsewhere [Johnson, 1972;

Thomas, 1977; Dresner and Johnson, 1980; Spencer, 1985; Neytzell-de Wilde et al., 1988; Spencer, 1989].

3.5 SUMMARY

The technology of making inorganic membranes has seen commercialization first in the microfiltration applications. Commercial ceramic membranes for ultrafiltration begin to emerge. Many of the commercial inorganic membranes are composite in nature while a few of them have homogeneous pore structures. Commonly used methods for producing those membranes are slip casting (dip or spin coating) of fine particle dispersions (e.g., from sol-gel process), phase separation/leaching and anodic oxidation. A wide variety of other preparation techniques have also been surveyed in this chapter. In contrast to porous inorganic membranes, dense membranes have limited commercial uses due to their costs and low permeabilities. Palladium and palladium-based dense membranes are the only significant dense inorganic membranes in the marketplace.

Production technology of porous inorganic membranes is progressing toward smaller pore sizes to meet more challenging requirements for such applications as gas separation and membrane reactors. Research in this area has intensified in recent years. The trend of exploring new combinations of materials and processes for finer pore membranes can be highlighted in Table 3.5 based on the literature.

TABLE 3.5
Pore size limits of various processes for making porous inorganic membranes

Process	Membrane material(s)	Smallest pore diameter (nm)
Suspension/dip coating	Al_2O_3	100
Sol-gel/dip coating	Al_2O_3, TiO_2, ZrO_2, CeO_2, RuO_2/TiO_2	2.5
Phase separation/leaching	Glass, glass ceramics	2
Pyrolysis of polymers	SiO_2, molecular sieve C, C, Al_2O_3/cordierite	1
Pore modifications	Al_2O_3/Fe_2O_3, SiO_2/CeO_2, Al_2O_3MgO, TiO_2/V_2O_5, ZrO_2/TiO_2, ZrO_2/Fe_2O_3, Al_2O_3/ZrO_2, Al_2O_3/silicate	0.3
Anodic oxidation	Al_2O_3, ZrO_2	10
Track etching	mica	6

It seems that while new nanoscale technology has long term promises an approach that is relatively simple to implement has attracted much attention. This widely investigated technique to making inorganic membranes with sub-nanometer pore sizes is to modify pre-fabricated membranes by various deposition techniques such as sol-gel process, CVD, ion implementation, etc.

REFERENCES

Abe, F., and T. Fujita, 1987, U.S. Patent 4,689,150.

Agency of Ind. Sci. & Technol. (Japan), 1986, Japanese Patent 60,017,443.

Aksay, I.A., and C.H. Schilling, 1984, Colloidal filtration route to uniform microstructure, in: Ultrastructure Processing of Ceramics, Glasses, and Composites (L.L. Hench and D.R. Ulrich, Eds.), Wiley, New York, USA, p. 483.

Altemose, V.O., 1961, J. Appl. Phys. **32**, 1309.

Anderson, M.A., and L. Chu, 1993, U.S. Patent 5,194,200.

Anderson, M.A., and Q. Xu, 1991, U.S. Patent 5,006,248.

Anderson, M.A., M.J. Gieselmann and Q. Xu, 1988, J. Membr. Sci. **39**, 243.

Anderson, M.A., Q. Xu and B.L. Bischoff, 1993, U.S. Patent 5,215,943.

Arrance, F.C., C. Mesa and C. Berger, 1969, U.S. Patent 3,437,580.

Athavale, S.N., and M.K. Totlani, 1989, Met. Fin. **87**, 23.

Atomic Energy Comm. (France), 1971, French Patent 2,061,933.

Baker, R.A., G.D. Forsythe, K.K. Likhyani, R.E. Roberts and D.C. Robertson, 1984, U.S. Patent 4,446,024.

Balachandran, U., S.L. Morissette, J.T. Dusek, R.L. Mieville, R.B. Poeppel, M.S. Kleefisch, S. Pei, T.P. Kobylinski and C.A. Udovich, 1993, Development of ceramic membranes for partial oxygenation of hydrocarbon fuels to high-value-added products, in: Proc. Coal Liquefaction and Gas Conversion Contractors' Review Conf., Pittsburgh, 1993 (U.S. Dept. of Energy).

Beaver, R.P., 1986, European Patent 186,129.

Berardo, M., J. Charpin and J.M. Martinet, 1986, French Patent 2,575,396.

Boudier, G., and J.A. Alary, 1989, Poral new advances in metallic and composite porous tubes, in: Proc. 1st Int. Conf. Inorg. Membr. , Montpellier, France, p. 373.

Bradley, D.C., R.C. Mehrotra and D.P. Gaur, 1978, Metal Alkoxides, Academic Press, London.

Brinker, C.J., K.D. Keefer, D.W. Schaefer and C.S. Ashley, 1982, J. Non-Cryst. Solids **48**, 47.

Brinker, C.J., T.L. Ward, R. Sehgal, N.K. Raman, S.L. Hietala, D.M. Smith, D.W. Hua and T.J. Headley, 1993, J. Membr. Sci. **77**, 165.

Brinker, C.J., R. Sehgal, N.K. Raman, P.R. Schunk, S. Prakash, S.L. Hietala and S. Wallace, 1994, Sol-gel strategies for amorphous, inorganic membranes exhibiting molecular sieving characteristics, in: Proc. Third Int. Conf. Inorg. Membr., Worcester, MA, USA.

Brinkman, H.W., and A.J. Burggraaf, 1995, J. Electrochem. Soc. **142**, 3851.

Burggraaf, A.J., K. Keizer and B.A. van Hassel, 1989, Solid State Ionics **32/33**, 771.

Cao, G.Z., H.W. Brinkman, J. Meijerink, K.J. de Vries and A.J. Burggraaf, 1993, J. Am. Ceram. Soc. **76**, 2201.

Carolan M.F., P.N. Dyer, S.M. Fine, A. Makitka, III, R.E. Richards and L.E. Schaffer, 1994, U.S. Patent 5,332,597.

Chai, M., M. Machida, K. Eguchi and H. Arai, 1994, J. Membr. Sci. **96**, 205.

Charpin, J., and P. Rigny, 1989, Inorganic membranes for separativetechniques: from uranium isotope separation to non-nuclear fields, in: Proc. 1st Int. Conf. on Inorganic Membranes, Montpellier, France (Trans Tech Publ., Zürich) p. 1.

Chen, Y.D., and R.T. Yang, 1994, Ind. Eng. Chem. Res. **33**, 3146.

Chiang, T.H., A. Nakamura and F. Toda, 1989, Thin Solid Films **182**, L13.

Clark, R.A., M.F. Hall and J. N. Kirk, 1988, Br. Ceram. Proc. **43**, 77.

Croopnick, G.A., and D.M. Scruggs, 1986, U.S. Patent 4,608,319.

de Albani, M.I.D., and C.P. Arciprete, 1992, J. Membr. Sci. **69**, 21.

Diggle, J.W., T.C. Dowme and C.W. Goulding, 1969, Chem. Rev. **69**, 365.

Dimitrijewits de Albani, M.I., and C.P. Arciprete, 1992, J. Membr. Sci. **69**, 21.

Dislich, H., 1986, J. Non-Cryst. Solids **80**, 115.

Dobo, E.J., 1980, U.S. Patent 4,222,977.

Dresner, L., and J.S. Johnson, Jr., 1980, Hyperfiltration (reverse osmosis), in: Principles of Desalination (K.S. Spiegler, Ed.), Academic Press, New York, USA, p. 401.

Dwivedi, R.K., 1986, J. Mater. Sci. Lett. **5**, 373.

Elmaleh, S., K. Jaafari, A. Julbe, J. Randon and L. Cot, 1994, J. Membr. Sci. **97**, 127.

Fleming, H.L., 1988, Carbon composites: a new family of inorganic membranes, presented at 6th Annual Membrane Planning Conference in Cambridge, MA, USA.

Furnas, C.C., 1931, Ind. Eng. Chem. **23**, 1052.

Furneaux, R.C., W.R. Rigby and A.P. Davidson, 1989, Nature **337**, 147.

Gaeta, S.N., H. Zhang and E. Drioli, 1989, Preparation of polyphosphazene membranes for gas and liquid separations, in: Proc. 1st Int. Conf. Inorg. Membr., Montpellier, France, p.65.

Galán, M., J. Llorens, J.M. Gutiérrez, C. González and C. Mans, 1992, J. Non-Cryst. Solids **147/148**, 518.

Gallaher, G.R., Jr., J.C.S. Wu, T.E. Gerdes, R.L. Gregg, D.F. Flowers, D.R. Smith, B. Hart, J.D. Sibold and R. Kleiner, 1994, Development of H_2 selective membranes and their use as catalytic membrane reactors, presented at Worldwide Catal. Ind. Conf. (CatCon '94), Philadelphia, PA, USA.

Gavalas, G.R., C.E. Megiris and S.W. Nam, 1989, Chem. Eng. Sci. **44**, 1829.

Gillot, J., 1987, U.S. Patent 4,698,157.

Goldsmith, R.L., 1989, International Patent WO 89/11342.

Govind, K., and D. Atnoor, 1991, Ind. Eng. Chem. Res. **30**, 591.

Graham, T., 1866, Phil. Trans. Roy Soc. (London) **156**, 399.

Grosgogeat, E.J., J.R. Fried, R.G. Jenkins and S.T. Hwang, 1991, J. Membr. Sci. **57**, 237.

Gryaznov, V.M., 1992, Platinum Metals Rev. **36**, 70.

Guizard, C., N. Idrissi, A. Larbot and L. Cot, 1986a, Br. Ceram. Proc. **38**, 263.

Guizard, C., N. Cygankiewicz, A. Larbot and L. Cot, 1986b, J. Non-Cryst. Solids **82**, 86.

Guizard, C., F. Garcia, A. Larbot and L. Cot, 1989, An inorganic membrane made from the association of a zirconia layer with a stainless steel support, in: Proc. 1st Int. Conf. Inorg. Membr., Montpellier, France, p.405.

Ha, H.Y., S.W. Nam, S.A. Hong and W.K. Lee, 1993, J. Membr. Sci. **85**, 279.

Haag, W.O., and J.G. Tsikoyiannis, 1991, U.S. Patent 5,019,263.

Hackley, V.A., and M.A. Anderson, 1992, J. Membr. Sci. **70**, 41.

Hammel, J.J., W.P. Marshall, W.J. Robertson and H.W. Barch, 1987a, European Pat. Appl. 248,391.

Hammel, J.J., W.P. Marshall, W.J. Robertson and H.W. Barch, 1987b, European Pat. Appl. 248,392.

Hazbun, E.A., 1988, U.S. Patent 4,791,079.

Hsu, C., and R.E. Buxbaum, 1987, Palladium-coated zirconium membranes for oxidative extraction, in: Preprint Annual AIChE Meeting, New York.

Huling, J.C., J.K. Bailey and D.M. Smith, 1992, Mater. Res. Soc. Symp. Proc. **271**, 511.

Hwang, S.T., D. Li and D.R. Seok, 1989, U.S. Patent 4,828,588.

Hyun, S.H., and B. Kang, 1994, J. Am. Ceram. Soc. **77**, 3093.

Ibiden Co., 1986, Japanese Patent 61,091,076.

Ilias, S., and R. Govind, 1989, AIChE Symp. Ser. 268, **85**, 18.

Isenberg, A.O., 1981, Sol. St. Ionics **3/4**, 431.

Ishizaki, K., and M. Nanko, 1995, J. Porous Mat. **1**, 19.

Jain, M.K., and S.K. Nadkarni, 1990, U.S. Patent 4,973,435.

Jayaraman, V., Y.S. Lin, M. Pakala and R.Y. Lin, 1995, J. Membr. Sci. **99**, 89.

Jia, M.D., K.V. Peinemann and R.D. Behling, 1993, J. Membr. Sci. **82**, 15.

Jiang, S., Y. Yan and G.R. Gavalas, 1995, J. Membr. Sci. **103**, 211.

Johnson, J.S., Jr., 1972, Polyelectrolytes in aqueous solutions - filtration, hyperfiltration and dynamic membranes, in:Reverse Osmosis Membrane Research (H.K. Lonsdale and H.E. Pondall, Eds.), Plenum, New York, USA, p. 379.

Johnson, S.M., and R.H. Lamoreaux, 1989, Am. Ceram. Soc. Bull. **68**, 1431.

Kato, A., and H. Mizumoto, 1986, Ceramics Int. **12**, 95.

Kennedy, J.H., and A. Foissy, 1977, J. Am. Ceram. Soc. **60**, 33.

Kim, S., and G.R. Gavalas, 1995, Ind. Eng. Chem. Res. **34**, 168.

Klein, L.C., and D. Gallagher, 1988, J. Membr. Sci. **39**, 213.

Konno, M., M. Shindo, S. Sugawara and S. Saito, 1988, J. Membr. Sci. **37**, 193.

Koresh, J., and A. Soffer, 1977, J. Electrochem. Soc. **124**, 1379.

Koresh, J.E., and A. Soffer, 1983, Separation Sci. Technol. **18**, 723.

Koresh, J.E., and A. Soffer, 1987, Separation Sci. Technol. **22**, 973.

Kraus, M., and K. Murphy, 1987, U.S. Patent 4,710,204.

Kuratomi, T., 1988, Japanese Patent 63,100,930.

Kyocera Corp., 1985, Japanese Patent 60,051,676.

Lamond, T.G., J.E. Metcalf III and P.L. Walker, 1965, Carbon **3**, 59.

Langlet, M., D. Walz, P. Marage and J.C. Joubert, 1992, J. Non-Cryst. Solids **147/148**, 488.

Larbot, A., J. Alary, C. Guizard, J.P. Fabree, N. Idrissi and L. Cot, 1987, Thin layers of ceramics from sol-gel process (or inorganic membranes) for liquid separations, in:Mater. Sci. Monographs (P. Vincenzini, Ed.) **38**, 2259.

Larbot, A., J.P. Fabre, C. Guizard and L. Cot, 1989, J. Am. Ceram. Soc. **72**, 257.

Lee, K.H., and S.J. Khang, 1987, Ceram. Eng. Sci. Proc. **8**, 85.

Lee, K.H., and Y.M. Kim, 1991, Key Eng. Mat. **61/62**, 17.

Leenaars, A.F.M., and A.J. Burggraaf, 1985, J. Colloid Interface Sci. **105**, 27.

Leenaars, A.F.M., K. Keizer and A.J. Burggraaf, 1984, J. Mat. Sci. **19**, 1077.

Lin, C.L., D.L. Flowers and P.K.T. Liu, 1994, J. Membr. Sci. **92**, 45.

Lin, Y.S., and A.J. Burggraaf, 1992, AIChE J. **38**, 445.

Lin, Y.S., L.G.J. de Haart, K.J. de Vries and A.J. Burggraaf, 1989, Modification of ceramic membranes by CVD and EVD for gas separation, catalysis and SOFC application, in: Euro-Ceramics (Vol. 3), eds. G. de With, R.A. Terpstra and R. Metselaar (Elsevier, London) p. 3.590.

Livage, J., 1986, J. Solid State Chem. **64**, 322.

Mardilovich, P.P., A.N. Govyadinov, N.I. Mazurenko and R. Paterson, 1995, J. Membr. Sci. **98**, 143.

Mårtensson, M., K.E. Holmberg, C. Löfman and E.I. Eriksson, 1958, Proc. 2nd U.N. Int. Conf. on Peaceful Uses of Atomic Energy **4** (Production of Nuclear Materials and Isotopes), 395.

Masuda, S., 1989, Japanese Patent 01,254,212.

Masuda, H., K. Nishio and N. Baba, 1993, Thin Solid Films **223**, 1.

Matsushita Electric Ind., 1985, Japanese Patent 60,129,119.

Mazdiyasni, K.S., C.T. Lynch and J.S. Smith, 1967, J. Am. Ceram. Soc. **50**, 532.

McCaffrey, R.R., and D.G. Cummings, 1988, Sep. Sci. Tech. **23**, 1627.

McCaffrey, R.R., R.E. McAtee, A.E. Grey, C.A. Allen, D.G. Cummings, A.D. Appelhans, R.B. Wright and J.G. Jolley, 1987, Sep. Sci. Tech. **22**, 873.

McHugh, K.M., and J.F. Key, 1994, J. Thermal Spray Technol. **3**, 191.

McHugh, K.M., L.D. Watson, R.E. McAtee and S.A. Ploger, 1993, U.S. Patent 5,252,212.

Megiris, C.E., and J.H.E. Glezer, 1992a, Ind. Eng. Chem. Res. **31**, 1293.

Megiris, C.E., and J.H.E. Glezer, 1992b, Chem. Eng. Sci. **47**, 3925.

Miller, J.R., and W.J. Koros, 1990, Separation Sci. Tech. **25**, 1257.

Miszenti, G.S., and C.A. Nannetti, 1975, U.S. Patent 3,874,899.

Mitrovic, M., and L. Knezic, 1979, Desalination **28**, 147.

Mitsubishi Heavy Ind., 1984a, Japanese Patent 59,059,224.

Mitsubishi Heavy Ind., 1984b, Japanese Patent 59,107,988.

Mitsubishi Heavy Ind., 1984c, Japanese Patent 59,147,605.

Mitsubishi Heavy Ind., 1984d, Japanese Patent 59,179,112.

Murphy, R., 1988, TechLink - Membrane Separation, October Issue, SRI International .

Neytzell-de Wilde, F.G., C.A. Buckley and M.P.R. Cawdron, 1988, Desalination **70**, 121.

NGK Insulators, 1986, Japanese Patent 61,238,305.

Niwa, M., S. Kato, T. Hattori and Y. Murakami, 1984, J. Chem. Soc. Faraday Trans. 1, **80**, 3135.

Ohya, H., Y. Tanaka, M. Niwa, R. Hongladaromp, Y. Negismi and K. Matsumoto, 1986, Maku **11**, 41.

Okubo, T., and H. Inoue, 1989, J. Membr. Sci. **42**, 109.

Okubo, T., K. Haruta, K. Kusakabe, S. Morooka, H. Anzai and S. Akiyama, 1991, J. Membr. Sci. **59**, 73.

Partlow, D.P., and B.E. Yoldas, 1981, J. Non-Cryst. Solids **46**, 153.

Peterson, E.S., M.L. Stone, R.R. McCaffrey and D.G. Cummings, 1993, Sep. Sci. Tech. **28**, 423.

Pflanz, K.B., R. Riedel and H. Chmiel, 1992, Adv. Mater. **4**, 662.

Powers, R.W., 1975, J. Electrochem. Soc. **122**, 490.

Quinn, J.A., J.L. Anderson, W.S. Ho and W.J. Petzny, 1972, Biophysical J. **12**, 990.

Rai, K.N., and E. Ruckenstein, 1975, J. Catalysis **40**, 117.

Raman, N.K., and C.J. Brinker, 1995, J. Membr. Sci. **105**, 273.

Riedel, C., and R. Spohr, 1980, J. Membr. Sci. **7**, 225.

Rigby, W.R., R.C. Furneaux, and A.T. Thomas, 1987, European Pat. Appl. 242,204.

Rigby, W.R., D.R. Cowieson, N.C. Davies and R.C. Furneaux, 1990, Trans. Inst. Metal Finish. **68**, 95.

Saito, S., 1988, Japanese Patent 63, 171,617.

Sano, Y., F. Mizukami, H. Yagishita, M. Kitamoto and Y. Kyozumi, 1994, Japanese Patent 06,099,044.

Sato, M, and K. Yamaguchi, 1985, Japanese Patent 60,243,287.

Satterfield, C.N., 1980, Heterogeneous Catalysis in Practice (McGraw-Hill, New York).

Scherer, G.W., 1990, J. Am. Ceram. Soc. **73**, 3.

Schleinitz, H.M., 1981, U.S. Patent 4,251,377.

Schnabel, R., and W. Vaulont, 1978, Desalination **24**, 249.

Schnabel, R., P. Langer and E. Bayer, 1989, Application oriented surface treatment of inorganic membranes, in: Proc. 6th Int. Symp. Synth. Membr. Sci. Ind., Tübingen, Germany, p. 171.

Shafer, M.W., D.D. Awschalom, J. Warnock and G. Ruben, 1987, J. Appl. Phys. **61**, 5438.

Sheng, G., L. Chu, W.A. Zeltner and M.A. Anderson, 1992, J. Non-Crystl. Solids **147/148**, 548.

Shimizu, Y., T. Yazawa, H. Yanagisawa and K. Eguchi, 1987, Yogyo Kyokaishi **95**, 1169.

Shkrabina, R.A., B. Boneckamp, P. Pex, H. Veringa and Z.R. Ismagilov, 1995a, React. Kinet. Catal. Lett. **54**, 181.

Shkrabina, R.A., B. Boneckamp, P. Pex, H. Veringa and Z.R. Ismagilov, 1995b, React. Kinet. Catal. Lett. **54**, 193.

Shu, J., B.P.A. Grandjean, A. van Neste and S. Kaliaguine, 1991, Can. J. Chem. Eng. **69**, 1036.

Shu, J., B.P.A. Grandjean, E. Ghali and S. Kaliaguine, 1993, J. Membr. Sci. **77**, 181.

Siemens Aktiengesellschaft, 1975, U.K. Patent 1,402,206.

Simon, C., R. Bredesen, H. Raeder, M. Seiersten, A. Julbe, C. Monteil, I. Laaziz, J. Etienne and L. Cot, 1991, Key Eng. Mat. **61/62**, 425.

Smatko, J.S., 1973, U.S. Patent 3,711,336.

Soffer, A., J.E. Koresh and S. Saggy, 1987, U.S. Patent 4,685,940.

Spencer, H.G., 1985, Dependence of dynamic membrane performance on formation materials and procedures, in:Mater. Sci. of Synthetic Membr. (D.R. Lloyd, Ed.), Am. Chem. Soc. , Washington, D.C., USA, p. 295.

Spencer, H.G., 1989, Formed-in-place inorganic membranes: properties and applications, in: Proc. 1st Int. Conf. Inorg. Membr., Montpellier, France, p. 95.

Stemme, G., and G. Kittilsland, 1988, Appl. Phys. Lett. **53**, 1566.

Strawbridge, I., and P.F. James, 1986, J. Non-Crystalline Solids **86**, 381.

Strumillo, C., and T. Kudra, 1987, Drying Principles, Applications and Design (Gordon and Breach, New York)

Subbarao, E.C., ed., 1980, Solid Electrolytes and Their Applications, (Plenum Press, New York).

Sugawara, S., M. Konno and S. Saito, 1989 J. Membr. Sci. **44**, 151.

Sumitomo Electric Ind., 1973, Japanese Patent 48,007,190.

Sumitomo Electric Ind., 1981, Japanese Patent 56,008,643.

Suzuki, H., 1987, U.S. Patent 4,699,892.

Teijin, K.K., 1984, Japanese Patent 59,177,117.

Terpstra, R.A., and J.W. Elferink, 1991, J. Mat. Sci. Lett. **10**, 1384.

Terpstra, R.A., B.C. Bonekamp and H.J. Veringa, 1988, Desalination **70**, 395.

Thomas, D.G., 1977, Dynamic membranes: their technological and engineering aspects, in: Reverse Osmosis and Synthetic Membranes (S. Sourirajan, Ed.), National Research Council of Canada, Ottawa, Canada, p. 294.

Tiller, F.M., and C.D. Tsai, 1986, J. Am. Ceram. Soc. **69**, 882.

Tokyo Shibaura Electric Co., 1974, Japanese Patent 49,115,997.

Toyo Soda Mfg., 1985, Japanese Patent 60,187,320.

Trocha, M., and W.J. Koros, 1994, J. Membr. Sci. **95**, 259.

Tsapatsis, M., and G.R. Gavalas, 1992, AIChE J. **38**, 847.

Tsapatsis, M., S. Kim, S.W. Nam and G.R. Gavalas, 1991, Ind. Eng. Chem. Res. **30**, 2152.

Uemiya, S., Y. Kude, K. Sugino, N. Sato, T. Matsuda and E. Kikuchi, 1988, Chem. Lett. **10**, 1687.

Uemiya, S., H. Sasaki, T. Matsuda and E. Kikuchi, 1990, Nippon Kagaku Kaishi **6**, 669.

Van Praag, W., V. Zaspalis, K. Keizer, J.G. van Ommen, J.R. Ross and A.J. Burggraaf, 1989, Preparation, modification and microporous structure of alumina and titania ceramic membrane systems, in: Proc. 1st Int. Conf. Inorg. Membr., Montpellier, France, p.397.

Voncken, J.H.L., C. Lijzenca, K.P. Kumar, K. Keizer, A.J. Burggraaf and B.C. Bonekamp, 1992, J. Mat. Sci. **27**, 472.

Walker, P.L., L.G. Austin and S.P. Nandi, 1966, Chem. Phys. Carbon **2**, 257.

Walukas, D.M., 1992, GASAR Materials: A Novel Approach in the Fabrication of Porous Materials, DMK TEK, Inc. brochure.

Wefers, K., and C. Misra, 1987, Oxides and Hydroxides of Aluminum, Alcoa Technical Paper No. 19 (Revised), Aluminum Co. of America.

Winnick, J., 1990, High temperature membranes for H_2S and SO_2 separations, DOE Quarterly Progress Report DOE/PC/90293-T1.

Witte, J.F., 1988, European Pat. Appl. 272,764-A.

Xu, Q., and M.A. Anderson, 1991, J. Mater. Res. **6**, 1073.

Xu, Q., and M.A. Anderson, 1994, J. Am. Ceram. Soc. **77**, 1939.

Yazawa, T., H. Tanaka, H. Nakamichi and T. Yokoyama, 1991, J. Membr. Sci. **60**, 307.

Yoldas, B.E., 1975a, Ceram. Bull. **54**, 286.

Yoldas, B.E., 1975b, J. Mater. Sci. **10**, 1856.

CHAPTER 4

PHYSICAL, CHEMICAL AND SURFACE PROPERTIES OF INORGANIC MEMBRANES

The preparation and fabrication methods and their conditions described in Chapter 3 dictate the general characteristics of the membranes produced which, in turn, affect their performance as separators or reactors. Physical, chemical and surface properties of inorganic membranes will be described in detail without going into discussions on specific applications which will be treated in later chapters. Therefore, much of this chapter is devoted to characterization techniques and the general characteristics data that they generate.

Macroscopic features such as the shapes of membrane elements, modules and systems and general application characteristics will be covered in Chapter 5. Good understanding of the basic membrane characteristics should provide the much needed background and guidelines for design and operation of the membrane systems.

4.1 MICROSTRUCTURES

Membrane morphology and, in the case of porous membranes, pore size and orientation and porosity are vital to the separation properties of inorganic membranes. As the general characterization techniques evolve, the understanding of these microstructures improves.

4.1.1 Morphology of Membrane Structures and Surfaces

Surface and structural features can be revealed by several visibly comprehensible techniques including various microscopy methods. Those tools provide direct guidance for the optimization of materials and processing parameters during membrane preparation.

Scanning electron microscopy. The most widely used method of characterizing the morphology of membranes, particularly inorganic membranes, is scanning electron microscopy. A scanning electron microscope (SEM) is based on the principle that an image is formed when the electrons it emits interact with the atoms of the specimen. SEMs can provide higher-resolution images than reflected light microscopes. Because of this advantage and competitive pricing as a result of recent advanced technology, SEMs have become the workhorse of many surface and microstructural characterization tools in various material applications, including membrane separation.

The theoretical maximum magnification of SEMs can exceed 500,000x (with a resolution of 6 nm) but the practical limits are less than 100,000x. The less expensive models, especially when operated under less ideal conditions (e.g., vibration), often can only magnify up to 30,000x or even lower. The depth of field feature of a scanning electron microscope enables both researchers and users of membranes to examine the three-dimensional microscopic network and texture of the porous surfaces. For poorly conductive materials such as many oxides, gold plating is usually used to reveal those detail structures.

As briefly mentioned in Chapter 1, some inorganic membranes exhibit a symmetric structure. An example is given in Figure 4.1 which shows the microstructure of a porous glass membrane under an SEM. Some porous metal membranes have similar structures. While these types of membranes are effective in rejecting solutes or retentates according to their permselectivity characteristics, they are not effective in maintaining high fluxes because of their large thicknesses. The permeate flux, according to the simple Hagen-Poiseuille flow equation, is inversely proportional to the membrane thickness. As elucidated in Chapter 1, an effective way of designing membranes for high flux is to design a multiple-layer structure in which a thin layer of membrane with a small pore size is adhered to subsequent support layers with successively larger pore sizes.

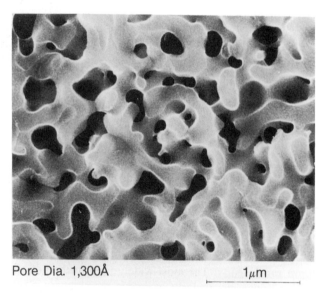

Pore Dia. 1,300Å   1μm

Figure 4.1 Cross-sectional microstructure of a porous glass membrane [Courtesy of Asahi Glass]

The principle involved in this type of multi-layered design is to ideally: (1) impart the retention function to a layer of the separating membrane as thin as possible (so long as it is sufficiently strong to possess structural integrity and is void of cracks and pinholes); (2) make the bulk support layer the source of mechanical strength (and negligible flow resistance) for the membrane/support composite; and (3) sandwich one or more intermediate support layers between the membrane and the bulk support.

The major reason for the intermediate support layers is the following. To make a bulk support with negligible flow resistance compared to that of the membrane layer, their pore sizes need to be widely different and so are the sizes of their corresponding precursor particles during synthesis. The small membrane precursor particles are likely to penetrate into the pores of the bulk support, thus decreasing the permeate flux through the membrane/support composite. A practical and proven way of surmounting this problem is to use one or more intermediate support layers with pores that are successively larger but not larger than that of the bulk support and not smaller than that of the membrane. Figure 4.2 shows such a commercial three-layer ceramic membrane. Typically in these membranes, the separating layer is the thinnest, on the order of 5 to 20 microns. The intermediate support layers are generally 10 to 50 microns thick compared to 1 to 2 mm thick bulk support. Well defined crystalline networking structure of the bulk support is evident in the figure. It is a result of sintering at elevated temperatures.

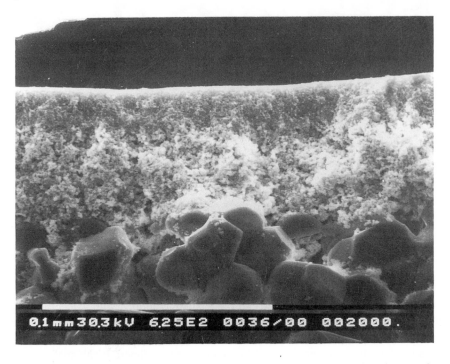

Figure 4.2 Cross-sectional SEM micrographs of a three-layer ceramic membrane [Courtesy of US Filters]

Multi-layered high-quality ceramic membranes do not have any appreciable pore blockage by the precursor particles from the adjacent layers. This is evidenced by the sharp and clean interface between layers shown in Figure 4.2. It should also be noted that the smoothness of the underlying layer is critical to forming a smooth layer on top of it.

One of the major reasons for the widespread interest of inorganic membranes in recent years is the ability to manufacture ceramic membranes with consistently uniform-sized pores and precursor particles on an industrial scale. Figure 4.3 reveals the surface morphology of a microfiltration grade ceramic membrane. It is an essential requirement in separation applications that a membrane is free of defects or cracks.

Figure 4.3 Surface morphology of a microfiltration grade ceramic membrane

Most SEMs today, except those low end models, are equipped with a spectrometer to perform X-ray microanalysis. This type of analysis, commonly known as energy-dispersive X-ray analysis (EDX or EDXRA), can reveal the elemental composition of a small spot (less than a square micron) in a membrane specimen. It can analyze a large number of elements except light elements such as carbon, nitrogen and oxygen. It is possible to carry out some analysis on some light elements with a light element detector.

EDX can be utilized to determine elemental composition of an inorganic membrane or membrane/support composite at different depth locations to guide membrane synthesis. In addition, EDX is capable of detecting the nature of any foulants that partially or completely block the passage of the permeate through membrane pores. Displayed in Figures 4.4(a) and (b) are examples of EDX data of an alumina membrane before and after it was fouled by some species containing barium, sulfur and strontium. Aluminum in the analysis result confirms that the membrane is made primarily from alumina and gold shows up as a coating material for SEM analysis.

Transmission electron microscopy. Using a higher accelerating voltage than SEM, transmission electron microscopy (TEM) provides higher resolution images than SEM does. Therefore it can be used to characterize membrane layers with pores in the ultrafiltration and the upper end of the gas separation range and their precursor particles or colloids. Contrary to SEM, TEM passes the image forming electrons through the specimens and, because of this, is commonly limited to specimens of a few hundred nanometers (or a few tenths of a micron) thick. More sophisticated models, however, can analyze samples with a thickness of a few microns and can provide a maximum resolution of 0.2 to 0.5 nm (or a magnification above 10,000,000x). As in the case with SEM, elemental analysis can be performed with EDX on a very small area on the order of a few angstroms in each dimension.

For TEM observation, more complicated and careful sample preparation procedures are involved. Extreme care must be taken during sample preparation to avoid cross-contamination. The sample sometimes needs to be embedded in a medium such as epoxy resin that cures at 50°C overnight before cutting by a microtome equipped with a diamond knife or thinning by electropolishing or ion beam.

Contrary to SEM, TEM does not provide the striking three-dimensional images. Shown in Figure 4.5 are some well dispersed crystallites of boehmite as the precursor to a thin, unsupported partially calcined alumina membrane the TEM image of which is given in Figure 4.6. It is estimated from the TEM that many of the crystallites appear to be smaller than about 50 nm. The ordered structure of the very thin, partially calcined alumina membrane is evident in Figure 4.6.

Atomic force microscopy. Scanning electron microscopes do not provide the necessary resolution for analyzing detail morphology of certain membrane layers of very fine pore sizes such as those suitable for ultrafiltration and gas separation. While transmission electron microscopes are capable of examining very small scale structures, the technique is limited to only very thin specimens. This makes sample preparation for a multi-layered membrane very difficult and tedious.

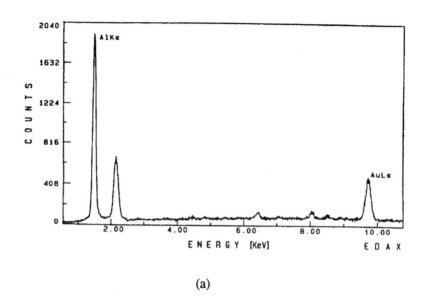

(a)

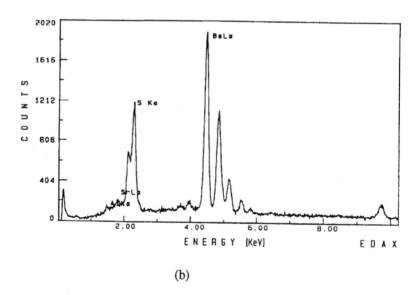

(b)

Figure 4.4 Energy dispersive X-ray analysis data of an alumina membrane (a) before fouling and (b) after fouling

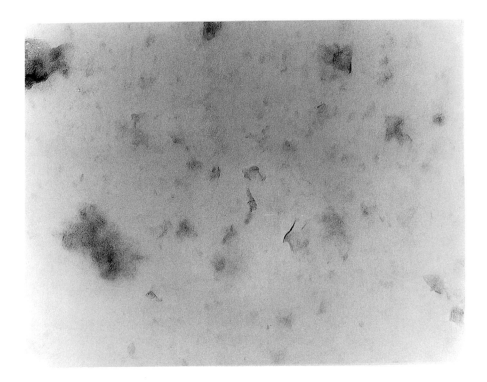

Figure 4.5 Transmission electron microscopy image of well dispersed boehmite particles as a precursor to fine-pore alumina membranes

A new alternative to solve this problem is atomic force microscopy (AFM) which is an emerging surface characterization tool in a wide variety of materials science fields. The method is relatively easy and offers a subnanometer or atomic resolution with little sample preparation required. The basic principle involved is to utilize a cantilever with a spring constant weaker than the equivalent spring between atoms. This way the sharp tip of the cantilever, which is microfabricated from silicon, silicon oxide or silicon nitride using photolithography, mechanically scans over a sample surface to image its topography. Typical lateral dimensions of the cantilever are on the order of 100 μm and the thickness on the order of 1 μm. Cantilever deflections on the order of 0.01 nm can be measured in modern atomic force microscopes.

Since it utilizes a force instead of a current, the technique is capable of imaging both conducting and nonconducting materials. AFM can make three-dimensional quantitative measurements with a higher resolution and on a wider variety of materials than any other surface characterization methods.

Figure 4.6 Transmission electron microscopy image of a very thin, partially calcined alumina membrane with nanometer-sized pores

AFM has been increasingly utilized to analyze organic polymeric membranes and has just been explored for studying inorganic membranes [Bottino et al., 1994]. Two gamma-alumina ultrafiltration membranes having pore diameters smaller than 5 nm and supported on three-layer alpha-alumina structures were characterized. Contrary to SEM, AFM does not require a conductive coating such as gold. AFM can be calibrated with a standard such as polystyrene latex spheres arrays and diffraction gratings.

Bottino et al. [1994] used silicon nitride tips and silicon nitride gold-coated cantilevers to obtain the images in the height mode (or constant force mode). The AFM technique has been compared quite well with SEM using alpha-alumina membrane samples. They found that AFM could determine the sizes of individual constituent gamma-alumina particles in the membranes and detect subtle differences in their dimension ratios. Based on the gray scale, information on the depth (and, therefore, roughness) of the membrane can be derived and used to construct a quantitative three-dimensional image of the membrane surface such as that shown in Figure 4.7. Surface roughness of a membrane may be used to explain its separation behavior in ultrafiltration [Capannelli et al., 1994].

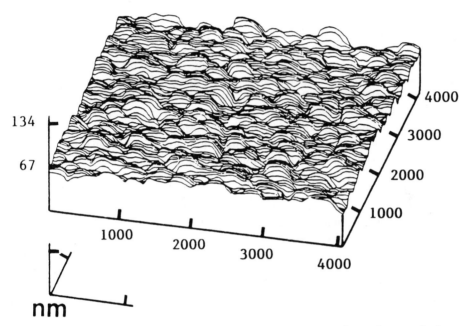

Figure 4.7 Three-dimensional image of a gamma-alumina membrane by atomic force microscopy [Bottino et al., 1994]

To obtain statistically significant depth data, the total membrane area inspected by the AFM should be larger than 100 times the area of a single gamma-alumina particle. For alpha-alumina particles, the area needed should exceed about 400 times. From the depth distribution data, it has been deduced that the particles in the gamma-alumina membranes should be plate-shaped which seems to agree with the idealized model for the microstructure of a gamma-alumina membrane [Leenaars et al., 1984]. For example, for a gamma-alumina membrane produced commercially, the dimensions of individual particles are approximately 300 nm in length or width and only 30 nm in height [Bottino et al., 1994].

The mean roughness of two gamma-alumina membranes prepared by the sol-gel process described in Chapter 3 is estimated to be about 30 to 40 nm. In contrast, the mean roughness of a commercial alpha-alumina membrane is about 340 nm. The values for the gamma-alumina membranes are close to that obtained by de Lange et al. [1995] as 40 nm. Interesting, when a gamma-alumina membrane is modified with an amorphous microporous layer of silica, the resulting maximum surface roughness of the modified membrane lies between 20 and 40 nm, which means that the modification only slightly smooths out the surface roughness [de Lange et al., 1995].

4.1.2 Pore Size

Separation properties such as permeability and permselectivity of a porous membrane depend, to a great extent, on its pore size. The pore size can be conveniently classified into various ranges: macropores (>50 nm), mesopores (2 to 50 nm), micropores (<2 nm) [Sing et al., 1985]. The micropores can be further grouped into supermicropores (1.4 to 2 nm) and ultramicropores (<0.5 nm) [Huttepain and Oberlin, 1990]. There are a variety of methods to characterize the pore size distribution (or its mean or median) of a membrane in those ranges. Among the more established methods, mercury porosimetry and nitrogen adsorption are particularly suitable to inorganic membranes which, owing to their mechanical and thermal stabilities, do not suffer from the problem of pore compaction or collapse typically associated with organic membranes due to high pressures used in the tests.

The pore size distribution or its mean value of a porous inorganic membrane can be assessed by a number of physical methods. These include microscopic techniques, bubble pressure and gas transport methods, mercury porosimetry, liquid-vapor equilibrium methods (such as nitrogen adsorption/desorption), gas-liquid equilibrium methods (such as permporometry), liquid-solid equilibrium methods (thermoporometry) and molecular probe methods. These methods will be briefly surveyed as follows.

Microscopic methods. While microscopic methods provide direct visual information on membrane morphology as discussed earlier, determination of pore size, especially meaningful pore size distribution, by this type of methods is tedious and difficult. Advances have been made on the electron microscopy techniques to visualize membrane surface pores. For example, Merin and Cheryan [1980] have developed a replica-TEM technique to observe membrane surface pores. Nevertheless, microscopic methods have remained primarily as a surface morphology characterization tool and not as a pore size determination scheme.

Bubble point test. For separation applications, it is critical that the membranes are free of defects such as hairline cracks or pinholes. A simple technique widely used in the membrane industry to detect any defect or very large pores in organic or inorganic membranes is the bubble point test.

The method is primarily used for determining the maximum pore size of a membrane. In the case of a defect, the pinholes or cracks are in essence the largest pores. It is based on the phenomenon that a pressure difference ΔP is required to apply to a pore with a diameter d to displace a liquid from the pore with a gas according to the Washburn equation

$$d = 4S \cos\theta / \Delta P \qquad (4\text{-}1)$$

where S is the surface tension of the liquid and θ its contact angle with the pore surface. The basic procedures call for saturating the membrane pores with the test liquid. The applied pressure is gradually increased until bubbles burst from the membrane when the test gas pushes the liquid out of the largest pores. The corresponding pressure difference is called the bubble point pressure from which the largest pore diameter can be calculated using Eq. (4-1).

Water is frequently the choice for the liquid medium used in the test. However, the surface tension of water at 25°C is 72 dyne/cm compared to 22 and 16 dynes/cm for ethanol and some fluorocarbons, respectively. Eq. (4-1) indicates that, for the same pore diameter to be determined, the required applied pressure needs to be only 1/3 or 1/4 approximately when an alcohol or a fluorocarbon liquid is used instead of water. Air or nitrogen is generally used as the gas medium.

Having been adopted by ASTM (procedure F316), this test utilizes an experimental setup similar to that shown in Figure 4.8 or a commercial bench top unit by Coulter Electronics [Venkataraman et al., 1988]. In principle, it can also be used to arrive at a rough estimate for the "average" pore size of the membrane provided there is no defect. The steps involved are the following: (1) First flow a gas through the membrane dry. The flow rate is generally linear with respect to the applied pressure difference; (2) Force the same test gas through the membrane pores again after saturating the membrane with the test liquid until the bubble point and determine the maximum pore size; (3) Continue to increase the pressure beyond the bubble point, as schematically shown in Figure 4.9; (4) The "wet" curve in Figure 4.9 will approach the "dry" curve at a high pressure, usually several bars. The "half-dry" curve is established by reducing the gas flow rate of the "dry" curve by half for a given pressure difference. The pressure at which the "half-dry" curve intersects with the "wet" curve can be used to calculate the "average" pore diameter according to Eq. (4-1).

Mercury porosimetry. Mercury porosimeters nowadays have become a very well established method for describing pore diameter distributions of porous materials down to about 3.0 nm. It is based on the observation that, for a non-wetting liquid such as mercury to enter a capillary, an external applied pressure is required to overcome the interfacial tension. The smaller the capillary diameter, the higher the required pressure. Thus, when applied to a membrane using mercury as the liquid, the equilibrium of the above two forces yields the following equation relating the pore diameter, d, to the corresponding applied pressure, ΔP, required for the liquid to enter pores of that size:

$$d = -2\sigma \cos\theta / \Delta P \qquad (4\text{-}2)$$

GA 35279.2

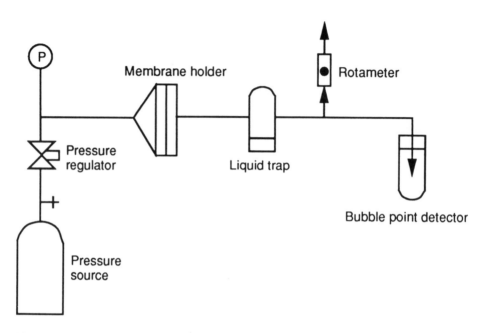

Figure 4.8 Schematic of a bubble point testing system

where σ is the surface tension of the liquid and θ the contact angle between the liquid and the pore wall. When mercury is used as the liquid, the surface tension is usually taken as 0.48 N/m and the contact angle with many inorganic materials varies only between 135 and 142°. Using an average wetting angle of 140°, Eq. (4-2) becomes

$$d = 15,000 / \Delta P \qquad (4-3)$$

where d and ΔP are in the units of nm and atm, respectively. Due to the approximate nature of the numerator in Eq. (4-3), other values ranging between 12,000 and 15,000 have been used.

Figure 4.10 shows the pore size distribution data of an alumina membrane by mercury porosimetry. This particular sample has a three-layered structure. The support has a relatively narrow pore size distribution but the membrane layer and the intermediate support layer do not show a clear distinction on the mercury porosimetry data. Typical mercury porosimetry analysis involves intrusion and extrusion of mercury. The intrusion data are normally used because the intrusion step precedes the extrusion step and complete extrusion of mercury out of the pores during the de-pressurization step may

take a substantially long time. It should be mentioned that the intrusion and extrusion
data usually show hysteresis.

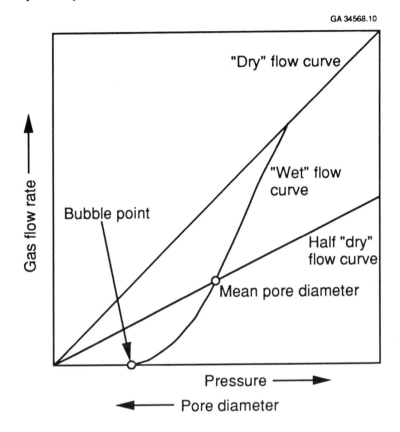

Figure 4.9 Determination of bubble point and mean pore diameter

Nitrogen adsorption/desorption. One of the most common techniques used for
analyzing size distributions of mesopores is the nitrogen adsorption/desorption method.
The method is capable of describing pore diameters in the 1.5 nm to 100 nm (or 0.1
micron) range. Pore size distribution can be determined from either the adsorption or the
desorption isotherm based on the Kelvin equation:

$$\ln (p/p_o) = (-2\sigma V \cos \alpha)/rRT \qquad (4\text{-}4)$$

where p_o and p are the vapor pressures above a planar surface and a curved surface the
radius of curvature of which is r, σ the surface tension of the test liquid, V the molar
volume of the liquid, α the contact angle, R the gas constant and T the absolute
temperature. Kelvin equation does not hold well for micropores with pore diameters less

than 1.5-2.0 nm. The distributions from the desorption branch are much narrower than those from the adsorption branch. An example is give in Figure 4.11 where the pore size distribution of a γ–Al_2O_3 membrane is calculated from both branches of the sorption isotherm [Anderson et al., 1988]. It is seen from the figure that the membrane pores are largely around 2 nm in radius (or 4 nm in diameter).

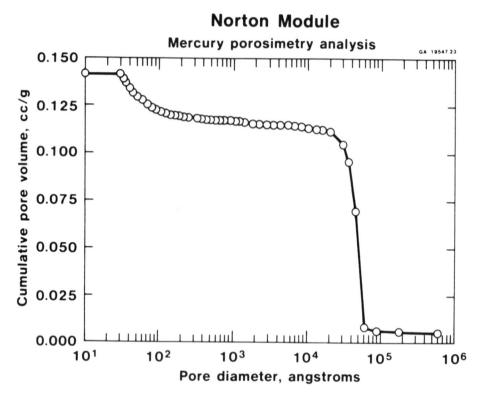

Figure 4.10 Pore size distribution of an alumina membrane by mercury porosimetry

There are differences in the opinion for the choice of adsorption versus desorption branches of the isotherms. On one hand, the desorption branch is more widely used although it overemphasizes the neck or narrow regions of the pore structure. The desorption branch data seem to correspond better with the mercury intrusion data. On the other hand, it is argued that the distribution based on the adsorption better represents the distribution of the entire sample. So far as permeability is concerned, the adsorption isotherm contains information on the large pores and, since solutes or retentates can pass through the membrane only via the largest pores available, it may be relevant to separation properties.

Multi-layered pore size distributions. The multi-layered structure of the membrane/support composites seen by the scanning electron microscopes as shown earlier is reflected in multiple "plateaus" of the cumulative pore size distributions (Figure 4.12). The sharp 'drops' in the distribution indicate that the pore size in each layer is quite uniform. The drop near 10 μm indicates the bulk support while the drop near 4 nm represents the selective membrane layer. The two drops in between are related to the two intermediate support layers.

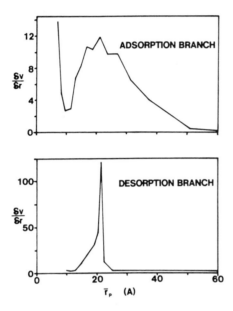

Figure 4.11 Differential pore size distribution of a γ-Al₂O₃ membrane by the nitrogen adsorption/desorption method [Anderson et al., 1988]

Many commercial ceramic membranes have two, three or even four layers in structure and their pore size distributions are similar to that shown in Figure 4.12. It should be noted, however, that determination of those multi-layered, broad pore size distributions is not straightforward. The major reason for this is the overwhelmingly small pore volume of the thin, fine pore membrane layer compared to those of the support layer(s) of the structure. It is possible, although very tedious, to remove most of the bulk support layer to increase the relative percentage of the pore volumes of the membrane and other thin support layers. Provided the amount of bulk support layer removed is known and the mercury porosimetry data of the "shaved" membrane/support sample is determined, it is feasible to construct a composite pore size distribution such as the one shown in Figure 4.12.

Permporometry. Permporometry is based on the well known phenomenon of capillary condensation of liquids in mesopores. It is a collection of techniques that characterize the interconnecting "active" pores of a mesoporous membrane as-is and nondestructively using a gas and a liquid or using two liquids. The former is called gas-liquid permporometry [Eyraud, 1986; Cuperus et al., 1992b] and the latter liquid-liquid permporometry [Capanneli et al., 1983; Munari et al., 1989]. The "active" pores are those pores that actually determine the membrane performance.

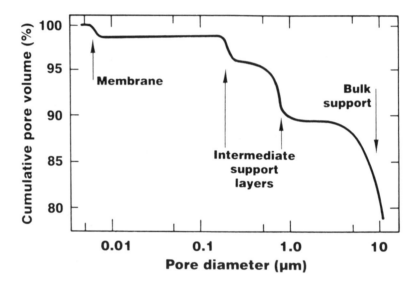

Figure 4.12 Cumulative pore size distribution of a four-layered ceramic membrane [Hsieh, 1991]

The gas-liquid permporometry combines the controlled stepwise blocking of membrane pores by capillary condensation of a vapor, present as a component of a gas mixture, with the simultaneous measurement of the free diffusive transport of the gas through the open pores of the membrane. The condensable gas can be any vapor provided it has a reasonable vapor pressure and does not react with the membrane. Methanol, ethanol, cyclohexane and carbon tetrachloride have been used as the condensable gas for inorganic membranes. The noncondensable gas can be any gas that is inert relative to the membrane. Helium and oxygen have been used. It has been established that the vapor pressure of a liquid depends on the radius of curvature of its surface. When a liquid is contained in a capillary tube, this dependence is described by the Kelvin equation, Eq. (4-4). This equation which governs the gas-liquid equilibrium of a capillary condensate applies here with the usual assumption of $\alpha=0$:

$$\ln (p/p_0) = (-2\sigma V)/r_k RT \tag{4-5}$$

The equation provides the corresponding pore radius for a given relative pressure, p/p_0. At the beginning of an adsorption process when the relative pressures is low, only a monolayer of the vapor molecules is adsorbed to the pore wall. This is followed by multilayer adsorption which is restricted to the so-called 't-layer' (adsorbed layer) with a maximum thickness on the order of a few molecules. As the relative pressure is increased, capillary condensation (i.e., the condensation of a vapor) occurs first in the smallest pores. Then progressively larger pores are filled according to Eq. (4-5). This continues until the saturation pressure is reached when the entire pore population is filled with the condensate.

Capillary condensation provides the possibility of blocking pores of a certain size with the liquid condensate simply by adjusting the vapor pressure. A permporometry test usually begins at a relative pressure of 1, thus all pores filled and no unhindered gas transport. As the pressure is reduced, pores with a size corresponding to the vapor pressure applied become emptied and available for gas transport. The gas flow through the open mesopores is dominated by Knudsen diffusion as will be discussed in Section 4.3.2 under Transport Mechanisms of Porous Membranes. The flow rate of the non-condensable gas is measured as a function of the relative pressure of the vapor. Thus it is possible to express the membrane permeability as a function of the pore radius and construct the size distribution of the active pores. Although the adsorption procedure can be used instead of the above desorption procedure, the equilibrium of the adsorption process is not as easy to attain and therefore is not preferred.

This method has been applied to ceramic membranes (e.g., gamma-alumina membranes) and compared to other methods such as nitrogen adsorption/desorption and thermoporometry (to be discussed next) in Figure 4.13. It can be seen that the mean pore diameter measured by the three methods agrees quite well. The pore size distribution by permporometry, however, appears to be narrower than those by the other two techniques. Similar conclusions have been drawn regarding the comparison between permporometry and nitrogen adsorption/desorption methods applied to porous alumina membranes [Cao et al., 1993]. The broader pore size distribution obtained from nitrogen adsorption/desorption is attributable to the notion that the method includes the contribution of passive pores as well as active pores. Permporometry only accounts for active pores.

Thermoporometry. Thermoporometry is the calorimetric study of the liquid-solid transformation of a capillary condensate that saturates a porous material such as a membrane. The basic principle involved is the freezing (or melting) point depression as a result of the strong curvature of the liquid-solid interface present in small pores. The thermodynamic basis of this phenomenon has been described by Brun et al. [1973] who introduced thermoporometry as a new pore structure analysis technique. It is capable of characterizing the pore size and shape. Unlike many other methods, this technique gives the actual size of the cavities instead of the size of the openings [Eyraud, 1984].

The essence of the technique is that the size of the transformed solid (from the condensate) which is confined by the pore walls is inversely proportional to the degree of undercooling (ΔT). Thus the finely dispersed transformed solid present in the membrane pores melts at temperatures below its ambient melting point when the liquid behaves as a bulk liquid outside the membrane pores. The change in the freezing (or melting) point of the capillary condensate, ΔT, allows the determination of the pore radius provided it is univocally related to the radius of curvature of the interface. The value of $1/\Delta T$ is linear with respect to the pore radius (r) with the following relationship:

$$r \ \text{(in nm)} = A_1/\Delta T + A_2 \tag{4-6}$$

GA 35278.4

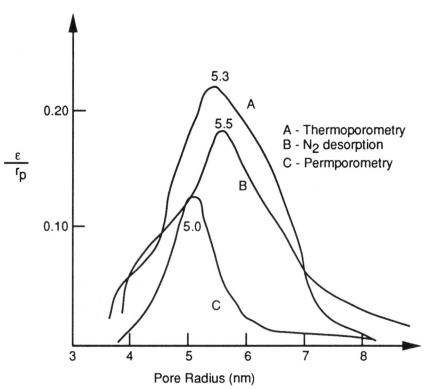

Figure 4.13 Comparison of pore size distribution data by permporometry, thermoporometry and mercury porosimetry [Eyraud, 1984]

where r is expressed in the unit of nm and the coefficients A_1 and A_2 depend on the liquid used. Implied in Eq. (4-6) is the assumption that the thickness of the condensate layer is 0.8 nm. Given in Table 4.1 are the values of these two coefficients for two liquids that have been used most often: water and benzene. The difference in the values for water from different sources may be due to the difference in the liquid-solid transition approach used, freezing versus melting, although in principle they should yield the same information. The melting transition approach is preferred over the freezing transition method [Cuperus et al., 1992a]. The above equation applies to cylindrical pores or spherical cavities. Depending on the nature and sensitivity of the calorimeter used, the amount of membrane sample required for the analysis is about 20 mg to 2 g. The weight of the condensate is determined. The sample holder is sealed and placed inside a differential scanning calorimeter. The temperature is programmed to decrease at a slow rate to a low temperature (e.g., close to -50°C) enough to effect freezing. The pore size distribution can then be deduced from the thermogram by an analysis procedure which has been automated [Eyraud, 1984].

TABLE 4.1
Values of coefficients for thermoporometry based on Eq. (4-6)

Liquid	A_1	A_2	Liquid-solid transition	Reference
Water	64.67	0.57	Freezing	Eyraud, 1984
Water	32.33	0.68	Melting	Cuperus et al., 1992a
Benzene	132	0.54	Freezing	Eyraud, 1984

This method has been found to provide pore size distributions of inorganic ultrafiltration membranes in good agreement with those using the mercury porosimetry and permporometry methods (see Figure 4.13 for their differential pore size distributions). It is applicable to a pore diameter range of 2 to about 150 nm. The lower limit is imposed by the assumptions made in the thermodynamic description of the process which are no longer valid below -40°C or so. The upper limit is set by the effect of the curvature on the freezing point depression. For pores larger than the upper limit, the shift from the ambient fusing point is so small that the effect can not be separated from the peak temperature associated with free or bulk liquid [Cuperus et al., 1992a]. For those asymmetric organic membranes with the top layer and the sublayer having pores of comparable sizes, it is sometimes difficult to determine which layer or both layers the measured pore size distribution corresponds to. For inorganic membranes, this is usually not the case.

Gas flow method. The above methods have a common characteristic. That is, they do not discriminate between the "active" pores and the "passive" pores. For separation applications, only the "active" pores are most relevant. A nondestructive technique for inferring the pore size distribution of a membrane is the gas flow method. Having been developed and tested at Oak Ridge Gaseous Diffusion Plant [Fain, 1989], it departs greatly from the above methods in that the pore size distribution is based on gas flow rate rather than volume. As such, it is believed to be more relevant to gas separation applications. The method is applicable to analyzing pore diameters in the range of 1.5-2.0 nm to about 0.1 μm where the Kelvin equation is valid.

This test method of flow-weighted pore size distribution measures the flow of a mixture of an inert gas such as nitrogen (or helium) and a condensable gas such as carbon tetrachloride (or cyclohexane) through the membrane pores of various sizes and is not sensitive to the amount of gas adsorbed. The basic procedures involved are the following: (1) The gas mixture is pressurized to the point that capillary condensation completely blocks the membrane pores; (2) The pressure is then incrementally decreased and the corresponding increase in the gas flow is measured in the large pores followed by small pores. The pressure is decreased until there is no longer increase in the gas flow; and (3) The change in gas flow rate relative to the pressure change can then be related to the pore size through the Kelvin equation for capillary condensation, Eq. (4-4).

The method performs most satisfactorily with a small pressure difference across the membrane and a low concentration of the condensable gas. The recommended pressure difference and the mole fraction are < 3 cm Hg and 0.05-0.1, respectively. The test can handle membrane samples of various shapes and sizes. It is generally very reproducible [Fain, 1989; Gallaher and Liu, 1994] and typically takes a few hours to complete.

For multi-layered asymmetric or composite membranes where the pore sizes between layers are widely different, the analysis gives the pore size distribution of the densest layer (membrane) even if they may be only a small volume or weight fraction of the total membrane/support structure. Shown in Figures 4.14 (a) and (b) are the pore size distributions of two very similar multi-layered alumina membranes prepared by the sol-gel process and two distinctively different alumina membranes made by the anodic oxidation process. The comparison of pore size distributions in each case reflects the similarities and differences of the membrane samples.

Gas permeability method. Like the gas flow method, this technique is also based on the measurement of the flow rate of a gas through a porous medium such as a membrane. The flow rate is monitored as a function of the pressure drop across the thickness of the membrane. But unlike the gas flow method, the gas used in this method is a pure, nonadsorbable and noncondensable gas.

The average radius of "active" pores can be estimated to be

$$r = 8.513 \, \mu \, (RT/M)^{0.5}(P_P P_K) \tag{4.7}$$

where μ is the fluid viscosity, T the absolute temperature, M the molecular weight of the gas, R the gas constant, and P_P and P_K the slope and the intercept of the flow rate vs. pressure drop curve, respectively.

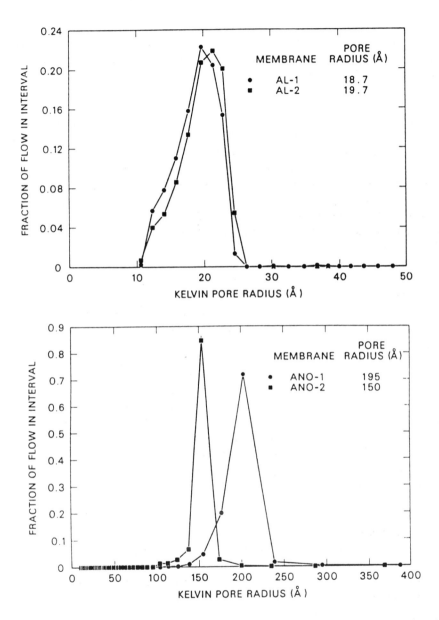

Figure 4.14 Pore size distributions by gas flow method of (a) two similar sol-gel alumina membranes and (b) two distinctively different anodic alumina membranes [Fain, 1989]

Although this technique has been frequently used to characterize microporous organic membranes, it has not been applied to microporous inorganic membranes. Moreover, this method provides only the average pore size but not the pore size distribution.

There is some indication that this method is influenced by the type of gas used. For example, nitrogen and carbon dioxide yield similar pore sizes which can differ substantially from that using helium [Altena et al., 1983].

Molecular probe method. When the membrane pores reach the molecular dimensions, the molecular sieving effect becomes operative and the membrane can discriminate gas molecules with a diameter difference as low as 0.02 nm. Some methods utilizing molecules of different sizes as molecular probes have been used to estimate the pore size of a membrane. In these cases, the membranes can be characterized by their permeabilities or accessible micropore volumes to different gases. The pore size can be inferred from the permeation rates or micropore volumes of different gases whose molecular dimensions are known or can be calculated.

The pore sizes of various carbon molecular sieve membranes prepared and modified under different activation conditions have been estimated by comparing the relative magnitudes of the measured permeability values of selected pure gases and the known molecular dimensions of those gases. The accumulated adsorption data of solid molecular sieves indicates that the molecules occluded during adsorption are preferably oriented so that their passage through the pores is determined by their smallest dimension. In other words, the smallest dimension (critical diameter) of non-spherical molecules is critical in controlling their penetration into the ultramicropores in molecular sieve materials.

Given in Table 4.2 are the shape-corrected critical diameters of some common gases based on this argument [Koresh and Soffer, 1980; Soffer et al., 1987]. Using these critical diameter values, one can infer the pore diameter of a given membrane exhibiting certain gas permeabilities. Shown in Table 4.3 are the permeability and calculated ideal selectivity values for a given pair of gases. Also included in the table are the calculated selectivity values assuming the predominant transport mechanism is Knudsen diffusion which will be discussed in Section 4.3.2 dealing with transport mechanisms in porous membranes. It can be seen that the first membrane (designated as DA1) should have a pore diameter close to about 0.36 nm because oxygen (0.328 nm) has a much higher permeability than nitrogen (0.359 nm) and sulfur hexafluoride (0.52 nm) permeates very little through. Thus, molecular sieving occurs with this membrane under this condition. The second membrane (DA2) is very different. Apparently the membrane pores of the second membrane have been enlarged substantially. The measured selectivity values for the O_2/SF_6 and O_2/N_2 gaseous mixtures are essentially equal to those based on the Knudsen flow assumption. Therefore it can be concluded that the pore diameter has been increased beyond the 1 nm range where molecular sieving is operative and into the Knudsen flow regime (i.e., likely larger than 5 nm).

In an attempt to estimate the resulting pore diameters of a series of porous silica-modified membranes, Lin et al. [1994] used nitrogen and neopentane gases as the molecular probes. The two gases have a kinetic diameter of 0.36 and 0.62 nm, respectively. From the nitrogen/neopentane selectivity data, one can distinguish those membranes with a pore diameter less than 0.36 nm, between 0.36 and 0.62 nm and greater than 0.62 nm.

TABLE 4.2
Critical diameters of various molecules

Molecule	Critical diameter or shaped corrected width (nm)
CO_2	3.10
O_2	3.28
C_2H_2	3.33
H_2	3.44
N_2	3.59
Ar	3.60
CH_4	4.00
SF_6	5.02

(Adapted from Koresh and Soffer [1980] and Soffer et al. [1987])

TABLE 4.3
Permeability and permselectivity of oxygen, nitrogen and sulfur hexafluoride through modified carbon molecular sieve membranes

Membrane treatment	Gas	Permeability x 10^8 (cm^3(STP)-cm/s-cm^2-cm Hg)	Measured selectivity	Knudsen selectivity
DA1	O_2	11.4	>38	2.1
DA1	SF_6	<0.3		
DA2	O_2	228	2.0	2.1
DA2	SF_6	115		
DA1	O_2	11.4	8.1	0.94
DA1	N_2	1.4		
DA2	O_2	228	0.92	0.94
DA2	N_2	248		

Note: DA1 = Dynamic activation-first step; DA2 = Dynamic activation-second step
(Adapted from Soffer et al. [1987])

A similar technique is based on the theory of micropore volume filling. It states that the total microporous volume accessible to a given adsorbate can be obtained from the Dubinin-Radushkevich equation as a function of the temperature, relative pressure, and characteristic energy of adsorption. When this procedure is applied to a few linear or spherical molecules (as probes) of different but known sizes, the adsorption isotherms of these gases at the same temperature can be employed in combination with their minimum molecule dimensions (such as the minimum kinetic diameters) to arrive at the pore size distributions. This method has been used to characterize microporous carbons [Garrido et al., 1986] and partially pyrolyzed polysilastyrene membranes [Grosgogeat et al., 1991]. Examples of the probe molecules are carbon dioxide, propane and sulfur hexafluoride. The minimum kinetic diameters of these molecules (as given in Table 4.4 along with those for other molecules) have been used as the index to monitor the evolving pore size distributions of the partially pyrolyzed polysilastyrene membranes, Figure 4.15 [Grosgogeat et al., 1991]. The accessible micropore volume vs. minimum kinetic diameter data for each probe molecule shown is an indication of the cumulative pore size distribution. For example, the membrane pyrolyzed at 644°C appears to have a pore diameter distribution smaller than 0.55 nm but including 0.33 and 0.43 nm. The membrane heat treated at 735°C does not exhibit pores larger than 0.43 nm. On the contrary, the molecular sieving effect is observed in the 557°C heat-treated membrane sample which should contain at least those pores in the diameter range of 0.33 to 0.55 nm.

TABLE 4.4
Mean free path and kinetic diameter data of various gases

Molecule	Molecular weight	Mean free path (nm)	Kinetic sieving diameter (nm)
He	4	180	0.26
H_2	2	112	0.289
O_2	32	64.7	0.346
Ar	40	63.5	0.34
N_2	28	60.0	0.364
CO	28	58.4	0.376
NH_3	17	44.1	0.26
CO_2	44	39.7	0.33
C_2H_4	28	34.5	0.39

(Adapted from Itoh [1990])

Other methods. Other techniques of measuring membrane pore size distributions, especially those non-destructive types, have been pursued. One such method, NMR spin-lattice relaxation measurement [Glaves et al., 1989], can analyze pore diameters from less than 1 nm to greater than 10 μm. The basic requirement of the method is to carefully membrane composite such that only the fine pores of the membrane (separating) layer control the moisture contents in the various layers of a multi-layered are saturated with water, while those larger pores in the support layers are slightly adsorbed with water. The underlying principle of this technique is the observation that the spin-lattice relaxation decay time of a fluid such as water in a pore is shorter than that for the same fluid in the bulk. The pore volume distributions can be calculated based on the measured relaxation data. The method is limited by the homogeneity of the magnetic field and, at current development stage, means the membrane sample can not exceed 10 cm long or so for practical reasons.

Stoeckli and Kraehenbuehl [1984] have proposed a relationship between the heat of immersion and the microporous volume of a porous solid, applicable to materials having a wide range of external surface areas. This allows a rapid determination of the pore size distribution below 0.8 to 1 nm. The technique, however, requires a non-porous standard material of surface composition similar to the membrane material.

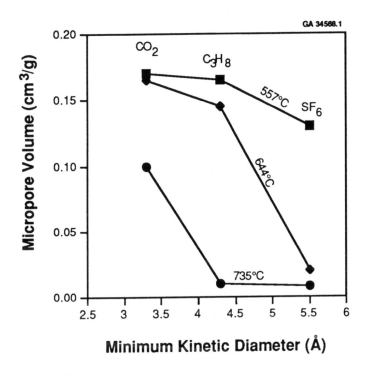

Figure 4.15 Use of three molecules to determine the evolving pore size distributions of partially pyrolyzed polysilastyrene membranes [Grosgogeat et al., 1991]

4.1.3. Porosity and Tortuosity

The openness (e.g., volume fraction) and the nature of the pores affect the permeability and permselectivity of porous inorganic membranes. Porosity data can be derived from mercury porosimetry information. Membranes with higher porosities possess more open porous structure, thus generally leading to higher permeation rates for the same pore size. Porous inorganic membranes, particularly ceramic membranes, have a porosity ranging from 20 to about 60% in the separating (membrane) layer. The support layers have even higher porosities.

As discussed in the previous chapter, the network of the inter-connected membrane pores formed during preparation and fabrication may be tortuous or nearly straight depending on the synthesis and subsequent heat treatment methods and conditions. The microstructure of a membrane, particularly the type with tortuous pores, is too complicated to be described by a single parameter or a simple model. Due to the relatively poor knowledge of flow through porous media, an empirical term called "tortuosity" has been introduced and used by many researchers to reflect the relative random orientation of a pore network and is based on the Kozeny-Carmen equation for the membrane flux, J:

$$J = \varepsilon^3 \Delta P \, / \, (1-\varepsilon)^2 \beta \mu \, L_m A_v{}^2 \tag{4-8}$$

where ΔP is the pressure difference across the membrane pore length (L_m), A_v the surface area per unit volume of the pores, ε the void fraction, μ the permeate viscosity and β the Kozeny-Carmen constant. β is the product of two factors k_s and k_t which depend on the shape and tortuosity of the pores, respectively [Leenars and Burggraaf, 1985b]:

$$\beta = k_s k_t \tag{4-9}$$

Basically it represents the relative average length of the flow path of a fluid element from one side of the membrane to the other. The tortuosity factor, k_t, calculated from flux measurements and Eqs. (4-8) and (4-9) has been used frequently to characterize laminar flow through a porous medium such as a membrane. There is evidence that the tortuosity thus determined agrees reasonably well with the value obtained by ionic conductivity [Shimizu et al. 1988]. If the pore network of the membrane consists of regular packing of more or less isometrically shaped particles, β is approximately equal to 5 [Carman, 1937]. This value is commonly used to characterize many porous media. Leenaars and Burggraaf [1985b], on the other hand, have determined that β has a value of 9 to 23 for boehmite gel layers and 13.3±2 for γ-alumina membrane calcined at 800°C. The shape-related factor, k_s, for the γ-alumina membrane pores is accepted to be 2.6 and, therefore, the tortuosity factor, k_t, becomes 5.3. This higher than normal values of β and k_t, can be explained mostly by the relatively large and thin plate-shaped alumina crystallites forming the membrane structure. Thus, the resulting pore network has a high value of the tortuosity factor.

However, the complexity of the manner in which tortuosity is determined and the general belief that it is not an intrinsic parameter directly related to material or transport properties have limited its use only in the scientific literature discussions but not by the membrane industry.

4.2 TRANSPORT MECHANISMS

The mechanisms by which various components in a liquid or gaseous feed stream to the membrane system are transported through the membrane structure determine the separation properties of the membrane. These transport mechanisms are quite different in liquid and in gas or vapor phases. So are their effects on permeate flux (or permeability) and retention (or rejection) coefficient or separation factor in the case of gas separation.

4.2.1 Dense Membranes

A widely accepted transport mechanism for certain gases through dense metal, solid electrolyte or ceramic membranes, similar to that for organic counterparts, is the solution-diffusion type.

Metal membranes. Since Thomas Graham [1866] discovered that metallic palladium absorbs a large amount of hydrogen, the metal hydride system of which palladium is an important example has been extensively studied. Hydrogen dissolved in a metal hydride system (according to the mechanism to be described below) is considered to behave in an atomic or ionic form which is more reactive than molecular hydrogen in a gas phase.

Barrer [1951] proposed that the permeation of hydrogen through metals (such as palladium) entails three processes: (1) dissociative chemisorption of hydrogen on the membrane surface followed by dissolution of the atomic hydrogen in the structural lattice of the metal.; (2) diffusion of the dissolved hydrogen in the membrane; and (3) desorption of combined hydrogen atoms as molecules. This can be illustrated in Figure 4.16 where it is shown that only hydrogen undergoes the above solution-diffusion type transport while other molecules are rejected right at the surface of the membrane. The driving force for the diffusion across the dense palladium membrane is the concentration difference of the dissolved hydrogen in the atomic form. The rate-limiting step is often the bulk diffusion as in the case of palladium membranes. The transport of oxygen through dense metal (e.g., silver) membranes also follows the same process and generally can be described by a solution-diffusion type model in terms of Fick's first law (Lewis, 1967):

$$J = \mathbf{P}\,[P_1{}^n - P_2{}^n]/\,L \tag{4-10}$$

where J is the permeate flux, $\mathbf{P}$ the permeability, L the membrane thickness and P_i the gas pressure on the membrane surface, i (i = 1 or 2 representing the upstream or downstream side). The constant n in Eq. (4-10) is often accepted to be 0.5 assuming that the diffusion through the bulk of the palladium membrane is rate-limiting compared to hydrogen dissociation and that the concentration of dissolved hydrogen is proportional to the square root of hydrogen pressure. The square root expression in the above equation (Sieverts law) reflects the atomic nature of the gas species in the membrane. Considerable experimental evidence, however, suggests a value higher than 0.5 for the exponent n in Eq. (4-10). Hydrogen permeation flux through a palladium membrane may be better correlated to an exponent of 0.68-0.8 of hydrogen pressures across the membrane [Uemiya et al., 1991].

The temperature-dependent permeability, as in the case of dense organic polymeric membranes, can be expressed in the form of a solubility term multiplied by a diffusivity term as follows:

$$\mathbf{P} = (K / K_s) * (D_\alpha) \tag{4-11}$$

where D_α is the Fick's diffusion coefficient and K and K_s are the adsorption and the Sieverts' constants, respectively, for the alpha-phase Pd. Eq. (4-10) is the result of several simplifying assumptions. More complicated models of gas transport through dense Pd-based membranes can be found elsewhere [Shu et al., 1991]. It is noted that Eq. (4-10) applies to the more stable alpha-phase of Pd. For a beta-phase Pd membrane, the hydrogen flux is described by

$$J = (D_\beta K/B) \ln (P_1/P_2) / L \tag{4-12}$$

where D_β is the Fick's diffusion coefficient and B which is temperature-dependent is a sorption constant for the beta-phase Pd. For a palladium membrane consisting of both phases, a mixing-rule type equation applies [Nagamoto and Inoue, 1983].

Solid electrolyte/ceramic membranes. Transport through solid electrolyte membranes differs from the dense metal membranes primarily in the form of the diffusing species in the membrane matrix. Contrary to dense Pd- or Ag-based membranes, solid electrolyte membranes have charged, rather than atomic species, present in the membrane matrix.

There are certain dense ceramic membranes such as silica that utilize pressure gradient for transporting selective gases. For example, very thin silica membranes have been deposited by chemical vapor techniques on porous Vycor glass or carbon/Vycor glass support to make composite membranes which show high selectivities of hydrogen [Gavalas et al., 1989; Megiris and Glezer, 1992].

An emerging family of dense ceramic membranes that are permselective to oxygen are made from dense fast ionic conducting oxides. A major advantage of these membranes is that they do not require electrodes and external electric connectors. Oxygen permeates these membranes by ionic conduction involving lattice vacancies and interstitial atoms at high temperatures under the influence of a pressure or electric field. These materials exhibit both the ionic and electronic conductivities and are called mixed conductors. Transport of oxygen species through these membranes involves surface electrochemical reactions on the membrane-gas interface and solid state diffusion of charged species in the bulk oxide.

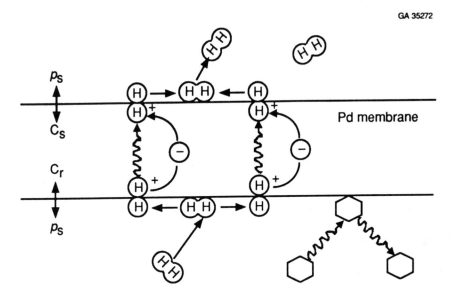

Figure 4.16 Transport mechanism of hydrogen through palladium membrane [Itoh, 1990]

The permeation flux of oxygen through a mixed oxide membrane described above depends on the oxygen partial pressures across the membrane, membrane thickness and temperature. The dependence, however, is embedded in a complicated implicit equation [Lin et al., 1994]. Only in special cases the permeation flux shows a pressure dependence similar to that for palladium membranes as given in Eq. (4-10). For example, when electronic conductivity predominates, the value of the exponent, n, is equal to 0.5 for thin membranes and 0.25 [Dou et al., 1985; Itoh et al., 1993] for thick oxide films. If the oxide membrane is essentially an ionic conductor and the surface reaction is the rate-limiting step, n takes on a value of 0.5.

4.2.2 Porous Membranes

Transport and therefore separation mechanisms in porous inorganic membranes are distinctively different from and more varying than those prevailing in dense membranes. Because of the variety of the mechanisms that can be operative, these membranes in principle are capable of separating more varieties of compound mixtures. Compared to dense membranes, these porous membranes generally exhibit permeabilities of one or two orders of magnitude higher. For example, Pd-based membranes typically have a hydrogen permeability of 10^4-10^5 barrers while many porous ceramic membranes show a permeability of at least 10^5-10^6 barrers. One barrer is defined to be 10^{-10} (cm^3 (STP)-cm/cm^2-s-cmHg). The majority of porous membranes, however, suffer from lower permselectivities.

Possible transport mechanisms in a fluid system through the membrane pores are multiple. They vary to a great extent with the membrane pore size and, to a less extent, with chemical interaction between the transported species and the membrane material. Under the driving force of a pressure gradient, permeants (whether in the form of solvents, solutes or gases) can transport across a membrane by one or more of the mechanisms to be discussed below. The degree by which they affect permeability and permselectivity depends on the operating conditions, membrane characteristics and membrane-permeating species interactions in the application environment.

Laminar and turbulent flows, which occur in large pores where the mean pore diameter (say, larger than 1 μm) is larger than the mean free paths of the fluid molecules involved, do not discriminate components in fluid mixtures. The same applies to molecular diffusion which occurs at relatively high pressures in large pores. Therefore, no separation is effected when only laminar or turbulent flow or bulk diffusion occurs in the pores. The permselectivity of a porous membrane will then have to rely on other transport mechanisms.

Liquid separation. Separation can take place between solvents and solutes, macromolecules or particles or between species in liquid media by the effect of size exclusion. That is, those molecules or colloids larger than the size of the membrane pores will be retained or rejected while those smaller ones can pass through the membrane. The size exclusion mechanism predominates in pressure driven membrane processes such as microfiltration, ultrafiltration and even nanofiltration which has a molecular selectivity on the order of one nanometer.

Microfiltration and ultrafiltration processes are primarily based on convective (Poiseuille) flows. By combining non-equilibrium thermodynamics and the pore theory, Sarrade et al. [1994] carried out a theoretical analysis of solvent and solute transport through a gamma-alumina membrane (on a titania support) with a mean pore diameter of approximately 1.2 nm in the nanofiltration range. The pore theory assumes the solute flux to be contributed by two components: convection which is associated with molecular entrainment by the solvent and diffusion which is due to a concentration gradient. With the theoretical findings agreed by experimental data using the uncharged solute molecules of polyethylene glycols, the authors concluded that, regardless of the size of the solute molecule, convective transport is always more important than diffusive transport.

The above mechanisms assume that the solutes, macromolecules or particles are not charged. In reality, interactions between feed components and the membrane/pore surfaces invariably exist. Preferential adsorption of a particular component on the pore or membrane surface, for example, can lead to effective separation. Thus, the membrane/pore surfaces can play a significant role in separating a given liquid system depending on their surface charge effects. This will be treated in more detail under the section 4.6 Surface Properties.

Gas separation. As mentioned earlier, not all transport modes occurring in the membrane pores result in separation of species. The schematic diagram in Figure 4.17 shows six possible transport mechanisms inside the pores of an inorganic membrane. The viscous flow does not separate species. Other five mechanisms can be effective in gas separation. As in the case of liquid separation, size exclusion (or molecular sieving) is an important factor but is usually operative only inside those pores of the molecular dimensions. As discussed in Chapter 3 and in Section 4.1.2 some zeolitic and carbon molecular sieve membranes have been prepared and characterized in the laboratory that effectively have very small and narrow pore size distributions. These membranes are capable of separating gases that even differ only by 0.02 nm in molecular dimension. By tailoring the membrane pore size through synthesis and any subsequent modification conditions, it is possible to make molecular sieve membranes that show very high permselectivities for a gas mixture as illustrated in Table 4.3.

In addition, Knudsen diffusion, surface diffusion, multi-layer diffusion and capillary condensation or some of their combinations may be responsible for separating gaseous mixtures. Knudsen diffusion (or called Knudsen flow), unlike bulk diffusion which is a result of predominantly intermolecular collisions, is based on collisions between molecules and the pore surface. As the pore diameter decreases or the mean free paths of the species molecules increase (as it would happen with a low density gas, reduced pressure or increased temperature), the permeating species collide with the pore wall more frequently than with other gas molecules. The contribution of Knudsen diffusion is evident generally when the pore diameter measures about 5-10 nm under pressure or 5-50 nm in the absence of a pressure gradient. As a general rule of thumb, when the pore size is approximately smaller than 1/10 that of the mean free path of the diffusing species, Knudsen diffusion may become dominant. The separation factor, which is an indicator for the separation efficiency and will be defined in Chapter 7, has been shown to depend on the square root of the molecular weight ratio of the gases separated:

$$F_A / F_B = (M_B / M_A)^{0.5} \qquad (4\text{-}13)$$

where F and M represent the permeation fluxes and molecular weights of the gas species A and B. In many cases the ratio involved is not substantially different from 1.0 which is the case of no separation. Laminar flow very often occurs in the presence of Knudsen diffusion. The contribution of Knudsen diffusion relative to that of lamina flow is determined by the Knudsen number, Kn, which is defined as the ratio of the mean free path (λ) of the gas to the radius of the membrane pore (r) as follows:

GA 34561.5

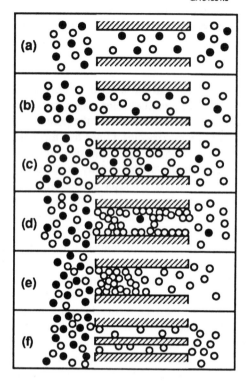

Figure 4.17 Transport mechanisms for gaseous mixtures through porous membranes: (a) viscous flow; (b) Knudsen diffusion; (c) surface diffusion; (d) multi-layer diffusion; (e) capillary condensation; and (f) molecular sieving [Saracco and Specchia, 1994]

$$Kn = \lambda / r \qquad\qquad\qquad (4\text{-}14)$$

The value of λ depends on the pressure and the gas according to the following equation (assuming the ideal gas law applies):

$$\lambda = (1/\sqrt{2}\sigma)\,(kT / p) \qquad\qquad\qquad (4\text{-}15)$$

where σ is the collision cross-section of the gas molecule which can be taken as πd^2 (with d being the molecule diameter) and $k = 1.38 \times 10^{-23}$ J/K. It is noted that the mean free path is proportional to the absolute temperature and inversely proportional to the pressure. For selected commonly handled gases, their mean free paths under the "standard condition" of 273°K and 1 atm are given in Table 4.4 [Itoh, 1990]. When the pores are small in diameter and/or the gas density is low (and therefore λ is large), Kn >>1. It means that, when λ is large and/or r is small, the molecules collide with the pore wall much more frequently than with each other, thus the Knudsen diffusion predominates over other transport mechanisms including the laminar flow. On the other hand, transition regimes of the two types of flow occurs when $10 \geq Kn \geq 0.01$ [Remick and Geankoplis, 1973].

Thus Eq. (4-13) implies that Knudsen diffusion is practical only when those gases with large differences in their molecular weights are to be separated. For applications where this mechanism poses as a severe limitation, other more effective separation mechanisms would be necessary. Two such possibilities are surface diffusion (and multi-layer diffusion) and capillary condensation, both of which are dependent on the chemical nature of both the membrane material and pore size and the species to be separated.

Surface diffusion becomes important when the molecules of a gas component are adsorbed in a significant amount on the pore surface. As the adsorbed molecules accumulate more and more on the pore wall, first to form a monolayer, the preferentially adsorbed component causes a difference in the surface concentration gradient and diffuses faster than the nonadsorbed ones under a pressure gradient. The surface adsorption/diffusion effects a difference in permeability and consequently leads to separation of components. This phenomenon begins to dominate at a pore diameter of 1-10 nm or when the specific surface area is very large. More adsorption, however, does not necessarily mean more transport of the adsorbed phase since the heat of adsorption can be a barrier [Uhlhorn et al., 1989]. The adsorbed phase can extend beyond a monolayer and start to form multiple layers. Surface diffusion with multilayer adsorption has been described by a random hopping model [Okazaki et al., 1981]. Some examples of surface diffusion-driven separation is carbon dioxide through the pores of gamma-alumina membranes.

At a relatively low temperature (e.g., near 0°C), some gases undergo capillary condensation and become liquid occupying the pores. The associated transport is then governed by pseudo-liquid phase diffusion. When other gases do not dissolve in the condensed component(s), separation of gases occurs. Two examples are SO_2/H_2 [Barrer, 1965] and H_2S/H_2 [Kameyama, et al. 1979] systems in which SO_2 and H_2S, respectively, condense in the pores and diffuse across the membrane while H_2 in both cases is blocked from the pores as it does not dissolve in either liquid SO_2 or H_2S.

4.3 SEPARATION PROPERTIES

Permeability and permselectivity of a membrane depend on its pore size distribution. But equally important, they are applications specific and determined by the interactions between the process stream and the pore or membrane surface. However, for general characterization purposes, some model permeants (solvents and molecules) are often used to obtain a "generic" permeability and permselectivity for that membrane. Water is often used as the model permeant for liquid-phase characterization while air, nitrogen, hydrogen and argon are employed for gas-phase characterization.

4.3.1 Permeability

For a given combination of a membrane and a model permeant such as water, the membrane permeability depends to a great extent on the mean pore diameter of the membrane. This is illustrated in Figure 4.18 for a family of experimental alumina membranes. The water permeability of these alumina membranes monotonically increases with increasing mean pore diameter. In addition, permeability varies with temperature due to the viscosity effect. For liquids, the viscosity generally decreases, and therefore the permeability increases, with increasing temperature (Figure 4.19). For gases, however, the opposite can be true. Shown in Figure 4.20 is permeability of carbon dioxide through an alumina membrane with a median pore diameter of 4 nm at two different temperatures: 20 and 450°C. The permeability at the higher temperature is lower due to the higher viscosity of carbon dioxide at the higher temperature [Hsieh et al., 1991].

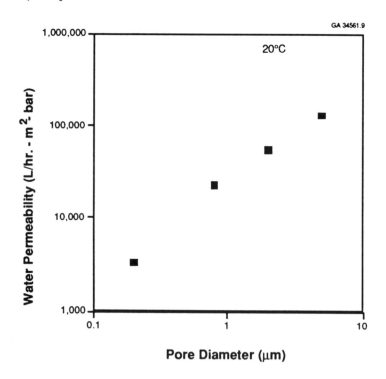

Figure 4.18 Water permeability of a family of alumina membranes having different mean pore diameters [Hsieh et al., 1991]

In real applications, build-up of solutes, macromolecules or particles near the membrane surface of a porous inorganic membrane can exert significant influence over the permeate flux. Generally speaking, the build-up layer becomes lessened with increasing crossflow velocity. This beneficial effect is shown in Figure 4.21 for a monolithic 19-channel alumina membrane processing oil-water mixtures [Bhave and Fleming, 1988].

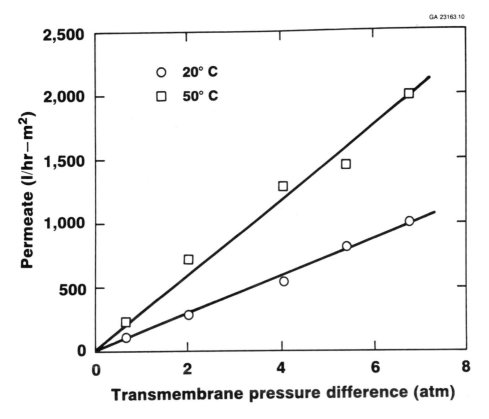

Figure 4.19 Effect of temperature on water flux through an alumina membrane

4.3.2 Permselectivity

The generic permselectivity of a membrane can be described by the retention coefficient for liquid phase or the separation factor for gas phase. Separation factor will be defined and discussed in Chapter 7. In the case of liquid-phase membrane separation, the retention coefficient of the membrane can be characterized by some commonly used model molecules such as polyethylene glycol (PEG) polymers which have linear chains and are more flexible or dextrans which are slightly branched. The choice of these model molecules is due to their relatively low cost. They are quite deviated from the generally implied idealization of perfect spheres and this is reflected in the typically broad molecular weight cutoff (MWCO) curves as shown in Figure 4.22. The MWCO curve provides an indication of how "selective" the pore size distributions is with respect to dextrans which are used as the model molecules. It is imperative to specify what model molecule is used when reporting a retention or rejection coefficient. Other materials such as latex particles are alternative model molecules but they are seldom chosen because of their prohibitive costs.

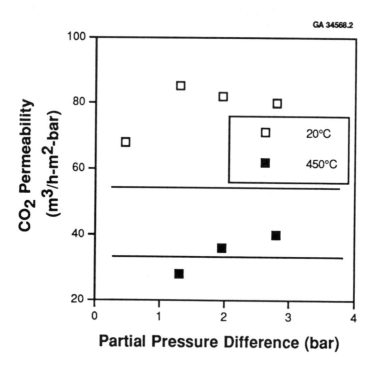

Figure 4.20 Permeability of carbon dioxide through an alumina membrane at two different temperatures [Hsieh et al., 1991]

4.4 THERMAL AND HYDROTHERMAL STABILITIES

Although inorganic membranes are known to be more stable thermally than organic counterparts, sintering or phase transformation can still alter the microporous structures of inorganic membranes at elevated temperatures during applications or cleaning between applications. One of the most critical requirements for these membranes to be useful in many practical applications, particularly when high temperatures and/or severe conditions are involved, is their thermal stability, in dry or wet environments.

Thermal stability of inorganic membranes not only depends on the membrane/support materials used but also on the environment and conditions. To illustrate this point, Table 4.5 provides the upper temperature limits for a number of inorganic membrane materials based on the recommendations of manufacturers. It is obvious that the temperature limits depend on the nature of the environment with the temperature limits being higher under the reducing atmosphere than the oxidizing atmosphere. The temperatures shown only reflect the thermal stability of the membrane materials only. In many applications, however, the maximum operating temperatures are actually dictated by the temperature limits of the sealing, gaskets or other accessory materials that are required to provide leakage-free operation.

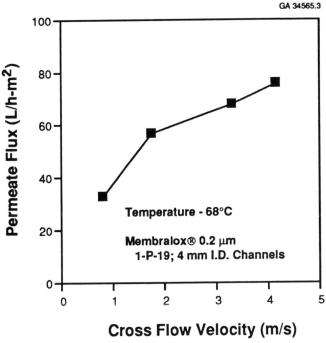

Figure 4.21 Effect of crossflow velocity on water flux of an alumina membrane for oil-water separation [Bhave and Fleming, 1988]

Thermal and hydrothermal exposures can change the pore size and its distribution, porosity and tortuosity of a porous membrane which in turn influence the separation properties of the membrane such as permeability and permselectivity. Several ceramic membranes have been investigated for their responses to thermal and hydrothermal environments.

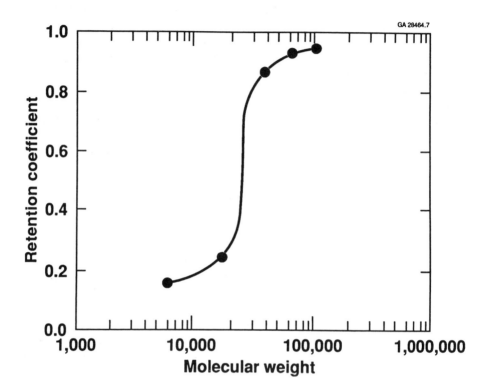

Figure 4.22 Molecular weight cutoff curve of a 5 nm pore diameter zirconia membrane using dextrans of varying molecular weights as test molecules [Hsieh et al., 1991]

Alumina membranes. It has been established that several phases of alumina exist and a particular phase of alumina is determined not only by the temperature it has experienced but also by the chemical path it has taken. For commercial membrane applications, the alpha- and gamma-phases of alumina are the most common. Alpha-alumina membranes are well known for their thermal and hydrothermal stabilities beyond 1,000°C. In fact, other transitional forms of alumina will undergo transformation towards the thermodynamically stable alpha-alumina at elevated temperatures beyond 900°C. On the contrary, commercial gamma-alumina membranes are typically calcined at 400-600°C during production and are, therefore, subject to potential structural changes beyond 600°C. Moreover, alumina chemistry reveals that phase transition also occurs beyond that temperature [Wefers and Misra, 1987].

TABLE 4.5
Upper temperature limits (°C) of membrane materials recommended by manufacturers

Material	Oxidizing atmosphere	Reducing atmosphere	Product brochure reference
Stainless steel	400	540	Mott
Inconel	590	810	Mott
Hastelloy X	780	930	Mott
Gamma-alumina	400	400	Anotec Separations
Zirconia/titania	350	350	TECH-SEP
Alumina/titania	400	400	TECH-SEP
Alumina	700	700	Holland Ind. Ceramics
Carbon		165	Le Carbone-Lorraine

The presence of water at elevated temperatures is also a critical factor in the performance of gamma-alumina membranes. Water has been known to promote the aforementioned phase transition of alumina at a lower temperature.

If simply heated to a higher temperature, gamma-alumina will be converted to alpha-alumina but the pore size and porosity change significantly. For example, by increasing the heating temperature from 800°C (gamma-alumina) to 1,000°C (alpha-alumina), the pore diameter increases from 4.8 to 78 nm and the corresponding porosity decreases from 55 to 41% [Leenaars et al., 1986]. A technique has been developed to avert these undesirable changes upon the conversion from gamma- to alpha-alumina. It involves adding a small amount of fine crystalline seed particles which primarily consists of alpha-alumina into the sol for deposition of the membrane [Hay, 1990]. These seed particles are used at a level of 0.05 to 1% by weight of the hydrated alumina in the sol and have a diameter on the order of 15 to 45 nm (with a surface area of 35 to 100 m^2/g). The resulting pore diameters are in the 30 to 50 nm range.

In a systematic characterization of a commercial gamma-alumina membrane with a nominal pore diameter of 4 nm, Gallaher and Liu [1994] have found that exposure of the membrane to a temperature below 400°C has negligible effect on the membrane permeability. However, thermal treatment at 500 and 600°C results in a near doubling of the membrane permeability although the membrane maintains its structural integrity. The permeability change occurs mostly in the first few hours and it is not accompanied by a significant modification of the membrane pore size as shown in Table 4.6 which provides the resulting data on pore diameter, nitrogen permeance (in the unit of m^3/m^2-h-atm) and calculated porosity/tortuosity ratio. It is seen that thermal treatment at 640°C for 100 hours results in an increase of nitrogen permeance from 52 to 110 m^3/m^2-h-atm while the pore diameter only has a slight increase from 4.0 to 4.3 nm. The increase in permeance far exceeds that accountable by the slight increase in pore size and is attributed to an appreciable change in apparent porosity due to desorption of adsorbed moisture and possibly other species during thermal treatment. This is supported by the substantial increase of the porosity to tortuosity ratio given in Table 4.6. The ratio is calculated from the measured permeance assuming Knudsen diffusion according to the following equation [Gallaher and Liu, 1994]:

$$P_m = (2r/3) \cdot (8RT/\pi M)^{1/2} \cdot (\varepsilon/\tau) \cdot (1/RTL_m) \tag{4-16}$$

It has also been found that the gamma-alumina membranes thermally treated in the absence of steam loses the permeance upon a long storage period, but, when the thermal treatment is repeated, the permeance is regained. Apparently the process involved is reversible.

TABLE 4.6
Characteristics of γ-Al_2O_3 membranes as a function of thermal exposure at 640°C

Exposure time (h)	Pore dia. (nm)	N_2 permeance $(m^3/m^2\text{-h-atm})$	Porosity/tortuosity (γ/τ) calculated
0	19.8	52.3	0.127
100	21.6	110.0	0.244

(Adapted from Gallaher and Liu [1994])

Thermal treatment in the presence of steam, however, results in irreversible changes frequently involving pore enlargement. Water is known to accelerate the phase transformation process and reduce the temperature at which the transformation takes place. Most likely, the phase transition results in pore changes. Similarly to thermal treatment without steam, hydrothermal exposure triggers a rapid increase in the membrane permeance. The majority of the pore size changes take place in the first few hours. Unlike pure thermal treatment, however, hydrothermal treatment does not reach a permeance plateau right away but continues to increase to a higher level. Given in Table 4.7 are the membrane characteristics before and after exposure to two levels of steam concentrations: 5 and 90% [Gallaher and Liu, 1994]. It is evident that both the pore diameter and the nitrogen permeance become substantially greater after steam exposure at 640°C. By assuming Knudsen diffusion, the extent of permeance increase can be fully explained by the corresponding pore increase. The pores are enlarged from about 3.7-3.8 nm to 4.9 and 6.4 nm, respectively, by having been immersed in a gas stream with 5 and 90% steam at 640°C. The higher steam concentration appears to have a mineralizing effect on the pore structure. The pore enlargement effect is also confirmed by the essentially constant ratio of porosity to tortuosity.

Glass and silica membranes. Porous Vycor glass membranes have been in the marketplace for some time although with a very limited sales volume. Porous glass membranes can have some variations in their compositions and consequently in their thermal and hydrothermal stabilities as well. Furthermore, dense or ultramicroporous silica membranes are found to be highly selective to hydrogen. Some understanding of their thermal and hydrothermal stabilities are obviously quite important to the design and operation of the membrane systems.

As discussed in Chapter 3, a porous glass membrane can be made by the phase separation and leaching method. The pore size of the resulting glass membrane is regulated by the annealing temperature which normally varies between 500 and 800°C. Porous glass membranes have the tendency to undergo separation of ingredients particularly on the surface when the membranes are exposed to a temperature above the annealing temperature. Some ultramicroporous, spongy glass membranes with a pore diameter less than 2 nm are not stable above 400°C [Elmer, 1978].

TABLE 4.7
Characteristics of γ-Al_2O_3 membranes as a function of hydrothermal exposure at 640°C

Steam content	Exposure time (h)	Pore dia. (nm)	N_2 permeance $(m^3/m^2$-h-atm)	Porosity/ tortuosity (γ/τ) calculated
5%	0 (after thermal)	3.70	109	0.282
	95	4.92	148	0.289
	11.8	4.90	152	0.298
90%	0 (after thermal)	3.76	109	0.278
	198	6.46	188	0.279

(Adapted from Gallaher and Liu [1994])

Although steam sterilizable and chemically and biologically inert, porous glass membranes generally have a maximum working temperature of 500 to 700°C. They are usually unstable above that temperature range.

Silica undergoes significant densification above 800°C. Apparently at a temperature away from that limit, such as 600°C, silica structure is not altered. Wu et al. [1994] prepared and characterized silica membranes on porous alumina membranes with a nominal pore diameter of 4 nm as the supports. When those membranes were exposed to nitrogen at 600°C for 130 hours, the helium and nitrogen permeation rates at room temperature essentially did not experience any changes. It was inferred then that the silica membranes are thermally stable at 600°C. Larbot et al. [1989] have prepared mesoporous silica membranes by the sol-gel process and calcination at a temperature of 400 to 800°C. If the membranes so obtained are subject to heat treating at, say, 900°C for an extended period of time (e.g., 10 hours or longer), the membrane pores begin to close.

Silica membranes are not thermally stable if water is present even at 600°C. For example, upon exposure to 20% water at 600°C, the same silica membranes Wu et al. [1994] have characterized appear to have undergone some permanent structural densification. As stated earlier, silica structure normally densifies above 800°C. The presence of water may have accelerated the densification process. This structural change leads to a significant decrease in the helium permeance above 300°C (Table 4.8). This change is most pronounced during the first two hours or so (not shown in the table) but continues at a slower rate over a period of close to 100 hours. The decrease of nitrogen permeance at 450 and 600°C as a result of the hydrothermal exposure are not as drastic as helium. The ideal separation factor (ratio of He to N_2 permeance) decreases following the hydrothermal exposure at 600°C for 95 hours. It has been hypothesized for the hydrothermally treated silica membrane that helium and hydrogen transport through the micropores of the membrane by hindrance diffusion while nitrogen through the mesopores by Knudsen diffusion and that as silica densifies following hydrothermal exposure the associated shrinkage creates microcracks [Wu et al., 1994]. These hypotheses seem to be supported by the observation that the nitrogen permeance decreases while the helium permeance increases with increasing temperature and that the He/N_2 selectivity drops after hydrothermal exposure.

TABLE 4.8
Effect of hydrothermal exposure[a] of silica-modified alumina membrane on gas permeance

Temperature (°C)	Permeance[b] (m^3/m^2-h-atm)					
	Before			After		
	He	N_2	He/N_2	He	N_2	He/N_2
300	7.29	0.319	22.9	1.80	0.408	4.4
450	9.36	0.441	21.2	2.68	0.311	8.6
600	11.0	0.655	16.9	3.67	0.238	15.5

Note: a - after exposed to 20 mol% water in N_2 flow at 600°C for 95 h; b - permeance was measured at a reject pressure of 5.3-8.8 psig and a permeate pressure of 0.0 psig (Adapted from Wu et al. [1994])

Zirconia membranes. Along with alumina membranes, zirconia membranes have been a major driving force among various inorganic membranes for serving the separation market. The zirconia membrane layer in a composite membrane element is fired at a temperature between 400 and 900°C to enhance its thermal stability. Zirconia exists as a metastable tetragonal phase up to about 700°C. Above that temperature, phase transition to the monoclinic form occurs. As the temperature increases, the pore diameter of the zirconia membrane gradually increases. If the zirconia membrane is subjected to a temperature higher than 950-1230°C, transformation to the high-temperature tetragonal phase takes place. Upon cooling, however, the membrane can undergo an appreciable volume expansion leading to cracks [Stevens, 1986].

Titania membranes. Due to its excellent chemical resistances and unique catalytic properties, titania membranes have received considerable interest in recent years. Titania membranes show the anatase phase until above 350-450°C when the material is transformed into the rutile phase. The transformation often generates cracks due to a significant volume change. This problem can be circumvented by adding a proper stabilizing agent such as sulfate. The pore diameter slowly increases with increasing temperature in the range of 500 to 1,100°C.

Metallic membranes. Among the metallic membranes, palladium and its alloys offer the most promise and are used and studied most extensively. Metallic membranes often suffer from the problem of structural degradation after repeated cycles of adsorption and desorption at an elevated temperature above, say, 200°C.

4.5 CHEMICAL STABILITY

The issue of chemical resistance is relevant not only during applications but also in membrane cleaning procedures which often specify strong acids and bases and sometimes peroxides. Moreover, nonoxide ceramic membrane materials are prone to reaction upon extended exposure to oxidizing environments.

For lack of definitive and quantitative chemical resistance data for various types of inorganic membranes, some simple dissolution test can be and has been employed to provide comparative corrosion rate by a specific corrosive chemical agent such as an acid or base. A generic version of the test is given as follows. The membrane samples to be compared are subject to ultrasonic cleaning with Freon for a period of, say, 5 minutes followed by drying at 200°C for 2 hours. The above steps are repeated once more with Freon replaced by demineralized water. The samples are then immersed in a constant-temperature bath of the corrosive chemical solution (e.g., 35% hydrochloric acid); care should be exercised not to trap air bubbles inside the membrane pores. The dissolution test continues for a fixed duration (e.g., one week). The membrane samples are removed from the bath and ultrasonically cleaned with demineralized water and dried using the same procedure before the dissolution step. The weights of the cleaned membrane samples before and after the dissolution test are compared to estimate the relative corrosion rate which is typically expressed as % weight loss per unit area exposed per day. For example, the very stable alpha-alumina membrane undergoes hydrochloric acid (35% strength) corrosion at a relative rate on the order of 10^{-5} wt. loss % per unit cm^2 per day which is an order of magnitude slower than the highly resistant plastic, polytetrafluoroethylene (PTFE), and three or four orders of magnitude slower than stainless steel.

Ceramic materials have compact crystal structure, strong chemical bonding and high field strengths associated with the small and highly charged cations. Consequently they are in general more stable chemically, mechanically and thermally than organic polymeric membranes. Thermodynamically, many ceramic materials having a large free energy and total energy of formation are very stable. Some examples are yttria and thoria. Alumina, titania and zirconia are also known for their chemical stability. However, these metal oxides are not resistant to such corrosive chemicals as hydrogen fluoride, ammonium fluoride and concentrated hydrochloric and sulfuric acids.

As a general rule, the more acidic metal oxides or ceramics show greater resistance towards acids but are more prone to attack by bases while the more basic ones are more resistant to bases but have less resistance to acids. For comparison purposes, the diagram illustrated in Figure 4.23 can be used [Lay, 1979]. For example, alumina or zirconia membranes generally are more stable than silica when exposed to caustic solutions. On the other hand, silica membranes have better acidic resistance than most of other metal oxide membranes. For those ceramic membranes which contain two oxides, the chemical resistance towards acids and bases often lies between those of the constituents. For example, spinel ($MgAl_2O_4$) has a greater resistance to basic solution than alumina but lower than magnesia. Similarly, zircon ($ZrSiO_4$ or $ZrO_2 \cdot SiO_2$) is more resistant to acidic solutions than zirconia but not as much as silica.

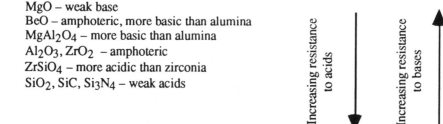

MgO – weak base
BeO – amphoteric, more basic than alumina
$MgAl_2O_4$ – more basic than alumina
Al_2O_3, ZrO_2 – amphoteric
$ZrSiO_4$ – more acidic than zirconia
SiO_2, SiC, Si_3N_4 – weak acids

Figure 4.23 Relative comparison of acidic and basic resistances of various metal oxides and ceramics [Lay, 1979]

Even within a given metal oxide, the chemical resistance varies with the particular phase. Consider the most commonly used ceramic membrane material: alumina. The hexagonal alpha-alumina crystals are very stable chemically even towards strong acids and bases. In comparison, the cubic crystalline transition-phase alumina which is capable of making membranes with nano-sized pores has been known to be subject to some attack under high and low pH.

Porous glass membranes made by the phase separation method, as in the case of commercial glass membranes, suffer from the problem of partial silica dissolution upon long exposure to water. This can result in some structural change. Although these membranes show excellent resistance to acids, they undergo some degree of silica elution in the presence of strong caustic solutions especially at high temperatures. An effective means of preparing porous glass membranes that are resistant to water and alkali is to make mixed oxide glass via the sol-gel process. Using $Si(OC_2H_5)_4$ and $Zr(OC_3H_7)_4$ as the precursors, Yazawa et al. [1991] found that the 70% SiO_2-30% ZrO_2 composition overall offers the best defect-free porous glass membranes that can withstand water and alkali attack. Compared in Table 4.9 are the water permeation rate and the sodium chloride retention coefficient data for the essentially ZrO_2-free porous glass membrane by the phase separation process and the 30%-ZrO_2 glass membranes by the sol-gel process. In all cases the membranes have been soaked in high-pH solutions for 5 hours at 30°C. The salt retention coefficient of the original membrane prior to alkali exposure is 77%. It is evident that the ZrO_2-loaded porous glass membranes are superior to the commercial ones in their alkali stability. The salt retention is much lower and the water permeation rate is much higher due to the partial dissolution of the silica pore structure in alkaline solution.

One way to improve chemical stability of a ceramic membrane is to introduce another oxide to the system as mentioned previously. On the other hand, even a small quantity of an ingredient present in the membrane composition may appreciably change the resulting chemical resistance in an undesirable direction. An example is calcia-stabilized zirconia which is likely to offer less resistance to acids than pure zirconia.

TABLE 4.9
Effect of alkali exposure on membrane permeability and permselectivity

Membrane constituent	pH of soaking solution	Salt retention coefficient (%)	Water permeation rate(cm^3-cm^{-2}-hr^{-1})
SiO_2	13.0	5	6.23
SiO_2-ZrO_2	11.0	74	0.125
SiO_2-ZrO_2	11.5	79	0.135
SiO_2-ZrO_2	12.0	75	0.190
SiO_2-ZrO_2	12.5	73	0.115
SiO_2-ZrO_2	13.0	75	0.140

(Adapted from Yazawa et al. [1991])

The kinetics of chemical corrosion for inorganic membranes can be quite different from that for the same materials in dense monolithic form. This is particularly true with porous inorganic membranes where the porous network resembles interconnected finely divided particles. Compared to dense bulk bodies, fine particles normally are dissolved more rapidly in corrosive chemicals.

The chemical environment during applications can have profound impact on the membrane characteristics through phase transformation. Thus phase transformation is not just caused by thermal energy. Balachandran et al. [1993] have demonstrated that a La-Sr-Fe-Co-O material with a certain stoichiometry exhibits a cubic perovskite phase in an oxygen-rich (20%) atmosphere at 850°C but converts to an oxygen-vacancy-ordered phase when the oxygen partial pressure is reduced to less than 5% at the same temperature. This results in a substantial volume expansion that could exceed the effects of thermal expansion. An implication of this observation is that, when the two sides of a relatively thick La-Sr-Fe-Co-O membrane with a certain stoichiometry are exposed to appreciably different oxygen atmospheres, it is possible that the membrane have different phase constituents across its thickness. The mismatch of the expansion coefficients due to phase transition near where the phases meet can cause the membrane to fracture. It should also be mentioned that direct contact of some La-Sr-Fe-Co-O membranes with hydrogen (and possibly CO as well) could also result in the undesirable phase change which could lead to structural decomposition.

4.6 SURFACE PROPERTIES

It has been widely recognized that both organic and inorganic membranes experience decline in the permeate flux, usually referred to as fouling, due to strong adsorption of solutes or macromolecules to the surface of pores or the membrane itself. Fouling is often enhanced by the presence of ionic species in the feed streams. Surface charges of a membrane are known to greatly influence its separation properties such as permeability and permselectivity. The liquid feed and the membrane interact by various mechanisms such as van der Waals and ionic interactions and hydrogen bonding. Benefits have been realized recently from surface modifications of membranes through better understanding of these interactions. Increasing research and development efforts in this area [Schnabel and Vaulont, 1978; Shimizu et al., 1987, Keizer et al. 1988; Yazawa et al., 1988] have extended the range of applications of inorganic membranes by limited tailoring of membrane surface properties which have led to more desirable separation properties in selected cases.

Electrokinetic phenomena such as the zeta potential and streaming potential can alter the surface properties of many porous inorganic membranes. The electrochemical properties of membrane surfaces can exert profound influence on the nature and magnitude of the interactions between the membrane and the liquid feed, thus affecting the permeating fluxes of the solvent and solute (or macromolecule/particle) through the membrane pores.

A common method of determining the surface electrochemical properties is to measure the zeta potential of the membrane. Bowen et al. [1991] have determined the membrane zeta potential from electro-osmotic flow rate data. The membrane is held at the end of a "dipped cell" and a constant current is applied between an electrode positioned behind the membrane and a counter electrode outside the cell. This induces electro-osmotic flow of electrolyte into the cell which is measured by an electronic balance. They found that the zeta potential of alumina membranes in contact with NaCl solution is substantially greater than those of polymeric membranes. Shown in Figure 4.24 is the zeta potential data of an anodic alumina membrane (having a pore diameter of 0.2 μm) in 0.01 M NaCl over a wide range of pH [Bowen et al., 1991]. It is noted that the isoelectric point (where the zeta potential reaches zero) for 0.01 M NaCl is 4.04 while that for 0.001 M NaCl is 4.00. It should be mentioned that most ceramic and organic polymer ultrafiltration membranes carry a net negative charge in aqueous solutions at the neutral pH, as shown in Figure 4.24. By manipulating the membrane zeta potential, it is possible to effect a significant increase in the membrane selectivity of protein mixtures [Saksena and Zydney, 1994].

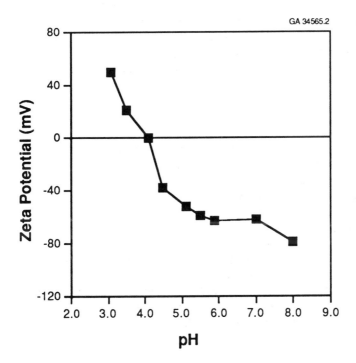

Figure 4.24 Dependence of membrane zeta potential on solution pH [Bowen et al., 1991]

Several investigators measure the zeta potential of the membrane by measuring that of the membrane precursor particles or the particles obtained from grinding the membrane.

Thus alumina membranes surface modified with silanes and sulfone [Shimizu et al. 1987] and with trimethyl chlorosilane TMS [Shimizu et al. 1989] and glass membranes adsorbed with surfactants [Busscher et al. 1987] have been studied this way. The results show that surface treatments alter the zeta potentials. Shimizu et al. [1989] have also demonstrated that under normal operating conditions the zeta potentials of alumina membranes do not change over time even for a period of two to three years. The isoelectric point for alumina particles thus determined is close to 4.00 as determined by direct measurement of membranes.

The streaming potential is also often used to characterize electrochemical properties of a membrane surface in the presence of a liquid. The pressure-drive flow of an electrolyte solution through a charged membrane tends to induce an electric field and creates the streaming potential. Thus, the streaming potential is an electrokinetic measurement indicative of the result of liquid flow through capillaries or pores under a pressure gradient. It should be mentioned that for a straight capillary the streaming potential can be directly related to the electro-osmotic flow and hence the zeta potential. Although the measurement is not easy and sometimes not very consistent, Nystrom et al. [1989] have developed a sophisticated device to measure the streaming potential which can then be used to determine the apparent zeta potential by applying the Helmholtz-Smoluchowski equation. The method has been found to give reliable results for predicting trends in adsorption, separation properties and fouling especially for solutions of ionic macromolecules.

Both the zeta potential and the streaming potential of a membrane in the presence of an aqueous solution have been theoretically analyzed as functions of the pore size distribution of the membrane. Assuming the membrane pore size can be described by a log-normal distribution, Saksena and Zydney [1995] have studied the effects of the pore size distribution on the solvent flux, membrane zeta potential and induced streaming potential. The solvent flow rate increases with an increase in the breadth of the pore size distribution. The rate increase is much more pronounced for pressure-driven flows than electrically-driven flows. The pore size distribution can dramatically change the streaming potential and the calculated values of the membrane zeta potential. For the same mean pore diameter and ionic strength, the absolute value of the normalized zeta potential decreases while the normalized streaming potential increases as the pore size distribution becomes broader. The induced streaming potential causes a relatively large reverse flow through the smallest pores in the distribution. The total magnitude of this reverse flow is much less pronounced for those membranes with broader pore size distributions.

Knowledge of the charge density of the membrane surface and membrane pores is also essential to characterizing the surface electrochemical properties of the membranes. Many attempts have been made to provide quantitative measures of the membrane surface charge density. The surface charge density can be determined by pH surface titration. A typical surface titration curve for an anodic alumina membrane in 0.01 M NaCl solution is given in Figure 4.25 which shows the titratable surface charge of the alumina membrane as a function of the solution pH [Bowen et al., 1991]. The alumina membrane tested has high surface charges. The sharp inflection points near pH 4 and 7 and the broad inflection point near 6.5 in the figure can be identified with the aluminum oxide surface groups, and the oxalate and phosphate groups incorporated during the anodic oxidation manufacturing process.

From the zeta potential and surface charge density, Bowen et al. [1991] have constructed an electrochemical model based on electrical double layer theory where the membrane surface consists of a three-dimensional array of various charged groups (gel) with both protons and counterions being able to penetrate the gel region. The model provides estimates for the thickness of the gel-layer, the total number of charged groups and the ratio of aluminum oxide groups to the incorporated anion groups.

GA 35275.2

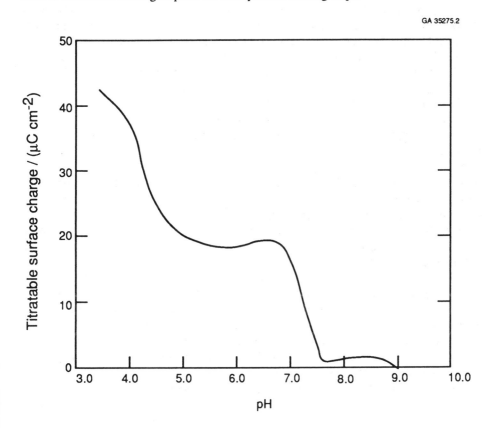

Figure 4.25 Titratable surface charge of an alumina membrane as a function of solution pH [Bowen et al., 1991]

From the chemical reaction point of view, some inorganic membranes having functional groups at their surfaces can be modified by linking them to or reacting them with some other functional groups or polymers. The surface of microporous glass membranes, for example, are covered with many silanol groups. These groups tend to be reactive and lend themselves to a wide range of chemical surface modifications such as the coupling of enzymes, or the attachment of microorganisms [Langer and Schnabel, 1989]. Microporous glass membranes as-is have a strong tendency to adsorb proteins. Their modifications with various silanes, however, deactivate the silanol groups and make the membrane surface much less prone to protein adsorption. This type of surface modification often involves crosslinking to such extent that the original glass surface is effectively shielded from the surrounding application medium and can result in hydrophilic or hydrophobic characteristics and unique separation properties for bioseparation applications. For example, alcoholic OH-groups can most effectively modify the surface of glass membranes to achieve a low level of nonspecific protein binding. The resulting protein adsorption can be as low as 1/1000 of that due to unmodified glass membranes [Langer and Schnabel, 1988]. Another intriguing surface modification of porous glass membranes is interaction with NH_2-groups which allow the coupling of enzymes through glutaraldehyde as the coupling agent [Langer and Schnabel, 1989]. This is particularly useful for enzymatic reactors.

Similarly suitable functional groups can be attached to the surfaces of other metal oxide membranes to impart improved separation properties or chemical stability to the membranes. Phosphoric acid and organic acids such as phosphonic and phosphinic acids, carboxylic acids or boronic acids have been chemically bonded to the reactive sites on the surface of inorganic membranes, particularly ceramic membranes [Martin et al, 1990; Wieserman et al., 1990; Martin et al., 1992]. The membranes thus treated have better pH stability. Titania and zirconia membranes have also been modified with phosphoric acid and alkyl phosphonic acids to improve separation properties [Randon et al., 1995]. Phosphate adsorption on the titania membranes changes the polar hydrophilic character of the titania or zirconia membranes to a nonpolar hydrophobic character. As shown in Figure 4.26, the phosphate-treated membranes offer higher values of the rejection (retention) coefficient of bovine serum albumin (BSA) protein than untreated membranes having the same mean pore diameter of 18 nm. The rate of the flux decline is slower and the asymptotic flux value is higher as the phosphate treatment time increases.

4.7 SUMMARY

The general characteristics of inorganic membranes that are not application specific but affect separation performance and their determination methods are reviewed in this chapter. Those application-specific characteristics will be treated in Chapters 6 through 11.

Morphology and qualitative elemental composition of membrane structures and surfaces have been routinely examined by scanning electron microscopy and, as needed, by transmission electron microscopy. Atomic force microscopy is an emerging technique capable of measuring surface topology.

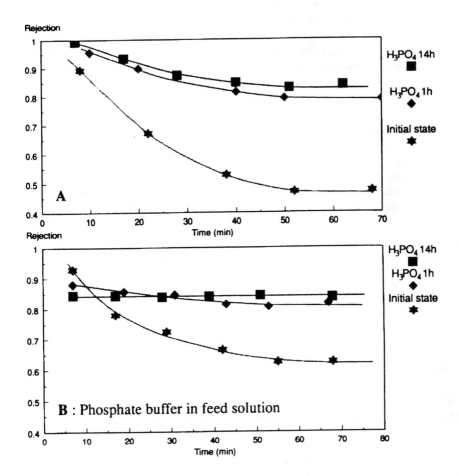

Figure 4.26 Rejection coefficient of bovine serum albumin (BSA) protein as a function of time for untreated titania membrane and two titania membranes treated with 0.1 M phosphoric acid at room temperature for 1 and 14 hours, respectively. (A) without phosphate buffer; (B) with 0.01 M Na_2HPO_4/NaH_2PO_4 buffer [Randon et al., 1995]

The critically important pore size and its distribution can be determined by a host of measurement techniques. The bubble point test is used primarily for detecting any defects or hairline cracks and for estimating the average pore size. Traditionally mercury porosimetry and nitrogen adsorption/desorption have been the workhorse for determining the pore size distribution of a porous membrane. They are, however, usually limited to a pore diameter range of >3 nm and 1.5-100 nm, respectively. Determination of multi-layered and broad pore size distributions is not straightforward; it often requires tedious and multiple measurements to construct the entire pore size distribution. Newer pore size measurement techniques such as permporometry, thermoporometry, gas flow method and gas permeability method are being developed. While some of them are able to detect pore diameters as small as 1.5 nm, the gas flow method appears to be the choice because of its robustness and simplicity. The molecular probe method using various molecules of known kinetic diameters in the permeation test, in some cases, can infer the pore diameters of ultramicroporous membranes. While it is possible to determine the porosity of an inorganic membrane, the determination of tortuosity is indirect and not frequently used.

Transport mechanisms for dense metal and solid electrolyte membranes are of the solution-diffusion type. The major difference is that the diffusing species in the dense metal and solid electrolyte membranes are in atomic and charged form, respectively. There are a multitude of transport mechanisms for porous membranes that effect separation of species. Size exclusion is often dominating in microfiltration and some ultrafiltration applications. In addition, Knudsen diffusion, surface diffusion and capillary condensation are capable of assisting separation under proper conditions. General separation characteristics of an inorganic membrane can be described by the permeability and permselectivity. The permeability of a porous membrane in the microfiltration and ultrafiltration range is largely determined by the pore size and temperature.

Thermal and hydrothermal stabilities of inorganic membranes depend on the particular phase of the membrane material which is determined during the membrane synthesis. Generally speaking, hydrothermal exposure of the membrane at elevated temperatures causes irreversible phase changes and/or microstructural alterations leading to deteriorated separation performance. As a general rule, the more acidic metal oxides or ceramics show greater resistance to acids and more basic materials have higher stabilities towards bases. The chemical stability of a membrane also depends on its chemical phase. Furthermore, like thermal energy, the chemical environment can cause phase transformation and affect the separation properties of the membrane.

Finally, zeta and streaming potentials are often used to describe the surface characteristics of a membrane and can substantially influence the permeation rates of the permeant as well as the solute, macromolecular or particulate species. Both potentials can be affected by the pore size distribution.

REFERENCES

Altena, F.W., H.A.M. Knoef, H. Heskamp, D. Bargeman and C.A. Smolders, 1983, J. Membr. Sci. **12**, 313.
Anderson, M.A., M.J. Gieselmann and Q. Xu, 1988, J. Membr. Sci. **39**, 243.

Anotec Separations (now under Whatman Sci.), 1988, Anotec inorganic membrane filtration, Product Brochure.

Balachandran, U., S.L. Morissette, J.T. Dusek, R.L. Mieville, R.B. Poeppel, M.S. Kleefisch, S. Pei, T.P. Kobylinski and C.A. Udovich, 1993, Development of ceramic membranes for partial oxygenation of hydrocarbon fuels to high-value-added products, in: Proc. Coal Liquefaction and Gas Conversion Contractors' Review Conf. (U.S. Dept. of Energy), Pittsburgh, USA.

Barrer, R.M., 1951, Diffusion in and through Solids (Cambridge University Press, London).

Barrer, R.M., 1965, AIChE Symp. Ser. **1**, 112.

Bhave, R.R., and H.L. Fleming, 1988, AIChE Symp. Ser. **84** (No. 261), 19.

Bottino, A., G. Capannelli, A. Grosso, O. Monticelli, O. Cavalleri, R. Rolandi and R. Soria, 1994, J. Membr. Sci. **95**, 289.

Bowen, W.R., D.T. Hughes and H.A.M. Sabuni, 1991, Key Engineering Mat. **61 & 62**, 117.

Brun, M., A. Lallemand, J.F. Quinson and C. Eyraud, 1973, J. Chim. Phys. **70**, 973.

Busscher, H.J., H.M. Uyen, G.A.M. Kip and J. Arends, 1987, Colloids and Surfaces **22**, 161.

Cao, G.Z., J. Meijerink, H.W. Brinkman and A.J. Burggraaf, 1993, J. Membr. Sci. **83**, 221.

Capannelli, G., F. Vigo and S. Munari, 1983, J. Membr. Sci. **15**, 289.

Capannelli, G., A. Bottino, S. Munari, D.G. Lister, G. Maschio and I. Becchi, 1994, J. Food Eng. **21**, 473.

Carman, P.C., 1937, Trans. Inst. Chem. Eng. **15**, 150.

Cuperus, F.P., D. Bargeman and C.A. Smolders, 1992a, J. Membr. Sci. **66**, 45.

Cuperus, F.P., D. Bargeman and C.A. Smolders, 1992b, J. Membr. Sci. **71**, 57.

Dou, S., C.R. Masson and P.D. Pacey, 1985, J. Electrochem. Soc. **132**, 1843.

De Lange, R.S.A., J.H.A. Hekkink, K. Keizer and A.J. Burggraaf, 1995, J. Membr. Sci. **99**, 57.

Elmer, T.H., 1978, Ceramic Bull. **57**, 1051.

Eyraud, C., 1984, Thermoporometry, presented at European Soc. Membr. Sci. & Tech.- Summer School (Cadarache, France).

Eyraud, C., 1986, Applications of gas-liquid permporometry to characterization of inorganic ultrafilters, in: Proc. Europe-Japan Congr. Membr. Membr. Processes, Stresa, Italy (1984), p. 629.

Fain, D.E., 1989, A dynamic flow-weighted pore size distribution, in: Proc. 1st Int. Conf. Inorg. Membr., Montpellier, France, p.199.

Gallaher, G.R., and P.K.T. Liu, 1994, J. Membr. Sci. **92**, 29.

Garrido, J., J.M. Martin-Martinez, M. Molina-Sabio, F. Rodriguez-Reinoso and T. Torregrosa, 1986, Carbon **24**, 469.

Gavalas, G.R., C.E. Megiris and S.W. Nam, 1989, Chem. Eng. Sci. **44**, 1829.

Glaves, C.L., P.J. Davis, K.A. Moore, D.M. Smith and H.P. Hsieh, 1989, J. Colloid and Interface Sci. **133**, 377.

Graham, T., 1866, Phil. Trans. Roy. Soc. (London) **156**, 399.

Grosgogeat, E.J., J.R. Fried, R.G. Jenkins and S.T. Hwang, 1991, J. Membr. Sci. **57**, 237.

Hay, R.A., 1990, U.S. Patent 4,968,426.

Holland Ind. Ceramics, 1993, Holland Ind. Ceramics problem solving with separation technology, Product Brochure.

Hsieh, H.P., 1991, Catal. Rev. - Sci. Eng. **33**, 1.

Hsieh, H.P., P.K.T. Liu and T.R. Dillman, 1991, Polymer J. (Japan) **23**, 407.

Huttepain, M., and A. Oberlin, 1990, Carbon **28**, 103.

Itoh, N., 1990, Sekiyu Gakkaishi **33**, 136 (in Japanese).

Itoh, N., M. A. Sanchez C., W.C. Xu, K. Haraya and M. Hongo, 1993, J. Membr. Sci. **77**, 245.

Kameyama, T., M. Dokiya, K. Fukuda and Y. Kotera, 1979, Sep. Sci. Technol. **14**, 953.

Keizer, K.R., J.R. Uhlhorn, R.J. van Vuren and A.J. Burggraaf, 1988, J. Membr. Sci. **39**, 285.

Koresh, J., and A. Soffer, 1980, J. Chem. Soc. Faraday Trans. (I), **76**, 2473.

Langer, P., and R. Schnabel, 1988, Chem. Biochem. Eng. **Q2**, 242.

Langer, P., and R. Schnabel, 1989, Porous glass UF-membranes in biotechnology, in: Proc. 1st Int. Conf. Inorg. Membr., Montpellier, France, p.249.

Larbot, A., J.P. Fabre, C. Guizard and L. Cot, 1989, J. Am. Ceram. Soc. **72**, 257.

Lay, L.A., 1979, The resistance of ceramics to chemical attack, Report No. Chem 96, National Physical Laboratory, Teddington, Middlesex, U.K.

Le Carbone-Lorraine, 1993, CFCC Cross-flow microfiltration membranes and modules, Product Brochure, .

Leenaars, A.F.M., K. Keizer and A.J. Burggraaf, 1984, J. Mater. Sci. **19**, 1077.

Leenaars, A.F.M., and A.J. Burggraaf, 1985, J. Membr. Sci. **24**, 245.

Leenaars, A.F.M., K. Keizer and A.J. Burggraaf, 1986, Chemtech **16** , 560.

Lewis, F.A., 1967, The Palladium Hydrogen System, Academic Press, New York, USA.

Lin, C.L., D.L. Flowers and P.K.T. Liu, 1994, J. Membr. Sci. **92**, 45.

Lin, Y.S., W. Wang and J. Han, 1994, AIChE J. **40**, 786.

Martin, E.S., and L.F. Wieserman, 1990, U.S. Patent 4,962,073.

Martin, E.S., L.F. Wieserman, K. Wefers and K. Cross, 1992, U.S. Patent 5,124,289.

Megiris, C.E., and J.H.E. Glezer, 1992, Ind. Eng. Chem. Res. **31**, 1293.

Merin, U., and M. Cheryan, 1980, J. Appl. Pol. Sci. **25**, 2139.

Mott Metall. Corp., 1991, Precision porous metals - engineering guide, Product Brochure DB 1000.

Munari, S., A. Bottino, P. Moretti, G. Panannelli and I. Becchi, 1989, J. Membr. Sci. **41**, 61.

Nagamoto, H., and H. Inoue, 1983, On a reactor with catalytic membrane permeated by hydrogen, in Proc.: 3rd Pac. Chem. Eng. Congr., Seoul, S. Korea, vol. 3, p. 205.

Nystrom, M., M. Lindstrom and E. Matthiasson, 1989, Colloids and Surfaces **36**, 297.

Okazaki, M., H. Tamon and R. Toei, 1981, AIChE J. **27**, 271.

Randon, J., P. Blanc and R. Paterson, 1995, J. Membr. Sci. **98**, 119.

Remick, R.R., and J. Geankoplis, 1973, Ind. Eng. Chem. Fund. **12**, 214.

Saksena, S., and A.L. Zydney, 1994, Biotechnol. Bioeng. **43**, 960.

Saksena, S., and A.L. Zydney, 1995, J. Membr. Sci. **105**, 203.

Saracco, G., and V. Specchia, 1994, Catal. Rev.-Sci. Eng. **36**, 305.

Sarrade, S., G.M. Rios and M. Carlès, 1994, J. Membr. Sci. **97**, 155.

Schnabel, R., and W. Vaulont, 1978, Desalination **24**, 249.

Shimizu, Y., T. Yazawa, H. Yanagisawa and K. Eguchi, 1987, Yogyo Kyokaishi **95**, 1067 (in Japanese).

Shimizu, Y., K. Matsushita, I. Miura, T. Yazawa and K. Eguchi, 1988, Nippon Seramikkusu Kyokai Gakujutsu Ronbunshi **96**, 556 (in Japanese).

Shimizu, Y., K. Yokosawa, K. Matsushita, I. Miura, T. Yazawa, H. Yanagisawa and K. Eguchi, 1989, Nippon Seramikkusu Kyokai Gakujutsu Ronbunshi **97**, 498 (in Japanese).

Shu, J., B.P.A. Grandjean, A. van Neste and S. Kaliaguine, 1991, Can. J. Chem. Eng. **69**, 1036.

Sing, K.S.W., D.H. Everett, R.A.W. Haul, L. Moscou, R.A. Pierotti, J. Rouquerol and T. Siemieniewska, 1985, Pure Appl. Chem. **57**, 603.

Soffer, A., J.E. Koresh and S. Saggy, 1987, U.S. Patent 4,685,940.

Stevens, R., 1986, Zirconia and Zirconia Ceramics (2nd ed.), Magnesium Elektron, U.K.

Stoeckli, H.F., and F. Kraehenbuehl, 1984, Carbon **22**, 297.

TECH-SEP, 1993, TECH-SEP: The exact solution in cross-flow filtration, Product Brochure.

Uemiya, S., N. Sato, H. Ando and E. Kikuchi, 1991, Ind. Eng. Chem. Res. **30**, 585.

Uhlhorn, R.J.R., K. Keizer and A.J. Burggraaf, 1989, J. Membr. Sci. **46**, 225.

Venkataraman, K., W.T. Choate, E.R. Torre, R.D. Husung and H.R. Batchu, 1988, J. Membr. Sci. **39**, 259.

Wefers, K., and C. Misra, 1987, Oxides and hydroxides of aluminum, Alcoa Tech. Paper 19 (rev.), Aluminum Co. of America, Pittsburgh, PA, USA.

Wieserman, L.F., K. Wefers, K. Cross, E.S. Martin, H.P. Hsieh and W.H. Quayle, 1990, U.S. Patent 4,957,890.

Wu, J.C.S., H. Sabol, G.W. Smith, D.L. Flowers and P.K.T. Liu, 1994, J. Membr. Sci. **96**, 275.

Yazawa, T. H. Nakamichi, H. Tanaka and K. Eguchi, 1988, Nippon Seramikkusu Kyokai Gakujutsu Ronbunshi **96**, 18 (in Japanese).

Yazawa, T., H. Tanaka, H. Nakamichi and T. Yokoyama, 1991, J. Membr. Sci. **60**, 307.

CHAPTER 5

COMMERCIAL INORGANIC MEMBRANES

The year 1980 marked the entry of a new type of commercial ceramic membrane into the separation market. SFEC in France introduced a zirconia membrane on a porous carbon support called Carbosep™. This was followed in 1984 by the introduction of alumina membranes on alumina supports, Membralox™ by Ceraver in France and Ceraflo™ by Norton in the U.S. With the advent of commercialization of these ceramic membranes in the eighties, the general interest level in inorganic membranes has been aroused to a historical high. Several companies involved in the gas diffusion processes were responsible for this upsurge of interest and applications.

Although some inorganic membranes such as porous glass and dense palladium membranes have been commercially available for some time, the recent escalated commercial activities of inorganic membranes can be attributed to the availability of large-scale ceramic membranes of consistent quality. As indicated in Chapter 2, commercialization of alumina and zirconia membranes mostly has been the technical and marketing extensions of the development activities in gas diffusion membranes for the nuclear industry.

Inorganic membranes, particularly ceramic membranes, are rapidly making inroads into food and beverage processing, waste water and water-oil treatments, biotechnology separations and other applications. The world market of inorganic membranes not including systems (i.e., elements and modules only) was estimated to be about 12 and 35 million U.S. dollars in 1986 and 1990, respectively, and projected to be about 100 million dollars by 1995 [Crull, 1991]. The estimate for the period of 1986-1988 appears to be independently confirmed by an estimate of about 10,000-12,000 m^2 inorganic membranes sold worldwide, mostly the ceramic type [Gester, 1986; Guyard, 1988]. This market size does not include the application of uranium isotope separation.

There are a variety of porous inorganic membranes in the market today. Highlighted in Table 5.1 are selected major commercial inorganic membranes according to their material type. So far the most widely used inorganic membranes are alumina membranes, followed by zirconia membranes. Porous glass and metal (such as stainless steel and silver) membranes have also begun to attract attention.

Some details of the microstructures and physical, chemical and surface properties of inorganic membranes (particularly the porous ones) have been described in Chapter 4. In this chapter, those properties related to membrane performance during and between applications and general features of commercial membrane elements, modules and systems will be discussed along with application characteristics and design and operating considerations.

TABLE 5.1
Selected commercial inorganic membranes

Manufacturer	Trade name	Membrane material	Support material	Pore diameter	Element geometry
Asahi Glass		Glass	None	8 nm-10 μm	Tube & plate
Astro Met	AmPorMat	Metal alloys, C, Al_2O_3, ZrO_2, WC	None	>5 μm	Plate
CARRE (du Pont)		$Zr(OH)_4$	S.S.	0.2 - 0.5 μm	Tube
Ceramem	CeraMem®	Ceramic oxides	Ceramic oxides	5 nm-0.5 μm	Monolith & tube
Ceramesh		ZrO_2	Inconel	0.1 μm	Plate
Ceram-Filtre	FITAMM	Ceramics	SiC	0.1-10 μm	Monolith
Corning	Celcor®	Cordierite, Al_2O_3, TiO_2, mullite	None	0.035-35 μm	Monolith
	Vycor®	Glass	None	4-20 nm	Tube & plate
DMK TEK		Ni, Cu, Mg, S.S., Al_2O_3, glass	None	>5 μm	Monolith
Du Pont	PRD-86	Al_2O_3, mullite, cordierite	None	0.06-1 μm	Tube
Fairey	Strata-Pore®	Ceramics	Ceramics	1-10 μm	Tube & plate
Fuji Filters		Glass	None	4-90 nm; 0.25-1.2 μm	Tube
Gaston County Filtration Systems	Ucarsep®	ZrO_2	C	4 nm	Tube
Golden Technologies (Coors)		Al_2O_3	Al_2O_3	0.2 μm; 0.8 μm	Monolith & tube
Holland Industrial Ceramics	HepHi-matic®	Al_2O_3	Al_2O_3	0.1-0.5 μm	Tube
Kubota		Al_2O_3	Al_2O_3	0.1-10 μm	Tube

TABLE 5.1 - CONTINUED
Selected commercial inorganic membranes

Manufacturer	Trade name	Membrane material	Support material	Pore diameter	Element geometry
Le Carbone (Pechiney)		C	C	0.01-10 µm	Tube
Millipore		Ag	None	0.45 µm; 0.8 µm	Plate
Mott	Hypulse®	S.S., Ni, Au, Ag, Ti, Pt, etc.	None	0.2-100 µm	Tube & plate
NGK	Cefilt®	Al_2O_3	Al_2O_3	4 nm; 0.1-5 µm	Monolith & tube
	Cefilt®	TiO_2	Al_2O_3	5-50 nm	Monolith
Nihon Cement		Al_2O_3	Al_2O_3	0.4-8	Tube & disk
NOK		Al_2O_3	Al_2O_3	0.4-12	Tube
Osmonics	Hytrex®	Ag	None	0.2-5 µm	Plate
	Ceratrex®	Ceramics	Ceramics	0.1-25 µm	Tube & plate
	Duratrex®	S.S.	None	0.2-100 µm	Tube
Pall		S.S., Ni, etc.	None	0.4-35 µm	Tube & plate
Poretics		Ag	None	0.2-5 µm	Plate
		Ceramics	None	0.1-12 µm	Plate
Schott Glass		Glass	None	10 nm; 0.1 µm	Tube
TDK	Dynaceram®	ZrO_2	Al_2O_3	10 nm	Tube
Tech-Sep (Rhone-Poulenc)	Carbosep®	ZrO_2/TiO_2	C	4 nm-0.14 µm	Tube
	Kerasep™	ZrO_2/TiO_2	Al_2O_3/TiO_2	4 nm-0.45 µm	Monolith
TOTO		Al_2O_3	Al_2O_3	0.2-0.4 µm	Monolith, tube & disk
		ZrO_2	Al_2O_3	10-60 nm	Monolith, tube & disk
U.S. Filter	Membralox®	Al_2O_3	Al_2O_3	4 nm-5 µm	Monolith & tube
		ZrO_2	Al_2O_3	20-100 nm	Monolith & tube

TABLE 5.1 - CONTINUED
Selected commercial inorganic membranes

Manufac-turer	Trade name	Membrane material	Support material	Pore diameter	Element geometry
Whatman	Anopore™	Al_2O_3	None	0.02-0.2 μm	Plate

An effective membrane separation system obviously requires much more than just an efficient membrane. It consists of membrane modules, process pipings, vessels, control instrument, pressurizing sources, heat exchange installations and membrane regeneration equipment. The important building block of the membrane system is the membrane modules which as units can be combined in many different ways to achieve the desired separation performance. Each module which contains one or many membrane elements are equipped with permeate collection chambers and backflushing hardware. The membrane elements are connected to pipings and housings to form a module. Inorganic membrane elements, modules and systems are discussed in the following sections.

5.1 MEMBRANE ELEMENT

Several possible microstructural features of an inorganic membrane in the direction of its thickness have been reviewed in Chapter 4. A few commercial inorganic membranes such as microporous glass, stainless steel and palladium membranes having more or less symmetric (or homogeneous) pore structures are made using the phase separation/leaching and pressing or extrusion processes, respectively. Some inorganic membranes possessing symmetric pore structures with essentially straight pores are prepared by the anodic oxidation technique.

Most of the other inorganic membrane elements have a multi-layered structure and consist of the membrane (or separating) layer and the underlying support layer(s). Depending on the pore size of the membrane layer, the number of support layers in a commercial membrane element varies from none to three. The available filtration area per unit volume of the support varies from 300 to 2,000 m^2/m^3. Each layer contains different pore size and porosity. The support, made of alumina, zirconia, titania, silica, spinel, aluminosilicate, cordierite or carbon, typically has a pore diameter of about 1 to 20 μm and a porosity of 30 to 65%. Any additional intermediate support layers have progressively smaller pore sizes than the underlayer of support. The intermediate support layers are typically 20-60 μm in thickness and 30-40% in porosity. The bulk support is generally manufactured by extrusion or sometimes by pressing and intermediate support layers by dip or spin coating. For the membrane layer in many commercial ceramic membranes, the processes employed are dip or spin coating, a special case of which is the sol-gel process for making nanometer pore size membranes.

The membrane material varies from alumina, zirconia, glass, titania, cordierite, mullite, carbon to such metals as stainless steel, palladium and silver. The resulting pore diameter ranges from 10 μm down to 4 nm and the membrane thickness varies from 3 to 10 μm. The membrane porosity depends on the pore size and is 40-55%.

A distinctively different concept in fabricating inorganic composite membranes is the use of woven metal mesh as an integral support for a microporous ceramic membrane [Gallagher, 1990]. This is in contrast to the general approach of employing a monolithic support. The ceramic membrane is formed by coating a slurry of ceramic precursor particles within the apertures of the woven metal mesh. Thus a thin, lens-shaped microporous ceramic membrane is suspended in the apertures of the mesh. Yttria-stabilized zirconia membranes with a mean pore diameter of 0.2 μm supported on a nickel-chrome alloy, Inconel 600, have been marketed recently. An advantage of this mesh-supported membrane element is that the ceramic membrane is under biaxial compression (due to differential contraction of the mesh and the ceramic after sintering) and the individual cells of the membrane are self-contained. These features reduce crack initiation and propagation which is associated with the common monolithic ceramic structures.

The overall membrane element shape comes in different types: sheet, single tube, hollow fiber, and multi-channel monolith. Photographs of some commercial membrane elements are shown in Figure 5.1. The use of disks (or sheets) has been confined to medical, pharmaceutical and laboratory applications, while tubes and monoliths are employed in larger-scale applications ranging from removal of bacteria from wine and beer fermentation to oil-water separation to waste water treatment.

Obviously a monolithic multi-channel honeycomb shape provides more filtration area per unit volume than either a sheet or a single tube. The former represents a technical evolution from the single-tube or tube-bundle geometry. It offers the potential advantages of high mechanical strength, savings in installation and repair/maintenance costs, and equally important, a higher membrane packing density. Multi-channel honeycomb membrane elements basically all have the main body in a cylindrical form, the cross section of which can vary from a circle to an octagon to other shape. The cross section of the open channels that are parallel to the axis of the element may appear as a circle, square or other shape. The number of open channels in a commercial membrane has been as high as 60.

A generalization of this concept of a monolithic multi-channel honeycomb structure is described in a patent by Hoover and Roberts [1978]. An integral support of porous ceramic material has a multiplicity of parallel passageways (or open channels). These passageways are substantially uniformly spaced. On the surface of these channels are coated with a permselective membrane layer. The feed stream flows inside the channels. The membrane, being the first layer in direct contact with the process stream, is selective to one or more species in the stream. In the normal cases of properly wetted membrane pores, the permeate under a driving force will transport through the membrane, any intermediate layers of support, and the bulk support (Figure 5.2) before leaving the membrane element for the permeate collection chamber in a membrane module.

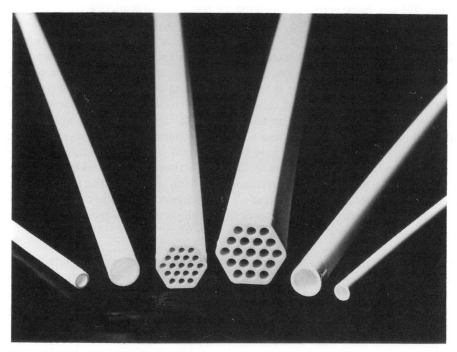

Figure 5.1 Photograph of alumina membrane elements in tubular and monolithic honeycomb shapes [Courtesy of U.S. Filters]

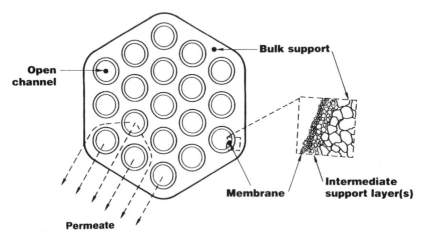

Figure 5.2 Schematic diagram showing permeate passage through a multi-layered, multi-channel membrane element [Hsieh et al., 1988]

Inorganic membranes in the form of hollow fibers give an even higher packing density. The technology of making hollow fiber ceramic membranes is available for producing silica-rich glass membranes [Beaver, 1986; Hammel et al., 1987a and 1987b]. Commercial hollow fiber inorganic membranes, however, are not available currently. To achieve a high membrane area per unit volume, the organic membrane industry has developed and widely used another membrane element shape. It is called spiral wound which, although having a ratio of filtering area/volume roughly only one-tenth that of the hollow fiber shape, in many cases achieves comparable separation performance. A porous, hollow-fiber ceramic tubule (having an outside diameter of 1.5-4.6 mm) spirally wound to form a self-supporting ceramic membrane module has been developed [Anonymous, 1988]. The membrane materials can be alumina, mullite, cordierite or a combination of these materials. Due to the complexity in construction, spiral-wound inorganic membranes have not been commercialized.

Very often a compromise has to be made for the highly compact (high area/volume ratio) membrane elements. They generally require more pre-treatments to lower the risk of fouling the membranes.

5.2 MEMBRANE MODULE

Large-scale industrial applications often require that a membrane module consist of multiple membrane elements, as many as a thousand tubes or a hundred honeycomb-like monoliths each of which may contain many channels. An example is an alumina membrane module consisting of 36 elements with 19 channels in each element. New product lines move towards smaller open channel diameters and more channels in an element. A schematic side-view of a membrane module consisting of several membrane elements is given in Figure 5.3. An important operational advantage of a module having multi-channel monolithic elements is that installation and maintenance become much easier. The membrane tubes or monoliths are bundled to form the multi-element module by the use of header plates at the ends and, in some cases, in the middle of the module length. Sometimes the bundles of membrane tubes or monolithic elements are reinforced by a few longitudinal tie rods. Some commercial membrane modules consist of multiple membrane tubes. Figure 5.4 shows photographically how the membrane elements (in this case, tubes) are assembled to form modules. In some cases, the number of membrane tubes in a module is even on the order of a hundred.

5.3 MEMBRANE SYSTEMS

A membrane module may be a stand-alone loop which may be operated in batch or continuous mode. Often a few membrane modules are combined in parallel or in series, sometimes even in a single vessel, to become a continuous loop or stage. When a system of multiple stages or loops are required to achieve the target permeation rate and final concentrations, the stages or loops may be configured in parallel or in series, as shown in Figure 5.5 to provide maximum capacity with the minimum number of loops.

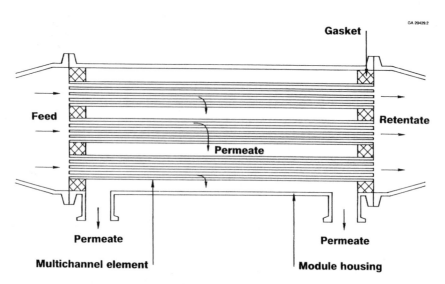

Figure 5.3 Schematic side-view of a membrane module consisting of several membrane elements [Hsieh et al., 1988]

Figure 5.4 Photograph displaying how multi-tube modules are assembled [Courtesy of NGK]

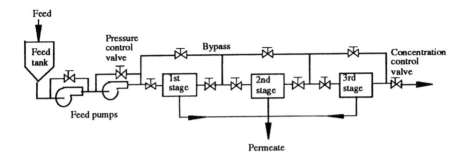

Figure 5.5 Example of a multi-stage or multi-loop membrane system

Figure 5.6 Photograph of a production-scale alumina membrane system [Courtesy of U.S. Filter]

In addition to membrane modules, various types of processing equipment may be required. Examples are feed and recirculation pumps, control valves, heat exchangers, and backflushing systems to be discussed later in this chapter. Shown in Figure 5.6 is a photograph displaying a production-scale membrane system using alumina membranes.

5.4 COMMERCIAL MEMBRANES

Both the number and the variety of inorganic membranes commercially available have multiplied in recent years. The majority of the commercial inorganic membranes have been highlighted in Table 5.1 where the chemical and physical natures and the shapes of their elements are summarized. As the inorganic membrane market is still evolving, the participating companies change hands rather frequently. This is largely due to the very competitive nature of the separation business and the realization by some inorganic membrane manufacturers that to be effective in market penetration in that business requires close integration of manufacturing, engineering design, distribution network and technical service. For reference purposes, the names and addresses of many current manufacturers of inorganic membranes are listed in the Appendix.

Although porous glass membranes have been around for some time, alumina membranes are finding more uses than other inorganic membranes. Zirconia membranes are also receiving much attention. Porous metal membranes are available and are used to less extent due to their unit costs. New inorganic membranes such as titania membranes are emerging.

Four parameters related to the membrane, feed stream and operating conditions determine the technical as well as economic performance of an inorganic membrane system. They are the transmembrane flux, permselectivity, maintenance of the permeating flux and permselectivity over time and stability toward the applications environment. These parameters are the primary considerations for all aspects of the membrane system: design, application, and operation.

5.5 DESIGN CONSIDERATIONS

Separation performance of inorganic membranes largely depends on their material and engineering designs. Engineering design to be discussed will focus on three aspects: the crossflow configuration, available membrane filtration area per unit volume of the module and backflushing design to maintain the membrane flux. Material design concerns the material and methodology for end sealing and module assembling.

Moreover, from the packaging and maintenance perspective, an efficient inorganic membrane system in an industrial environment should possess the following characteristics:

- an element and module with high filtration area per unit volume
- simple assembly and maintenance procedures

- the capability of regenerating effective permeation area

5.5.1 Crossflow Configuration

As mentioned in Chapter 1, there are two major operating configurations of a membrane process: through-flow (or dead-end) and crossflow (see Figure 1.1). In the dead-end or through-flow filtration, there is no exit for the retentate, the rejected species (particles, macromolecules or colloids) keep accumulating on the feed side near the membrane. In most of the pressure-driven membrane separation processes, however, the crossflow configuration is preferred. This applies to both organic and inorganic membranes. In crossflow membrane systems, the feed stream flows parallel to the membrane surface. This parallel flow results in shear forces and in some cases turbulence that help sweep away particles or colloids loosely blocking the membrane pores. Thus crossflow is considered to be a more effective fluid management strategy than the through-flow mode.

The permeate is continuously removed through the membrane from the entrance to the exit of the flow channel of the membrane (in frame-and-plate, tubular or multichannel monolithic elements) as the process stream flows across the membrane surface. Consequently the tangential velocity decreases along the length of the membrane. The velocity at the entrance, however, normally differs from that at the exit only by a small amount since in one pass the amount of permeate removed is relatively small. The operating pressure determines the tangential velocity.

To increase the tangential velocity which generally also increases the transmembrane flux, the pressure will need to be increased but its magnitude should not exceed the recommended maximum operable pressure of the membrane element. Inorganic membranes in general are mechanically stronger than organic counterparts; nevertheless, they are subject to some limits such as burst strength and other mechanical properties (see Section 5.6.2 Mechanical Properties of Membrane Elements) which depend on the material and its microstructures. Another way of increasing the tangential velocity is to decrease the cross sectional area of the flow channel. Typically the inside diameter of the flow channel is in the range of a few millimeters or smaller. Decreased channel diameter, however, increases the pumping energy required. Thus the design choice of a suitable channel diameter is a trade-off between the tangential or crossflow velocity (and therefore the transmembrane flux), the energy required and the mechanical strength consideration.

The design for the majority of the industrial applications of crossflow inorganic membranes system contains all or most of the basic elements shown in Figure 5.7.

The feed stream enters the membrane system which, in one pass, separates the feed into two streams: the permeate and retentate (or concentrate depending on the applications). Very often the separation efficiency is not sufficient to raise the concentrate of the retentate to the economically acceptable level. Therefore, part of the retentate is recycled to be processed again. Sometimes the stream temperature increases appreciably as a result of recirculation, for example, when processing viscous solutions or slurries, and, when necessary, heat exchangers may be used to maintain the system temperature particularly for those applications where the processed material is heat sensitive. An important feature of many inorganic membranes (e.g., ceramic and metal membranes) is their ability to be regularly cleaned by backflushing as part of the operational process. This will be explained in some details later.

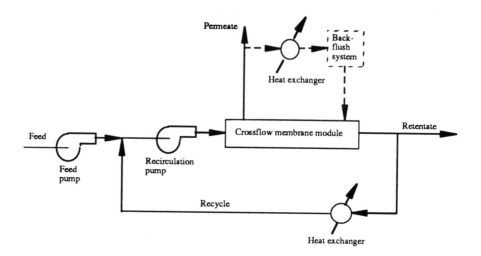

Figure 5.7 Basic crossflow membrane system

It is understandable that many types of fluid metering, indicating and controlling equipment are an integral part of the membrane system. This is one of the reasons why the investment cost of the membrane module is not a predominant fraction of the total capital costs.

5.5.2 Packing Density

An important distinction among various module shapes is the wetted (or available filtering) area per unit volume of the membrane module called membrane packing density. A higher packing density has more specific filtration area and means less capital equipment requirement. It is also likely to incur lower operating cost due to less maintenance to attend. Given in Table 5.2 are the packing density of various geometries of organic membrane modules in general [Leeper et al., 1984]. Other process-related characteristics such as typical feed channel diameters, method of membrane replacement, replacement labor, concentration polarization tendency and typical levels of suspended solids build-up are also included for general comparison.

Shown in Table 5.3 is the packing density of the commercially available inorganic membrane modules. The single plate geometry has a very low packing density. Single tubes typically also show a low packing density unless the membrane tube and the module both have small diameters which are not practical other than for laboratory

TABLE 5.2
Comparison of process-related characteristics of various module configurations of organic polymer membranes

Characteristics	Module type			
	Plate & Frame	Spiral-Wound	Tube-in-shell	Hollow-fiber
Packing density (m^2/m^3)	200-400	300-900	150-300	9,000-30,000
Feed channel diameter (mm)	5	1.3	13	0.1
Method of replacement	As sheets	As module assembly	As tubes	As entire module
Replacement labor	High	Medium	High	Medium
Concentration polarization	High	Medium	High	Low
Suspended solids buildup	Low to medium	Medium to high	Low	High

(Adapted from Leeper et al. [1984])

TABLE 5.3
Packing densities of modules containing various inorganic membrane element geometries

Membrane element geometry in module	Packing density (m^2/m^3)	Note
Single plate	30-40	
Single tube	35-280	High values only for lab units
Shell-and-tube	120-300	
Multi-channel monolith (single element in a module)	300-540	
Multi-channel monolith (multiple elements in a module)	100-200	

evaluation purposes. An improvement over the geometry of a single plate or a single tube is a bundle of parallel tubes held together by headers or spacers at both ends and possibly in-between. Like the typical shell-and-tube heat exchanger arrangement, this geometry has been used in some applications. In extreme cases, the number of tubes in the shell-and-tube geometry can exceed a thousand. The packing density employed in most of the inorganic membrane modules of the shell-and-tube type lies in the same range as the organic membrane type, 150 - 300 m^2/m^3. In contrast to organic polymer membranes, no commercial inorganic membranes are currently available in the high-density geometries such as spiral wound and hollow fiber yet. However, monolithic membrane elements of the honeycomb geometry fill the gap between those two high packing-density configurations and the tube-in-shell shape. Their packing density for modules containing single membrane elements is in the range of 300 - 450 m^2/m^3. However, as the number of monolithic membrane elements increases in a module, particularly when they are not closely packed, the packing density tends to be lower in the 100 - 200 m^2/m^3 range.

5.5.3 Backflushing

Crossflow configuration is a good fluid management technique to maintain permeate flux in the majority of membrane filtration applications. However, the permeate flux is limited by a layer of particles, colloids or macromolecules that deposit on the membrane surface, rather than by the resistance to flow through the membrane itself. This built-up layer makes the permeate flux diminishing with time and is called fouling. For some microfiltration applications, the reduction in the permeation rate can be severe enough to undermine the inherently higher permeability associated with MF membranes. The flux decline can often be quite severe during the initial filtration period. The initial flux decline can be described by the following equation [Redkar and Davis, 1995]:

$$J = J_o (1+t/\tau)^{-0.5}$$ (5–1)

where J and J_o are the flux with and without fouling, t the elapsed time from the beginning of the filtration cycle and τ the time constant related to the growth of the foulant and can be measured by the procedure outlined by Redkar and Davis [1993]. The above equation is valid only until t is equal to t_s when the flux reaches a steady state value of J_s. Figure 5.8 illustrates actual deterioration of the permeating flux of an alumina membrane unit with time due to fouling [Cumming and Turner, 1989]. Dornier et al. [1995] applied the neural network theory to dynamically model the rate of membrane fouling as a function of time for a raw cane sugar syrup feed stream. The effects of both constant and variable transmembrane pressure and crossflow velocity were successfully modeled.

Some more drastic measures than increasing crossflow velocity need to be taken to help reduce the flux decline over an economically acceptable period of operating time. One such technique frequently adopted in many porous inorganic membrane systems today is backflushing. This unique feature design leads to a more constant throughput rate of the permeate and a higher energy efficiency particularly for those cases suffering from fouling. This in-process regeneration technique is also called backpulsing, backwashing or blow-back from the permeate side of the membrane.

Several variations of backflushing exist and they all involve temporary reverse flow of either the permeate or air. Thus, the permeate can be delivered through the membrane from the permeate side to the feed side by applying a momentarily higher pressure on the permeate side or suction on the feed side. Alternatively, air can be pushed through the membrane in a similar way. Matsumoto et al. [1988] have concluded from a study on the separation of yeast that the reverse flow of permeate by either pressurizing or suction is more effective than the use of air. The pulse type of reverse permeate flow is more effective than the steady reverse flow type. In this flux enhancement method generally employed in microfiltration, a counter pressure is applied on the permeate side of the membrane to deliver controlled pulses of permeate back through the membrane pores in the opposite direction to the normal flow of permeate. This backpulsing often dislodges the deposits on the membrane surface and the loosened deposits are then carried away by the retentate flow. The process of backflushing can be schematically represented in Figure 5.9. The dashed curve indicates the flux behavior without backflushing while the solid curve represents the results of applying several reverse filtration or backflushing. The reverse direction of the backpulse is reflected in a negative flux. Within each backflushing cycle, the flux will start to decline and a fresh backpulse is imposed to restore the flux level.

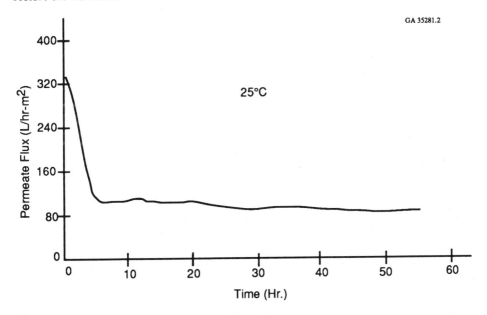

Figure 5.8 Flux decline of a porous alumina membrane system with time due to fouling [Cumming and Turner, 1989]

Two of the important design parameters of a backflushing system are the frequency and duration of the pulses employed. Galaj et al. [1984] has developed a simple mathematical model to describe membrane fouling by particles and membrane regeneration by backflushing. They assume that the particles of a uniform size in a suspension enter and clog up the pore entrances on the surface of the membrane first to form a single layer of particles before all the pores are covered after which particles can

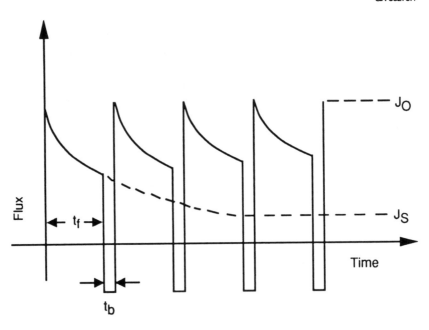

Figure 5.9 Flux vs. time for crossflow membrane filtration with backflushing (solid curve) and without backflushing (dotted curve) [Redkar and Davis, 1995]

form multiple layers. They also assume that particles can be expelled from the surface by backflushing. It can be shown that the reduction of permeate flux is given by

$$\Delta F = 1-(F/F_o) = \Omega/\{1+ n(1 - \Omega)\} \qquad (5\text{-}2)$$

which is depicted in Figure 5.10. The number of layers of fouling particles is described by n. The permeate flux decreases slowly as a small number of pores are blocked (i.e., when Ω is small) but drastically when the fraction of pores blocked is high (i.e., large Ω). It has also been shown that the membrane is totally regenerated (i.e., all pores are open) when

$$\Delta P_B/\Delta P_T = (n+1)/\{1+n(1- \Omega)\} \qquad (5\text{-}3)$$

where ΔP_B and ΔP_T are the pressure difference from the permeate to the feed side during the backflushing cycle and the operating transmembrane pressure difference during the normal cycle, respectively.

From Eq. (5-3) it is clear that, when the membrane pores are completely blocked (i.e., $\Omega=1$),

$$\Delta P_B = \Delta P_T (1 + n) \qquad\qquad (5\text{-}4)$$

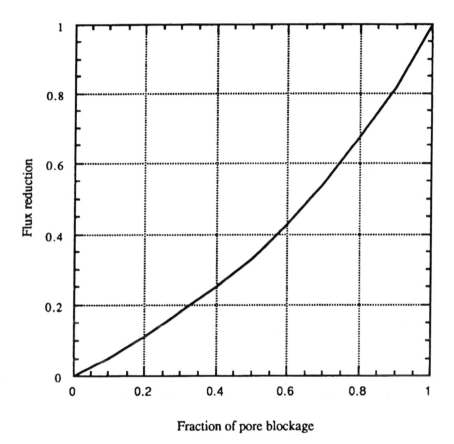

Fraction of pore blockage

Figure 5.10 Flux decline as a function of the fraction of membrane pores blocked by a monolayer of particles

which indicates that the pressure required for backflushing to dislodge particles can be several times that of the transmembrane pressure difference and increases with increasing number of particle layers accumulated on the membrane surface. A large n or

Ω implies less frequent pulses of backflushing. According to Eq. (5-3), this causes the required backflushing pressure to be higher. Therefore, an effective strategy of design and operation of backflushing should be to apply more frequent short high-pressure pulses. This has been supported by experimental data of yeast separation by an alumina membrane with a nominal pore diameter of 1.6 μm (Figure 5.11). It is evident that as the backflushing interval decreases (or as the frequency increases), the resulting permeate flux improves [Matsumoto et al., 1988].

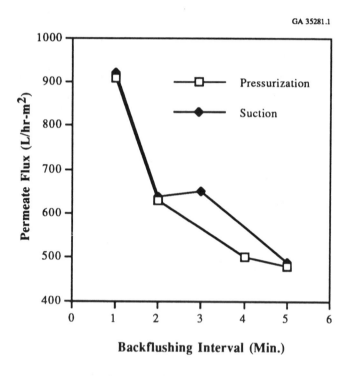

Figure 5.11 Effect of backflushing interval on permeate flux [Matsumoto et al., 1988]

An example of the details of a backflushing system is shown in Figure 5.12. It uses an air-driven piston mechanism mounted on the permeate port of the membrane housing. At the end of the period of the backflushing cycle, the mechanism is activated. The first segment of the piston stroke closes the permeate port, followed by the second segment which drives the permeate trapped between the piston and the membrane backward through the membrane as a result of the higher piston pressure than the membrane tube side pressure.

In practice, air pressure or pumps can be used for generating backpulses. Air pressure may be preferred since pumps sometimes generate particles which can possibly plug membrane pores on the permeate side [Norton Co., 1984]. The backflushing system contains a timer for setting the frequency and duration of the pulses, a controller, valves, backpulse override switch and a backpulse reservoir. Normally the permeate is used for backpulsing. Typically the amount of permeate used in the backpulses is small. Therefore, the reservoir capacity does not need to be large. The reservoir interface can be simply air or water contact (Figure 5.13). If the direct contact of the permeate with air or water dilution is not desirable, a flexible interface diaphragm can be used or the air may be replaced with inert gases.

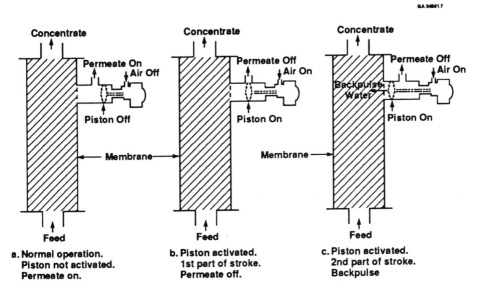

Figure 5.12 Details of a backflushing system using piston backpulses [Lahiere and Goodboy, 1993]

Although most of the discussions on backflushing so far use the crossflow configuration as an example, backflushing also can be and has been employed in through-flow membrane separation applications.

5.5.4 Membrane End Sealing

The porosity at both ends of a tubular or monolithic honeycomb membrane element can be a potential source of leakage. These extremities need to be made impervious to both permeates and retentates so that the two streams do not remix. Typically the end surfaces and the outer surfaces near the ends of a commercial membrane element are coated with some impervious enamel or ceramic materials.

5.5.5 Module Assembly

Two types of connection (and therefore sealing) are involved in assembling membrane modules. The first type connects tubular or monolithic membrane elements in bundles using header plates at the ends and, in some cases, in the middle of the module length. The second type provides sealing between the plates and the module housing.

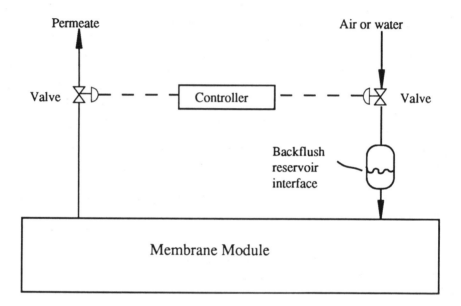

Figure 5.13 Schematic diagram of an air- or water-actuated backflushing reservoir system

The element-plate connection in commercial inorganic membranes often calls for the use of gaskets or potting materials. Commercial porous glass membranes are usually potted with silicon rubber in a housing made of a borosilicate glass. The potting compound is biocompatible and can withstand steam sterilization at a temperature at or below 135°C. Most of the commercial ceramic membranes today come equipped with gaskets made of silicon, neoprene, Buna-N (a butadiene-acrylonitrile copolymer), ethylene propylene, EPDM (a terpolymer elastomer made from ethylene-propylene diene monomer) and a family of fluorinated polymers. Some of the fluorocarbon elastomers are Viton (a series of fluorocarbon elastomers based on the copolymer of vinylidene fluoride and hexafluoropropylene or tetrafluoroethylene), PTFE (polytetrafluoroethylene), and Kalrez (a fluorocarbon elastomer based on the copolymer of tetrafluoroethylene and perfluoro(methyl vinyl ether)). The fluorocarbon elastomers are superior to hydrocarbon elastomers in their thermal stabilities. The above elastomeric or polymeric gaskets are particularly suitable for sealing the gap between ceramic membrane elements and metal housing when the application requirements call for an ultraclean (e.g., microelectronics and food and beverage uses) and easy to change system to be operated at a low temperature (<100°C).

For those applications where steam sterilization and processing temperatures are below 130°C, the element-to-plate sealing often calls for the use of a one-piece elastomer as the sealing ring. In cases where there is appreciable temperature rise or sufficient compression on the sealing material by the plates that sandwich the sealing material, the elastomer tends to be "extruded" or squeezed in the direction of the membrane length into clearances formed between the membrane element and the plates. This accelerates the deterioration of the sealing material and, consequently, the loss of the fluid-tightness. A design solution to this problem is to use a three-layered sealing ring in place of the one-layered elastomer. Alternatively, the elastomer has a stratified structure with varying properties in its thickness. In either approach, the central idea is to use a softer layer in the middle and two harder layers sandwiching the soft layer [Bathelet, 1994]. The two harder outer layers prevent the softer intermediate layer from deforming into the clearances, thus preserving the physical integrity of the sealing layer.

The connections between the elements and the plates in special cases can also be made by chemical reaction. For example, carbon membrane tubes can be connected to carbon sheets through cross linking at high temperatures and circumvent the need for gaskets.

The plate-to-housing connection is made by the use of gaskets or "O-rings". The considerations for the choice of gasket materials are the same as those for the element-to-plate connection.

The module housing or casing is usually made of plastics such as PVC, CPVC, PTFE or polypropylene, 304 stainless steel, PTFE-lined steel or, in cases where sanitary design is required, 316L stainless steel. The quality of the finish can be mill finish or AAA sanitary. The fittings for the housing are available in three major types: sanitary, flanged or threaded fitting. Sanitary grade fittings call for the use of clamps while the industrial grade uses flanges or threads. Threaded connections which provide a positive seal without the need for gaskets usually can be used in a broader range of temperature and chemicals.

5.6 APPLICATION CONSIDERATIONS

Besides higher unit costs, in both initial investment and replacement, another factor limiting inorganic membranes from wider usage is the control of their pore size distributions. Recent advances, however, such as the sol-gel process and anodic oxidation have started to be implemented in commercial production and have made significant impacts on the current and future market shares of the inorganic membranes.

It is well known that the mean pore diameter of a membrane determines its applicability in the separation spectrum. Microfiltration is generally considered to cover the approximate range of 0.1 to 10 μm while ultrafiltration spans from 0.1 μm down to about 3 nm. Ceramic membranes, for example, started their industrial applications in microfiltration and gradually expand toward the ultrafiltration market with new commercial membranes made by the sol-gel and anodic oxidation processes. The market pull for even finer pore membranes is evident as the potentials for high-temperature gas separation and membrane reactor applications are enormous. At the same time, the general expectation for these inorganic membranes to be commercialized in a few years has been quite high. Whether this challenge can be met in the time frame hinges on close interactions between the membrane manufacturers and the end users.

5.6.1 Material Selection for Chemical Compatibility

For a given application where the process fluid and temperature are known, material selection is important as it is related to chemical compatibility not only between the process fluid and the membrane but also between the fluid and any gasket, header plate and housing materials used. While the possible combinations are endless, useful guidelines can be developed from compiling manufacturers' and literature data. Some general guidelines are given in Table 5.4 where most of the data are based on static exposure at room temperature for 48 hours.

Included in the table are some porous inorganic membrane materials as well as the materials for their gaskets, header plates and housing. A material is considered to be resistant toward a chemical when essentially no change is observed in the flux or the bubble point of the membrane. "Not resistant" means that the membrane is not stable upon its exposure to the chemical and appreciable change in its permeability occurs. The membrane shows limited resistance when only minor changes in the separation properties of the membrane are associated with short term exposure to the chemical. The information provided in the table only serves as a guide and it should be emphasized that chemical compatibility can be influenced by many other factors such as time, temperature, pressure, condensation, etc. Testing under actual conditions is often required to assess the long term chemical stability of the membrane. Table 5.4, however, provides overall information on the potentials of inorganic membranes. Their generally good resistance to chemicals make them candidates in those applications involving a wide variety of acids, alkalines, solvents, detergents and many other hydrocarbon compounds during applications or treatments between applications.

There are special applications where it is required that essentially no dissolution or reaction takes place between the membrane/module material and the process stream. One such application area is food and beverages. For these uses, not only the membrane material but also the housing and gasket materials need to pass certain tests for sanitary reasons (e.g., FDA approval). Stainless steel (especially the 316L type) is typically used as the casing material and fluorinated polymers, EPDM, silicon or other specialty polymers are selected as the gasket materials for assembling modules. To comply with regulations, no material comprising of the membrane element or module can be extracted or leached out into the process stream beyond a certain limit (e.g., 50 ppb in some cases).

TABLE 5.4
Chemical compatibility of construction materials of inorganic membrane, gaskets, header plates and housing with various chemical reagents

Chemical reagent	α-Al_2O_3	γ-Al_2O_3	Glass	ZrO_2	C	Ag	S.S.	Teflon	Buna N	Viton
HCl (conc.)	LR	NR	R	LR	R		NR	R	R	R
HCl (dil.)	R		R	LR	R		LR	R	R	R
HF	LR-NR	NR	NR	LR-NR	LR					
HNO_3 (conc.)	R	R	R	LR	NR	NR	R	R	NR	R
HNO_3 (dil.)	R		R	LR	LR	NR	R	R	NR	R
H_3PO_4 (conc.)	LR-R			LR	R		R	R	LR	R
H_2SO_4 (conc.)	LR	NR	R	NR	LR	NR	NR	R	NR	R
H_2SO_4 (dil.)	R		R	LR	R	NR	LR	R	NR	R
NH_4OH	R						R	R	NR	LR
KOH	R			R			LR	R	R	R
Na_2CO_3	R				R		R	R	R	R
NaOH	R	NR		R	R		R	R	R	R
Benzene	R	R				R	R	R	NR	R
Toluene	R	R				R	R	R	NR	R
Xylene	R	R				R	R	R	NR	R
Carbon tetra-chloride	R	R					R	R	R	R
Chloro-form	R	R					R	R	NR	R
Methyl-ene chloride	R	R					R	R	NR	R
Trichloro-ethylene	R	R					R	R	LR	R

TABLE 5.4 – CONTINUED
Chemical compatibility of construction materials of inorganic membrane, gaskets,
header plates and housing with various chemical reagents

Chemical reagent	α-Al_2O_3	γ-Al_2O_3	Glass	ZrO_2	C	Ag	S.S.	Teflon	Buna N	Viton
Acetone	R	R			R		R	R	NR	NR
Cyclo-hexanone	R	R					R	R	NR	NR
Methyl ethyl ketone	R	R					R	R	NR	NR
Ethanol	R	R					R	R	R	LR
Iso-propanol	R	R					R	R	R	R
Methanol	R	R					R	R	R	NR
Ethyl acetate	R	R					R	R	NR	NR
Isopropyl acetate	R	R					LR	R	NR	NR
Tetrahy-drofuran	R	R					R	R	NR	NR
Ethylene glycol	R	R					R	R	R	R
Glycer-ine	R	R					R	R	R	R
Dimethyl forma-mide	R	LR					R	R	R	NR
Formal-dehyde	R	LR					R	R	LR	NR
Hexane (dry)	R	R					R	R	R	R
Phenol	R	LR					R	R	NR	R

Note: R=resistant; LR=limited resistant; NR=not resistant; S.S.=stainless steel; conc. =
concentrated; dil. = diluted

5.6.2 Mechanical Properties of Membrane Elements

Most inorganic membranes are mechanically stronger than their organic counterparts.
However, installation, repair and maintenance, mechanical vibration, pressurization and
other application conditions sometimes can cause problems. This is especially true with

ceramic membranes. Therefore, considerations of mechanical properties of inorganic membranes are vital to ensure crack-free and trouble-free operation. Quantitative information, however, is scarce and not yet standardized even for commercial membranes. Most of the methods used to arrive at limited available mechanical properties of inorganic membranes, particularly the porous type, are based on the ASTM procedures for nonporous shaped solids.

Based on available product brochure information of commercial membranes and the literature data of developmental membranes, Table 5.5 provides some indication of the various mechanical properties of a few inorganic membrane elements. It should not be used for serious comparison as the testing conditions are usually not given or sketchy at best. In addition, the mechanical properties of a membrane element depend, to a great extent, on its shape, membrane thickness, porosity and pore size. While overall the majority of the data appears to be consistent, there is some strength data that seems to be off the trend line.

TABLE 5.5
Mechanical property data of various inorganic membrane elements

Bulk material (porous)	Element shape	Bend. S. (bar)	Burst S. (bar)	Comp. S. (bar)	Reference
α-Al$_2$O$_3$	Honeycomb monolith	50-450	>60-100		
α-Al$_2$O$_3$	Tube		>30		
α-Al$_2$O$_3$	Tube (30-40 % porosity)	250-350	70	4,000-5,000	Auriol & Gillot, 1985
α-Al$_2$O$_3$/TiO$_2$	Honeycomb monolith		50		
Cordierite (high porosity)	Honeycomb monolith			80-330	
Carbon	Tube		60		Veyre, 1986
Glass	Tube	3-300	40	2,050	Schottglas
SiC		180			Ibiden Co., 1986
Silver	Disk			1,000	Jordan and Greenspan, 1967
Stainless steel	Tube (15-45 % porosity)		150-400		Boudier & Alary, 1989

Note: Bend. S.= Bending strength; Burst S.= Burst strength;
Comp. S.=Compressive strength

The effect of porosity on the mechanical properties of ceramic or metallic bodies are complex. It not only decreases the cross sectional area to which the load is applied but also acts as a stress concentrator. Details such as the pore shape and orientations can significantly influence the mechanical properties. It is generally expected that the ceramic strength decreases with increasing porosity. The form of the dependency, however, may vary from one mechanical strength to another. In general, the rate of decline in strength has been found experimentally to be higher at a lower porosity. An empirical relationship describing such a behavior has been developed [Ryskewitsch, 1953]:

$$\sigma = \sigma_o \exp(-n\, \emptyset) \qquad\qquad\qquad\qquad\qquad (5\text{-}5)$$

where σ and σ_o are the strengths of the porous and dense parts of the same ceramic material, respectively, and the constant n ranges from 4 to 7 and $\emptyset$ is volume fraction of the porosity. It appears that this equation is applicable to at least some commercial ceramic membranes.

It should be cautioned that, for a given membrane porosity, it is entirely possible that two membranes can have different pore sizes or pore size distributions which, in turn, can exhibit varying strength properties. Given in Table 5.6 are two mechanical properties of homogeneous porous stainless steel membranes as functions of the membrane pore diameters. It is evident that, as the pore size increases, both the yield strength and the minimum ultimate tensile strength deteriorate drastically. The rate of decline is particularly pronounced for pore diameters less than 5 μm. Therefore, the use of Eq.(5-5) should be limited to crude estimation in the absence of test data and the potential interdependence of pore size and porosity can not be overlooked.

TABLE 5.6
Yield strength and minimum ultimate tensile strength of porous stainless steel membranes as functions of membrane pore diameter

Pore diameter (μm)	Yield strength (bar)	Minimum ultimate tensile strength (bar)
0.2	1,795	2,070
0.5	1,450	1,620
2	910	1,220
5	635	920
10	520	725
20	395	485
40	240	310
100	200	270

(Adapted from Mott Corp. [1991])

5.7 OPERATING CONSIDERATIONS

Successful performance of inorganic membranes depend on three types of variables and their interactions. The first type is related to the characteristics of the feed stream such as the molecular or particulate size and/or chemical nature of the species to be separated and concentration of the feed to be processed, etc. The second type is membrane dependent. Those factors are the chemical nature and pore size of the membrane material and how the membrane and its accessory processing components are constructed and assembled. The third type is processing conditions such as pressure, transmembrane pressure difference, temperature, crossflow velocity and the way in which the membrane flux is maintained or restored as discussed earlier in this chapter.

There are certain general operating considerations that are common in many applications. These considerations will be discussed here and leave those factors specific to individual applications in Chapter 6.

5.7.1 General Effects of Operating Variables

Among the operating parameters of a membrane system, the fluid temperature, transmembrane pressure (TMP) and feed concentration play important roles in determining the permeate flux.

Transmembrane pressure. The driving force for membrane permeation is the transmembrane pressure or transmembrane pressure difference (Δp_T):

$$\Delta p_T = (p_1 + p_2)/2 - p_3 \qquad (5\text{-}6)$$

where p_1, p_2 and p_3 are the inlet (feed) and outlet (retentate) pressures on the feed side and the permeate side pressure, respectively (Figure 5.14). When only a pure solvent (e.g., water) is filtered, the permeate flux is essentially a linear function of TMP. On the other hand, when a fluid-fluid or fluid-solid mixture is processed, the permeate flux of the solvent is lower than that in the case of filtering the pure solvent due to concentration polarization. Nevertheless, the permeate flux increases nonlinearly with increasing TMP. However, as the TMP becomes large and exceeds some threshold value, the permeate flux is more or less independent of the TMP. The region where the flux is independent of the TMP due to gel formation or fouling is called the mass-transfer controlled region. The other region where the flux is an increasing function of the TMP is called pressure-controlled region. This is schematically illustrated in Figure 5.15. In severe cases, the permeate flux may even begin to decline with further increase in the TMP.

If the pressure drop along the membrane length is large such that the inlet and outlet pressures p_1 and p_2 on the feed side are significantly different, the arithmetic mean should be replaced by the logarithmic mean.

Temperature. The effect of operating temperature on the permeate flux is usually significant due to the change of fluid viscosity with temperature. In liquid-phase processing, the permeate flux increases as the temperature increases primarily due to viscosity or solid solubility. Liquid viscosity generally decreases as the temperature increases. Furthermore, the solubility of suspended solids also typically increases with increasing temperature.

GA 35292.1

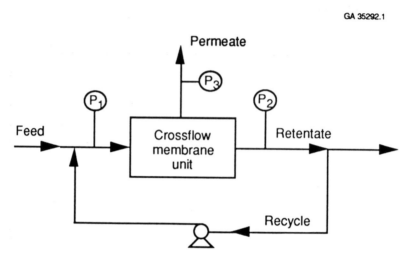

Figure 5.14 Schematic diagram of a simple crossflow membrane system

On the contrary, the viscosity of gas at low density increases with increasing temperature. Therefore, as the temperature rises, the gas permeation rate becomes lower.

While a higher operating temperature may be desired for liquid-phase membrane separation, the associated higher energy cost needs to be balanced with the savings in pumping energy or a higher throughput.

Feed concentration. As expected, the higher the feed concentration is, the higher propensity of mass-transfer limiting phenomena such as concentration polarization, gel formation or fouling becomes. The resulting permeate flux is lower. A higher feed concentration also raises the viscosity.

Perhaps the most important operating consideration related to membrane processes is how to enhance, maintain and restore membrane flux (or permeability) and permselectivity. They can be approached by the following techniques.

5.7.2 Feed Pretreatment

There are many reasons why in some cases the feed stream needs to be pretreated prior to entering the membrane system. When the membrane has great fouling potential with respect to the feed components, or when the membrane pore size is close to or significantly larger than the size of the species to be retained, both the permselectivity and permeability may be compromised.

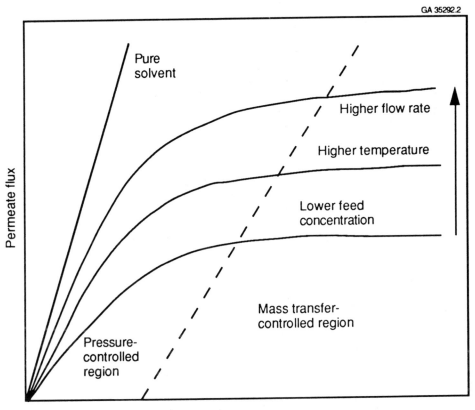

GA 35292.2

Figure 5.15 General trends of effects of operating variables on permeate flux

Several pretreatment techniques have been practiced in conjunction with the use of organic polymeric membranes. The choice of a particular technique to apply depends to a great extent on the chemical nature of the feed, membrane resistance toward the feed stream and the product requirements. The characteristics of the feed such as the particle size or molecular weight distribution, effective fluid viscosity and pH (and therefore the zeta potential as discussed in Chapter 4) are the primary factors involved in developing an effective pretreatment procedure. It is quite common to introduce chemical additives such as calcium hydroxide, alum and polyelectrolytes into the feed stream before the membrane filtration step.

Many times the prime objective of pretreatment is to produce a fluffy and cohesionless floc and thereby considerably reduce fouling of membrane pores. Thus, in those applications involving particulate separation, flocculants may be added to the feed stream prior to membrane filtration to form larger particles for ease of filtration and sometimes for improving the permselectivity. It is desirable that these large particles retain their physical integrity under the prevailing operating conditions including high shear rates.

In other applications such as oil-water separation and electrocoat paint recovery, it may be necessary to adjust the feed pH to prevent precipitation or adsorption of certain species on the membrane surface or into the pores.

5.7.3 Crossflow Velocity

As most of the membrane separation processes are operated in the crossflow mode for the reasons discussed earlier (Section 5.5.1 Crossflow Configuration), the crossflow velocity has marked effects on the permeate flux. A higher crossflow velocity typically results in a higher flux. The rate of flux improvement as a function of the crossflow velocity usually can be described by the following equation with specific units:

$$Log F = m_{cf}(Log v_{cf}) + q_{cf} \tag{5-7}$$

where F is the flux (L/hr-m^2) and v_{cf} the crossflow velocity (m/s). The constants m_{cf} and q_{cf} are the slope and the intercept of the straight line in the log-log plot of F vs. v_{cf}. The value of the slope m_{cf} appears to have a value of near 1.0 with a range of 0.8 to 1.2 from various sources for different applications.

Although there are no clear-cut guidelines as to the recommended crossflow velocity for inorganic membranes to be used in general applications, Lajiere and Goodboy [1993] suggested without specific reasoning a minimum crossflow velocity of 0.9 m/s for microfilters and 4.6 m/s for ultrafilters. Generally speaking, a crossflow velocity of 5 to 7 m/s is considered high for inorganic membrane systems.

5.7.4 Flux Enhancement

Fouling is one of the most prevalent operational problems associated with microfiltration and ultrafiltration applications. The high mechanical strength and chemical as well as structural stabilities of many inorganic membranes (especially the ceramic types) and their bonding with the underlying support layer(s) allow them to be periodically regenerated from the fouled condition. Membrane regeneration can be achieved by several methods some of which are part of the operation cycle while others are performed during shut-down. Two in-process techniques are discussed here. Detailed information is generally provided by the manufacturers of inorganic membranes.

Fast flushing and flow reversal. One of the simplest methods to dislodge or remove the deposits from the membrane surface is to increase the recycle rate to a level equivalent to a linear velocity greater than the recommended maximum crossflow velocity of, say, 5-7 m/s. This "fast" flush procedure is carried out with the permeate ports closed. At high recycle rates the transmembrane pressure difference varies considerably from the inlet to the outlet of the module with the outlet pressure approaching the atmospheric pressure. This usually causes some backflow of permeate, particularly in the downstream portion of the module. As a result, the upstream may be subject to a relatively greater fouling compared to the remainder of the module. One remedy to this problem is to perform a momentary flow reversal toward the inlet, still in the crossflow mode. Flow reversal can be effective and is sometimes adopted in some microfiltration and ultrafiltration applications such as waste water treatment and food and beverage processing.

Backflushing. By incorporating this mode in the normal operation cycle of the membrane system, membrane fouling can be reduced substantially between consecutive applications of backflushing. The membrane flux can thus be restored periodically. The efficiency of this technique, however, often depends on both the type of suspension to be filtered and the type of fouling that occurs. The frequency and amplitude of the backpulse can affect the sustainable membrane flux. Short and frequent back pressure pulses (e.g., once every 1 to 30 minutes, more often 3 to 5 minutes, for a duration of 0.5 to 2 seconds) help maintain high permeate flux [Galaj et al., 1984]. The backflushing pressure is typically 20 to 40 psi higher than the feed pressure or 2 to 3 times the transmembrane pressure. This key operational feature of backflushing offers one of the main advantages of ceramic and metal membranes for a variety of separation applications where fouling is a constant problem.

Galaj et al. [1984] made a detailed experimental investigation and proposed a model for backflushing on alumina membranes in the crossflow filtration mode. Their model relates permeate flux to a number of key membrane and operational parameters: the number of membrane/support layers, total number of pores, number of pores plugged, pressure loss across the membrane layer and particles concentration in the feed. It was found that during backflushing the recycle flow rate only has a secondary effect as long as it is sufficiently large to prevent redeposition of the rejected particles or species on the membrane surface.

Redkar and Davis [1995] have made a theoretical and experimental study on a high-frequency backflushing system to maximize the global average permeating flux. For every backflushing duration, there is an optimum normal (forward-filtration) operation cycle time that gives the maximum global average flux. The optimum average flux increases with decreasing backflushing duration and feed concentration. The same flux essentially does not depend on the crossflow velocity and the reverse filtration transmembrane pressure.

When operating the backflushing system optimally the resulting flux with rapid backflushing can be 20 to 30 times higher than the long-term flux in the absence of backflushing [Redkar and Davis, 1995].

For the reason of possible blocking of the membrane pores on the permeate side, only permeate or other prefiltered cleaning agent should be used in the backpulses.

Flux enhancement by backflushing can control fouling in inorganic membranes and consequently reduce the operating costs. Most of the cost savings lie in the decrease of the membrane replacement costs [Muralidhara, 1991].

5.7.5 Membrane Cleaning

Generally speaking, all microfiltration and ultrafiltration applications require some form of periodic, disruptive cleaning to remove the foulants and thereby restore the membrane flux close to its initial level. The cleaning solutions and the procedures used industrially vary with the type of fouling encountered.

Chemical cleaning. Prior to chemical cleaning, the membrane element should be separated from the module housing and fittings or gaskets. They should also be degreased or drained of the process fluid. The membrane is then immersed in an appropriate cleaning solution. The cleaned membrane is then rinsed with water. The cleaning cycle may require the use of different solutions with rinsing between different solutions.

As a general guideline, particulate foulants require acid cleaning while oily foulants call for alkaline cleaning agents. Tough protein deposits on the membrane surfaces can be cleaned effectively by the use of suitable enzyme cleaners. In industrial practice, alkaline cleaning may be sufficient. But it is not uncommon to see that sequential applications of concentrated alkaline and acidic cleaning solutions, sometimes alternately, are used to effectively clean severely fouled membranes. In other situations, organic solvents and strong oxidizing agents may be necessary as a precleaning step. The commonly used alkaline and acid solutions are sodium hydroxide and nitric acid, at an approximate strength of 0.5-3 % by weight for ceramic membranes and higher (to 14-17 %) for metal (e.g., stainless steel) membranes. An example of the oxidizing agents is sodium perchloride with 200 to 300 ppm free chlorine. It is generally recommended to minimize the transmembrane pressure drop during cleaning by closing or restricting the permeate outlets.

In cases where a strong oxidant followed by an alkaline solution and an acid solution (can be in either order), are sequentially used in a cleaning cycle, water rinsing at room temperature is often performed between each type of cleaning solution. The recommended temperature for performing the cleaning procedure is in the 50 to 95°C range (lower end for ceramic membranes and higher end for metal membranes) except for strong oxidizing agents which are recommended to be used at room temperature. Many inorganic membranes, however, may undergo chemical as well as microstructural changes resulting from the use of strong oxidizing agents. Alpha-alumina and zirconia membranes are some of the exceptions. Many membrane manufacturers provide the recommended cleaning procedures for their membrane systems.

The choice of the acid type and strength can be crucial to maintain the structural and chemical integrity of the membranes. For example, while silver membranes can withstand strong acids such as 10% hydrofluoric acid, they are not resistant toward nitric or sulfuric acids. On the contrary, hydrofluoric acid is not recommended for alpha-alumina membranes.

Thermal cleaning. Some ceramic membranes such as alumina membranes can also be thermally cleaned by steam (or high pressure water) sterilization or autoclaving [Gillot et al., 1984]. This feature can be important in many bioseparations, pharmaceutical and food applications. Prior to the sterilization or autoclaving treatment, the membranes should be chemically cleaned to remove foulants. Steam or water sterilization can be performed at a temperature of approximately 120-130°C. Heating or cooling in the sterilization or autoclaving treatments should progress slowly (e.g., at a rate of 10°C or lower) to avoid thermal shock. This problem is more pronounced for ceramic membranes.

For those membranes that can withstand temperatures of 250°C or higher, organic foulants, when present in the feed, can be burned off by the ignition cleaning process to restore the membrane surface. The membranes are subject to a high temperature treatment in a furnace. Care, however, should be exercised not to change the membrane pore size and morphology as a result of heating. Considerations should also be given to the thermal stabilities of the gasket materials. For example, when elastomers are used, the maximum operating temperatures for thermal cleaning are limited to approximately 130°C. Silver membranes can be heated in a furnace for ignition cleaning as long as the temperature is below 425°C for membrane pores larger than 1.2 µm and below 205°C for membranes with a pore diameter in the 0.2 to 0.8 µm range [Poretics, 1990].

5.8 SUMMARY

Currently commercial inorganic membranes are dominated by porous ceramic membranes. They are available in the shape of disk, tube or multi-channel monoliths, particularly the monolithic form. The material for the separating membrane layer varies from alumina, zirconia, glass, titania, carbon, stainless steel to other ceramics or metals. Many of them exhibit a multi-layered composite microstructure with each underlying layer having a progressively larger pore size. The bulk support which may be of a different material than the membrane layer provides the mechanical strength and has the largest pore size. The smallest pore diameters of commercial inorganic membranes are about 4 nm.

A membrane system consists of many membrane modules which, in turn, are made of several membrane elements. Both ends of a membrane element are sealed with such materials as enamels or ceramic materials. The connections between elements and between elements and the housing or pipings are typically made from plastics or elastomers for liquid phase applications.

To improve or sustain the separation performance of an inorganic membrane, design and operational factors should be carefully considered. The crossflow configuration with a crossflow velocity of up to about 7 m/s is commonly used to maximize the permeation rate. To slow down flux decline due to fouling, backflushing mechanisms are often installed in the inorganic membrane systems. One of the major advantages of porous inorganic membranes is that they are amenable to backflushing. High-frequency short-pulse backflushing is preferred.

A simplified material selection guide has been provided to determine the general chemical compatibility between the membrane (and its accessory components) and the process stream. The guide is primarily for relatively low temperature applications.

Finally, mechanical properties of a porous membrane can be significantly reduced by the presence of pores in the matrix. Both pore size and porosity have large impacts on the mechanical properties. Available mechanical properties of porous inorganic membranes are scarce and the strength measurement techniques are not specified. More systematically acquired mechanical properties of porous inorganic membranes are needed.

REFERENCES

Anonymous, 1988, North Am. Membr. Soc. Membr. Quarterly 3, 16.
Auriol, A., and J. Gillot, 1985, International Patent 85/01937.
Bathelet, P., 1994, U.S. Patent 5,366,624.
Beaver, R.P., 1986, European Patent 186,129.

Boudier, G., and J.A. Alary, 1989, Poral new advances in metallic and composite porous tubes, in: Proc. 1st Int. Conf. Inorg. Membr., Montpellier, France, p. 373.

Crull, A., 1991, Prospects for the inorganic membrane business, in: Proc. 2nd Int. Conf. on Inorganic Membranes, Montpellier, France, 1991 (Trans Tech Publ., Zürich) p. 279.

Cumming, I.W., and A.D. Turner, 1989, Optimization of an UF pilot plant for the treatment of radioactive waste, in: Future Industrial Prospects of Membrane Processes (L. Cecille and J.C. Toussaint, Eds.), Elsevier Applied Sci., London, U.K., p. 163.

Gallagher, P.M., 1990, Novel composite inorganic membranes for laboratory and process applications, presented at Annual Membrane Technology/Planning Conference, Newton, MA (USA).

Galaj, S., A. Wicker, J.P. Dumas, J. Gillot and D. Garcera, 1984, Le Lait **64**, 129.

Gerster, D., 1986, State of mineral membranes in the world, in: Proc. 4th Annual Membrane Technology/Planning Conf., Cambridge, MA, USA (Business Communications Co.), p.152.

Gillot, J., G. Brinkman and D. Garcera, 1984, New ceramic filter media for crossflow microfiltration and ultrafiltration, presented at Filtra'84 Conf. , Paris, France.

Guyard, C., 1988, Ind. Tech. Special **12**, 32.

Hammel, J.J., W.P. Marshall, W.J. Robertson and H.W. Barch, 1987a, European Pat. Appl. 248,391.

Hammel, J.J., W.P. Marshall, W.J. Robertson and H.W. Barch, 1987b, European Pat. Appl. 248,392.

Hoover, F.W., and R.E. Roberts, 1978, U.S. Patent 4,069,157.

Hsieh, H.P., R. R. Bhave and H. L. Fleming, 1988, J. Membr. Sci. **39**, 221.

Ibiden Co., 1986, Japanese Patent 61,091,076.

Jordan, G., and R.P. Greenspan, 1967, Tech. Quart. (Master Brewers Assoc. Am.) **4**, 114.

Lahiere, R.J., and K.P. Goodboy, 1993, Environ. Prog. **12**, 86.

Leeper, S.A., D.H. Stevenson, P.Y.C. Chiu, S.J. Priebe, H.F. Sanchez and P.M. Wikoff, 1984, Membrane technology and applications: an assessment, U.S. Dept. of Energy Report No. DE84-009000, p. 35.

Matsumoto, K., M. Kawahara and H. Ohya, 1988, J. Ferment. Technol. **66**, 199.

Mott Metall. Corp., 1991, Precision porous metals-engineering guide, Product Brochure DB 1000.

Muralidhara, H.S., 1991, Key Eng. Mat. **61&62**, 301.

Norton Co., 1984, Asymmetric ceramic microfilters -- Testing tubular crossflow modules, Product Brochure.

Poretics Corp., 1990, Microfiltration Products Catalog.

Redkar, S.G., and R.H. Davis, 1993, Biotech. Prog. **9**, 625.

Redkar, S.G., and R.H. Davis, 1995, AIChE J. **41**, 501.

Ryskewitsch, E., 1953, J. Am. Ceram. Soc. **36**, 65.

Veyre, R., 1986, Houille Blanche. **41**, 569.

CHAPTER 6

TRADITIONAL LIQUID-PHASE SEPARATION APPLICATIONS

Commercial as well as potential uses of inorganic membranes multiply rapidly in recent years as a result of the continuous improvement and optimization of the manufacturing technologies and applications development for these membranes. Most of the industrially practiced or demonstrated applications fall in the domains of microfiltration or ultrafiltration. Microfiltration is applied mostly to cases where the liquid streams contain high levels of particulates while ultrafiltration usually does not involve particulates. While their principal separation mechanism is size exclusion, other secondary mechanisms reflecting the solution-membrane interactions such as adsorption are often operative. Still under extensive research and development is gas separation which will be treated in Chapter 7.

Inorganic membranes have been adopted in industrial processing or evaluated for new applications in a number of technology areas. The choice of inorganic membranes can be for different reasons depending on the uses as shown in Table 6.1. So far inorganic membranes have enjoyed the first large-scale commercial success in the food and beverages market. Significant application developments are also occurring in a number of other market areas: biotechnology/pharmaceutical, petrochemical, environmental control, electronic, gas separation and other process industry applications. Some have advanced to more maturing stages than others. These, with the exception of gas separation, will be surveyed in this chapter.

TABLE 6.1
Major advantages of inorganic membranes for various application areas

Application area	Major advantage(s) of inorganic membranes
Food and beverages	Resistance to alkaline cleaning solutions; being steam sterilizable; consistent pore size
Biotechnology	Being steam sterilizable; consistent pore size
Petrochemical processing	Thermal stability; solvent resistance
Textile processing	Resistance to acid, alkaline and peroxide solutions; thermal resistance.
Paper and pulp processing	Thermal staiblity
Metal cleaners treatment	Resistance to cleaning solutions

Emphasis will be placed on the application aspects for the discussions in this and the next chapter. Some details are given to the application description, advantages and limitations of inorganic membranes for the applications and the critical issues involved.

6.1 FOOD PROCESSING

Membrane filtration has been increasingly applied to food processing. It is estimated that the total membrane area for food, beverage and biotechnology applications as of 1989 is about 100,000 m^2 in which inorganic membranes constitute 12% [Merin and Daufin, 1989]. The main usage (80%) of inorganic membranes is in the dairy industry. The remaining major applications are in the beverage, protein concentration and biotechnology.

Many food and beverage applications require thorough cleaning and sanitizing including thermal cleaning (e.g., steam sterilization) and chemical cleaning. Furthermore, some of them involve handling of viscous liquid feed streams that exert relatively high shear stress on the membrane surface. A number of inorganic membranes, such as the ceramic type, due to their inherent resistance to heat and many chemicals and their mechanical strengths, are particularly suitable for these uses. Except isotope separations, food and beverage are the two most successful commercial application areas for inorganic membranes. Manufacturing plants employing inorganic membranes with a total membrane area of several thousand square meters are now in operation. Compared to their organic counterparts, inorganic membranes currently are more expensive and are not widely available in some high packing density geometries such as hollow fiber and spiral wound modules. In addition, their costs per unit membrane area under the present market conditions are higher than those of organic membranes without regard to their longer service life.

6.1.1 Applications in Dairy Industry

The first industrial-scale food application of inorganic membranes was reportedly the concentration of whey proteins in 1980 followed by milk protein standardization [Merin and Daufin, 1989]. In recent years they have been heavily used for milk and whey processing as microfilters and for high viscosity fluids applications (e.g., milk at high volume concentration factor) as ultrafilters. Most ultrafiltration membranes for dairy applications retain the majority of proteins in milk or whey, but permit smaller sized ingredients such as lactose, soluble salts and non-protein nitrogen fractions to pass through the membranes with the permeate. Inorganic membranes possess better chemical resistance during the acid/alkali treatment between operations and withstand higher operating pressures than organic membranes. Thermal stability of the membranes is probably not a major factor here because above 65°C irreversible denaturation of whey proteins occurs [Muir and Banks, 1985].

Almost invariably when concentrating milk or whey, a major factor that can limit the membrane flux is the issue of fouling which has been mentioned in Chapter 5. Fouling not only significantly reduces the permeation rate but also can deleteriously affect the quality of the permeate (for purification or separation applications) or the retention of the product in the retentate (for concentration applications). In the dairy industry, major constituents of foulants are proteins, lipids and calcium salts [Daufin et al., 1989; Vetier et al., 1988]. Fouling very much depends on the nature of the interaction between the feed components and the membrane. Protein adsorption is a well known cause of fouling. Despite the operational problems that fouling has caused, the phenomenon is still not well understood. In the chemical cleaning cycle to remove these and other foulants, nitric acid (e.g., at a strength of 0.5%) or phosphoric acid are most often used while the standard alkaline solution for cleaning is sodium hydroxide (e.g., at 0.1%). Other chemical agents such as hydrogen peroxide, sodium hypochlorite and peracetic acid are also applied in some cases. Sterilization is also important in this application and, therefore, many steam sterilizable inorganic membranes are preferred.

It has been proven over the years that the effect of fouling can be lessened to some extent for the application of whey concentration by pretreating the feed streams for the ultrafilters. Whey contains many insoluble solids such as casein fines, lipoprotein complex, mineral precipitates, free fats and microorganisms. Clarification of these debris helps reduce fouling potential during ultrafiltration. In addition, it is quite evident that calcium phosphate minerals in whey are not stable and their precipitation in the membrane pores often results in flux decline. Demineralization of whey before ultrafiltration helps maintain high permeate flux considerably [Muir and Banks, 1985].

Bacteria removal from milk. This is a relatively new high-flux application for ceramic membranes with a pore diameter of 0.1 μm or larger. In this case, the desired product is the filtered milk. Given in Table 6.2 are the approximate diameters of the various milk constituents. There is a clear distinction in size between the low-molecular-weight components such as water, ions and lactose and the high-molecular-weight components such as fats, proteins and bacterias. It can be seen from the table that bacterias are typically larger than 0.2 μm and fat globules range in size from 0.1 to 2 μm. It is, therefore, expected that the quality of skimmed milk will not be changed after the bacteria removal but that of whole milk will be affected because most of the fats will be removed along with bacterias. For example, the first documented pilot study of microfiltration of whole milk has shown that even 1.8 μm alumina membranes can reduce the bacteria counts by two orders of magnitude but also skim 98% of the fats. The proteins, however, remain in the filtered milk [Merin and Daufin, 1989].

Other demonstration tests using alumina membranes with a mean pore diameter of about 1.4 μm and with a cocurrent flow of permeate (to maintain a constant transmembrane pressure, TMP) to process skimmed milk by the cold pasteurization process have witnessed a flux exceeding 700 L/hr-m^2 and 99.7% bacteria removal [Malmberg and

Holm, 1988]. The optimal TMP is 0.3 to 0.8 bar. The membranes are subject to heavy
fouling which requires regeneration after 5 to 10 hours of normal operation.

TABLE 6.2
Approximate diameters of major milk constituents

Component	Particle diameter (nm)
Water	0.3
Cl⁻, Ca⁺⁺	0.4
Lactose	0.8
Whey proteins	3-5
Casein micelles	25-300
Fat globules	100-2,000
Bacterias	>200

[Merin and Daufin, 1989]

Concentration of raw or pasteurized whole milk. The purpose of this commercial
operation using ceramic membranes is to prepare a liquid precheese material with a high
protein content (higher than about 20-22%) as a precursor to soft or semi-hard cheese
such as Saint-Paulin cheese. Precheeses are subsequently cooled to about 30°C and
added with either classical lactic starters or freeze-dried lactic starters and also sodium
chloride. This is followed by molding and demolding, hardening, brining, drying and
ripening before semi-hard cheeses are made, as schematically shown in Figure 6.1
[Goudédranche et al., 1980].

The ultrafiltration process is operated in a batch mode at a temperature of about 50°C.
Ceramic membranes with 0.1 or 0.2 μm pore diameter enable processing of this highly
viscous and concentrated raw or pasteurized whole milk due to their inherent mechanical
strength. The viscosity of the concentrate has been found to increase exponentially with
the rise of protein content in the precheese. Polymeric membranes have also been
considered not suitable for this process in view of their structural compaction under
pressure and their difficulty of cleaning.

Today zirconia membranes are used for making milk concentrates with a volume
concentration factor of 5 and for desalting and delactosing by the combination of
ultrafiltration and diafiltration [Merin and Daufin, 1989]. The resulting permeate flux is
about 30 L/hr-m² (under a crossflow velocity of 4.2 m/s at 52°C) and the total solids
content of the concentrate can reach 45% [Goudédranche et al., 1980]. The precheese
viscosity is at its minimum when the pH is about 5.5. Therefore, it is very desirable to
filter high protein precheeses by membrane at this pH. The increase in cheese yield due
to the use of inorganic membrane ultrafiltration is close to 19% compared to the
traditional process. The general chemical cleaning procedure outlined in Chapter 5 has
been found to be effective. Basically it involves acid cleaning (with phosphoric acid at

50°C for 10 minutes) and alkaline cleaning (with sodium hydroxide with EDTA at 74°C for 30 minutes) with water rinsing before and after each type of cleaning.

GA 35279.1

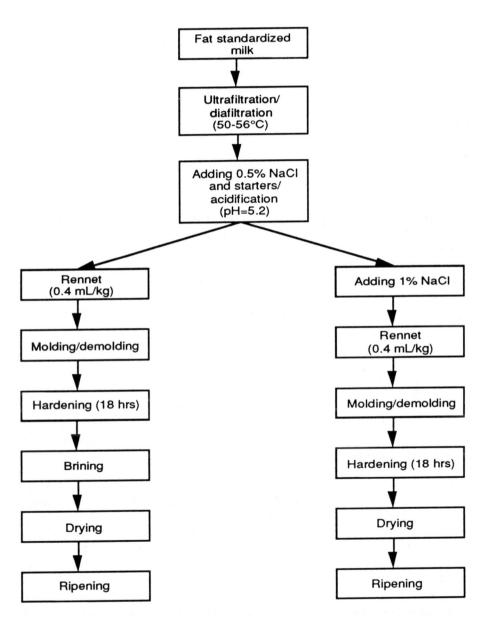

Figure 6.1 Schematic process diagram of semi-hard cheese manufacture [Adapted from Goudédranche et al., 1980].

Milk protein standardization (concentration of pasteurized skimmed milk). Milk protein standardization is designed to maintain the protein level in the milk constant all year round for automated cheese making. It basically involves concentration of pasteurized skimmed milk. It has been one of the major commercial successes in using inorganic membranes for food applications. In commercial production, microfiltration alpha-alumina and zirconia membranes with a pore diameter of 0.1 to 0.7 μm (mostly 0.2 μm) are used. Skimmed milk pasteurized at 70°C is typically concentrated to a volume concentration factor of 2 to 5 [Attia et al., 1988; van der Horst et al., 1994]. The volume concentration factor is the ratio of the initial feed volume to the retentate volume. Thus the higher the factor is, the more concentrated the product becomes.

A high crossflow velocity of 5-7 m/s and a low TMP helps slow down fouling which is a serious problem in many microfiltration applications of dairy streams, causing low flux and poor permselectivity. As a general trend, the flux increases as the crossflow velocity increases or the concentration factor decreases. The protein rejection depends on the pore size. For example, the apparent rejection decreases with increasing flux (but the intrinsic rejection based on the concentration at the membrane surface is essentially constant) for a 0.4 μm membrane while it is independent of the flux (at approximately 1%) for a 0.7 μm membrane [van der Horst et al., 1994]. It is noted that the transmembrane pressure (TMP) can be variable or uniform over the length of a membrane module. It was found that, for this application with the 0.4 μm membrane, the use of a proper uniform TMP was better than a variable TMP for achieving a high overall module capacity. However, this was not the case with the 0.7 μm membrane. The effects of a variable or uniform TMP needs to be studied more systematically before definitive conclusions can be drawn.

Milk protein standardization for continuous cheese making can also be done by ultrafiltration using ceramic membranes. Zirconia membranes with an average molecular weight cut-off (MWCO) of 70,000 daltons on carbon supports have been used for this purpose. The objective for this application is to concentrate either the whole volume of the milk to a volume concentration factor of 1.3 to 1.6 or just a fraction of the feed volume to a volume concentration factor of 3 to 4 followed by mixing the concentrate with raw milk to reduce the requirement of milk storage space [Merin and Daufin, 1989].

Fouling is one of the most serious problems associated with membrane filtration for various types of milk. In these cases fouling is the consequence of one or more of the following phenomena: (1) adsorption of proteins, fats or bacterias on the membrane surface or pore wall; (2) pore blockage by particulates or macromolecules; and (3) concentration polarization. Attia et al. [1991a; 1991b] studied fouling characteristics and mechanisms of filtering skimmed milk and acidified milks with 0.2 and 0.8 μm alumina membranes, using SEM and electrophoresis measurements. External deposit layers are formed on both membranes that function as the true filtering (ultrafiltration) media as a result of strong adsorption of proteins and in some cases salts on the alumina particles in the membrane surface layer. The dense fouling substance appears to consist of a coalescence of juxtaposed casein micelles [Attia et al., 1991a]. This limits the steady state flux to 69 and 35 L/hr-m^2 for 0.2 and 0.8 μm membranes, respectively, under the condition of 50°C, TMP of 5 bars and a crossflow velocity of 3 m/s. This is shown in Figure 6.2 for membrane filtration of a reconstituted skim milk at a pH of 6.62 [Attia et al., 1991b].

The pore size of the membrane apparently plays an important role in determining if internal fouling (i.e., plugging of particles inside the pores) occurs. The 0.2 μm membranes do not experience internal fouling while the 0.8 μm membranes suffer from internal casein fouling which makes the larger pore membranes unsuitable for this application. It is noted that the casein micelles are approximately 0.02 to 0.3 μm in size and it is quite possible that they enter the membrane pores. This is reflected in the lower flux for the 0.8 μm membrane.

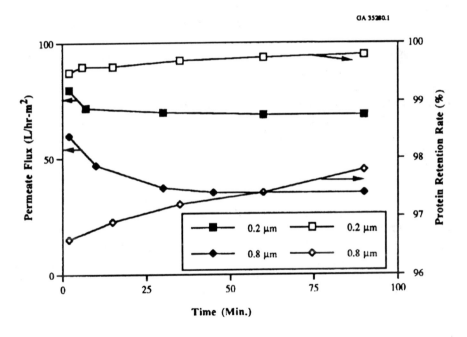

Figure 6.2 Effect of membrane pore diameter on permeate flux of skimmed milk and protein retention rate [Adapted from Attia et al., 1991b]

Also shown in Figure 6.2 is the protein retention rate for the above two membranes. As expected, the smaller pore (0.2 μm) membrane shows a higher retention capacity than the larger pore membrane.

Concentration of acidified milk. Fresh unripened, textured cream cheese can be produced by acidifying milk. This, for example, can be done by ultrafiltering the acidified milk (at a pH of about 4.4) after lactic fermentation. This pH level makes calcium salts fully soluble and therefore preserves their contents and the organoleptic qualities of the prepared cheese like those of the traditionally prepared cheeses. The use of membrane filtration eliminates the need for curd centrifuge separation. Industrial plants employing inorganic membrane filtration are currently in operation. One of the most severe processing problems lies in the high viscosity (often greater than 200 cp) of the process stream which requires the use of inorganic membranes. Ultrafiltration zirconia membranes on carbon supports with a MWCO of about 70,000 daltons have been reported to give a flux of 60 L/hr-m^2 at 40°C with a volume concentration factor of about 2.3 [Mahaut et al., 1982].

Microfiltration-grade alumina and zirconia membranes with a pore diameter of 0.2 and 0.1 μm, respectively, have also shown promising results for concentrating acidified milk [Merin and Daufin, 1989]. Furthermore, larger pore size (0.2 and 0.8 μm pore diameter) alumina membranes have been investigated for filtering acidified milks [Attia et al., 1991a; 1991b]. Both 0.2 and 0.8 μm membranes give comparable, satisfactory fluxes and protein retention coefficients.

The pH value of the milk is critical in the occurrence of internal micellar fouling. There is indication of membrane pore blockage when the milk pH is above 5.6. At a pH below 5.6, no internal micellar fouling is evident. This transition is also reflected in the permeate flux. For example, the permeate fluxes at a pH of 4.4 (acidified milk) and 6.62 (milk), under otherwise identical conditions, are 130 and 35 L/hr-m^2, respectively, for an 0.8 μm alumina membrane.

Parametric studies of the effects of TMP, temperature and crossflow velocity on the permeate flux and protein retention rate have been conducted using 0.8 μm alumina membranes at a pH of 4.4. The maximum steady state flux is observed at a TMP of 3 bars. As expected, a higher crossflow velocity increases the steady state permeate flux, as illustrated in Figure 6.3 under the condition of 50°C, TMP of 5 bars and pH of 4.40 [Attia et al., 1991b]. The protein retention rate also improves with the increase in the crossflow velocity. The permeate flux reaches 175 L/hr-m^2, accompanied by a protein retention rate of 97.5% when the crossflow velocity is at 3.8 m/s. This improvement in the flux corresponds to a reduction in the thickness of the external fouling layer.

Shown in Figure 6.4 are the effects of the process stream temperature (35 to 50°C) on the permeate flux at a crossflow velocity of 3 m/s and a TMP of 5 bars for a feed pH of 4.40 [Attia et al., 1991b]. The steady state flux increases from 20 to 130 L/hr-m^2 as the temperature increases from 35 to 50°C. Correspondingly, the protein retention rate decreases from 98.6 to 95.6 %. This increase in the flux and decrease in the protein retention may be explained by the reduced viscosity, increased diffusion coefficients and increased solubility of the constituents in the membrane and solution as a result of raising the temperature.

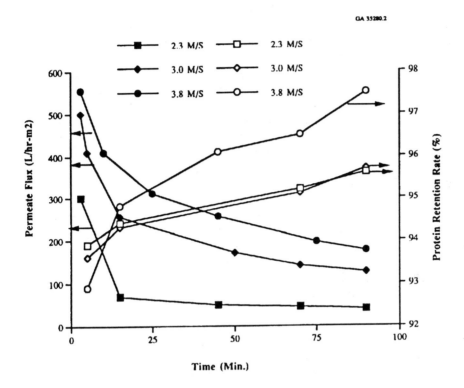

Figure 6.3 Effect of crossflow velocity on permeate flux and protein retention for an 0.8 μm alumina membrane [Adapted from Attia et al., 1991b]

Acidified goat milk has also been processed similarly but with a lower flux of about 20 L/hr-m^2 at a volume concentration factor of 6.7. The corresponding total solids is 40-45% and fat 18%, with a cheese composition similar to that obtained by the traditional processes.

Concentration of whey proteins. As mentioned earlier, microfiltration can be used to remove bacterias. In addition, they are capable of separating phospholipids, fats and casein fines of sweet whey or sour (acidified) whey. Ultrafiltration of whey has been well proven to provide an array of protein products of diverse compositions and properties. Inorganic membrane filtration can be used at different stages of the process to make whey protein concentrates (WPC) in powder form with a protein content reaching 50%.

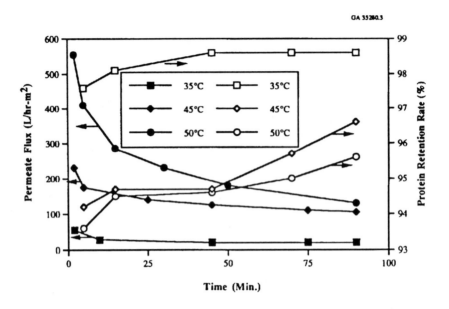

Figure 6.4 Effect of temperature on permeate flux and protein retention for an 0.8 μm
alumina membrane [Adapted from Attia et al., 1991b]

First, microfiltration-grade inorganic membranes can be utilized to remove residual fats
and phospholipids and improve the quality of the feed to the subsequent ultrafiltration
step for producing the WPC [Merin and Daufin, 1989]. To further increase the
microfiltration flux of cheese whey, thermocalcic aggregation can be applied first. And
the flux increase can be substantial. The treated feed to the ultrafilter which can also be
an inorganic membrane has been claimed to enhance the subsequent UF permeate flux.
The resulting microfiltration flux and the retention rate of proteins such as β-
lactoglobulin and α-lactalbumin are strongly influenced by the crossflow velocity.

High-purity WPC (i.e., 70-95% proteins or total solids) can be produced by thermocalcic
aggregation, followed by microfiltration, ultrafiltration and diafiltration of whey
proteins. Ultrafiltration has been practiced since early 1970s. It appears that zirconia
membranes on carbon supports with a MWCO of 10,000 to 20,000 daltons and zirconia
membranes on alumina supports with a pore diameter of 0.05 to 0.1 μm are suitable for
this purpose. A permeate flux of as high as 60 L/hr-m^2 for processing acid whey to a
protein content of 25 to 37% using a zirconia membrane with a MWCO of 10,000
daltons has been reported [Merin and Daufin, 1989].

6.1.2 Other Food Processing Applications

Protein recovery and concentration are other major applications for inorganic membranes in the food processing area.

Protein recovery in fish processing. A huge quantity of soluble proteins are carried in the effluents of fish products processing plants. This poses two issues. It represents a significant economic loss and at the same time raises the pollution level of the fish pulp washwater. Therefore, the recovery of soluble proteins is of practical interest.

Watanabe et al. [1986] pioneered the use of inorganic membranes for this application. Dynamic membranes having a nominal pore diameter of 0.05 μm are formed on porous ceramic supports to recover essentially 100% of the protein with a molecular weight of 10,000 daltons or higher from a fish jelly processing plant. The dynamic membranes are composed of the water-soluble protein itself. The membrane modules consist of single membrane tubes or multiple tubes. The permeate flux increases with increasing the TMP or crossflow velocity before reaching asymptotic values. The rejection rate is essentially constant at 95% independent of the TMP or crossflow velocity. The optimal operating condition has been found to be the following:

temperature = 15°C
crossflow velocity = 1.4 m/s
transmembrane pressure = 4.5 bars.

Like all other membranes, dynamic membranes experience flux decline over time to low steady state values. The permeate flux can be restored to its initial high value by applying a 500 to 1,000 ppm of sodium hypochlorite solution to wash out the fouled dynamic membrane. Porous ceramic supports were used instead of organic polymer membranes due to their resistance to chemicals and heat during the cleaning cycle.

Swafford [1987] has studied the use of ceramic membranes for processing the washwater of minced Alaska pollack. The membrane with a nominal MWCO of 10,000 daltons yields a membrane flux about 20 to 30% lower than that of a membrane with a MWCO of 20,000 daltons.

Inorganic membranes, particularly ceramic membranes, provide the operational advantages of not being subject to compaction as in the case of organic membranes under relatively high pressures and the ease of regeneration by such techniques as backflushing or various means of cleaning as discussed in Chapter 5.

One of the most challenging issues in this application is the adsorption of proteins on both inorganic and organic membranes. This leads to fouling which in turn limits the membrane flux as shown in Table 6.3 [Quemeneur and Jaouen, 1991]. The hydraulic resistance due to the layer of the adsorbed protein, R_a, can be significant when compared to the total resistance, R_a+R_m, where R_m represents the intrinsic hydraulic resistance due to the composite membrane. In all the cases studied, R_a exceeds 50% and can reach 80% in some cases. The effects of protein adsorption on the alumina and zirconia membranes tested are less than that on the sulfonated polysulfone membrane but appreciably higher than that on the regenerated cellulose membrane. The retention of proteins by both types of ceramic membranes can easily reach 100%.

TABLE 6.3
Effects of fish protein adsorption on membrane flux

Membrane designation	Membrane material	Membrane pore dia. or MWCO	Water flux[1] $(L/hr-m^2)$	$R_a/(R_a+R_m)$[2] (%)
TECH-SEP PSS 3026	sulfonated polysulfone	10,000 daltons	320	98
S.C.T. T1-70	alumina	0.2 μm	1,080	80
TECH-SEP M5	zirconia	10 kD[3]	45	51
TECH-SEP M4	zirconia	20 kD	72	57
TECH-SEP M1	zirconia	50-80 kD	104	57
TECH-SEP M6	zirconia	0.08 μm	102	53
Millipore PLGC	regenerated cellulose	10 kD	50	0

Note: (1): water flux at 30°C under 1 bar; (2) R_a = hydraulic resistance due to the adsorbed layer and R_m = intrinsic hydraulic resistance of the membrane; (3): kD = 10,000 daltons
[Quemeneur and Jaouen, 1991]

As in many other applications, the crossflow velocity and the transmembrane pressure (TMP) can significantly affect the membrane flux. In Figure 6.5 the water flux is expressed as a function of the crossflow velocity and TMP. A higher crossflow velocity invariably yields a higher permeate flux. At a given crossflow velocity, the flux monotonically increases with increasing TMP until a threshold point beyond which the flux levels off to a constant value characteristic of the crossflow velocity, temperature and feed concentration. This asymptotic flux value is rather insensitive to the pore size of the membrane [Quemeneur and Jaouen, 1991]. The permeate flux does not differ much among the various membranes studied despite the wide spectrum of pore diameter involved. This is probably the result of protein adsorption leading to an additional dynamic barrier layer on the original membrane. This dynamic layer becomes the dominant separation barrier regardless of the structure of the underlying layer.

Quemeneur and Jaouen [1991] have made an economic assessment of utilizing ceramic membranes for recovering the fish proteins and recycling the membrane treated washwater. They concluded that the extra process and associated investments for the protein recovery will become economically justified when the concentrated proteins demand a high market value.

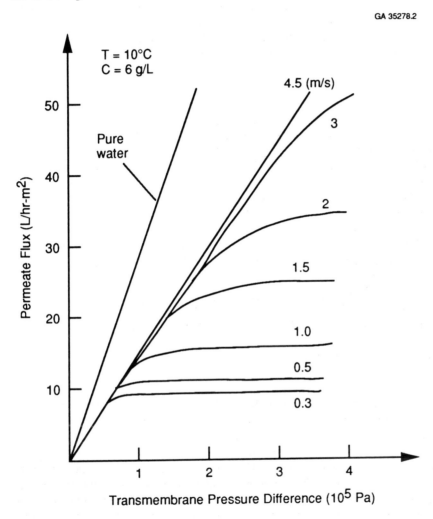

Figure 6.5 Permeate flux of fish processing effluent stream as a function of crossflow velocity and TMP [Quemeneur and Jaouen, 1991]

Protein concentration. Concentration of proteins of several animal and plant products is practiced in food applications. Inorganic membranes have been used in many of these cases [Merin and Daufin, 1989]. For example, separation and concentration of egg white proteins from yolk using 20,000 dalton zirconia membranes increases the protein solids from 11-12% to 32-35%. Zirconia membranes with a MWCO of 70,000 daltons raises the whole egg protein solids concentration from 24% to 40-50%. Blood proteins from the slaughterhouse are sometimes used in some food products (e.g., meat delicatessen). Blood plasma has been ultrafiltered using 20,000 dalton ceramic membranes followed by drying to prepare dried powder products. Zirconia membranes with a MWCO of 20,000 daltons have also been employed to concentrate soybean proteins.

Rios and Freund [1991] investigated the prospect of applying an electro-ultrafiltration (EUF) process for protein concentration. The underlying principle is that the permeate flux and permselectivity of an ultrafiltration membrane may be improved by the following phenomenon. If, when processing electrically charged solutes, an electric field is superimposed upon the applied pressure to act on and move the retained solutes away from the membrane surface, the concentration polarization or fouling can then be controlled. A tubular alumina membrane with a mean pore diameter of 0.2 μm was used to filter gelatin solutions. Shown in Figure 6.6 is an EUF module consisting of a membrane exposed to the solution, an anode placed at the center of the membrane tube and a cathode located outside the tube. In this arrangement, electrophoretic migration will draw gelatin molecules away from the alumina membrane. Both permeate flux and protein retention increase when the electric field is applied such that the electrophoretic force tends to move macromolecular species away from the membrane. Contrary to purely pressure-driven ultrafiltration, the crossflow velocity in an EUF process is preferably as low as possible. There is an optimal TMP that offers the maximum permeate flux. It was found that switching on the voltage before circulating the fluid is a more effective procedure than applying the electric field after starting ultrafiltration.

6.1.3 Wastewater Treatment in Food Processing

Due to their chemical stability and structural rigidity, ceramic membranes have been tested, and in some cases, used for the pretreatment of industrial waste waters.

Vegetable oil liquid wastes. A liquid waste generated in the production of olive oil is called "green water." Transition alumina membranes have been studied for their feasibility of reducing the turbidity of the "green water" from olive oil processing [Fabiani et al., 1989]. At 30°C and a TMP of 2 to 6 bars, the performance of an alumina membrane having a pore diameter of 38-46 nm for improving clarity of the permeates is shown in Table 6.4. The resulting steady state permeate flux of the treated waste waters under these conditions averages about 18 L/hr-m^2. This flux is relatively independent of the TMP which is indicative of the presence of a gel layer on the membrane surface. The turbidity index of the permeate is about 0.62-0.67 (compared to 0.60 for potable water) which is drastically reduced from a value of 57 for the "green water." The membrane permeability can be restored to 80% of its initial water permeability with a single backflush and an alkali detergent solution.

GA 35286

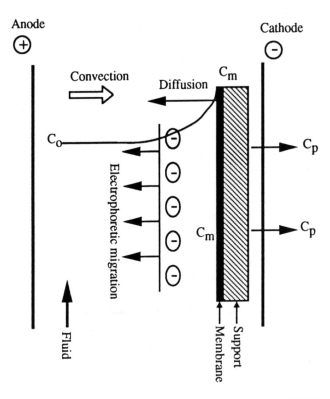

Figure 6.6 A schematic diagram of an electro-ultrafiltration (EUF) module for improving separation performance [Rios and Freund, 1991]

TABLE 6.4
Permeate flux and turbidity of waste waters from olive oil processing ("green water") and from washing operations in dairy processing

TMP (bars)	Waste water from olive oil processing		Waste water from wash operation in dairy processing	
	Permeate flux $(L/hr-m^2)$	Permeate turbidity index	Permeate flux $(L/hr-m^2)$	Permeate turbidity index
2	19.1	0.62	6.84	0.82
4	18.0	0.64	6.12	0.67
6	16.9	0.67	5.76	0.61

Note: Turbidity index of potable water, waste water from olive oil processing and waste water from dairy processing are 0.60, 57 and 57, respectively.
[Adapted from Fabiani et al., 1989]

Dairy liquid wastes. Clarification of the waste water from the washing operations in the dairy processing has also be investigated by Fabiani et al. [1989]. Also given in Table 6.4 are the separation properties of an alumina membrane (38-46 nm in pore diameter) for treating the dairy waste water. The permeate flux is quite insensitive to the TMP (over the range of 2 to 6 bars) and at 30°C it is at a level of 5.8-6.8 L/hr-m^2. The turbidity index decreases from 57 for the feed stream to 0.61-0.82 for the permeate. Again, for comparison, the turbidity index for potable water is about 0.60.

6.2 PROCESSING OF BEVERAGES AND DRINKING WATER

Besides food processing, alcoholic and non-alcoholic beverages and drinking water are the next largest commercial applications of inorganic membranes in recent years. For those cases involving particulates, inorganic membranes with larger pore sizes are used as microfilters. In other cases where only liquid streams are processed, ultrafiltration-grade inorganic membranes are the choice.

Compared to their applications in food processing, the use of inorganic membranes in beverage processing has not been as widely documented. For these applications as well, fouling can be a serious limiting factor for the permeate flux as a result of the choices of the membrane pore size and material, feed composition, operating pressure and temperature. One of the commonly observed foulants is polysaccharides in the use of ceramic membranes for beer brewing and wine processing [Poirier et al., 1984].

6.2.1 Clarification of Fruit Juices

Microfiltration and ultrafiltration have been used on a production scale for clarifying pressed or prefiltered fruit juices. Traditionally juice processing steps involve comminution or pulverizing of the fruit, pressing, coarse or rough filtration and polishing. These steps call for additional processing agents: press aids, filter aids, to name just a few. Ultrafiltration can combine pressing and filtration in a single unit operation. Organic ultrafiltration membranes, however, have been found to alter flavor and color of the membrane-treated juice. The increasingly accepted ceramic membranes used for this application replaces diatomaceous earth in some applications.

Clarification of apple juice. Inorganic membranes have been utilized in two ways in the production of apple juice. One is to clarify pressed or prefiltered apple juice and the other is to extract clarified apple juice directly from apple puree or pomace. Microfiltration of apple juice has been one of the most successful commercial applications of inorganic membranes.

To maximize the permeate flux using 0.2 μm zirconia membranes on carbon supports, a TMP of 3.5 bars and a crossflow velocity of 5 to 6.5 m/s at 30 and 50°C are recommended [Alvarez et al., 1994]. Apple contains pectins, sugars and polyphenols as some of its major constituents. The retention of pectins and sugars remains almost constant at any volume concentration factor value, but polyphenyl retention increases sharply beyond a volume concentration factor of 3 at which point the pectins retention is near 100%. Zirconia membranes of a pore diameter of 0.14 μm have also been implemented with a resulting flux of 125 L/hr-m^2 and intense color of the clarified apple cider. Finer pore zirconia membranes with a MWCO of 20,000 and 150,000 daltons have been commercially used [Merin and Daufin, 1989].

When extracting apple juice from apple puree, two related problems surface. One is fouling presumably due to accumulation of the constituents of the puree including pectins, colloids, starches and phenolics. Fouling increases with increasing concentration. The other is high viscosity of the high-solids, high-concentration retentate particularly near the end of the membrane module. Fouling or reduction of the permeate flux is not solely determined by the stream viscosity as has been demonstrated by Thomas et al. [1987]. Acceptable viscosity modifiers are used to increase the permeate flux as well as reduce the pumping energy (or pressure drop) required in a food-grade metallic oxide membrane formed-in-place within the matrix of a sintered stainless steel tube. Liquefaction enzymes such as cellulase and pectinase are found to be effective for reducing viscosity of the apple puree and break down the pectins. Even at an enzyme concentration of as low as 0.0065% by weight of the puree, the viscosity can be reduced by 70-80 % within approximately 40 minutes. To significantly increase the permeate flux, however, the required enzyme treatment needs to be higher at 0.044% for one hour at 50°C. The permeate flux increases with increasing dosage of the enzyme but apparently levels off at an optimal amount. Shown in Figure 6.7 is an optimization study on the proper dosage of the enzyme amount to maximize the permeate flux of dynamic metallic oxide membranes [Thomas et al., 1987]. It appears that fouling, and consequently the permeate flux, is strongly determined by the amount of total pectin present in the apple purees. Apparently pectins form a compressible gel layer which can lead to fouling. The optimal dosage of 0.044% enzyme corresponds to the minimum amount of the total pectin.

Using a two-stage, single-pass module of food-grade dynamic metal oxide membrane on a sintered stainless steel support tube under a TMP of approximately 20 bars and a temperature of 50°C, Thomas et al. [1987] have successfully clarified apple puree with a juice yield of 86% and a steady state flux of about 85 L/hr-m^2. The quality of the clarified apple juice by this dynamic or formed-in-place membrane is excellent -- sparkling looking and flavor retained. As much as 65% reduction in the total pectins has been attained.

Clarified apple juice has been commercially produced by the above dynamic metal oxide membrane process under the trade mark of Ultrapress^{TM} by Carre in the U.S. This process of combining pressing and ultrafiltration in a single-pass membrane filtration of enzyme-treated fruit puree has been patented [Thomas et al., 1987b]. It offers the potential for reducing process time and making microbiologically stable juice which bypasses the need for heat pasteurization. The molecular weight cut-off (MWCO) of the membranes is usually about 30,000 to 40,000 daltons. The membrane module can be as long as 60 m. Under a typical TMP of about 20 to 27 bars, the juice yield reaches 85 to 88%. Reportedly this ultrafiltration membrane process, as a result of reducing the processing steps required, offers a processing savings of 10-15¢/gal juice produced. The clarification system can be cleaned in place by flushing with a caustic solution followed by an acid rinse. It can also be sterilized by steam [Swientek, 1987].

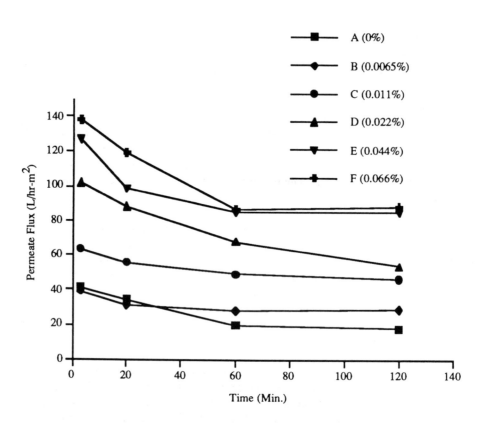

Figure 6.7 Determination of the optimal enzyme dosage for maximizing permeate flux of apple juice through a metal oxide dynamic membrane [Thomas et al., 1987a]

Clarification of other fruit juices. Clarification of cranberry juice has also been practiced commercially using ceramic membranes recently. Through a series of tests with alumina microfilters, Venkataraman et al. [1988] determined that the optimal pore diameter is about 0.45 μm with respect to the permeate flux and the clarity of the permeate. The optimal TMP appears to be about 3.5 bars and crossflow velocity about 7 m/s at 45°C. The fouled membranes can be regenerated by applying caustic soda and sodium hypochlorite at a pH of 11. The use of ceramic microfilters replaces the use of diatomaceous earth. The turbidity can be reduced to less than 1 NTU compared to about 10 NTU when diatomaceous earth is used.

Several fruits and vegetables such as apples, pears, pineapples, peaches, carrots and beets have been pilot tested with promising results [Swientek, 1987]. In addition, strawberry and kiwifruit juices have also been clarified using ceramic membranes.

6.2.2 Clarification and Purification of Drinking Water

Suspended solids in groundwater or surface water can be removed to produce potable water by some conventional technology (e.g., filtration) in combination with other processing steps such as coagulation, flocculation, decantation, disinfection, etc. Traditional filters like media filters are used and normally perform satisfactorily for feed water with a low concentration of colloidal and particulate suspensions, at a turbidity level lower than about 20 nephelometric turbidity unit (NTU). In some cases due to snow melting or heavy rainfall or other factors, the turbidity in the groundwater can be quite high, exceeding 100 NTU. The traditional filters become unacceptably inefficient as a result of fouling.

In addition, bacterias need to be removed from well, river or lake water before water can be rendered potable. A properly selected microfiltration ceramic membrane is effective for bacteria decontamination. When used preceding a reverse osmosis or an ion-exchange unit, the microfiltration membrane protects the downstream separation process from bacteria contamination and possible fouling due to colloids.

Furthermore, ultrafiltration ceramic membranes have been found to be effective in removing some heavy metal pollutants.

Crossflow microfiltration ceramic membranes provide a viable alternative to treating various types of water with high suspended solids. The filtrate flux of 0.2 μm alumina membranes between two consecutive cleaning cycles can be maintained at a relatively high level, reaching 800 L/hr-m^2, with efficient backflushing [Guibaud, 1989]. The turbidity can be significantly reduced from 120 down to 0.5 NTU. Ultrafiltration ceramic membranes are also capable of processing the groundwater in some special cases. Membrane fouling occurs easily even with low solids contents in the groundwater. Although the crossflow velocity can help reduce the fouling propensity, the associated pumping energy cost may not be justified in view of the generally perceived low cost product: drinking water. Therefore, a low pressure (about 1 bar or so) and a low crossflow velocity (much less than 1 m/s) are preferred for this application. Economic assessment indicates that the crossflow ceramic membrane process is more attractive than other conventional processes for a processing capacity of less than 120 m^3/hr.

Surface water can also be processed to become drinking water but it requires some pretreatment prior to the microfiltration step. A filtrate flux of 1,000-1,500 L/hr-m^2 can be realized [Guibaud, 1989].

Both microfiltration (0.2 μm) and ultrafiltration (4 nm) alumina membranes are very effective in removing bacterias. For example, the bacteria level of a lagoon water is reduced from 1,000-5,000/cm^3 to 0.03-0.4/cm^3 and 0.03-0.1/cm^3 with the microfiltration and ultrafiltration membrane, respectively [Castelas et al., 1984]. The total coliform level drops from 50-500/cm^3 to zero for both types of membranes. The accompanying permeate flux is 600-1,200 L/hr-m^2 for 70 hours for the microfiltration membrane when the water contains a low level of colloids and only 200 L/hr-m^2 for 20 hours when the concentrations of colloids and organic materials are high. The ultrafiltration flux varies between 100 and 250 L/hr-m^2 for 1,000 hours of operation.

In addition to bacteria decontamination, the ultrafiltration membranes also reduce some ions such as sulfates (118 to 1.5 mg/L) and nitrites (0.22 to 0.02 mg/L). This is attributable to complexation of these ions with the humic and fulvic acids which are retained by the membranes. On the other hand, the non-complexed ammonium and chloride ion concentrations are essentially the same before and after ultrafiltration.

6.2.3 Clarification in Wine Making

Conventional separation techniques in wine making to stabilize wines consist of decantation, fining, cold treatment and filtration. Decantation is used after natural clarification separates the solids from the juice or wine. Decantation can be accelerated by centrifuging. Suspended colloidal particles of grape or yeast proteins, peptides, pectins, gums, dextrans, unstable grape pigments, tannins, and other compounds cause clouds and hazes in wines [Thoukis, 1974]. To enhance sensory and clarity properties of the wines, they must be removed. Their separation can be achieved by the judicious use of a small quantity of fining agents which adsorb on the particles or neutralize their electric charges causing them to agglomerate and settle. This results in a clear wine. Fining can be done at several stages in the wine making process: for phenolic removal in press juice, decreasing proteins in the wine after fermentation and reducing astringency of young wines prior to bottling [Long, 1981]. Different fining materials serve different purposes. For example, bentonite, probably the most widely used, removes proteins or peptides. Other fining materials such as gelatin (animal protein), casein (milk protein), isinglass (fish protein), activated carbon and polyvinylpolypyrrolidone facilitate the removal of phenolic and clarification of the wine. Cold treatment can precipitate crystals of potassium bitartrate and calcium tartrate. Traditional filtration steps involve one or more of the following: coarse filtration using diatomaceous earth, fine filtration and organic membrane separation as a sterile filtration step. Crossflow microfiltration and ultrafiltration in general have been intensively evaluated in recent years as a modern one-step separation process to replace clarification, sterilization and stabilization of crude wines.

Microfiltration and ultrafiltration of lees or crude wines. More specifically, crossflow micro- and ultra-filtration ceramic membranes have the potential for replacing all the above separation steps except cold treatment [Castelas and Serrano, 1989]. When using inorganic membranes for removing bacterias, yeasts or suspended particles, the choice of the pore size is very important in determining the filtrate flux and the rejection performance of these materials from wines.

Given in Table 6.5 is the data of filtering lees (settled solids in a tank) of a red wine using three alumina membranes with different pore diameters with a crossflow velocity of 4.5 m/s and a TMP of 2 bars. Lees typically entrain juice or wine and can account for an appreciable percentage of the original juice. Most of the bacterias lie in the range of 0.5 to 7 μm and yeast in the 1 to 3 μm range while proteins and condensed tannins are smaller, 0.01 to 0.1 μm, as indicated in Table 6.6. It is, therefore, expected that the 0.2 μm membrane is very effective in removing all the lees components, while the larger pore membranes (1.4 and 2.0 μm) are too large to effectively retain bacterias and yeasts. It is, however, surprising to find that the flux is higher for the 0.2 μm membrane (85 L/hr-m^2) than the other two (42 and 64 L/hr-m^2 for the 1.4 μm and 2.0 μm, respectively). A plausible explanation for this is that some of the components plug the pores of the larger pore membranes. The starting contents of bacteria, turbidity and particles are 1,000/ml, 5025 NTU and 8%, respectively. Their corresponding levels after filtration by the three alumina membranes used can be as low as 0.01/ml, 0.22 NTU and <0.02%. It appears that in the application of filtering lees, the 0.2 μm membrane is the preferred choice.

Typical membrane flux data using a 0.2 μm alumina membrane for clarifying a red and a white wine is displayed in Figure 6.8. The TMP is 2 bars and the crossflow velocity is 4.5 m/s. The flux declines in the first two hours or so before reaching steady state values.

TABLE 6.5
Quality of filtrates from microfiltration of lees in a red wine by three alumina membranes having different pore sizes

Membrane pore diameter (μm)	0.2	1.4	2.0
Permeate flux (L/hr-m^2)	85	42	64
Turbidity (NTU)	0.22	0.73	0.86
Yeast concentration (count/ml)	0.01	0.04	8.25
Bacteria concentration (count/ml)	0.01	2.10	27.7
Particles (% by volume)	<0.02	<0.02	<0.02

[Adapted from Castelas and Serrano, 1989]

Chapter 6

TABLE 6.6
Sizes of various particles typically present in wines

Particle size (μm)	Types of particles
0.01-0.1	Peptides, proteins, polysaccharides, gums, dextrins, pectins, condensed tannins, leucocyanins and anthocyanins, and phenolic compounds
0.5-1.0	Tartrate cyrstals
1.0-3.0	Yeasts (Saccharomyces sp., Hensenula, Torulopsis, Debaromyces, Candida, Kluyveromyces, Pichia and Brettanomyces
0.5-7.0	Bacteria (Leuconostoc, Acetobacter, etc.)
50-100	Diatomaceous earth, fibers and debris

[Kilham, 1987]

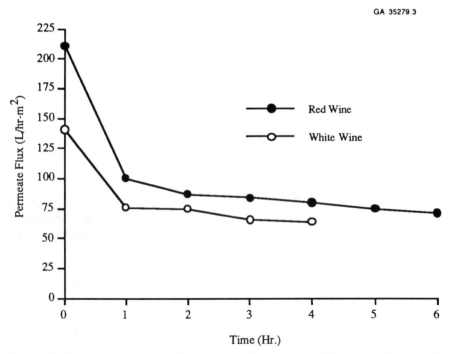

Figure 6.8 Permeate flux of a white and a red wine using 0.2 μm alumina membrane [Castelas and Serrano, 1989]

The characteristics of the flux decline depends, to a great extent, on the nature of the wine. For example, Belleville, et al. [1990] studied crossflow microfiltration of two different red wines: one obtained by destemming and crushing before fermentation followed by post-maceration (wine A) and the other made by destemming, crushing, rapid heating at 70°C before pressing and fermentation (wine B). The permeate flux data in Figure 6.9, obtained at 20°C under a crossflow velocity of 4.5 m/s and a TPM of 3 bars, shows that wine A exhibits the typical flux decline result which drops off sharply initially and quickly establishes a steady state. This agrees with a model proposed by Bennasar and Tarodo de la Fuente [1987]. The flux of the filtrate from wine B, however, does not stabilize even after the initial sharp drop and continues to decrease gradually without reaching a steady state. It has been postulated [Belleville, et al., 1990] that in wine A the dominant polysaccharides of high molecular weights accumulate at the membrane surface, forming another filtration layer which prevents small molecules from penetrating the membrane pores. On the other hand, with wine B, either internal fouling (penetration into the pores) due to the relative loss in high molecular weight polysaccharides or structurally different molecules from wine B interact more strongly with the membrane, thus forcing the membrane to gradually lose its permeability.

An important concern for microfiltration of wines is how much the organoleptic characteristics (e.g., color intensity, tint, polyphenols and polysaccharides) of the wines will be altered due to retention of some wine components by the membrane. This can be illustrated by a case study of microfiltration of two red wines using 0.2 μm alumina membranes (Table 6.7). The operating temperature, TMP and crossflow velocity are 20°C, 3 bars and 4.5 m/s, respectively. The effects on the concentrations of polysaccharides and polyphenols are rather pronounced. Other organoleptic characteristics such as color intensity and tint can also undergo appreciable changes. The color intensity is the sum of D_{420}, D_{520} and D_{620} and the hue is the ratio of D_{520} over D_{420} where D's are the optical densities in the light absorption spectra at 420, 520 and 620 nm. The color intensity has undergone a major decrease but the percent relative distribution of various color components is not changed. The change in hue is not as significant. As a result of the microfiltration, the discolored colloids (total soluble proteins, polysaccharides and residual polyphenolics) are lowered by 52-74%, proteins by 43-72% and polysaccharide by 54-75%. The relative neutral sugar distribution is also altered. Polysaccharides, along with polyphenolics, can adversely affect the taste of wines.

The impact of membrane filtration is not limited only to the quality of red wines. It is also observed with white wines. Given in Table 6.8. are the microfiltration results of a white and a red wine using 0.2 μm alumina membranes [Castelas and Serrano, 1989]. The TMP and crossflow velocity are 2 bars and 4.5 m/s, respectively. The changes on the wine quality are very obvious in almost every property.

However, not all wines suffer from the loss of organoleptic quality as severely as those shown above as a result of membrane filtration. Given in Table 6.9 are the organoleptic properties for other white and red wines which only exhibit slight or essentially no changes.

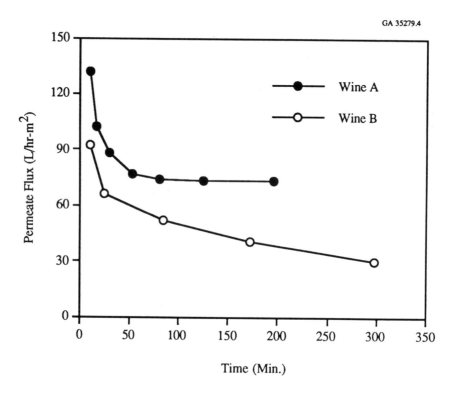

GA 35279.4

Figure 6.9 Decline of permeate fluxes of two red wines made in different conditions [Adapted from Belleville, et al., 1990]

Where membrane filtration is applied in the wine production process can also have long terms effects. Care should be taken to ensure that long term colloidal stability as a result of membrane filtration is not a problem. It may take several months for the colloids to reestablish new equilibrium. Proteins, anthocyanins, tannins and tartrate salts may precipitate.

Besides capital investments, there are two important technical issues. The first, related to the throughput as in many other applications, is the permeate flux. A major obstacle to establishing microfiltration in enology is the relatively low filtration rate compared to competitive processes. Economic viability determines that the flux should be at least 250 L/hr-m^2 after 18 hours [Guibaud, 1989] although this may seem to be higher than the best values reported in the literature. Among various factors, fouling can seriously affect the flux values. The second important issue is the potential color retention by the membrane as just discussed earlier. The color retention should be low enough to preserve the quality of the wine processed. Alumina membranes reportedly have been operated in some wineries in Italy and California [Guibaud, 1989].

TABLE 6.7
Changes in organoleptic and other filtrate qualities as a result of crossflow
microfiltration of two red wines using 0.2 μm alumina membranes

Property	Red wine A		Red wine B	
	Crude	Filtered	Crude	Filtered
Turbidity (abs. 650 nm)	0.295	0.210	0.348	0.138
Absorbance				
420 nm	0.502	0.391	0.454	0.221
520 nm	0.735	0.658	0.597	0.293
620 nm	0.163	0.118	0.153	0.061
Color intensity	1.40	1.17	1.21	0.58
Hue	0.68	0.59	0.76	0.75
Discolorized colloids	929	443	545	140
Proteins	63	36	50	14
Polysaccharides	683	314	356	88
Acidid	158	80	90	61
Neutral	525	234	266	27

[Adapted from Belleville et al., 1990]

TABLE 6.8
Changes in organoleptic and other filtrate qualities as a result of crossflow
microfiltration of a white and a red wine using 0.2 μm alumina membranes

Property	White wine before filtering	White wine after filtering	Red wine before filtering	Red wine after filtering
Total polyphenol index	65	38	63	38
Polysaccharides (mg/L)	207	26	810	627
D_{420}	0.62	0.08		
Color intensity			3.3	0.56
Tannins (g/L)			2.63	2.43
Turbidity (NTU)	393	0.35	1550	0.10
Yeast conc. (count/ml)	3,000	<0.01	800	<0.01
Bacteria conc. (count/ml)	4	<0.01	55	<0.01
Flux (L/hr-m^2)	94		96	

[Adapted from Castelas and Serrano, 1989]

TABLE 6.9
Effects of crossflow microfiltration on organoleptic characteristics of white and red
wines using alumina membranes

Type of wine[1]	Color intensity[2]	Tint[3]	Polysaccharides
White (11°) not filtered	0.138	2.73	610
White (11°) filtered--0.2 μm memb.	0.099	2.80	424
White (11°) filtered--0.8 μm memb.	0.098	2.77	392
Rosé (12°) not filtered	0.725	1.35	557
Rosé (12°) filtered--0.2 μm memb.	0.656	1.32	438
Rosé (12°) filtered--0.8 μm memb.	0.661	1.35	417
Red (11°) not filtered	0.399	0.79	860
Red (11°) filtered--0.2 μm memb.	0.366	0.76	451
Red (11°) filtered--0.8 μm memb.	0.388	0.76	440
Red (12°) not filtered	0.445	0.92	853
Red (12°) filtered--0.2 μm memb.	0.411	0.91	621
Red (12°) filtered--0.8 μm memb.	0.414	0.91	426

Note: (1): sugar content expressed in ° Brix; (2): sum of light absorptions at 420 nm
(D_{420}) and 520 nm (D_{520}) ; (3): ratio of D_{420} to D_{520}
[Adapted from Castelas and Serrano, 1989]

6.2.4 Clarification and Separation in Beer Brewing

Inorganic membranes have also been used in the clarification of other fermented
alcoholic beverages such as beer and vinegar in recent years. Two important applications
of membrane filtration for beer production are the removal of bacterias and beer
recovery from the so-called tank bottoms. They are treated in the following.

Bacterias/microorganisms removal. Bacterias or microorganisms are known to spoil
beer and reduce its shelf life. Traditionally the filtration process is applied to remove
microorganisms as well as the yeast. This method, however, is not efficient enough to
produce beer of consistent quality with a very low level of bacterias. Pasteurization,
therefore, is normally used before tapping to ensure the removal of bacterias or
organisms. This added heat treatment, however, can cause oxidation of some aroma
compounds as a result.

Microfiltration with inorganic membranes is a promising alternative. As early as 1964
porous silver membranes (composed of permanently molecular bonded pure silver
particles) in the disk form were commercially available. Silver membranes with a
maximum pore diameter of 1.2 μm were tested successfully on the pilot scale for cold
sterilization of beer to remove any organisms that can cause spoilage in closed

containers during normal storage [Jordan and Greenspan, 1967]. The resulting permeate flux of beer is about 12,200 L/hr-m^2 for a single membrane and the yeast cells in the feed stream was reduced to essentially zero.

Crossflow microfiltration with a porous inorganic membrane is a feasible alternative to the pasteurization process with the potential advantages of avoiding heat exposure of the beer. In principle, by choosing the proper pore size of the membrane, those particles or organisms otherwise passing through the filtration process can be captured by the microfiltration membrane, thus improving the quality of the clarified beer. The desired quality of the membrane filtered beer demands that essentially no changes in the color, bitter flavor and some proteins (such as foam stabilization proteins) should occur [Trägårdh and Wahlgren, 1989].

Two alpha-alumina membranes of 0.2 and 0.5 μm pore diameters have been tested at 20°C for filtering several types of light, low-alcohol (1.8% alcohol by weight) beer. The test samples of beer were prepared by adding the Lactobacillus type microorganisms to a pasteurized beer at a controlled level of 100-1,000 microorganisms/ml [Trägårdh and Wahlgren, 1989]. This range of microbial counts is equivalent to a major contamination at a brewery.

Taking an analogy from a study of milk microfiltration, Malmberg and Holm [1988] employed the concept of a constant transmembrane pressure (TMP) where the pressure difference across the membrane is essentially constant along the length of an alumina membrane. This has the advantage that the membrane flux can be sustained at a high level over a long period of time. With a crossflow velocity of 7 m/s and under a TMP of 0.55-1 bar, the 0.2 μm membrane retains too much nitrogen (12%) and color (30%) which are not acceptable. In contrast, the 0.5 μm membrane has reduced the microorganisms content of the beer to less than 1 count per 10 ml while retaining only 3% color and insignificant amounts of nitrogen and total solids in the retentate.

As usual, the permeate flux increases with increasing TMP in the 0.09-0.6 bar range. High fluxes were observed initially but fouling results in flux decline over time. This flux decline appears to require an optimized start-up procedure and TMP and a chemical cleaning procedure more effective than that used in their study: cleaning with a detergent and a 1-2 % nitric acid augmented by a rinsing step with deionized water before and after each chemical cleaning step.

Beer recovery from tank bottoms and yeast recovery. In beer production, yeast is recovered after fermentation. This is normally done by centrifuging but potentially can be accomplished by dead-end or crossflow microfiltration inorganic membranes.

Meunier [1990] tested the feasibility of this new approach with 0.8 μm alumina membranes under a transmembrane pressure of 3 bars and a crossflow velocity of 3 m/s at 5°C. The initial permeate flux declines to a level of about 20 L/hr-m^2.

"Tank bottoms" is another potential area that can utilize inorganic membranes for processing beer. Tank bottoms refers to a coagulated liquid extract that contains solids, fining materials and some yeast. It contains 90 to 99% beer. There are two types of tank bottoms: the fermentation vessel tank bottoms and the maturation vessel tank bottoms. Their solids contents are 10-15 g/L and 40-50 g/L, respectively. Traditionally, tank bottoms is treated with such techniques as vacuum filters, filter press and centrifuges. Microfiltration has become an alternative in recent years in view of its potential savings in operating costs. It is capable of directly producing bright beer without the use of filter aids and with low dissolved oxygen as the membrane process is not exposed to the air. To realize significant savings on polishing and clarifying chemicals (such as kieselguhr, various forms of a diatomaceous earth widely used in beer brewing, and fining materials), it is advantageous to recover beer extract from the tank bottoms by membrane filtration after yeast fermentation. The clarified beer can be blended in the process and amounts to about 5% of the total beer produced [Guibaud, 1989]. The retentate contains a concentrated yeast cream. Processing of the tank bottoms will also reduce the effluent problems.

A serious issue in the microfiltration of tank bottoms by ceramic membranes at a low temperature is the adsorption of the fining materials to the membrane surface which leads to fouling. Therefore, a number of microfiltration tests have been conducted in the absence of finings [e.g., Finnigan et al., 1987; Meunier, 1990]. Alumina membranes with a pore diameter in the range of 0.12-1.8 μm have been tested at 0-4°C and ambient temperature under an average TMP of 4 bars. Microfiltration at the lower temperatures helps preserve the key components (for appearance and taste) in a quality beer. The concentration due to microfiltration increases from an initial value of 2-6% to about 26% by weight and the resulting steady state flux is about 20 L/hr-m^2, similar to that in the separation of yeast. Backflushing can be used to minimize the fouling layer and maintain the permeate flux. In another study, Shackleton [1987b] has obtained a steady state permeate flux of about 60 L/hr-m^2 with backflushing and a dry solids content of 25%. Beer clarification or recovery of beer from the tank bottoms can be done in a two-stage process using monolithic, multi-channel ceramic membranes [Shackleton, 1987a]. A portion of the retentate from the first membrane filtration is directed to the second membrane filtration for further recovery of beer. The feed to the second membrane has a solids concentration of about 4%.

Beer clarification by ceramic membrane crossflow microfiltration has not been commercially practiced yet. But the ongoing application research and development efforts in this area and the potential savings in the diatomaceous earth cost will undoubtedly drive the technology to commercialization.

6.3 BIOTECHNOLOGY AND PHARMACEUTICAL APPLICATIONS

A number of important applications of inorganic membranes in biotechnology are related to bioreactors for enzymatic and microbial conversion processes. These will be discussed in Chapter 8 which covers the use of inorganic membranes both as the separators and reactors or catalysts.

Most of other major applications in biotechnology have to do with the separation of plasma and separation and harvesting of enzymes or microorganisms. A clear advantage of many inorganic membranes for these biotechnology applications, as for the food applications, is their biocompatibility, narrower pore size distribution than organic polymeric membranes and chemical and thermal stabilities to meet the requirements of the cleaning cycle which may include sterilization and/or chemical cleaning. Another reason why inorganic membranes are attractive for biotechnology applications is that many liquids in these applications have high viscosity and many ceramic membranes can withstand the resultant high shear stress imposed by these liquids when they are pumped past the membrane surface. Higher permeate flux, longer production runs between cleaning cycles and shorter cleaning times have been some of the major reasons that ceramic membranes have the edge over organic polymeric membranes [Shackleton, 1987].

Many processes in this field have been developed only in recent years and their details are often well protected as a competitive edge. However, it is evident that ceramic and glass membranes are increasingly being evaluated for biotechnology applications. The generic layout of a fermentation plant using ceramic membranes is described in Figure 6.10. The entire system is fully steam sterilizable. During the fermentation run, a portion of the liquid can be circulated through the membrane to remove some of the product so that nutrient can be fed continuously to the fermentation tank. At the end of the fermentation run, the membrane is operated at a full capacity to separate biomass from the spent medium. The retentate can be recirculated until the total solids reaches a very viscous state and is discharged automatically from the system.

6.3.1 Filtration of Fermentation Broths

Due to their greater chemical and thermal stabilities and narrower pore size distributions compared to polymer membranes, ceramic membranes are attractive in a number of filtration applications related to the fermentation broths. They can be used for either upstream or downstream processing.

Separation of microorganisms. In wastewater treatment utilizing bioreactor systems, there are needs to concentrate and separate microorganisms. During or after the fermentation processes, bacterias or cell debris need to be removed from the system. Crossflow filtration has been investigated for these purposes. An example of this is to treat the evaporator condensate discharged from a Kraft pulp mill.

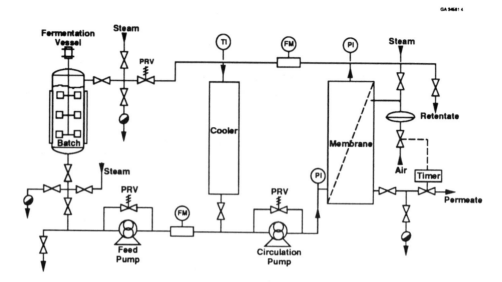

Figure 6.10 A generic layout of a fermentation plant using ceramic membranes [Shackleton, 1987]

To separate microorganisms from such a thermophilic methane fermentation broth, Imasaka et al. [1989] have employed a different approach where a normal liquid phase flow, gas-liquid flow or gas-liquid-solid flow is circulated through vertically installed alumina membranes in multistages. The idea of the multiphase flows is to promote turbulence which should reduce the permeate resistances attributed to concentration polarization, gel layer and fouling. The flow patterns can encompass slug flow, froth flow, annular flow and gas-liquid-solid three-phase flow. Symmetric (homogeneous) alumina membranes having a pore diameter ranging from 0.2 to 0.57 μm and a composite alumina membrane with a mean pore diameter of 0.2 μm for the separating layer were tested for concentrating microorganisms in the methane fermentation broth. The microorganisms are successfully retained by these membranes as further evidenced by scanning electron micrographs showing deposit of microorganisms on the surface of but no penetration into the membranes. The microorganism used has a size distribution that is all larger than 0.4 μm and has a mean particle diameter of about 0.8 μm. The flocs of the microorganism are significantly larger at 4 μm or so.

As expected, a composite membrane gives a higher permeate flux than a homogeneous membrane having the same pore size, membrane thickness and operating conditions. The multiphase flow approach mentioned above, however, has not provided a higher flux than the normal liquid flow case. A permeate flux of 100 L/hr-m^2 can be obtained with a crossflow velocity of about 2 m/s or when the feed gas velocity is about 3 m/s (under the standard condition of 0°C and 1 atm), both under a TMP of 1 bar using a 0.2 μm membrane [Imasaka et al., 1989]. While increasing the crossflow velocity helps improve the permeate flux in many cases, too high a crossflow velocity may not be desirable. Apparently, intensive shear force may be generated by an excessive crossflow velocity or feed gas flow rate and/or solids content in the multiphase flow inside the membrane tubes. When this occurs, the dynamic gel layer that acts as the effective self-rejecting membrane may be removed or becomes so thin that plugging of membrane pores may begin to take place. This consideration is also important in maintaining a sustainable flux over an extended period of time.

In another study, similar fluxes have been observed. Through the use of microfiltration grade alumina membranes, broth clarification has concentrated the bacterias to a solids content of 21% with an average permeate flux of 120 L/hr-m^2 [Guibaud, 1989].

Yeast separation and concentration. Extraction of ethanol from biomass requires several separation steps, traditionally by centrifuging, sedimentation and cake filtration. First, after the fermentation, the yeast is removed from the fermentation broth and may be recycled. Additionally, after ethanol is stripped from the fermentation broth by steam followed by the removal of solid fractions, the remaining material called the thin stillage can be clarified for reuse upstream as the process water.

Alumina and other ceramic membranes of various microfiltration pore sizes have been used for the separation of yeast (saccharomyces cerevisiae) from the broth and the clarification of thin stillage [Cheryan, 1994]. A typical flux of 110 L/hr-m^2 can be obtained with a crossflow velocity of 4 m/s and a transmembrane pressure of 1.7 bars. The crossflow velocity is found to markedly affect the membrane flux. Concentration factors (ratios of final to initial concentrations) of 6 to 10 for both the broth and the stillage can be achieved. Backflushing with a frequency of every 5 minutes and a duration of 5 seconds helps maintain the flux, particularly in the initial operating period. The permeate flux for both types of separation reaches steady state after 30 to 90 minutes.

Similarly, Redkar and Davis [1993] studied crossflow microfiltration of yeast cell suspensions in tubular alumina membranes having a mean pore diameter of 0.2 μm. Under a TMP of 0.35-2.1 bars and a crossflow velocity of 0.1-2.6 m/s, the steady-state permeate flux varies from 108 to 270 L/hr-m^2 without backflushing. Like in many other microfiltration tests, the initial flux decline follows dead-end filtration theory. The steady-state flux increases with increasing shear rate (or crossflow velocity) and decreasing feed concentration, and they are nearly independent of the TMP. The permeate flux of washed yeast cells is 1.5-3.5 times that of unwashed yeast cells. The difference is attributed to the presence of extracellular proteins and other macromolecules in the unwashed yeast suspensions. These biopolymers cause higher cell adhesion and greater resistance in the cake layer, thus reducing the permeate flux.

Another study utilizing microfiltration zirconia membranes with a pore diameter of 0.14 μm for separation in the production of alcohols from biomass indicates that the yeast concentration can reach 300 kg/m^3 (dry weight) with a permeate flux of 400 L/hr-m^2 under an average TMP of 1.8 bars [Lafforgue et al., 1987]. Langer and Schnabel [1989] have also studied the use of porous glass membranes (0.04 μm) for the separation of yeast from fermentation broth with a permeate flux of up to 80 L/hr-m^2 under 0.3 - 0.6 bar using backflushing. Initially the permeate flux drops to a level of about 50% but backflushing is effective in restoring the flux to its original value.

Filtration of yeast cell debris using a composite ceramic/metal-mesh membrane with a pore diameter of 0.2 μm has been studied [Gallagher, 1990]. Shown in Figure 6.11 is the permeate flux data of a 1% (by weight) bakers yeast suspension under a TMP of 1 bar. The flux appears to have a square root dependence on the crossflow velocity. This dependence appears to be deviated from the normal range of Eq. (5-8). The exponent of 0.5 is smaller than the lower end of the range. The flow is believed to be in the laminar flow regime and the concentration polarization, not particle fouling, controls the permeate flux.

The flux is affected by the solids content of the suspension. Their relationship is depicted in Figure 6.12 which shows that, while the flux decreases with increasing solids content when the suspension is dilute, it is relatively insensitive to the solids concentration in the 1 to 10% (by weight) range.

Filtration of biomass solids. Microfiltration of methane fermentation broth has been demonstrated with alumina composite membranes having pore diameters of 0.05 to 0.8 μm [Narukami et al., 1989]. The feed to the membranes is a fermentation broth incubated at 37°C using heat treated liquor of sewage sludge as a substrate. Under the condition of a low crossflow velocity of 0.5 m/s and a transmembrane pressure of 1 bar, the permeate flux for filtering 4.3 - 8.1 g/L suspended biomass solids using the 0.8 μm membrane is approximately 30 L/hr-m^2 without backflushing. But the flux drops off to approximately 20 L/hr-m^2 in 5 hours and stays relatively stable for 50 hours. The membrane layer lies on the outer surface of the membrane tube and the driving force of the membrane filtration comes from suction on the inner surface of the membrane tube. It has been found that in this configuration the flux becomes 40 to 80% lower if, instead of suction from the inner surface, a pressure is applied on the outer surface. This difference is attributed to the notion that the biomass is compacted under the pressure mode but only forms a loose cake layer under the suction mode. The SEM observation and gel permeation chromatography results seem to support this hypothesis.

Yeast can be separated from an ethanol fermentation broth by porous ceramic membranes with backflushing [Matsumoto et al., 1988]. Tubular alumina membranes with a nominal pore diameter of 1.6 μm were demonstrated to be effective for this application with a maximum permeate flux of 1,100 L/hr-m^2 with backflushing. The permeate flux increases with increasing feed rate (or crossflow velocity) and TMP and with decreasing yeast concentration. Various backflushing techniques were investigated and the reverse flow of filtrate (instead of air) either by pressure from the permeate side or suction from the feed side was found to be most effective. The backflushing interval (the reciprocal of the backflushing frequency) should be as small as possible.

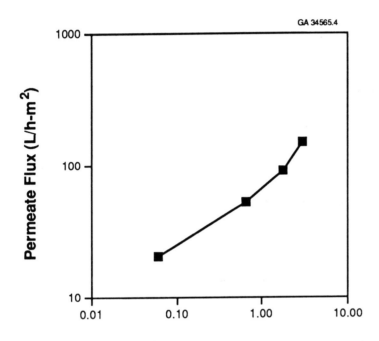

Cross Flow Velocity (m/s)

Figure 6.11. Effect of crossflow velocity on permeate flux of a bakers yeast suspension through a ceramic/metal mesh membrane [Gallagher, 1990]

Separation of lactic and propionic acids. The lactose fraction in the sweet whey permeate from cheese whey ultrafiltration can be fermented to produce lactic acid. In conjunction with the fermentation step, inorganic membranes have been tested in a continuous process to separate the lactic acid. This approach improves the productivity and reduces energy consumption compared to a conventional fermentation process. In addition, it produces a cell-free product. In a conventional process, some cells, although immobilized, are often detached and released to the product. Zirconia membranes with a MWCO of 20,000 daltons were operated at 42°C and a crossflow velocity of 2-5 m/s for this purpose [Boyaval et al., 1987]. The resulting permeate flux is 12-16 L/hr-m^2. To obtain a relatively high sustainable permeate flux, the transmembrane pressure should be increased slowly to reduce the immediate impact of fouling, thus prolonging the filtration cycle. The high cell concentration from the fermentator can cause fouling. Therefore, periodic cleaning is required and involves circulation of an alkaline solution at 70°C and nitric acid at 50°C with water rinsing before and after each treatment.

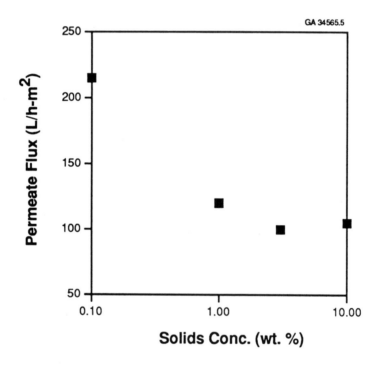

Figure 6.12 Effect of solids concentration on permeate flux of a bakers yeast suspension through a ceramic/metal mesh membrane [Gallagher, 1990]

Propionic acid and its derivatives are used in food, perfume and plastic applications. Traditional processes for making this compound, however, have limited productivity due to the low growth rate of the propionic bacteria and the inhibitory effect of the acid on the fermentation. The cheese whey permeate can be an inexpensive source of propionic acid. Propionic acids can be produced by fermentation of sweet whey permeate in a stirred tank reactor with cells separated from the medium and recycled back to the reactor by an ultrafiltration ZrO_2 membrane on a carbon support [Boyaval and Corre, 1987]. This arrangement reduces the propionic acid concentration and increases the volumetric productivity. High TMP can cause severe fouling of the membrane. The highest volumetric productivity was 14.3 g/L-h for a propionic acid concentration of 25 g/L. This value of the productivity was high compared to 5 g/L-h for a comparable system using a 0.14 μm zirconia-titania membrane on a carbon support with the propionic acid concentration at 17 g/L [Blanc and Goma, 1989]. Nevertheless, both values are significantly higher than other methods of producing propionic acid, by an order of magnitude difference in most cases.

Other filtration applications. Another application is the recovery of polysaccharide from sugar fermentation broth. Good clarification results have been obtained with a flux of 200 L/hr-m^2 [Guibaud, 1989].

6.3.2 Separation and Purification of Blood Components

Separation of plasma from blood can be used to remove toxic substances with high molecular weights in body fluid which is important in the treatment of many fatal diseases [Nose et al., 1983] and to collect plasma for blood banks to produce plasma fractionates [Ikeda et al., 1986]. Organic polymeric membranes with a mean pore diameter finer than 0.5 μm have been employed to some extent for these purposes. However, their wide pore size distributions and protein adhesion problems make their permeate flux quickly decline and their separation efficiency low. In addition, polymeric membranes generally can not withstand sterilization by autoclaves or chemical cleaning.

In contrast, microporous glass and ceramic membranes have the potential of being well suitable for plasma separation. They can be sterilized by autoclave prior to each use, repeatedly regenerated after fouling by heating in furnaces or rinsing with chemicals. Microporous glass and alumina membranes with a mean pore diameter of 0.2 to 0.8 μm offer practically the same level of permeate flux as organic polymeric membranes with an equivalent pore size [Ozawa et al., 1986] for treating bovine plasma and 0.1% aqueous solution of bovine albumin at 37°C. Fouled membranes can be regenerated many times by heating at 1,000°C for two hours (alumina) or by rinsing with 500 ppm sodium hypochlorite solution at 37°C for 30 minutes (glass). The ability to essentially completely regenerate the fouled membranes is definitely an advantage of inorganic membranes. Besides filtration and regeneration characteristics, another important issue in plasma separation is any appreciable elution of the membrane constituents. Aluminum is not detected. However, some silicon and boron are eluted in an amount below 100 ppb. In terms of protein sieving and the permeate flux, the 0.4 μm glass membrane was found to be the most acceptable with a flux of 20 L/hr-m^2 and a protein sieving of 93%. The preferred porosity should exceed 40% and membrane thickness less than 100 μm.

Plasma separation for preparing plasma fractionates should be performed at low temperatures to prevent bacterial contamination and protein denaturation. Sakai et al. [1989] found that organic polymer membranes, when used for low-temperature plasma

separation, creates hemolysis at a TMP of 0.067 bar while microporous glass
membranes produce no hemolysis even at a TMP of 0.133 bar. Hemolysis occurs when
red blood cells enter small membrane pores of several microns in diameter while
pressurized during membrane separation and the stress applied to cell membranes
exceeds a threshold value over a certain period. Adequate permeate flux and protein
sieving for the low-temperature (10-37°C) plasma separation can be obtained with
porous glass membranes having a mean pore diameter of 1 to 1.5 μm at a TMP of 0.033
to 0.067 bar. The permeate fluxes for a bovine plasma (concentration of 65 kg/m^3) and a
bovine blood (protein concentration of 65 kg/m^3 and hematocrit concentration of 30%)
are shown in Figure 6.13. Scanning electron microscopy reveals the presence of red
blood cells at cavities located on the membrane surface but not in the permeate after
plasma separation. This indicates that no red blood cells pass through the glass
membranes with a pore diameter in the 1.0 to 1.5 μm range. The amount of red blood
cells adhered to the membrane surface is equivalent to about 30 g/kg which is essentially
the same level when polymer membranes are used for the same application. If
membranes with a pore diameter larger than 1.5 μm are used, red blood cells tend to
plug the pores resulting in a lower flux.

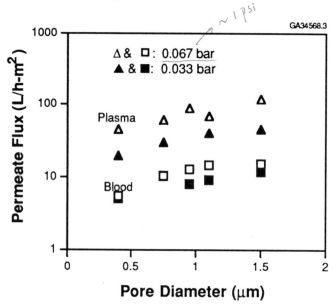

Figure 6.13 Dependence of permeate flux of bovine plasma and bovine blood through
microporous glass membranes on pore diameter under a TMP of 0.033 and 0.067 bar
and at 37°C [Sakai et al., 1989]

Blood fractionation has been commercially practiced using microfiltration ceramic
membranes. A multi-stage system is used where alumina membranes with varying pore

sizes effect the separation and purification of the main blood components [Guibaud, 1989].

6.3.3 Water for Injection

To produce acceptable water for injection, pharmaceutical manufacturers are required to use only the approved technologies. Most of the pharmaceutical companies have been using distillation which involves phase change allowing only less than 0.25 endotoxin units/ml. The water so produced is considered "endotoxin-free." The endotoxins, in non-aggregated form, have a molecular weight of about 10,000 daltons [Collentro, 1994]. Reverse osmosis has also been approved for making the water for injection in the past decade although only a few pharmaceutical manufacturers actually use the new unit operation. This is a result of several concerns associated with reverse osmosis: the fragile nature of organic membranes (in relation to the high transmembrane pressures required), improper brine seal design, and membrane scaling and fouling. Ultrafiltration, particularly using inorganic membranes, offers an attractive alternative to reverse osmosis or distillation. In fact, Japan has adopted ultrafiltration as an acceptable method for producing water for injection. The Pharmaceutical Manufacturer's Association in the U.S. has proposed changes that will allow the use of "any validatable process." This will present opportunities for inorganic membranes in the ultrafiltration range due to their inherent characteristics.

6.3.4 Other Biotechnology and Pharmaceutical Applications

Inorganic membranes have been mentioned in other usage in the biotechnology and pharmaceutical industries. For example, zirconia and alumina-based ceramic membranes have been incorporated in the following operations [Cueille and Ferreira, 1989]: purification and concentration of antibiotics, vitamins, amino-acids, organic acids, enzymes, biopolymers and biopeptides for the fermentation steps in the more conventional applications; human blood derivatives, vaccines, recombinant proteins, cells culture and monoclonal antibiotics in newer applications; and pyrogen removal for ultrapure water.

6.4 WASTE OIL TREATMENT

6.4.1 Oil-water Separation

The sources of oily wastes in the industry are abundant. Industrial wastewaters from metal fabricating and plating, food processing, etc. contain oily contaminants such as motor oils, greases and vegetable oils. They are required to be treated for removing oils, greases and some solids before discharge into sewers or open water storage. Compared to other treatment technologies such as adsorption, cartridge filtration and chemical treatment, membrane separation offers the advantage of relatively low energy consumption. Organic membranes have been used for removing oily contaminants by ultrafiltration and reverse osmosis. In those applications, however, organic membranes are prone to fouling caused by some other components of the wastewaters such as microorganisms and bacterias and corrosion attack by such chemicals as ketones, benzene and toluene. The inherently better chemical and thermal resistances and the ability to be periodically backflushed have motivated feasibility studies of inorganic membranes for processing oils and greases in the wastewaters.

Some examples of wastewater samples from a lubricating oil producing plant and a vegetable oil processing plant are profiled in Table 6.10 [Bhave and Fleming, 1988]. The waste lubricating oil from the slop oil stream does not require pretreatment. Since there is an appreciable fraction of solids finer than 0.2 μm in the lubricating oil, ultrafiltration alumina membranes with a mean pore diameter of 4 and 50 nm were used in their study. In contrast, the vegetable oil-containing sample from a soybean oil refinery needs a pretreatment in conjunction with membrane separation. First, the pH of the vegetable oil plant sample is adjusted to approximately 3 using nitric acid and the sample is then added with a high molecular weight cationic polymer as a flocculant at an optimal dosage of 50 mg/L. The pretreated vegetable oil-containing sample has the added advantage that it can be processed with a microfiltration alumina membrane which results in a greater permeate flux than an ultrafiltration membrane. Since the mean particle size of the flocs is about 1.5 to 2 μm, an alumina membrane with a mean pore diameter of 0.2 μm can be used with a gain in the flux.

TABLE 6.10
Compositions of wastewater samples from a lube oil plant and a vegetable oil plant

	Lube oil plant sample (mg/L)	Vegetable oil plant sample (mg/L)
BOD	7-50	500-2,000
COD	200-1,000	1,500-10,000
TSS	500-1,000	400-1,600
TDS	1,000-2,500	1,100-2,900
Oil & grease	80-120	200-700
pH	9-11	4-6
Total phosphorus (as PO_4)	0.01-1	1,000-2,500
Iron	--	1-6
Silicon (as SiO_2)	--	1-16
Mn	--	0.01-0.25

Note: BOD = biological oxygen demand; COD = chemical oxygen demand; TSS = total suspended solids; TDS = total dissolved solids.
[Adapted from Bhave and Fleming, 1988]

Alumina membranes having a mean pore diameter of 4 and 50 nm reduce the oil and grease concentration in the wastewater containing lubricating oil from 80-120 mg/L in the feed to about 2-4 ppm in the permeate. This is equivalent to a retention efficiency of about 96% [Bhave and Fleming, 1988]. The attainable permeate flux is close to about 90 L/hr-m^2. The permeate fluxes for the two membranes are shown in Figure 6.14 where they are compared to corresponding clean water fluxes.

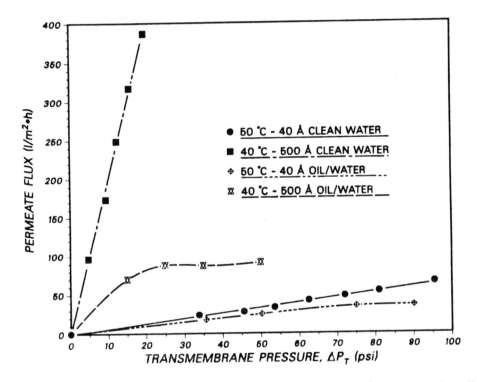

Figure 6.14 Permeate fluxes of ultrafiltration alumina membranes for processing oil-water samples [Bhave and Fleming, 1988]

As in many other membrane separation applications, the clean water flux is essentially proportional to the transmembrane pressure difference (TMP). When solutes, macromolecules or particulates are to be separated from the solvent (e.g., water), the permeate flux is first a linear function of the TMP and is in the pressure controlled regime. Although similar to the behavior of water flux, the permeate flux is nevertheless lower. Beyond a "threshold pressure," the permeate flux is insensitive to TMP due to concentration and gel polarization near the membrane surface. This behavior is so-called mass transfer controlled. It appears that the larger pore membrane, 50 nm in pore diameter, reaches the threshold pressure sooner than the finer pore membrane, 4 nm in pore diameter. There is a significant advantage of operating the membranes at a higher temperature to increase the permeate flux. Figure 6.15 compares the permeate flux at 20 and 50°C for the oil/water mixture and clean water. As a result of a lower viscosity at a higher temperature for a typical dilute solution or dispersion, the permeate flux increases. For example, by raising the temperature from 20 to 50°C and keeping other conditions constant, the permeate flux for a 4 nm alumina membrane approximately doubles. While this may be a process benefit, the trade-off for a higher energy cost needs to be considered.

As indicated in Table 6.11, the 0.2 μm alumina membranes treating the oily wastewater from the vegetable oil plant can produce a very high permeate flux (several hundred L/hr-m^2), particularly when a high dosage of the flocculant is used in the pretreatment step. The oil and grease level is reduced from 200-700 mg/L in the feed to about 5-10 mg/L in the permeate with a corresponding volume reduction of more than 90% [Bhave and Fleming, 1988]. The alumina membrane system in that study uses backflushing (with a backflush pressure of 3.5 to 4 bars) for a duration of 0.5 to 1 second and a frequency of once every 3 to 5 minutes. Flux decay for the periods between two consecutive backflushing operations has been found to be about 7 to 10%.

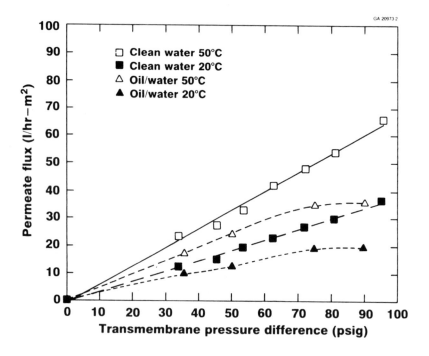

Permeate Flux of Oil/Water Emulsion

Figure 6.15 Effect of operating temperature on the permeate flux of an alumina membrane [Bhave and Fleming, 1988]

TABLE 6.11

Permeate flux of an 0.2 μm alumina membrane for removing oil/grease from a wastewater containing vegetable oil

Flocculant	Flocculant conc. (ppm)	TMP (bar)	Temp. (°C)	Flux (L/hr-m^2)
None	0	0.9	60	27
Percol 778	10	0.3	65	42
Percol 778	50	0.7	58	430
Percol 778	250	0.5	59	550

Note: TMP = transmembrane pressure difference; crossflow velocity is about 3 m/s; pH = 5.5 to 6.0
[Adapted from Bhave and Fleming, 1988]

A process involving ceramic membranes has reportedly regenerated oil-water emulsions without breaking the emulsion. An application of this is to remove metal chips, bacterias and oil and grease from the oil-water machine coolant [Guibaud, 1989]. The bath regenerated by the ceramic membranes can last three months instead of the usual three weeks.

6.4.2 Filtration of Waste Oils

There are other types of oily wastes that are not mixed with water and can be processed by membranes for possible reuse. Purification of waste lubricating oil, for example, can be achieved by using crossflow ultrafiltration ceramic membranes. Residues consisting of suspended solids, ash and heavy metals are removed as the retentate while purified oil is generated as the permeate. Porous ceramic membranes on honeycomb cordierite supports have been successfully tested at 150 to 260°C for converting waste oils into usable "clean" fuel for combustion [Higgins et al., 1994]. The membrane flux greatly depends on two factors: temperature and membrane-feed stream interaction. The flux increase with the temperature can be explained by the temperature dependence of the oil viscosity. Membrane fouling, leading to flux decline, strongly depends on the nature of the membrane surface and the constituents in the feed stream. The fouling is particularly profound above 250°C most likely due to the thermal decomposition reaction of the oil.

6.5 PETROCHEMICAL PROCESSING

Many separation problems in the petroleum or petrochemical industry occur in gas or vapor phase and at high temperatures. These gas separation applications will be treated in the next chapter. Only those applications involving liquid phase are reviewed here.

The generally inherent thermal, chemical and mechanical stabilities of inorganic membranes, particularly ceramic ones, make them amenable to several liquid-phase separation processes in petrochemical applications including wastewater treatment. Three examples related to petroleum residues processing will be given first: solvent recovery from deasphalted oil, recovery of catalyst from converted oil and direct deasphalting of petroleum residues. These will be followed by the use of ceramic membranes for treating wastewaters containing precipitated heavy metals, solids and oils.

6.5.1 Petroleum Residues Processing

__Solvent recovery from deasphalted oil__. A large proportion of residues from petroleum refining by distillation has been used as heavy oil. More stringent environmental regulations have driven the need for catalytic processing of these heavy fuels. The major problem associated with this refining process is the presence of asphaltenes which tend to poison the catalysts. The asphaltenes can be precipitated from the residues by adding a light paraffinic hydrocarbon solvent such as pentane. The deasphalted oil is then removed from the solvent by ultrafiltration. The pentane can be recycled and the deasphalted oil is sent to catalytic cracking or hydrocracking. Figure 6.16 shows the separation results using alumina and zirconia membranes with pore diameters ranging from 2 to 15 nm at 147°C with a crossflow velocity of 4 m/s and a feed concentration of 250 g/L [Deschamps et al., 1989]. As expected, the flux increases with the pore size while the permselectivity (rejection rate) decreases from 100% for a pore diameter of 2 nm to less than 30% for 15 nm. Separation of the deasphalted oil and the solvent appears to occur mainly through a dynamic layer of the oil molecules at the membrane surface. The process does not appear to be economically competitive when compared to supercritical extraction.

__Recovery of catalyst from converted oil__. Another way to process the residues is to add hydrogen to effect hydroconversion which avoids the formation of a large quantity of asphalt. Solid catalyst is formed afterward by reaction. Membrane filtration is used to separate the converted oil from the catalyst. This makes it possible to partially recycle the catalyst to the reactor. Alumina and zirconia membranes with pore diameters ranging from 30 to 600 nm have been tested for this application. The membrane with a pore diameter of 30 nm yields a stable flux and a catalyst retention better than 98% [Deschamps et al., 1989]. Concentration polarization is significant and requires a high crossflow velocity and temperature to overcome it.

__Direct deasphalting of petroleum residues__. Ultrafiltration zirconia membranes with a pore diameter of 6.3 nm on carbon support have been used to remove asphaltenes from a long residue at a temperature of 150°C [Guizard et al., 1994]. With a higher than normal crossflow velocity of 11.5 m/s, no membrane fouling is evident over a period of 500 hours. The permselectivity can be higher than 75% with a flux as high as 40 L/hr-m^2 under a transmembrane pressure (TMP) of 8 bars. At higher TMP's, a dynamic membrane layer is formed and controls the separation mechanism. In this case, the effects of the material and pore size of the membrane become diminishing. In contrast, at low TMP's, the interactions between the residue and the membrane surface can be quite significant. Compared to other ceramic membrane materials, zirconia exhibits weaker interactions with the petroleum residues due to its lower adsorption potential. It has also been found that the structure and the molecular weight distribution of the asphaltene exert great influence over the performance of the ceramic membranes.

GA 35278.3

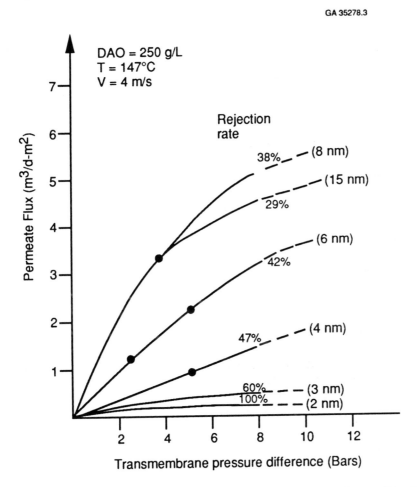

Figure 6.16 Separation of refining residues using ceramic membranes [Deschamps et al., 1989]

In the manufacturing processes of polymer resins, insoluble impurities in the form of gels and other high molecular weight materials (e.g., greater than 1×10^6 daltons) are often produced as well. They can cause undesirable product quality such as haze in the product. These impurities are currently removed from the resin by employing precoat filters. Recently ceramic membranes are being investigated for replacing the precoat filters [Muralidhara and Bhave, 1994].

6.5.2 Removal of Precipitated Heavy Metals, Solids and Oils

Ceramic membranes are attractive alternatives to conventional technologies for treating wastewater streams in petrochemical processing. They can function consistently without chemical attack for an extended period of time and can be backflushed and chemically cleaned in situ. And because of the latter reasons they eliminate atmospheric and worker exposure to hazardous and corrosive fluids containing, for example, aggressive chlorinated organics like EDC (1,2 dichloroethane). Ceramic membranes have been innovatively applied in pilot tests to treat different aqueous streams of varying pH and containing heavy metals, oils and solids at petrochemical manufacturing plants [Lahiere and Goodboy, 1993]. The pH of the waste stream and the nature and concentration of the various waste components are listed in Table 6.12. Three types of wastewaters were tested: vinyl chloride monomer (VCM) wastewater with heavy water, VCM oil/water/solids emulsions and linear alkylbenzene oily wastewater. It is seen that these characteristics of the waste streams can be very different. The membranes, however, respond to the upsets in the operating conditions and the feed concentrations rather robustly.

TABLE 6.12
Characteristics of three different petrochemical wastewaters and their components

	VCM wastewater with heavy metals	VCM oil/water/solids emulsions	Linear alkylbenzene oily wastewater
pH	0.5-1	12-13	4-12
Metal hydroxides	120 ppm by wt.	1-15% by volume	-----
Emulsified EDC	0	2-4% by volume	-----
Water-soluble EDC (wt. %)	0.3	0.3	-----
Oil & grease (ppm by wt.)	-----	-----	15-500
Other organics	water-soluble chlorinated organics	water-soluble chlorinated organics	1.5 ppm (by wt.) of benzene
Temperature (°C)	35-55	30-45	25-50

Note: VCM = vinyl chloride monomer; EDC = 1,2 dichloroethane
[Lahiere and Goodboy, 1993]

Removal of precipitated heavy metal solids. In the manufacture of VCM, an acidic process water stream with a low pH (e.g., 0.7) contains a high concentration of heavy metals such as iron, copper and nickel in hydroxide forms (120 ppm by weight as hydroxides). In addition, it is saturated with EDC at 0.3% by weight and also contains other chlorinated by-products of VCM production. Since incineration is required to dispose of the EDC-containing solids, a hazardous waste by itself, ceramic membranes can be used to filter and concentrate the solids to minimize the final volume before incineration. A two-stage membrane separation in series is employed for this purpose: the first stage to filter the solids in the stream and the second to concentrate the sludge [Lahiere and Goodboy, 1993].

First the process water is adjusted to a pH of 12 by the addition of caustic to precipitate the heavy metals as insoluble hydroxides, thus enhancing the separation by the membranes. Alpha-alumina membranes with a pore diameter of 0.8 μm and 1.4 μm are used as the first- and second-stage membranes, respectively. The resulting heavy metal hydroxide concentration in the permeate has been consistently lowered to about 1 ppm by weight from the original level of 120 ppm in the feed stream to the membrane operation. The sludge is concentrated to 17-20% by weight. The permeate flux values for the filtering stage and the sludge concentration stage are 630-920 and 160-230 L/hr-m^2, respectively, at a temperature in the 35-55°C range. These levels of the permeate flux are quite high which is required in petrochemical applications. About 80 and 99.5% of the water in the feed streams to the first and second membranes are recovered in their respective permeates.

It is noted that at the high solids concentration as in the second stage, only inorganic membranes, as opposed to organic polymeric membranes, are capable of processing such a highly viscous sludge without suffering from structural degradation. And because of the higher viscosity, the second stage membrane requires a higher crossflow velocity than the first stage membrane. Although the crossflow velocity was not specified, it is probably greater than about 1 m/s.

Removal of EDC emulsions. A commonly occurring alkaline (pH of 12-13) process waste stream in VCM plants contains emulsions of EDC, water and inorganic solids such as calcium/iron hydroxides. The emulsion which settles between the heavier (bottom) EDC layer and the lighter (top) water layer in a tank can create deficiency in the stripping and distillation towers by causing fouling and corrosion. As indicated in Table 6.12, the amount of emulsified EDC varies from 2 to 4% and that of metal hydroxides from 1 to 15% by volume.

Alumina membranes with a pore diameter of 0.2 μm have been tested successfully for removing emulsified EDC [Lahiere and Goodboy, 1993]. The soluble EDC, however, is not affected by the membrane treatment and remains at the same level before and after the operation. Pilot test results show that the water recovery is about 94%. The steady state permeate flux from the pilot test at 30-45°C is 1290 L/hr-m^2 which is about 4 to 6 times lower than that from the laboratory test. When the TMP required to maintain the same flux exceeds 2.1 bars, it is an indication that chemical cleaning is needed. The frequency and duration of the chemical cleaning cycle are approximately 5 days and one hour, respectively. Chemical cleaning consists of sequential applications of 5% hot caustic solution, water rinsing, hot 5% hydrochloric acid and another water rinsing.

Removal of oils from wastewater. Wastewater from a linear alkyl benzene plant contains 15-500 ppm by weight of aromatic and paraffinic oils, solids and dirt as shown in Table 6.12. Alumina membranes having a pore diameter of 0.2 and 0.8 μm have been proven in pilot tests for separating the above aromatic oils from the wastewater [Lahiere and Goodboy, 1993]. The oil and grease levels decrease from 15-500 ppm to less than 5 ppm by weight when chemical pretreatment is applied. The most effective pretreatment for yielding the highest permeate flux is the addition of 160 ppm HCl and 160 ppm ferric chloride by weight when 0.2 μm alumina membrane is used. The resulting flux is in the range of 1,250-1,540 L/hr-m^2. The finer pore membrane (0.2 μm) actually produces a higher flux than the coarser membrane (0.8 μm) probably as a result of the artifact of the pretreatment chemicals used which lead to a less densely packed gel layer. An ultrafiltration alumina membrane with a pore diameter of 0.05 μm has been found to be ineffective due to severe fouling.

6.5.3 Treatment of Waters Produced by Oil Wells

Water produced by offshore and onshore oil wells is called "produced water" and needs to be treated to render "clean water" with very low oil, grease and suspended solids concentrations. The current U.S. Environmental Protection Agency standards for those concentrations are in the tens parts per million (ppm) range. Future requirements are expected to be even more stringent. The treated "clean water" can be discharged overboard, re-injected or used as the feed water to the boilers. The current commercial technologies for removing oil, grease and suspended solids from the produced water include parallel-plate coalescers, gas flotation, granular media filtration, gravity separation, water clarification and emulsion breakers. It is desirable to develop new technologies which are more cost competitive, generate a lower volume of waste stream to be disposed of and are less sensitive to upsets or fluctuations in the feed conditions.

The oil/water emulsions with solids can be broken by the membrane process. The emulsions generated by both the waterflood and steamflood processes have been successfully treated using alumina membranes of different pore sizes at various locations including the U.S. and Canadian coasts, Gulf of Mexico and North Sea [Humphrey et al., 1989; Chen et al., 1991]. A schematic diagram of one such operation is shown in Figure 6.17.

Utilizing alpha-alumina membrane having a mean pore diameter of 0.5 to 1.5 μm, Humphrey et al. [1989] conducted extensive tests with 40 to 600 ppm oil and grease levels in the feed streams and reduced them to less than 8 ppm. The steady state permeate flux averages about 1,000 to 2,200 L/hr-m^2 under a crossflow velocity of 1.2 to 3.7 m/s. Using ceramic membrane modules for offshore applications has several advantages. One is the compact size and low weight. Second is the modular construction which makes it possible to easily expand. And the membrane operation appears to be able to accommodate upsets or fluctuations in the feed rates and compositions.

GA 34561.3

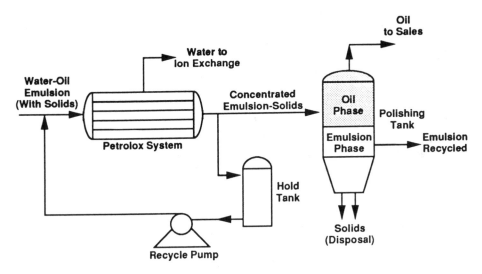

Figure 6.17. A ceramic membrane system used to separate water-oil-solids emulsions [Humphrey et al., 1989]

Similar results have been observed in an offshore application using alumina membranes [Guibaud, 1989]. With a flux of 1,600 L/hr-m^2, the membranes employed reduces the oil level in the treated water to less than 5 ppm. The filtered sea water can be either reinjected into the well or discharged to the sea.

Chen et al. [1991] further evaluated the above approach using alumina membranes with a pore diameter of 0.2, 0.5 and 0.8 μm in two onshore and two offshore pilot plants. By chemically pretreating the produced water prior to membrane filtration with backflushing and fast flushing, membrane fouling is significantly reduced, resulting in a higher sustainable permeate flux of up to about 3,000 L/hr-m^2. The oil and grease discharge level is reduced from 27-583 ppm to below 5 ppm and the suspended solids is lowered from 73-350 ppm to less than 1 ppm. The crossflow velocity varies from 0.9 to 4.6 m/s. The effective chemicals for pretreatment were not disclosed. It appears that the chemicals serve to flocculate a portion of the emulsified oil and suspended solids in the produced water. Thus, the flocculated solids loosely deposit on the membrane surface to form a hydrophilic dynamic separation barrier that prevents the oily substances and fine particles from penetrating the membrane pores to cause fouling. chemical cleaning is required after 66 or longer hours of continuous operations. The backflushing frequency is once every 2 minutes and lasts for 0.5 s. Fast flushing condition is 6 m/s, once every 2 minutes for 4 to 5 s.

6.6 OTHER PROCESS APPLICATIONS

6.6.1 Textile Industry

Inorganic membranes are attractive to textile processing due to the associated high temperatures and varying pH values of several process streams. The first application, currently commercialized, is to recover certain synthetic polymers such as polyvinyl alcohol as the warp sizing agents from textile waste streams. The second application under development is the removal of coloring dyes from the finishing and dyeing effluents before the wastewater can be discharged.

Recovery of polyvinyl alcohol. Zirconia membranes on porous carbon supports have been successfully used on a commercial scale in the textile industry since 1973. They are used to recover polyvinyl alcohol from the waste streams and reduce the "slashing" costs. The recovering rate is approximately 95%. The sizing agents are not biodegradable; therefore, recycling of the sizing polymers also minimizes the potential need for waste disposal. The recycled sizing chemicals although having changed in color can be reused again and do not affect the quality of the finished woven cloth. Strong sulfuric acid, caustic and hydrogen peroxide solutions are circulated through the membrane system at temperatures approaching boiling. These zirconia membranes appear to function well in these harsh environments and have been claimed to have a service life of 5 years or longer [Gaston County, 1979]. Each shell-and-tube module of the carbon-supported zirconia membranes may consist of about 1,000 membrane tubes which are subject to a transmembrane pressure of 7 to 10 bars. A permeate flux in the order of 100-150 L/h-m^2 under a transmembrane pressure has been reported. Two or more modules form a continuous loop through which the process stream is recirculated. Often a series of the ultrafiltration loops (e.g., 14 loops) are required to reach the target concentrations.

A schematic for the membrane-based process for recovering polyvinyl alcohol (PVA) as the sizing agent is depicted in Figure 6.18. In this process, the waste water containing the sizing chemical from the desize washer is first removed of any lint and trash through a coarse, mechanical filter. From the filter the stream flows into a buffer tank to be normalized prior to entering the ultrafiltration system which may consist of a series of loops to achieve the desired concentration level of the sizing agent. The reclaimed sizing chemicals can be supplemented with fresh PVA before returning to the slasher. Other polymeric sizing agents have also been reclaimed for reuse. Polyester sizing is an example where the permeation rate is higher than that involving PVA and the sizing cost is an even greater incentive for recycling. A production-scale zirconia membrane system for this application is shown in Figure 6.19.

Wastewater treatment. The effluents from textile processing contain some pollutants and become highly colored due to soluble and insoluble dyes. At present, there is no economically feasible method for removing the dyes and other chemicals present in the effluents from the finishing and dyeing operations.

Microfiltration ceramic membranes were recently proposed for treating those effluents [Soma et al., 1989]. Alumina membranes having a mean pore diameter of 0.2 μm were found to be effective for reducing the dyes and chemical oxygen demand (COD). Under the laboratory conditions of a TMP of 1-5 bars and a crossflow velocity of 3-5 m/s, the membranes retain 98% of insoluble dyes and 20-80% of soluble dyes. The rejection of soluble dyes can be improved by adding certain surfactants and reaches 97%. The effluents from a textile plant often contain surfactants which are capable of incorporating dyes in micellar structures or forming a dynamic separation layer near the alumina membrane surface, thus potentially improving the separation of soluble dyes and COD from the wastewater. Actual plant tests showed an 80% rejection of the dyes and 40% removal of COD with a long-term permeance of 26-28 L/hr-m^2-bar.

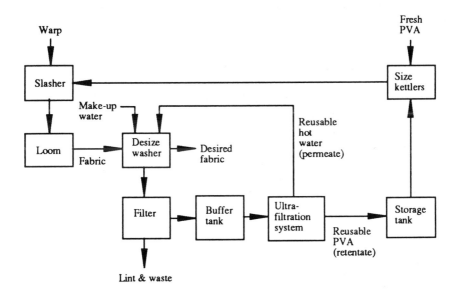

Figure 6.18 Schematic of the polyvinyl alcohol reclaim process by inorganic membrane ultrafiltration [adapted from Gaston County, 1976]

Zirconia membranes can also be utilized to remove wool oil and grease, resins, latex and high molecular weight dyestuffs larger than 20,000 daltons from process waste streams. The filtered hot detergent solutions can be reused in the washers.

While the formed-in-place or dynamic hydrous zirconium oxide membranes on porous stainless steel supports have been studied mostly for biotechnology applications, they have also demonstrated promises for processing the effluents of the textile industry [Neytzell-de-Wilde et al., 1989]. One such application is the treatment of wool scouring effluent. With a TMP of 47 bars and a crossflow velocity of 2 m/s at 60-70°C, the permeate quality was considered acceptable for re-use in the scouring operation. The resulting permeate flux was 30-40 L/hr-m^2. Another potential application is the removal of dyes. At 45°C, the dynamic membranes achieved a color removal rate of 95% or better and an average permeate flux of 33 L/hr-m^2 under a TMP of 50 bars and a crossflow velocity of 1.5 m/s.

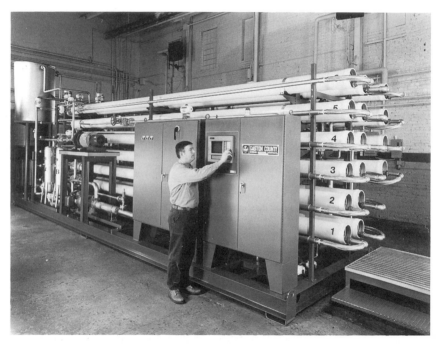

Figure 6.19 Photograph of a production-scale zirconia membrane system for recovering sizing agents in textile industry [Courtesy of Gaston County]

6.6.2 Paper and Pulp Industry

Many application studies have been conducted on the use of ultrafiltration and reverse osmosis to process "black liquors" in the paper industry. Inorganic membranes have the advantage of withstanding the relatively high process temperature, thus avoiding the need to lower the temperature in order to use some well known separation techniques such as organic polymer membranes. Zirconia membranes were the first inorganic membranes adopted for this field of applications.

In the manufacture of chemical pulp by the bisulfate process, residual "black liquor" contains, among other things, lignosulfonates (or called lignin sulfonates) and small organic molecules (especially sugar and mineral products). High molecular weight lignosulfonates can be separated by the use of polyvalent metal salts as the coagulants, but the coagulation treatment does not permit the resulting water to be reused. The use of inorganic ultrafiltration membranes enables separation and concentration of high molecular weight lignosulfonates in the effluents at the relatively high process temperatures (90-140°C) and, simultaneously, reduce those small organic molecules such as sugars so that the ratio of lignosulfonates to sugars increases. Additionally the relatively high viscosity and the pH value of the liquor to be processed can easily make organic membranes unfit for the application. The membrane-processed liquor can be recycled.

Barnier et al. [1987] tested two ceramic membranes at 85-115°C: one has alumina as the porous support for a metal oxide layer as the membrane with a molecular weight cut-off (MWCO) of 110,000 daltons and the other uses a porous carbon support for a metal oxide membrane with a MWCO of 70,000 daltons. To reduce fouling, high crossflow velocities are used to generate turbulent flow with a Reynolds number exceeding 15,000. Uninterrupted operations longer than a month have been observed with a slight decrease in the permeate flux. Chemical cleaning with acid, alkali, oxidation, reduction and detergent can restore the membrane flux to 90-100% of its initial value. Data in Table 6.13 reveals that the lignosulfonates concentration can be increased from about 105-124 g/L to 280-300 g/L and the lignosulfonates/sugars ratio increased more than two- or three-fold. The permeate flux ranges from 43 to 60 L/hr-m^2 and the energy consumption from 70 to 220 kwh/t of concentrate produced.

The low-molecular-weight lignosulfonates (LMW-LS) and metal salts remaining in the residual black liquors require a more refining separation scheme to remove them from the liquor. Reverse osmosis (or hyperfiltration) is a strong candidate for desalinating the liquor. The pH value of the liquor (either as-is or as a result of coagulation which leads to a pH of 3.0-3.5), however, essentially rules out most of the organic membranes. Reverse osmosis, however, can occur with dynamic membranes of colloidal iron hydroxide [Badekha et al., 1986]. It has been found that the highest retention of the LMW-LS is obtained when the dynamic membranes are formed from the iron hydroxide sol particles prepared through the oxidation of $FeCl_2$ by hydrogen peroxide. The membranes are formed dynamically on cellulose acetate ultrafilters. Rejection of sodium and calcium salts depends to a great extent on the LMW-LS concentration which may act as an organic polyelectrolyte with negatively charged functional groups. An LMW-LS concentration greater than 40 mg/L drastically reduces the retention of the salts by the membranes. In addition, pretreatments prior to the membrane filtration can significantly affect the rejection of LMW-LS and the salts. The best overall treatment for their removal has been found to be coagulation of $FeCl_3$ followed by oxidation of clarified liquor by hydrogen peroxide in the presence of Fe^{+3} and then membrane filtration. The resulting decreases of LMW-LS and salts concentrations in the liquor, as shown in Table 6.14, are accompanied by a surprisingly high permeate flux in the range of 45 to 60 L/hr-m^2.

TABLE 6.13
Concentration of lignosulfonates (LS) and reduction of sugars (S) in residual black liquors from chemical pulp production by ceramic membrane ultrafiltration

Membrane	A	A	B
Temperature (^oC)	110-115	85	85
Transmembrane pressure (bar)	7.5	6.7	5.0
Permeate flux (L/hr-m^2)	43	60	45
Starting LS conc. (g/L)	105	124	127
Final LS conc. (g/L)	300	280	292
Starting LS/S ratio	3	6	7
Final LS/S ratio	10	14	16
Volume concentration factor	5	6	7

[Barnier et al., 1987]

TABLE 6.14
Removal of low molecular weight lignosulfonates and sodium and calcium salts from residual black liquors

Separation method	LMW-LS conc. (mg/L)	Ca^{2+} conc. (mg/L)	Na^+ conc. (mg/L)	pH of treated liquor
Before treatment	300	0.2	0.23	10
Coagulation of $FeCl_3$ followed by membrane filtration	20-30	0.025-0.03	0.09-0.1	2.8-3
Coagulation of $FeCl_3$ followed by oxidation of clarified liquor by H_2O_2 in presence of Fe^{2+} and then membrane filtration	1	0.01-0.02	0.05-0.06	2.8-3.5

[Badekha et al., 1986]

6.6.3 Metal Industry

The conventional metal cleaning processes have been vapor degreasing, solvent immersion cleaning and cold cleaning. The solvents commonly used in these processes, unfortunately, are harmful to human health or environment. With the expected increasingly stringent regulations regarding those organic solvents, water-based metal cleaners are being used or developed. Recycling of these aqueous cleaners is preferred over their disposal or replacement because of the costs involved. Prior to recycling, however, oil, grease and particulates need to be separated from the cleaners. Crossflow ultrafiltration with ceramic membranes has been developed with an aim to reduce operating costs and improve operational reliability [Bhave et al., 1993; Chen, 1994]. The membrane separation does not need to be operated continuously and is used only when the cleaner bath quality reaches a threshold level.

Spent aqueous alkaline metal cleaners for an aluminum fabricating plant contains oil and grease from its machining operation and suspended solids (mainly aluminum oxide) from its deburring operation. The removal of these contaminants can be accomplished by employing an ultrafiltration ceramic membrane such as a zirconia membrane module with a mean pore diameter of 50 nm [Chen, 1994]. Seven 19-channel membrane elements are contained in a module. An average permeate flux of 40 L/hr-m^2 was obtained. The ultrafiltration system required chemical cleaning every 53 to 67 hours of actual operation. The chemical cleaning procedure involved circulating a 2% alkaline solution and a detergent at 71 to 82°C for one hour followed by 2 to 3.5 hours of water rinsing. Effective cleaning can restore the membrane permeability to 96% of its theoretical value. It is imperative to keep the feed tank agitated and/or heated to avoid solidification of the cleaner and formation of gel-like materials which may lead to membrane fouling.

The ultrafiltration zirconia membrane system described above removes about 92% of the total suspended solids (TSS), 47% of the total dissolved solids (TDS), 12% of chromium and 13% of copper. However, the membrane also rejects some alkaline salts (about 9%) and consequently make-up with fresh cleaner is required.

The above process for recycling spent aqueous alkaline cleaners for metal manufacturing plants can utilize other ultrafiltration ceramic membranes with a mean pore diameter of 5 to 100 nm, although zirconia membranes are preferred [Bhave et al., 1993]. A crossflow velocity of 3 m/s and a TMP of less than 5 bars are recommended.

6.6.4 Other Industries

<u>Radioactive waste treatment</u>. Radioactive liquid wastes can be treated by such conventional methods as precipitation of ferric hydroxide followed by sedimentation or ion exchange. This approach, however, suffers from the drawback of a relatively low decontamination efficiency and a large volume of sludge resulting from the use of flocculants. Furthermore, it has been found that organic membranes used to remove the active nuclides are degraded by exposure to high levels of γ radiation [Gutman et al., 1986]. Those membranes may become brittle and their separation efficiency is reduced, particularly under alkaline conditions. Ultrafiltration or microfiltration by inorganic membranes have been successfully tested in a pilot plant and reportedly chosen to treat some liquid wastes at a commercial facility [Cumming and Turner, 1988]. From the operational as well as economical points of view, it is very desirable to run the membrane treatment units over a long period of time (e.g., on the order of a year). Zirconia ultrafiltration and alumina microfiltration membranes appear to meet the needs.

Cumming and Turner [1988] used a zirconia membrane (MWCO of 20,000 daltons or equivalent to a nominal pore diameter of 2 nm) on a carbon support and a 0.2-μm alumina membrane on an alumina support to remove α, β, and γ activities from low level (about 10^{-5} Ci/m^3) active liquid wastes. The membrane modules were operated at a crossflow velocity of 4.5 m/s and a TMP of 2 to 5 bars. The best α activity removal, from 0.0068 Bg/ml down to below the detection limit of 0.0015 Bg/ml, is achieved at a pH of 4.5. The removal rate through crossflow filtration is typically at least 5 times higher than that by the ferric floc treatment. The removal can be enhanced by adding hydrolyzed titanium tetrachloride at a concentration of 0.01 Ti/L. Given in Table 6.15 are the reductions of β and γ (as emitted by ^{137}Cs and ^{60}Co) activities by processing the wastes with a 0.2 μm alumina membrane. Because these species are soluble, they require some additives such as copper ferrocyanide and zirconia phosphate at low dosages in the filtration circuit to enhance the separation. The overall removal of β and ^{137}Cs are consistently high while the separation of ^{60}Co varies more.

TABLE 6.15
Redution of radioactivities in low level wastes using a 0.2 μm alumina membrane

Batch	Activity (Bg/ml)		
	β	^{137}Cs	^{60}Co
1	0.394 → 0.06	0.23 → <0.002	0.12 → 0.055
2	0.131 → 0.045	0.177 → <0.002	0.07 → 0.012
3	0.215 → 0.065	0.098 → 0.002	0.06 → 0.03
4	0.18 → 0.081	0.06 → <0.002	0.041 → 0.014
5	0.143 → 0.046	0.054 → <0.002	0.037 → 0.025
6	0.15 → 0.035	0.066 → <0.002	0.028 → 0.011
7	0.585 → 0.47	0.034 → 0.005	0.49 → 0.38

Note: 20 mg/l of copper ferrocyanide and 10 mg/l of zirconia phosphate are added to the wastes [Cumming and Turner, 1988]

Some results with a 0.2 μm alumina membrane pilot plant containing 29 membrane tubes indicate that increasing the ferric ion concentration from 0.04 to 0.1 g/L effectively raises the steady state permeate flux from 125 to 210 L/hr-m^2 and prolongs the time for the flux decline to reach the steady state from about 6 hours to more than 20 hours, as shown in Figure 6.20 [Cumming and Turner, 1988]. Decreasing the ferric concentration below 0.04 g/L does not cause any further decline in the flux and raising it above 0.1 g/L does not enhance the flux.

The zirconia membranes used can be chemically cleaned by circulating a sodium hydroxide solution and a nitric acid solution through the filtration circuit, each at 50°C for one hour. The strengths of the alkali and acid can be 0.5 to 1 M. Alternatively, a new technique under development called direct electrical membrane cleaning may be used. It is based on the principle that the surface fouling materials can be periodically removed from a conductive membrane through the in-situ electrolytic generation of microscopic bubbles by the application of a short current pulse [Cumming and Turner, 1988]. The dislodged materials are carried away in the retentate. The effectiveness of this method is enhanced by a transient reduction in the TMP during current application. It is capable of maintaining a high permeate flux at a low crossflow velocity.

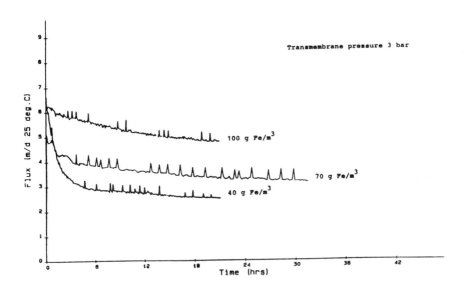

Figure 6.20 Permeate flux of an alumina membrane module processing low level activity wastes dosed with three levels of ferric nitrate at pH 5 [Cumming and Turner, 1988]

The above discussion applies to low level activity liquid wastes. Inorganic membranes have also been tested for treating medium level liquid wastes containing approximately 100 times the activity of the aforementioned low level wastes [Gutman et al., 1986]. The same type of zirconia membranes with a MWCO of 10,000 daltons used in treating low level wastes does not perform as well (in terms of the permeate flux and decontamination factor), but can function acceptably when processing medium level wastes. The problem is particularly pronounced if there are some defects in the membranes which tend to allow sufficient bacteria to flow through for a colony to grow on the permeate side of the membrane, leading to serious flux decline and poor quality of the filtrate. An adequate dosage of a biocide such as chlorine helps prevent the problem.

Calcium electrowinning in plutonium production. A potential application of inorganic membranes in radioactive waste treatment is in the industrially practiced direct oxide reduction process. In this process plutonium oxide is calciothermically reduced to plutonium in the presence of calcium chloride according to the following reaction:

$$PuO_2 + 2Ca = Pu + 2CaO \qquad\qquad (6-1)$$

The effluent salt is the radioactively contaminated calcium chloride saturated with calcium oxide which can not be discarded as-is. Thus it is desirable to cathodically reduce calcium oxide to calcium which can then be recycled back to the direct oxide reduction process. The primary difficulty in obtaining a cathodic calcium deposit is due to the evolution of carbonaceous anodic gases which encourage various reverse reactions in the cell leading to loss of deposited calcium.

A plausible approach to the problem is to separate the cathodic and anodic products, thus keeping the deposited calcium away from the anode. A porous ceramic (such as MgO) membrane can be used which provides effective transport of oxygen ions and yet separate all other anodic and cathodic products [Mishra et al., 1994]. The choice of the ceramic membranes is determined by other consideration factors as well: sufficient mechanical integrity to withstand the thermal shock and agitation, high corrosion resistance in the high temperature (about 900°C) molten salt environment, inertness toward the ionic species and electrical insulation. The cathode and the anode used are steel and graphite, respectively. The electrolytic recovery rate of calcium depends on the oxygen ionic diffusion through the porous membrane. As the flux of oxygen ions increases, the current density increases and so does the calcium recovery efficiency. The diffusion coefficient is affected by the temperature as well as the porosity of the ceramic membrane. The desired porosity is found to be over 30%. Thin and high surface-area membranes are also preferred.

Integrated circuit manufacture. In the manufacture of integrated circuits a light-sensitive coating called "photoresist" is applied to an electronic substrate to act as a mask for selective etching of the microcircuits. The quality of this fluid is imperative as it dictates the yield of the final integrated circuits. That is, the fluid should be free of all particulate matters and gels. Currently microfiltration is commonly used for purifying the photoresists.

Organic polymer membranes such as Teflon® membranes may be used but fiber shedding and medium migration are potential concerns. Silver membranes do not suffer from this problem and have been tested for removing particles and gels in the photoresists. Some of the photoresists can be processed this way with as fine a membrane pore diameter as 0.2 μm without causing observable damage to some critical low-molecular-weight branched polymers which are key to the photosensitive properties of the coating [Minneci and Paulson, 1988]. This may be attributed to the more orderly pore structures of the silver membranes than those of the organic polymer membranes.

Inorganic membranes are less prone to dissolution or leaching. This advantage can be put to use in some critical applications where valuable process liquids or liquid aids are filtered through an inorganic membrane such as silver or ceramic membrane provided the membrane matrix is chemically compatible with the liquids involved. Silver membranes have been reportedly used for purifying cleaning fluids in the microelectronics manufacturing [Minneci and Paulson, 1988]. In this case the low adsorptive propensity of the silver membrane minimizes any loss of the valuable fluids.

Separation of ethanol/water mixtures. Pervaporation is a developing membrane process which has elements of reverse osmosis and gas separation. In pervaporation, a liquid mixture is brought in contact with one side of a membrane while the permeate is removed as a vapor from the other side. The driving force is the difference between the partial pressure of the feed stream and the vapor pressure on the permeate side.

Certain inorganic membranes, particularly of the zeolite type, have shown great promises for separating ethanol/water mixtures by pervaporation [Sano et al., 1992; Sano et al., 1994]. For example, pure polycrystalline zeolite membranes of silicalite have been prepared hydrothermally on a porous stainless steel or alumina support and calcined at 500°C to decompose the organic amine occluded in the channels of silicalite as a template for the synthesis of the silicalite membranes. While pores in the order of 500 nm exist in the middle and the outer layer on the support side, it appears that there are no pores originated from silicalite grains in the selective layer on the solution side. A high separation factor, α, of ethanol/water exceeding 60 for a 5% by volume of ethanol solution at 30°C has been obtained. The separation factor here is defined as

$$\alpha = (C_{ethanol}/C_{water})_{permeate}/(C_{ethanol}/C_{water})_{feed} \qquad (6\text{-}2)$$

where $C_{ethanol}$ and C_{water} are the volume fractions of ethanol and water, respectively.

As in the case of other pervaporation processes, the underlying transport mechanism is by sorption-diffusion. In view of the small molecular sizes of ethanol and water (both smaller than 0.5 nm which is the estimated pore size of the membrane), the high separation factor is attributed to the dominant selective sorption of ethanol into the silicalite membrane. The data may also suggest that there are no cracks and pores between the silicalite grains within the membrane. The issue of cracks and/or pores between grains has been a major challenge associated with making integral zeolite or molecular sieve membranes. Any stress originating from the removal of volatile organic materials from the zeolite structure and from the difference in the thermal expansion coefficients between the silicalite crystals and the support can easily cause cracks during synthesis.

Increasing temperature has been found to decrease the separation factor only slightly while the concentration of ethanol in the mixture has a notable impact on the separation factor. The separation factor reaches a peak value at a feed ethanol concentration of about 3% by volume and then drops off rapidly with increasing ethanol concentration.

It is noted that only the calcined silicalite membranes give high separation factors and high permeate fluxes as indicated in Table 6.16. The air or vacuum dried membranes offer substantially lower flux and separation factor values. The results appear to suggest that the silicalite membrane changes from a water-selective membrane (with an α less than 1) to an ethanol-selective one (with an α greater than 1) by decreasing the amount of template occluded in the zeolite framework and that ethanol is separated from water by transport through the zeolite channels [Sano et al., 1994].

TABLE 6.16
Effect of membrane treatment conditions on the permeate flux and separation factor of a 5% ethanol/water mixture at 60°C by silicalite membranes

Treatment	Treatment temp. (°C)	Treatment time (h)	Permeate flux (kg/m^2-h)	Separation factor, α
Drying in air	100	12	0.00303	0.38
Drying in vacuum	300	6	0.00840	0.58
Drying in vacuum	380	6	0.0394	7.8
Calcination	500	20	0.760	58

Note: Feed concentration of ethanol at 5% by volume
[Sano et al., 1994]

6.7 LABORATORY APPLICATIONS

The majority of commercial as well as potential applications of inorganic membranes are for pilot or production scale operation. There are, however, limited uses of some inorganic membranes in laboratories.

6.7.1 Removal of Particulates from Samples for HPLC

One area is analytical separations by inorganic membranes as part of sample cleanup. For example, it is imperative that a high performance liquid chromatography (HPLC) system be void of particulate contaminants or the system components may be damaged. Organic polymer membranes have been employed in some cases but suffer from such problems as low resistance to organic solvents, low thermal stability and broad pore size distributions. Although radiation-etched polycarbonate membranes possess tight pore size distributions, their porosity is low. Therefore, inorganic membranes with sharp pore size distributions such as those made by anodic oxidation can provide a more acceptable alternative. In fact, they have demonstrated the feasibility for this application. Compared to those membranes with tortuous pores, these membranes are closer to being a true sieve rejecting particles larger than the diameter of the rather uniformly sized pores.

The aforementioned alumina membranes have proven to be resistant to all the solvents used in HPLC. This avoids contamination of the filtered solutions by organic extractables. The presence of any leached inorganic ions is limited to less than 0.05 parts per billion (ppb) except aluminum ions at a level of 6 ppb [Hoffman, 1989]. Anopore™ alumina membranes which are made by the anodic oxidation method come in the form of disposable syringe filters and filter disks. Contained in a hexagonal polypropylene housing, the disposable syringe filters with a mean pore diameter of 0.02 to 0.2 µm are used in filtration devices. The filter disks, also available from 0.02 to 0.2 µm, are suitable for filtering chromatographic mobile phases and fit standard vacuum filtration devices. These small sized filters find uses in food and beverage manufacturing for sample cleanup prior to analysis by HPLC. Their application in ion chromatography has also been mentioned.

Molecularly bonded silver membranes have also been used to purify aggressive HPLC solvents [Minneci and Paulson, 1988]. Similar to the anodic alumina membranes, silver membranes are preferred for this application due to the following reasons: high chemical and thermal stabilities leading to no or negligible amount of dissolved or leached membrane constituents in the solvents.

6.7.2 Analysis of Soils and Clays

Inorganic membranes can function as the capturing medium of clays, soils and other particulate contaminants in fluids. The inorganic membranes, such as silver membranes, containing the deposit of the contaminants or minerals are then used as the X-ray diffraction (XRD) substrate for material analysis of the particulates. In some cases, they are part of the established test procedures as a National Institute of Occupational Safety and Health (NIOSH) standard XRD substrate.

Silver membranes have been used as filters to collect clays or soils in a dispersion medium and then as an XRD substrate. In a typical analysis for mineral contents, a clay or soil is dispersed in water and vacuum filtered on the membrane. The silver membrane in this case has an advantage over ceramic membranes in that the XRD patterns of the minerals are distinctively different from that of silver. In contrast, the patterns of the filtered clays may be difficult to distinguish from those of the filter medium if ceramics is used.

6.7.3 Organics in Water Analysis

Two important ecological characteristics of natural waters are dissolved organic carbon (DOC) and suspended organic carbon (SOC). The United States Geological Survey (U.S.G.S.) has developed a standard analytical procedure for determining the two types of organic carbons. The procedure calls for the use of a 0.45 μm silver membrane. The organic carbon that passes through the membrane with the rated pore size is defined as DOC while that is retained by the membrane is SOC [Minneci and Paulson, 1988]. Analysis of humic and fulvic acids uses a standard that is prepared by a silver membrane, a method also developed by the U.S.G.S. In this application, up to about 11,500 L of water is filtered from specific rivers at certain times of the year using silver membranes. The collected material is used as a comparative standard.

6.7.4 Other Laboratory Applications

Another specific application is the use of 0.1 μm syringe filters for the removal of mycoplasma in tissue culture work. They are capable of removing 99.99% of three common human mycoplasma species (M. hominis, M. salivarium and M. fermentans) and two common contaminants of fetal calf serum (M. arginini and Acholeplasma Laidlawii) [Hoffman, 1989].

6.8 SUMMARY

Porous inorganic membranes have been used commercially or tested successfully in selected liquid-phase industrial production and analytical applications. More notable examples are dairy and fruit juice processing, wine and beer clarification, filtration of potable water, wastewater treatment in food and petrochemical processing, biotechnology applications, paper and textile processing and laboratory uses for material analysis. Depending on the application requirements, membranes of various ceramic and metallic materials with different pore sizes function as either microfiltration or ultrafiltration membranes. These applications came about because of unique properties of inorganic membranes such as those listed in Table 6.1. At the same time, they pose specific technical as well as economic challenges. Among the various technical issues listed in Table 6.17 for different applications, some common ones are fouling and initial investment costs of the membranes. As the type and number of applications of inorganic membranes proliferate, these issues are expected to be lessened or resolved.

TABLE 6.17
Major technical challenges of inorganic membranes for various application areas

Application area	Major challenge(s) of inorganic membranes
Dairy	Fouling (protein adsorption)
Beverages	Fouling; permeation rate; organoleptic quality of alcoholic beverages
Biotechnology	Fouling
Petrochemical processing	Investment costs
Waste oil treatment	Fouling
Textile processing	Permeation rate
Paper and pulp processing	Fouling; pretreatment required
Metal cleaners reuse	Permeate quality

REFERENCES

Alvarez, V., L.J. Andrés, F.A. Riera and R. Alvarez, 1994, Microfiltration of apple juice: a study of membrane rejection behavior, in :Proc. 3rd Int. Conf. Inorg. Membr., Worcester, MA, USA.

Attia, H., M. Bennasar and B. Tarodo de la Fuente, 1988, Le Lait **68**, 13.

Attia, H., M. Bennasar and B. Tarodo de la Fuente, 1991a, J. Dairy Res. **58**, 39.

Attia, H., M. Bennasar and B. Tarodo de la Fuente, 1991b, J. Dairy Res. **58**, 51.

Badekha, V.P., M.I. Medvedev, D.D. Kucheruk and A.T. Pilipenko, 1986, Sov. J. Water Chem. Technol. **8**, 36.

Barnier, H., A. Maurel and M. Pichon, 1987, Paperi ja Puu **69**, 581.

Belleville, M.P., J.M. Brillouet, B. Tarodo de la Fuente and M. Moutounet, 1990, J. Food Sci. **55**, 1598.

Bennasar, M., and B. Tarodo de la Fuente, 1987, Sci. Alim. **7**, 647.

Bhave, R.R., and H.L. Fleming, 1988, Removal of oily contaminants in wastewater with microporous alumina membranes, in: New membrane materials and processes for separation (K.K. Sirkar and D.R. Lloyd, eds.), AIChE Symp. Ser. No. 261, Vol. 84, p.19.

Bhave, R.R., S.P. Evans, A.S. Chen and H.J. Weltman, 1993, U.S. Patent 5,205,937.

Blanc, P., and G. Goma, 1989, Biotechnol. Lett. **11**, 189.

Boyaval, P., and C. Corre, 1987, Biotechnol. Lett. **9**, 801.

Boyaval, P., C. Corre and S. Terre, 1987, Biotechnol. Lett. **9**, 207.

Castelas, B., and M. Serrano, 1989, Utilization of mineral membranes for wine treatment, in: Proc. 1st Int. Conf. Inorg. Membr., Montpellier, France, p.283.

Castelas, B., R. Rumeau, L. Cot, C. Guizard and J.A. Alary, 1984, Application de la filtration tangentielle sur membrane minérale à la décontamination bacterienne des eaux, presented at 4 ème journées information Eaux, Poitiers, France.

Chen, A.S.C., 1994, Evaluating a ceramic ultrafiltration system for aqueous alkaline cleaner recycling, Report to U.S. EPA (Risk Reduction Eng. Lab., Cincinnati, Ohio) under contract 68-CO-0003.

Chen, A.S.C., J.T. Flynn, R.G. Cook and A.L. Casaday, 1991, SPE Prod. Eng. (May), 131.

Cheryan, M., 1994, Processing ethanol fermentation broths and stillage with ceramic membranes, presented at 3rd Int. Conf. Inorg. Membr., Worcester, MA, USA.

Collentro, W. , 1994, The use of ultrafiltration for producing water for injection, presented at 3rd Int. Conf. Inorg. Membr., Worcester, MA, USA.

Cueille, G., and M. Ferreira, 1989, Place of mineral membranes in the processes for bio-industry and food-industry, in: Proc. 1st Int. Conf. Inorg. Membr., Montpellier, France, p.303.

Cumming, I.W., and A.D. Turner, 1988, Optimization of an UF pilot plant for the treatment of radioactive waste, in: Proc. Future Ind. Prospects Membr. Proc. Conf., Brussels, Belgium, p.163.

Daufin, G., J.P. Labbe, A. Quemerais, F. Michel and C. Fiaud, 1989, Encrassement de la membrane Carbosep M_5 lors del'ultrafiltration de lactoserum clarifie, in: Proc. 1st Int. Conf. Inorg. Membr., Montpellier, France, p.425.

Deschamps, A., C. Walther, P. Bergez and J. Charpin, 1989, Application of inorganic membranes in refining processes of petroleum residues, in: Proc. 1st Int. Conf. Inorg. Membr., Montpellier, France, p.237.

Fabiani, C., C.A. Nannetti, R. Vatteroni and L. Bimbi, 1989, Alumina tubular membranes for gas and liquid separations, in: Proc. 1st Int. Conf. Inorg. Membr., Montpellier, France, p.497.

Finnigan, T., R. Shackleton and P. Skudder, 1987, Filtration of beer and recovery of extract from brewry tank bottoms using ceramic microfiltration, in: Proc. Filtech Conf., Utrecht, p. 533.

Gallagher, P.M., 1990, Novel composite inorganic membranes for laboratory and process applications, presented at Ann. Membr. Technol./Planning Conf., Newton, MA (USA), October 15-17.

Gaston County, 1976, Gaston County ultrafiltration systems (product brochure).

Gaston County, 1979, Textile size reclamation through ultrafiltration (product brochure).

Goudédranche, H., J.L. Maubois, P. Ducruet and M. Mahaut, 1980, Desalination **35**, 243.

Guibaud, J., 1989, Some applications of Membralox® ceramic membranes, in: Proc. 1st Int. Conf. Inorg. Membr., Montpellier, France, p.343.

Guizard, C., D. Rambault, L. Cot, D. Uhring and J. Dufour, 1994, Deasphalting of a long residue using ultrafiltration inorganic membranes, presented at 3rd Int. Conf. Inorg. Membr., Worcester, MA, USA.

Gutman, R.G., I.W. Cumming, G.H. Williams, R.H. Knibbs, I.M. Reed, P. Biddle, C.G. Davison, J.W. Sharps, M. Smith, J.A. Jenkins, J.A. Blackwell, T.E. Hilton and C.E. Barclay, 1986, Active liquid treatment by a combination of precipitation and membrane processes, Report EUR 10822 (Commission of the European Communities).

Higgins, R.J., B.A. Bishop and R.L. Goldsmith, 1994, Reclamation of waste lubricating oil using ceramic membranes, presented at 3rd Int. Conf. Inorg. Membr., Worcester, MA, USA.

Hoffman, J., 1989, Am. Lab. (August Issue), p.70.

Humphrey, J.L., K.P. Goodboy and A.L. Casaday, 1989, Ceramic membranes for the treatment of waters produced by oil wells, presented at Am. Chem. Soc. 197th National Meeting in Dallas, TX.

Ikeda, H., T. Tomono, A. Shimada, T. Hori, K. Ozawa, K. Sakai, H. Nakanishi and M. Inoue, 1986, Jpn. J. Artif. Organs **13**, 914.

Imsaka, T., N. Kanekuni, H. So and S. Yoshino, 1989, J. Ferment. Bioeng. **68**, 200.

Jordan, G., and R.P. Greenspan, 1967, Tech. Quart. (Master Brewers Assoc. Am.) **4**, 114.

Kilham, O.W., 1987, Wine separation: Membranes, applications. Presented at 5th Ann. Membr. Technol./Planning Conf., Cambridge, MA, USA.

Lafforgue, C., J. Malinowski and G. Goma, 1987, Biotechnol. Lett. **9**, 347.

Lahiere, R.J., and K.P. Goodboy, 1993, Environ. Prog. **12**, 86.

Langer, P., and R. Schnabel, 1989, Porous glass UF-membranes in biotechnology, in: Proc. 1st Int. Conf. Inorg. Membr., Montpellier, France, p.249.

Long, Z.R., 1981, White table wine production in California's north coast region, in: Wine Production Technology in the United States, ed. M.A. Amerine (Symp. Ser. 145, Am. Chem. Soc., Wash. D.C.) p. 29.

Mahaut, M., J.L. Maubois, A. Zink, R. Pannetier and R. Veyre, 1982, Tech. Lait. **961**, 9.

Malmberg, R., and S. Holm, 1988, North. Eur. Food Dairy J. **1**, 75.

Matsumoto, K., M. Kawahara and H. Ohya, 1988, J. Ferment. Technol. **66**, 199.

Merin, U., and G. Daufin, 1989, Separation processes using inorganic membranes in the food industry, in: Proc. 1st Int. Conf. Inorg. Membr., Montpellier, France, p.271.

Meunier, J.P., 1990, Use of crossflow filtration for process beer yeast and tank bottoms, in: Proc. 5th World Filtration Congr., Nice, France.

Minneci, P.A., and D.J. Paulson, 1988, J. Membr. Sci. **39**, 273.

Mishra, B., D.L. Olson and P.D. Ferro, 1994, Application of ceramic membranes in molten salt processing of radioactive wastes, in: Proc. 123rd Ann. Meeting of Minerals, Metals, and Mater. Soc., San Francisco, USA, p. 233.

Muir, D.D., and J.M. Banks, 1985, J. Soc. Dairy Technol. **38**, 116.

Muralidhara, H.S., and R. Bhave, 1994, A novel application of ceramic membranes to resin systems, presented at 3rd Int. Conf. Inorg. Membr., Worcester, MA, USA.

Narukami, Y., A. Kayawake, M. Shioyama, Y. Okamoto, K. Tokushima and M. Yamagata, 1989, Ceramic membrane filtration of methane fermentation broth, in: Proc. 1st Int. Conf. Inorg. Membr., Montpellier, France, p.267.

Neytzell-de-Wilde, F.G., R.B. Townsend and C.A. Buckley, 1989, Hydrous Zr(IV) oxide and Zr-polyelectrolyte membranes on porous stainless steel supports, in: Proc. 1st Int. Conf. Inorg. Membr., Montpellier, France, p.113.

Nose, Y., H. E. Kambic and S. Matsubara, 1983, Introduction to therapeutic apheresis, in: Plasmapheresis: Therapeutic applications and new techniques (Y. Nose, P.S. Malchesky, J.W. Smith and R.S. Krahauer, eds.), Raven Press, New York, p. 1.

Ozawa, K., H.B. Kim, H. Sakurai, S. Takesawa and K. Sakai, 1986, Novel utilization of ceramic membranes in plasma treatment, in: Progress Artificial Organs-1985 (Y. Nosé, C. Kjellstrand and P. Ivanovich, Eds.), ISAO Press, Cleveland, OH, U.S.A., p.913.

Poirier, D., F. Maris, M. Bennasar, J. Gillot, D.A. Garcera and B. Tarodo de la Fuente, 1984, Ind. Alim. Agric. 101, 481.

Quemeneur, F., and P. Jaouen, 1991, Key Eng. Mater. 61/62, 585.

Redkar, S., and R.H. Davis, 1993, Biotechnol. Progr. 9, 625.

Rios, G.M., and P. Freund, 1991, Key Eng. Mater. 61/62, 255.

Sakai, K., K. Ozawa, K. Ohashi, R. Yoshida and H. Sakurai, 1989, Ind. Eng, Chem. Res. 28, 57.

Sano, T., H. Yanagishita, Y. Kiyozumi, D. Kitamoto and F. Mizukami, 1992, Chem. Lett. 2413.

Sano, T., H. Yanagishita, Y. Kiyozumi, F. Mizukami and K. Haraya, 1994, J. Membr. Sci. 95, 221.

Shackleton, R., 1987, J. Chem. Tech. Biotechnol. 37, 67.

Swafford, T.C., 1987, Proc. Pacific Fish Technol. Meeting, February, Monterey, CA, USA.

Swientek, R.J., 1987, Food Proc. 48, 74.

Thomas, R.L., J.L. Gaddis, P.H. Westfall , T.C. Titus and N.D. Ellis, 1987a, J. Food Sci. 52, 1263.

Thomas, R.L., T.C. Titus and C.A. Brandon, 1987b, U.S. Patent 4,716,044.

Thoukis, G., Chemistry of wine stabilization: a review, in: Chemistry of Winemaking, ed.A.D. Webb (Adv. in Chem. Ser. 137, Am. Chem. Soc., Wash. D.C.) p. 116.

Trägårdh, G., and P-E Wahlgren, 1989, Removal of bacteria from beer using crossflow microfiltration, in: Proc. 1st Int. Conf. Inorg. Membr., Montpellier, France, p.291.

Van der Horst, H.C., M. Timmer and I. Piersma, 1994, Cross-flow microfiltration of skim milk, presented at 3rd Int. Conf. Inorg. Membr., Worcester, MA, USA.

Venkataraman, K., M.T. Giles and P.K. Silverberg, 1988, Ceramic membrane applications in juice clarification: a case study, presented at 2nd Annual Meeting of North Am. Membr. Soc., Syracuse, N.Y., USA (details given in: Inorganic Membr. Synth., Charact. Appl., R.R. Bhave, Ed., p. 239).

Vetier, C., M. Bennasar and B. Tarodo de la Fuente, 1988, J. Dairy Res. 55, 381.

Watanabe, A., T. Ohtani, H. Horikita, H. Ohyia and S. Kimura, 1986, Recovery of soluble protein from fish jelly processing with self-rejection dynamic membrane, in: Food Eng. & Process Appl.. (M. Le Maguer and P. Jelen, Eds.), Vol . 2, p. 225.

CHAPTER 7

GAS-PHASE AND NON-TRADITIONAL SEPARATION APPLICATIONS

In contrast to liquid-phase applications reviewed in Chapter 6, the current sales volume and application varieties of inorganic membranes in the gas-phase separation market are still quite limited. Their commercial usage in the gas- and vapor-phase environments is far from being a significant presence. Even the first largest gas-phase separation application, gas diffusion for uranium enrichment, discussed in Chapter 2 no longer requires any major production efforts.

However, technological developments in the gas-phase applications are accelerating at an ever increasing rate as is evident in the voluminous literature in this field. As will become obvious, the technology of inorganic membranes for gas separation is still at its early stage. It is, therefore, inevitable that most of the discussions in this chapter focus on the potentials, rather than proven utilization, of inorganic membranes. As science and technology of this field is evolving, it is virtually impossible to draw definitive conclusions at this point. Some discussions, therefore, may appear to be speculative because of this nature of the technology development.

First to be examined in this chapter are the uses of some inorganic membranes, particularly of the ceramic types, to remove particulates from gases or vapors. Some of these are commercially practiced. Next, separation of gas components in gaseous mixtures will be considered. This is a relatively new frontier except for isotope separation and, to some extent, hydrogen purification. Most of the examples of usage in this area are exploratory and require significant further materials and engineering developments before commercialization. Finally, some unusual uses of inorganic membranes for gas and vapor phases are discussed. These include their utilization in facilitated transport and gas and chemical sensors. Here again, technological gaps need to be closed prior to commercialization in most cases.

7.1 PARTICULATE FILTRATION OF AIR OR GAS

Some ceramic membranes are used for filtering particulate matters for reasons of process requirements or protection of equipment. Clean room applications constitute some major commercial usage of these membranes.

7.1.1 Clean Room Applications

Currently commercial installations of inorganic membranes in gas- and vapor-phase applications besides the isotope separation are mostly in the area of particulate filtration, although there have been very limited cases of separation of gaseous mixtures reported. Removal of fine particles from gas streams can be achieved by means of porous inorganic membranes. The major challenge, however, lies in the strict requirements in some of the commercial applications. This is the case in microelectronic packaging and some food and pharmaceutical processes where clean air is needed.

Until recently, air filtration for clean rooms uses dead-end fabric filters. They are not efficient in the particle diameter range of 0.1 to 0.5 μm and also suffer in many cases from two of the most important problems in clean room gases applications: particle shedding and gas reactivity (or called hydrocarbon outgassing). Some ceramic membranes such as alumina membranes have made a visible entry into the clean room market as in-line gas filters.

These ceramic membranes offer several advantages. First, essentially they are thermally stable and chemically inert in the application environments. For example, alumina membranes for clean room air filtration can be operated at up to 120°C and heated to 200°C during bake-out, a temperature range above the softening points of many organic polymer membranes. The rigid ceramic media does not introduce particles into the gas stream by shedding even under the harsh conditions of flow pulsation and mechanical vibration or shocks. And since typically they do not react with the process gases or generate gaseous impurities, no hydrocarbon outgassing problem occurs. Some of the corrosive or reactive process gases include silane, chlorine, hydrogen chloride, hydrogen bromide, arsine, ammonia and boron trichloride. Second, their pore size distributions are narrower than those of fabric filters and, therefore, they can retain an extremely high percentage of particulates (e.g., a retention rating $\geq$ 99.99999% at 0.01 μm). Third, as in the case of many liquid-phase applications, membranes are subject to concentration polarization or fouling, but, because of their rigid and tightly bonded structures, alumina and some other porous ceramic membranes can be easily regenerated periodically by backflushing. Finally, many ceramic membranes can be steam sterilized essentially without changing their microstructural as well as chemical characteristics.

To ensure the cleanliness of the ceramic membrane gas filters, the housing typically made of stainless steel should have electropolished internal wetted surfaces and welded parts free of flux materials. In addition, considerations need to be given to the materials of construction for housing (e.g., stainless steel) and seals.

These ceramic membranes are relatively easy to operate for filtering particulates. The pressure drop across the thickness of a membrane element and the gas flow rate follows a linear or nearly linear relationship. It has been found, however, that not all inorganic membranes are suitable for clean room air filtration. Glass membranes, for example, suffer from the problem of particle shedding under mechanical shock conditions [Jensen and Goldsmith, 1987]. Sometimes high moisture content in the filtered air can be a problem. Some chemical treatments to ceramic membranes prior to their utilization as clean room filters have been found to be effective in lessening the problem [Grosser, 1989].

7.1.2 Gas/particulate Separation

Another important fine particle filtration application is the use of membranes for hot gas clean-up applications. They include such processes as removing dust contaminants from the flue gases or from the hot gases to gas turbines in electric power plants using coals, catalyst recovery and other chemical processing. Being operated at high temperatures, these processes can benefit from ceramic membranes in two major aspects. One is the reduced number of cyclones required and the other is, in comparison to organic membranes or some other separation media, the elimination of the need to ramp the temperature down before and then up after particle separation. Significant amount of research is being pursued in this general area.

A particulate filter comprised of a thin ceramic membrane coating on a cordierite honeycomb monolithic porous structure has been developed for the aforementioned particulate filtration applications [Bishop et al., 1994]. The filter is a dead-end microfilter by plugging alternate passageways on both ends of the honeycomb. Having a particle filtration efficiency greater than 99.9%, the filter surface can be easily regenerated by removing the filter cake through backflushing using clean gas. The honeycomb monolithic design offers the engineering advantage of having a high packing density of up to about 500 m^2/m^3 which is an order of magnitude greater than those of conventional bag filters. Other monolithic ceramic membrane filters have been commercially used or tested for hot gas filtration [Eggerstedt et al., 1993]. A particulate collection efficiency higher than 99% has been reported for a commercial-scale cross-flow ceramic membrane operated at 650-800°C under 11-18 bars [MPS Review, 1988].

Because of the severe operating conditions typical of hot gas clean-up applications, the ceramic membranes are subjected to the following problems: corrosion, erosion, sintering, phase change, binder slippage, delamination, grain growth, fusion and thermal shock [Eggerstedt et al., 1993]. The candidate materials for the membrane filters include mullite, cordierite, aluminum titanate, refractory concretes, silicon carbide and silicon nitride.

7.1.3 Analytical Applications

As mentioned in Chapter 6, some ceramic and metal membranes can have duel uses in some X-ray diffraction analysis for liquids or dispersions. Using the same concept, one can utilize, for example, a molecularly bonded silver membrane in the analysis of particulate contaminants in the air. The membrane with the deposit of the contaminants also serve as the X-ray diffraction (XRD) substrate for material analysis of the particulates. This has become a test method established by National Institute of Occupational Safety and Health (NIOSH).

The determination of quartz dust in the air samples in industrial workplace is an established procedure. Although capable of collecting the particulates, organic polymer membranes can not be employed as an XRD substrate since the diffuse diffraction lines at or near the 2θ angle of quartz makes polymeric membranes not suitable for this application [Minneci and Paulson, 1988]. It is possible to quantify as low as 0.005 mg quartz under well controlled conditions [Bumsted, 1973]. Similarly, silver membranes can also be used as a collecting medium and XRD substrate for measuring crystalline and amorphous silica, lead sulfide, boron carbide and chrysotile asbestos [Leroux and Powers, 1970].

Finally, in an established application, silver membranes have also been utilized to capture coal-tar pitch volatiles and fly ash directly from smoke stacks and used in subsequent analysis [Minneci and Paulson, 1988].

7.2 GAS SEPARATION

Inorganic membranes can be utilized to separate major components or remove minor or trace contaminant gas(es) from a gaseous mixture. Bulk separation of the valuable hydrogen from other major gas constituents and the separation of carbon dioxide and water vapor from other coal gasifier products to increase the Btu content of a low-Btu gas are just some industrially important examples of the first type of applications. The removal of acid gases (such as sulfur dioxide, hydrogen sulfide and carbon dioxide) from flue gases and the removal of hydrogen sulfide from fuel gases represent cases of the second type.

In either type of application, the inherent thermal and chemical stabilities of inorganic membranes, as in the case of particle filtration, potentially can avoid the energy waste due to the need of precooling the unseparated gas stream and reheating the separated gases as required when organic membranes are used. Many industrially important gas separation problems occur at high temperatures. Generally speaking, high temperature separations are those applications in which the feed stream enters at a temperature higher than 500 to 600°C. Eliminating the need for multiple gas-cooling and heating steps not only reduces the requirement for heat exchange equipment but in some cases also has profound implications. Take the case of coal conversion processes. If the pressurized coal-gasification streams must be cooled down because of the limitation of the gas separation method used, tars and organics may begin condensing into liquid wastes. This can cause significant problems in treating the wastewater [Dellefield, 1988].

Compared to liquid-phase applications, commercial gas- or vapor-phase applications of inorganic membranes have been limited. Due to their low permeabilities, dense inorganic membranes are utilized only in special and low-volume cases. Hydrogen purification and separation from other gases by dense palladium alloy membranes has been practiced on a commercial basis to a limited extent. As discussed in Chapter 2, some porous ceramic membranes have been used for uranium isotope enrichment in a production setting for three decades. Recently, porous ceramic membranes are used commercially in filtration of particulates from gases. In addition, the use of porous alumina membranes for dehydration applications in manufacturing plants has been reported [Mitsubishi Heavy Industries, 1986]. Specifically, water can be separated from air or such organic solvents as alcohols because water condenses in and migrates through the ultrafine pores of the alumina membranes. Separation of azeotropic mixtures becomes feasible.

While commercial uses of inorganic membranes for gas-phase applications are limited at the present time, their potentials for separating components in a gaseous mixture are huge and promising. Therefore, the interest in inorganic membranes, particularly porous inorganic ones, has surged in both the academia and the industry. They have attracted much research and development effort as evidenced by the very large volume of literature and the number of patents in recent years. Many of the investigations on the use of inorganic membranes for gas separation involve chemical reactions and, because of their unique technical characteristics, deserve a separate treatment. Inorganic membranes used for separation and reaction simultaneously will be reviewed in later chapters.

7.2.1 Performance Parameters in Membrane Separation of Gases

How well economically a gas separation membrane system performs is largely determined by three parameters. The first parameter is its permselectivity or selectivity toward the gases to be separated. Permselectivity affects the percentage recovery of the valuable gas in the feed. For the most part, it is a process economics issue. The second is the permeate flux or permeability which is related to productivity and determines the membrane area required. The third parameter is related to the membrane stability or service life which has a strong impact on the replacement and maintenance costs of the system.

A frequently used indicator of how much two gases (say, gas m and gas n) in a multicomponent gaseous mixture are separated with respect to each other through a membrane is called separation factor. It is defined as

$$\alpha_{mn} = (y_m/x_m)/(y_n/x_n) \tag{7-1}$$

where y and x represent the mole fractions of the gas components downstream and upstream of the membrane, respectively. It is essentially determined by their relative permeabilities through the membrane. For a binary gas mixture, the separation factor becomes

$$\alpha_{12} = (y_1/x_1)/[(1-y_1)/(1-x_1)] \tag{7-2}$$

It can be shown that

$$\alpha_{12} = \alpha^*_{12}[1-P_d\, y_1/P_u\, x_1]/[1-P_d\, (1-y_1)/P_u(1-x_1)] \tag{7-3}$$

where P_d and P_u are the downstream and upstream pressures, respectively, and α^*_{12} is the so called ideal separation factor which is the ratio of the pure component permeabilities (P_1 and P_2):

$$\alpha^*_{12} = P_1/P_2 \tag{7-4}$$

It is a theoretical overall selectivity of a membrane and is often referenced in the literature.

From Eqs. (7-1) and (7-2), it follows that the separation factor is purely based on the compositions of the entering and exit streams regardless of their flows. Another measure of the separation efficiency of a membrane process is the extent of separation proposed by Rony [1968]. In the context of applying this index of separation efficiency between two components, it is assumed that there is no difficulty in separating the third component. Thus the segregation fractions, Y_{ij}, are obtained from the molar flow rates of the permeate and retentate streams on the basis of only two components. The extent of separation is defined as the absolute value of a determinant of a binary separation matrix consisting of the segregation fractions as follows:

$$\xi = abs.\ det. \begin{vmatrix} Y_{11} & Y_{12} \\ Y_{21} & Y_{22} \end{vmatrix} \tag{7-5}$$

where $Y_{ij} = Q_i / (F_i + Q_i)$ when $j=1$ and $Y_{ij} = F_i / (F_i + Q_i)$ when $j=2$. F_i and Q_i are the molar flow rates of the component i in the retentate and permeate streams, respectively. Although the extent of separation has been used by some investigators, it has not been widely accepted in the literature of membrane technology.

Besides some measures of separation efficiency such as the separation factor and extent of separation defined above, some quantity indicative of the throughput rate of a membrane system is needed to compliment the permselectivity of the membrane. It is quite common and practical in the membrane technology to use a phenomenological expression to relate the permeate flux (J_A in the unit of cm^3 (STP)/s-cm^2) of a given gas (A) through the membrane to the driving force, the transmembrane pressure difference (Δp) as follows:

$$J_A = (P_A/\Delta l)(\Delta p) \tag{7-6}$$

where (Δl) is the membrane thickness and the parameter P_A is frequently called permeability of the gas A through the membrane. Therefore, P_A is essentially the flux normalized with the driving force and the membrane thickness according to the following equation:

$$P_A = J_A(\Delta l)/(\Delta p) \qquad (7\text{-}7)$$

The permeability provides an overall measure of the relative ease of gas permeation through the membrane. An often used unit for the permeability is barrer which is equal to 10^{-10} cm^3 (STP)-cm/s-cm^2-cmHg.

It is noted that when comparing permeabilities of various gases the variations in the membrane thickness and the TMP are taken into account. Sometimes it is difficult to accurately determine the membrane thickness and in those cases $(P_A/\Delta l)$ is lumped into a parameter called permeance which has the unit of cm^3 (STP)/s-cm^2-cmHg. The terminology used in expressing the permeation characteristics of a gas separation membrane is often confusing. It is advisable to distinguish various terms such as flux, permeability and permeance by their units. The permeability of a porous membrane is usually significantly greater than that of a dense membrane but, in many cases, at the expense of a lower permselectivity.

Inorganic membranes are in general inherently more stable than their organic counterparts in response to the application environments. The service life of a ceramic membrane, for example, is claimed to be typically three to five times that of an organic polymer membrane. More reliable comparisons will be available in the future after inorganic membranes have been used in the industry long enough to establish statistically significant data.

It is obvious that these performance parameters not only depend on the membrane characteristics but also the operating conditions. In addition, process design can result in profound impacts on the performance of membranes. For example, single-stage and multistage membrane designs can lead to vastly different performance. Generally speaking, the multistage design or a more complex design can be economically justified only when the value of the separated components increases and/or the size of the process expands [Spillman, 1989].

7.2.2 Potential Application Areas for Inorganic Membranes

Technical advances in the preparation, characterization and applications of membranes for the more established liquid-phase usage have led to similar developments for gas-phase operations. Since the early 1980s organic membrane-based gas separation technologies have seen commercialization at an ever increasing rate. It is emerging as a new unit operation potentially important in many areas. The success probably can be attributed to some of their attributes including low energy demand, ease of operation, modular in construction, low capital investments, mobility and compactness. Given in Table 7.1 are the commercially more significant gas separation applications for membranes [Spillman, 1989]. This list is almost exclusively based on the application experiences and their extensions of dense organic membranes such as polysulfones, cellulosic derivatives, polyamides and polyimides in two predominant configurations: hollow-fiber and spiral-wound. Nevertheless, they represent a reasonable reference target for market entry of inorganic membranes.

TABLE 7.1
Major gas separation application areas

Common gas separations	Application
O_2/N_2	Oxygen enrichment, inert gas generation
H_2/hydrocarbons	Refinery hydrogen recovery
H_2/CO	Syngas ratio adjustment
H_2/N_2	Ammonia purge gas
CO_2/ hydrocarbons	Acid gas treatment, landfill gas upgrading
H_2O/hydrocarbons	Natural gas dehydration
H_2S/hydrocarbons	Sour gas testing
He/hydrocarbons	Helium separations
He/N_2	Helium recovery
Hydrocarbons/air	Hydrocarbon recovery, pollution control
H_2O/air	Air dehumidification

[Spillman, 1989]

It has been pointed out earlier that the only major production scale separation process using inorganic membranes is gas diffusion by porous membranes although low flow-rate commercial applications of hydrogen purification by dense palladium alloy membranes do exist. Therefore, those applications identified in Table 7.1 will be used as the framework for discussions on potential gas separation applications here. When appropriate, certain applications will be emphasized for their greater potentials in this field. Other potential applications exist where the demanding process requirements can only be technically or economically matched by the unique properties of certain inorganic membranes. These niche application areas will be reviewed as well. The latter application areas differ significantly from those listed in Table 7.1 which, for the most part, must compete directly with other separation processes (e.g., cryogenics, pressure swing adsorption, liquid scrubbing such as amine treatment, and distillation) on the basis of overall economics.

Often in practice, several membranes for gas separation are cascaded with recycle streams to effect a multistage system for increasing the separation factor to an effective level, particularly when the individual gases in the mixture have similar permeabilities.

An example of a multistage gas separation membrane process is given in Figure 7.1 for natural gas treating [Spillman, 1989]. It is generally accepted that for a gas separation membrane to be cost effective, a separation factor of at least 5-10 (sometimes even higher) will be required.

As a general rule, gas separation by membranes is most attractive in applications where a product purity of 95% or lower is acceptable or the feed flow-rate is not too high. As the required purity approaches 100%, the membranes become less cost effective than other separation processes. This is particularly true with single-stage units. For more stringent applications, some traditional separation processes are preferred or required to integrate with the membrane system.

GA 34558.2

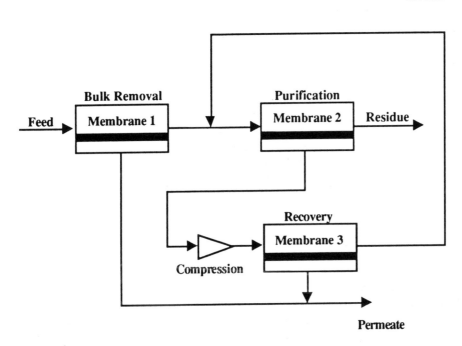

Figure 7.1 A multistage gas separation membrane process for natural gas treating [Spillman, 1989]

A major area of potential applications of inorganic membranes is high-temperature and/or high-pressure contaminant removal systems to protect process components and to control emissions from advanced coal-based power generation systems. The high temperature and pressure refer to 538°C (1000°F) and above 5-35 bars, respectively. The potential applications include both bulk separation of gases and the removal of trace contaminants (with less than 1 or 2% by volume) for the following processes [Dellefield, 1988]:

integrated gasifier combined cycles;
direct coal-fueled turbine systems;
pressurized fluidized-bed combustors;
coal gasifier fuel cell systems; and
direct coal-fueled diesels

The first two processes have received particular attention from the U.S. Department of Energy which has been sponsoring research and development encompassing various aspects of inorganic membranes. Some examples of trace contaminant removal are the removal of hydrogen sulfide from fuel gases or the removal of sulfur oxides and nitrogen oxides from combustion gases. Some common examples of the bulk separation of gases are separating the valuable hydrogen from other major gas constituents and the separation of carbon dioxide and water vapor from other coal gasifier products to increase the Btu content of a low-Btu gas or as a method of adjusting the ratio of gases prior to the synthesis of chemical compounds.

In the following sections, specific application areas, either commercially practiced or potentially important, will be reviewed individually. Current status of developments as reflected in the separation performance parameters, mainly the separation factor and permeability or its equivalent, will be discussed. Emphasis is placed on the key characteristics or requirements of each application, test data indicative of any promises and, wherever appropriate, major technical and, in some cases, economical challenges.

7.2.3 Manufacture of High-Purity Hydrogen

Diffusivity of hydrogen in metals is known to be 15 to 20 orders of magnitude higher than that of oxygen or nitrogen in the same metal. Dense palladium and its selected alloys exhibit extremely high permeability of hydrogen. For example, at 400°C under a differential pressure of 6.8 bars, the hydrogen permeability through a 25 μm thick dense Pd-8% Y, Pd-5.75% Ce and Pd-25% Ag membrane is about 4.1×10^4, 1.7×10^4 and 1.5×10^4 barrers, respectively [Hughes and Harris, 1978]. These permeability values are higher than those typical of organic polymer membranes. The hydrogen purity obtained through a Pd-Ag membrane can be guaranteed to 99.99995% or an impurity level of 0.5 ppm [Philpott and Coupland, 1988]. Commercial dense Pd-Ag membrane units have been available for many years for hydrogen purification on both laboratory and small production scales. Most of them come in the form of tubular membranes. They range in capacity from 1 L/min to 500 m^3/h and can be operated within a temperature range of -40 to 55°C. Some are designed to be operated at a temperature up to 500°C and under a pressure of 35 bars [Connor, 1962].

An existing industrial application of the ultrapure hydrogen separated by dense Pd alloy membranes is for electronics industry. In the fabrication of silicon chips, hydrogen acts as a carrier to transport small quantities of vaporized chemical compounds required to "dope" the chip to the surface of the silicon wafer [Philpott and Coupland, 1988]. The hydrogen used must be of a very high purity. The membrane units are used not only for gas purification but also for hydrogen recovery from hydrogen-rich gas streams.

Another application is on-site hydrogen production particularly at remote locations where hydrogen cylinders are not available or expensive to deliver. In a situation like this, a convenient source of any impure hydrogen is essential. For example, a 1:1 molar methanol-water mixture reacts to form hydrogen and carbon monoxide which upon contact with steam converts further to carbon dioxide and more hydrogen. Hydrogen so produced then can be separated from other gases involved and purified by the Pd-alloy membrane units. These mobile hydrogen generators have been used at weather stations in remote areas (e.g., in arctic and desert areas) for both military and civilian applications.

Other commercial uses of Pd alloy (primarily Pd-Ag) membranes include alternator cooling in power stations, tungsten heat treatment and removal of oxygen from hydrogen gas streams containing oxygen or hydrogen from oxygen gas streams [Philpott, 1985].

Similar permeability and hydrogen concentration levels as those above have been reported for Pd-Ag membranes supported on other porous materials. An ultrathin (about 1 μm thick) dense palladium-silver alloy membrane on a ceramic support has been fabricated for separating hydrogen from other gaseous mixtures, specifically for an ultrapure hydrogen purification system [Ye et al., 1994]. The hydrogen concentration in the permeate reaches 99.9995 to 99.9999% and the permeate (hydrogen) flux is about 1.9×10^4 barrers at 400°C and under a transmembrane pressure difference of 3 bars. In a case like this where a thin layer of a dense Pd alloy membrane is deposited on a porous support to improve the permeability, extreme care needs to be exercised to ensure no microcracks or defects on the membrane layer. When defects do occur, it is sometimes possible to make the repair in the composite membrane by a process such as the sol-gel or chemical vapor deposition process.

For many industrial bulk processing applications, the purity of the hydrogen required can not justify the use of dense palladium or its alloy membranes due to their low permeabilities. In these cases, porous inorganic membranes are more often considered.

In most cases, hydrogen has a higher permeability than many other gases. However, Ohya et al. [1994] prepared zirconia-silica composite membranes which allow only H_2O and HBr, but not hydrogen, to permeate. This particular gas separation is relevant to the thermochemical water decomposition process which uses high heat energy to produce hydrogen. If hydrogen can be separated from the product gases (containing H_2, H_2O and HBr) without phase change, the cost of hydrogen production is reduced as a result of the improved thermal efficiency. Hydrogen is not detected in the permeate and the separation factor of HBr over H_2O ranges from 6 to 36.

7.2.4 Separation of Oxygen and Nitrogen

Air separation has been one of the fastest growing applications of organic membranes in recent years. It has also been suggested to be one of the important applications of inorganic membranes for the refining industry [Johnson and Schulman, 1993]. The primary products of this application are nitrogen (or nitrogen-enriched air) and oxygen (or oxygen-enriched air). Using organic polymer membranes, this process enriches oxygen in the permeate and concentrates nitrogen in the retentate with an O_2/N_2 separation factor of 3.5 to 5.5. Today most of the commercial applications of air separation are for the production of nitrogen as a result of the practical limits of the current membrane technology. Current membranes produce nitrogen and oxygen with a highest purity greater than 99% and in the 30-50% range, respectively. They are, however, most cost effective for purifying nitrogen in the 95-98% range among various separation technologies. Since the feed to the process is essentially free, the added cost of recycle design is not justified. Therefore, air separation generally adopts the single-stage configuration. The fact that air has an essentially fixed composition and contains no impurities that will require any pretreatments makes the membrane process relatively simple.

The primary application for the nitrogen so produced by membranes is inert gas blanketing for storing and shipping of flammable liquids, fresh fruits and vegetables. Membranes are most competitive in this application when the volume and the acceptable purity level are both relatively low. An added advantage of membrane-based nitrogen separation process is that the nitrogen so produced, mostly in the retentate, is already pressurized. For those applications where high pressure gas is needed or convenient, air separation by membranes has additional values.

There is only limited utility of membranes for commercial-scale production of oxygen and it is primarily for medical uses. No large volume market exists today to support high volume production although enhanced combustion has the potential. Air is one of the major feedstock for coal processing. Oxygen-enriched air improves combustion efficiency of coal and is also a desirable feedstock for gasifiers. Oxygen currently produced by organic membranes has a purity limit of 50%.

There have been some scattered efforts testing various inorganic materials as membranes for separating oxygen and nitrogen in the air. Shown in Table 7.2 are some separation data taken from the open literature dealing with porous inorganic membranes for oxygen/nitrogen separation. It is interesting to note that, with oxygen having a molecular weight slightly greater than that of nitrogen, one would expect an O_2/N_2 separation factor of less than 1 (0.94 to be exact) in cases where Knudsen diffusion predominates. But some data in Table 7.2 shows that it is feasible to select inorganic membranes (e.g., molecular sieve carbon or zeolites) that are actually more selective to oxygen than nitrogen although the separation factor is not as high as one would desire. As molecular sieve materials including zeolites are being actively pursued by many researchers, it is reasonable to expect the achievable oxygen/nitrogen separation factor to be higher than those values shown in the table, perhaps close to the range where air separation may be feasible with inorganic membranes.

TABLE 7.2
Separation of oxygen and nitrogen by inorganic membranes

Membrane material (pore dia. in nm)	Temp(°C)	TMP (bar)	S.F.	P (barrer)	Reference	Note
Molecular sieve carbon (0.3-0.5)	25	<10	8*	1.1×10^3	Koresh & Soffer [1983a,b]	
Zeolite		9.8	2.7- 3.3	1.3×10^5	Matsushita Elect. Ind. Co. [1985]	on alumina support
Silicalite (4-5)					Sano et al. [1994]	
Glass	20-70		11		Ash et al. [1976]	

Note: S.F. = Separation factor; KD = Knudsen diffusion; P = Permeability (barrer or 10^{10} x (cm^3(STP)-cm/s-cm^2-cm Hg)); * indicates ideal separation factor; (S.F.)$_{KD}$ = 0.94

Not listed in the table are solid electrolyte membranes such as stabilized zirconia membranes. Currently oxygen production from these dense membranes is not as economical as cryogenic oxygen production for large scale oxygen plants. However, for a small plant below approximately 100 tons per day, solid electrolyte membranes could be economical [Dellefield, 1988]. These dense ceramic membranes could supply oxygen for small or modular coal gasification sites or for hospitals. Stabilized zirconia membranes require a temperature near 982°C to make them conductive to oxygen ions.

A recent development of a ceramic membrane appears to be promising for selectively removing oxygen from air. Multi-channel membrane elements have been fabricated for that purpose [Anonymous, 1995]. The membrane has the potential for reducing the cost of converting natural gas to synthesis gas.

7.2.5 Separation of Hydrogen and Hydrocarbon Gases

Recovery of hydrogen by membranes is important in the refining industry. Associated with the use of hydrogen at refinery plants, light hydrocarbon gases are produced which must be purged from the reaction system. The hydrogen value in the purge gas can be used as a fuel as in the past. However, two major factors have provided incentives for recovering hydrogen for refinery processing instead. One is that hydrogen in the off-gases, if recovered, can be as much as 20% of the total hydrogen generated at the refinery plants. The other factor is that the cost of hydrogen manufacture for various refining processes such as coking, hydrocracking and hydrodesulfurization is increasing. The cost of recovering hydrogen by gas separation via organic membranes depends on the permeate pressure and the % recovery of hydrogen but has been estimated to be less than $1.10 per 1,000 scf (standard cubic feet) of hydrogen compared to about $2.50 per 1,000 scf of hydrogen produced from natural gas [Johnson and Schulman, 1993]. The calculation is based on a "typical" case where the feed contains 75% hydrogen and 25% methane and the hydrogen purity in the permeate is 95% with a hydrogen recovery of 90%. It is advantageous to maintain as high a permeate pressure as practically possible, even if some loss of yield occurs.

In these applications hydrogen is separated from the purge gas containing an appreciable amount of hydrogen as well as hydrocarbon gases. The recovered hydrogen is returned to the plant to supplement hydrogen made by the expensive direct production route to improve the quality of fuels or the feed quality for various processes. Thus membrane separation in this case improves process efficiency. Membrane processing competes with other separation processes like pressure swing adsorption, cryogenics and oil scrubbing for these applications. Separation of hydrogen from methane has been identified as one of the promising applications of inorganic membrane technology for the refinery industry in the future [Johnson and Schulman, 1993].

Selected studies have been made to investigate separation of hydrogen from C_2-C_4 hydrocarbons as highlighted in Table 7.3. Porous alumina and silica membranes with pore diameters greater than 1 nm appear to fall within the Knudsen diffusion regime and offer a limited separation factor for the low molecular weight hydrocarbons. The actual or ideal separation factor ranges from 3.0 to 8.5 depending on the hydrocarbons involved. There is some indication that finer pore membranes such as the reportedly 0.3-0.8 nm silica membranes made by e-beam evaporation of quartz on porous glass or alumina supports may give a separation factor exceeding 8 for the hydrogen/isobutane gas mixture which has a Knudsen separation factor of 5.4 [Maier, 1993].

A major challenge is to develop inorganic membranes that offer higher separation factor and permeability values than the organic polymer membranes or provide the same level of separation performance at higher permeate pressures. The last point is related to the compression costs which can be very significant in determining if a membrane process is economical or not. On one hand, a lower permeate pressure increases the pressure driving force for the separation process. On the other hand, as the permeate pressure becomes lower, a penalty may have to be paid if hydrogen so separated needs to be recompressed. Figure 7.2 provides a quick estimate of the compression cost and it is seen that the compression cost can be approximated to be a linear function of the compression ratio (i.e., the ratio of the outlet to inlet pressures, both expressed in absolute pressures) [Johnson and Schulman, 1993].

For example, if hydrogen recovered as a permeate is at 20 bars (abs.) and needs to be recompressed to 40 bars (abs.) as a feedstock to a process, the required compression cost is approximately \$0.15 per Mscf (1×10^6 scf). Obviously, the trade-off between the recovery and the recompression requirements must be optimized for achieving acceptable process economics. Thus for inorganic membranes to be advantageous, they must achieve comparable separation of hydrogen as organic membranes at a higher permeate pressure. This is an ambitious target in view of the high hydrogen recovery (on the order of 90%) achievable with organic membranes today and requires much further development of inorganic membranes.

TABLE 7.3
Separation of hydrogen and hydrocarbons by inorganic membranes

Membrane material (pore dia. in nm)	Temp (°C)	TMP (bar)	S.F.	P (barrer)	Reference	Note
Al_2O_3 (10-20)	-196 to 97	0.23-1.05	4.1*	3.2×10^5	Itaya et al. [1984]	H_2/C_2H_6; $(S.F.)_{KD} = 3.88$
Al_2O_3 (10-20)	-196 to 97	0.23-1.05	5.0*	3.2×10^5	Itaya et al. [1984]	H_2/C_3H_8; $(S.F.)_{KD} = 4.69$
γ-Al_2O_3 (5)	200	0.02-0.13	7.0*		Okubo et al. [1991]	H_2/cyclohexane; $(S.F.)_{KD} = 6.5$
γ-Al_2O_3 (5)	200	0.02-0.13	5.9*		Okubo et al. [1991]	H_2/bezene; $(S.F.)_{KD} = 6.2$
Silica (0.3-0.8)		0.41-1.33	1.1-8.5		Maier [1993]	H_2/isobutane; $(S.F.)_{KD} = 5.40$; on glass support

Note: S.F. = Separation factor; KD = Knudsen diffusion; P = Permeability (barrer or 10^{10} x (cm^3(STP)-cm/s-cm^2-cm Hg)); * indicates ideal separation factor

As hydrogen is removed from a purge gas stream containing hydrocarbons, the dew point of these hydrocarbons will be reduced. If a significant amount of hydrogen is removed, further reduction in the dew point may trigger hydrocarbon condensation on the membrane surface. Although this condensation may not damage the membrane, it does reduce the gas separation efficiency [Spillman, 1989].

Bulk separation of hydrogen from other gases, in this case as in other cases, requires an acceptable level of hydrogen permeation rate or permeability. Typical hydrogen permeability through a dense palladium-based membrane is too low to be economically viable. Many porous inorganic membranes have a permeability in the order of 10^4-10^5 barrers. It is interesting to note that a permeance of 0.01 cm^3 (STP)/s-cm^2-cmHg has been set as a target for ceramic membranes to be effective for gas separation based on classified development work at the Oak Ridge Gaseous Diffusion Plant [Dellefield, 1988]. Assuming an effective membrane layer thickness of a few microns, the corresponding target permeability is in the 10^4 barrers range. Thus the permeability values of many inorganic membranes are within the economically viable range. What remains to be improved is the permselectivity (separation factor).

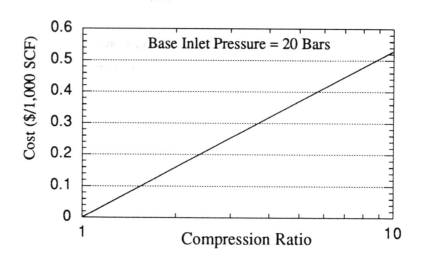

Figure 7.2 Compression cost as a function of compression ratio [Johnson and Schulman, 1993]

7.2.6 Separation of Hydrogen and Carbon Monoxide

Produced from natural gas, oil or coal using different processes, syngas (or synthesis gas) contains a mixture of mostly hydrogen and carbon monoxide plus low percentages of carbon dioxide and nitrogen. With transition metal catalysts, it is used in a number of reactions for making a wide variety of organic and inorganic compounds such as alcohols, aldehydes, acrylic acids and ammonia. The stoichiometry of the feed gas for these reactions must be adjusted according to the process requirements. The ratios of hydrogen to carbon monoxide and other gases in the feed streams to those chemical production processes can be varied by selectively removing hydrogen as a permeate through a membrane. In this case carbon monoxide comes out of the membrane separation process at high pressure which is an advantage over other separation processes such as cryogenics and pressure swing adsorption.

Another important application is the removal of carbon monoxide from the valuable hydrogen produced in steam reformed natural gas.

Alumina membranes made by anodic oxidation were tested for separating hydrogen and carbon monoxide at a temperature up to 97°C [Itaya, 1984]. The ideal separation factor only achieved 3.5. See Table 7.4. Wu et al. [1993] also examined the use of alumina membranes for this application at an even higher temperature of 300°C and obtain a maximum separation factor of only 2.2. The alumina membranes used in the above studies all have pore diameters larger than 4 nm. Their separation factors for the H_2/CO gas pair are no higher than the Knudsen diffusion values of 3.7 most likely as a result of the limitation imposed by the relatively large pores involved.

TABLE 7.4
Separation of hydrogen and carbon monoxide by inorganic membranes

Membrane material (pore dia. in nm)	Temp(°C)	TMP (bar)	S.F.	P (barrer)	Reference	Note
Al_2O_3 (10-20)	-196 to 97	0.23-1.05	3.5*	$3.2x10^5$	Itaya et al. [1984]	anodic oxide
Al_2O_3 (4)	20-300	1.3-17.4	1.2-2.2	(5.4 to 6.3)$x10^5$	Wu et al. [1993]	on multilayer support
Silica (0.3-0.8)		0.41-1.33	3.9-29.6		Maier [1993]	on glass support
Silica	25-70	21	31-62*	(0.18 to 2.8)$x10^2$	Way & Roberts [1992]	hollow fiber

Note: S.F. = Separation factor; KD = Knudsen diffusion; P = Permeability (barrer or 10^{10} x (cm^3(STP)-cm/s-cm^2-cm Hg)); * indicates ideal separation factor; (S.F.)$_{KD}$ = 3.7

Wu et al. [1993] have developed a mathematical model based on Knudsen diffusion and intermolecular momentum transfer. Their model applies the permeability values of single components (i.e., pure gases) to determine two parameters related to the morphology of the microporous membranes and the reflection behavior of the gas molecules. The parameters are then used in the model to predict the separation performance. The model predicts that the permeability of carbon monoxide deviates substantially from that based on Knudsen diffusion alone. Their model calculations are able to explain the low gas separation efficiency. Under the transport regimes considered in their study, the feed side pressure and pressure ratio (permeate to feed pressures) are found to exert stronger influences on the separation factor than other factors. A low feed side pressure and a low pressure ratio provide a maximum separation efficiency.

On the contrary, silica membranes with pore diameters in the 0.3-0.8 nm on porous glass supports have been claimed to exhibit an H_2/CO separation factor as high as 30. This is significantly higher than what Knudsen diffusion alone predicts. It is possible that some molecular sieving takes place in the reportedly small pores in the silica membranes. This postulation appears to be further reinforced by another study using hollow fiber silica membranes which shows rather high ideal separation factor of hydrogen over carbon monoxide [Way and Roberts, 1992]. Although no pore size data is provided in that study, it is believed that the membranes have very fine pore diameters judging from their low permeability values.

7.2.7 Separation of Hydrogen and Nitrogen

Ammonia is typically made from syngas at a temperature of about 500°C and a pressure of about 300 bars. Incomplete conversion to ammonia results in a need to recycle unreacted gases back to the reactor. Some gases inert to the reaction such as argon and unreacted methane will build up and require a continuous gas purge to manage these contaminants at an acceptable level. Like in some other applications, the purge gas contains valuable hydrogen and nitrogen as the major components and the two gases can be separated cost effectively by organic membranes in comparison to other traditional separation processes.

In this application hydrogen is separated as a permeate at the lower pressure side of the membrane. Consequently, it must be recompressed to the pressure of the reactor.

A number of inorganic membrane materials have been investigated for separating hydrogen from nitrogen at a wide range of temperatures, Table 7.5. Many of them predictably have not produced a separation factor higher than the Knudsen diffusion value of 3.7. In some studies, an ideal separation factor in the 140-300 range has been observed with silica membranes [Gavalas and Megiris, 1990; Megris and Glezer, 1992; Way and Roberts, 1992]. This potential for high separation factors seems to be further suggested by other investigations. Silica membranes with a pore diameter of approximately 0.3 nm show a separation factor of 7 [Maier, 1993]. Moreover, chemical vapor deposited silica membranes produce an actual separation factor as high as 72 [Wu et al., 1994]. While this is promising in the laboratory, large-scale testing is needed to confirm the findings.

TABLE 7.5
Separation of hydrogen and nitrogen by inorganic membranes

Membrane material (pore dia. in nm)	Temp (°C)	TMP (bar)	S.F.	P (barrer)	Reference	Note
Al_2O_3 (80)	20	0.4-2.0	3-4[*]		van Vuren et al. [1987]	$P/\Delta l = 3.3 \times 10^{-6}$
Al_2O_3 (10-20)	-196 to 97	0.23-1.05	4.1[*]	3.2×10^5	Itaya et al. [1984]	
Al_2O_3 (10-20)	20		3.7	2.5×10^5	Toyo Soda [1985]	anodic oxide
Al_2O_3 (4)	20-538	1.3-35.3	1.04-2.1	(5.4 to 6.3)$\times 10^5$	Wu et al. [1993]	on multilayered support

TABLE 7.5 – CONTINUED

Separation of hydrogen and nitrogen by inorganic membranes

Membrane material (pore dia. in nm)	Temp (°C)	TMP (bar)	S.F.	P (barrer)	Reference	Note
Al$_2$O$_3$ (4)	104-445	0.03-0.21	4-10		Lee et al. [1994]	impregnated with Pd; P/Δl = (6.0 to 9.6) x 10^{-3}
C	25-120		5*		Ash et al. [1976]	
Glass	-193 to 327		2.5-3.6	2.0x10^5	Tock & Kammermeyer [1969]	
Silica			3.3		Agency Ind. Sci. Tech. [1986]	P/Δl = 1.7x10^{-4}
Silica	200		11*		Okubo & Inoue [1989]	P/Δl =1.6x10^{-5}
Silica	100-600		4-300*		Gavalas & Megiris [1990]	P/Δl =4.0x10^{-5} to 1.2x10^{-4}
Silica	450		1000-5000*		Tsapatsis et al., 1991	P/Δl =7.3x10^{-7} to 7.7x10^{-5}
Silica (0.3-0.8)		0.41-1.33	3-7		Maier [1993]	on glass support
Silica	25-70	21	125-163*	0.18-2.8x10^2	Way & Roberts [1992]	hollow fiber
Silica/C	750	2	3.7-141*		Megiris & Glezer [1992]	amorphous membranes on glass support; P/Δl =(3.5 to 8.8)x10^{-5}
Silica	600	0.5-6	3.7-72		Wu et al. [1994]	P/Δl =1.0x10^{-5} to 6.4x10^{-3}
Titania	600		250*		Gavalas, 1991	P/Δl =4.0x10^{-5}
Zeolite silicalite	20	1	3.06*		Jia et al. [1993]	P/Δl =6.8x10^{-4}

Note: S.F. = Separation factor; KD = Knudsen diffusion; P = Permeability (barrer or 10^{10} x (cm^3(STP)-cm/s-cm^2-cm Hg)); P/Δl = Permeance (cm^3(STP)/s-cm^2-cm Hg); * indicates ideal separation factor ; (S.F.)$_{KD}$ = 3.7

7.2.8 Separation of Hydrogen and Carbon Dioxide

Carbon dioxide separation from hydrogen is important in the separation of the water gas shift reaction products. This is quite relevant, for example, in the production of hydrogen from coal, see Figure 7.3 [Pellegrino et al., 1988]. First syngas is produced by removing hydrogen sulfide and carbon dioxide from other gasification products, hydrogen and carbon monoxide. This step permits the use of sulfur sensitive catalysts for the shift reaction. After the water gas shift reaction where CO is converted to CO_2, the hydrogen and carbon dioxide so obtained can be separated by an inorganic membrane to concentrate hydrogen as a valuable product.

GA 34558.1

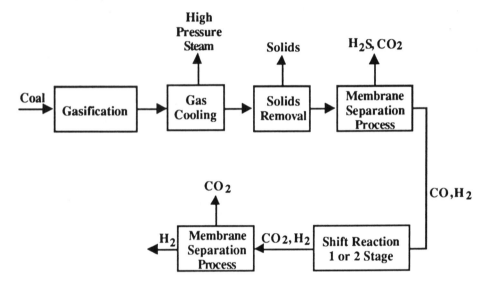

Figure 7.3 A schematic diagram of an entrained-flow coal gasification process to produce hydrogen from coal [Pellegrino et al., 1988]

Listed in Table 7.6 are some studies of inorganic membranes for the separation of hydrogen from carbon dioxide. Alumina membranes with as fine as 10 to 20 nm pore diameter have been investigated and a maximum separation factor of only 5 has been obtained. Obviously, new membrane materials with more promising separation results are required before this application can be tackled by inorganic membranes.

7.2.9 Separation of Hydrogen and Coal Gasification Streams

As mentioned elsewhere in this book, hydrogen is increasingly becoming a high-value product with a variety of potential uses. It can be applied to upgrade coal and oil shale-derived liquids, as a chemical feedstock and as an alternative fuel. Therefore, in an integrated coal gasification combined cycle unit, the process economics can be improved if hydrogen can be efficiently separated from the gasification stream. An approximate composition of a dry gasification stream is given in Table 7.7. It contains about 20% hydrogen, 13% carbon monoxide and 16% carbon dioxide. The remaining components are 47% nitrogen and only 4% methane. A conceptualized process for producing hydrogen from coal gasification has been shown earlier (Figure 7.3).

TABLE 7.6
Separation of hydrogen and carbon dioxide by inorganic membranes

Membrane material (pore dia. in nm)	Temp(°C)	TMP (bar)	S.F.	P (barrer)	Reference	Note
Al_2O_3 (10-20)	-196 to 97	0.23-1.05	5.0*	3.2×10^5	Itaya et al. [1984]	anodic alumina
Al_2O_3 (10-20)	20		5.0	2.5×10^5	Toyo Soda [1985]	anodic alumina
Al_2O_3 (4)	104-445	0.03-0.21	4-8		Lee et al. [1994]	impregnated with Pd; $P/\Delta l$ = (6.0 to 9.6) x 10^{-3}
SiO_2	25-70	21	4.7-5.4*	(0.18 to 2.8) $\times 10^2$	Way & Roberts [1992]	hollow fiber

Note: S.F. = Separation factor; KD = Knudsen diffusion; P = Permeability (barrer or 10^{10} x (cm^3(STP)-cm/s-cm^2-cm Hg)); $P/\Delta l$ = Permeance (cm^3(STP)s-cm^2-cm Hg); * indicates ideal separation factor; (S.F.)$_{KD}$ = 4.7

TABLE 7.7
Approximate composition of a dry gasification mixture

Gas	Concentration (%)
Hydrogen	20
Nitrogen	47
Carbon monoxide	13
Methane	4
Carbon dioxide	16

[McMahon et al., 1990]

A porous Vycor glass membrane and a porous gamma-alumina membrane, both having about the same pore diameter of 4 nm, were used to study the separation of hydrogen from other components in the gasification mixture [McMahon et al., 1990]. At a temperature up to 125°C, a maximum hydrogen concentration of about 40% in the permeate is observed under various feed pressures and pressure ratios. An accompanied modeling analysis which agrees well with the laboratory data reveals that even under the hot-gas cleanup conditions of 590°C and 20 bars, a hydrogen concentration of only 38% can be expected due to Knudsen diffusion as the prevailing separation mechanism. The purity of hydrogen can be improved if the pore size of the membrane used is smaller.

Potential benefits of having a highly selective (say, capable of removing 90% of hydrogen produced) ceramic membrane with a permeance of 0.01 cm^3 (STP)/s-cm^2-cmHg can be enormous [Dellefield, 1988]. It has been estimated that this type of membrane would only require one 7-foot diameter pressure vessel approximately 18 feet long to remove 90% of the hydrogen produced by oxygen gasification of 2,777 tons of coal per day in a Texaco gasifier which produces both hydrogen and electricity at 650-760°C. This compact vessel is capable of separating 16.6 million kcf (1x10^3 cubic feet) of hydrogen per year from the raw fuel gas produced by this gasifier system. The biggest challenge lies in the development of ceramic membranes with a sufficiently high separation factor. An alternative to the highly hydrogen-selective but low-permeability palladium-based membranes is proton-conducting ceramics. This type of ceramic membranes can be used not only in high-temperature gas separations (e.g., separation of high-purity hydrogen from fuel gases) but also in solid electrolyte fuel cells as will be discussed in Chapter 8.

7.2.10 Separation of Carbon Dioxide and Hydrocarbons

Carbon dioxide is pumped into depleting oil reserves at high pressures to drive residual oils to existing oil wells. Over an extended period of time, the carbon dioxide mixes with the natural gas associated with the wells and can reach as high as 95% [Spillman, 1989]. In this enhanced oil recovery application using current organic membranes, the CO_2 is separated as the permeate and, therefore, required to be recompressed in order to be reinjected at high pressures back into the reservoir. But, at high permeate pressures, the membrane efficiency is compromised. Thus, a proper balance needs to be established between the two considerations.

In some natural gas treatment, the issue of separating carbon dioxide from hydrocarbons (e.g., methane) is also relevant. Some natural gas contains impurity gases such as hydrogen sulfide and carbon dioxide which both are corrosive to the pipeline and must be removed. In addition, being non-combustible the impurity gases reduce the heating content of the fuel gas. Natural gas is typically produced at high pressures and is, therefore, very amenable to membrane processing. In this case, the product, methane, exits the membrane system as the retentate and suffers no appreciable pressure drop. The membrane separation operation can best be performed at the wellhead instead of a downstream central plant to avoid the problems of corrosion due to the acid gases and safety. A typical specification of the treated natural gas calls for a carbon dioxide concentration of less than 2% [Spillman, 1989]. Single-stage membrane units are recommended for low-flow applications while a recycle loop should be considered for higher flow rates to minimize the loss of hydrocarbons due to incomplete separation. Separation of carbon dioxide from hydrocarbons is most cost competitive at low flow rates, for high carbon dioxide concentrations or for offshore platforms. Organic membranes start to be applied in this application. To be competitive with organic membranes, the inorganic membranes need to provide higher separation factors and/or higher permeabilities.

Inorganic membranes can also be used to separate carbon dioxide from hydrocarbons in a reducing environment to increase the Btu content of the fuel gas by removing the non-combustible components of the gas stream such as CO_2. A further example of the need to separate carbon dioxide from hydrocarbons is landfill gas upgrading. Landfills generate methane that is mixed with approximately equal amount of carbon dioxide. Organic membranes are currently employed to separate the two gases [Spillman, 1989]. The use of membranes is attractive for those cases where it is desirable to have the product gas at a high pressure.

Separation of the polar gases such as carbon dioxide, hydrogen sulfide and sulfur dioxide behaves in many respects differently from other nonpolar gases. Their transport mechanisms through porous membranes are often different and therefore their separation performances can also be markedly different. This has been observed for separating carbon dioxide from other nonpolar, non-hydrocarbon gases.

Although some porous membranes, for example porous glass membranes, have been studied for removing carbon dioxide from hydrocarbons, dense inorganic membranes made of various polyphosphazenes clearly offer much higher separation factors as shown in Table 7.8. A separation factor exceeding 8 has been observed, compared to a value very close to 1 in other cases. These dense inorganic polymer membranes, varying in their substituent groups and therefore their capacities to solubilize various gases, operate in the solution-diffusion mode in contrast to predominantly Knudsen diffusion for porous membranes. Inorganic polymers such as polyphosphazenes, however, have a thermal stability range that is between those of organic polymers and ceramics or metals. They can not be operated for any extended periods of time at a temperature beyond 300°C.

TABLE 7.8
Separation of carbon dioxide and hydrocarbons by inorganic membranes

Membrane material (pore dia. in nm)	Temp(°C)	TMP (bar)	S.F.	P (barrer)	Reference	Note
Glass	-193 to 327		0.5-1.1		Tock & Kammermeyer [1969]	CO_2/C_3H_8; $(S.F.)_{KD} = 1.0$
Poly-hosphazenes (dense)	30-190	0.7-2.7	1.0-8.3	7-678	Peterson et al. [1993]	CO_2/CH_4; $(S.F.)_{KD} = 0.6$

Note: S.F. = Separation factor; KD = Knudsen diffusion; P = Permeability (barrer or 10^{10} x (cm^3(STP)-cm/s-cm^2-cm Hg)); * indicates ideal separation factor

There are, however, evidences that other more effective separating mechanisms such as surface diffusion and capillary condensation can occur in finer pore membranes of some materials under certain temperature and pressure conditions. Carbon dioxide is known to transport through porous media by surface diffusion or capillary condensation. It is likely that some porous inorganic membranes may be effective for preferentially carrying carbon dioxide through them under the limited conditions where either transport mechanism dominates.

7.2.11 Separation of Hydrogen Sulfide and Hydrocarbons

As mentioned earlier, hydrogen sulfide is present in some natural gases as a non-combustible impurity. In addition, hydrogen sulfide is toxic. Therefore, the purpose of natural gas treatment which is best performed at the wellhead is to remove hydrogen sulfide along with carbon dioxide.

Although an important subject, separation of hydrogen sulfide from hydrocarbons by inorganic membranes has not been documented much in the literature. Peterson et al. [1993] synthesized and evaluated the separation performances of a family of polyphosphazene membranes, all of which are nonporous in nature. Some have reached a separation factor of 22 for carbon dioxide/methane (at 30-190°C under a TMP of 0.7-2.7) which is very significant. The corresponding separation factor by the Knudsen diffusion mechanism alone would be 0.69. The permeability of hydrogen sulfide is at 13 to 228 barrers. It has been found that the transmembrane pressure difference does not exert much influence on the separation factor but the temperature does have a strong effect. As revealed in Table 7.9, the separation factor decreases with increasing temperature although the hydrogen sulfide permeability improves with temperature (Table 7.10). On the contrary, transmembrane pressure difference or the feed side pressure essentially does not affect the separation factor.

TABLE 7.9
Effect of temperature on separation performance of poly[bis(phenoxy)phosphazene] membranes for removing carbon dioxide from methane

Temperature (°C)	Separation factor
30	16
70	8.6
110	4.4
150	3.3
190	2.8

Note: Pressure: 1.4 bar
[Peterson et al., 1993]

TABLE 7.10
Effect of temperature on permeability of hydrogen sulfide in a 10% H_2S/CH_4 gaseous mixture through various polyphosphazene membranes

Variety of polyphosphazene	Permeability at 30°C (barrer)	Permeability at 80°C (barrer)
SO_3-PPOP	13	77
8%Br-PPOP	20	120
m-F-PPOP	29	176
8%COOH-PPOP	21	109
PPOP-Ethyl	20	92

Note: PPOP: poly[bis(phenoxy)phosphazene]; pressure: 1.4 bars
[Peterson et al., 1993]

The permeability data in Table 7.10 and other data show that the polarity of the substituent group on the polymer backbone (such as poly[bis(phenoxy)phosphazene] or PPOP) has a significant impact on the membrane permeability. The more polar gas (i.e., SO_2), the more easily it permeates a polar polymer (i.e., m-F-PPOP) and a less polar gas (i.e., CO_2) exhibits a lower permeability through a more polar membrane (i.e., SO_3-PPOP). This seems to provide a vast opportunity for chemically designing an inorganic polymer membrane for a particular separation application [Peterson et al., 1993].

It is commonly accepted that gas permeates through a dense membrane such as a polyphosphazene membrane by the solution-diffusion mechanism as mentioned in Chapter 4. It consists of sorption into the membrane followed by diffusion across the membrane thickness and desorption out of the membrane. The gas solubility into the membrane depends greatly on the gas-membrane and gas-gas interactions while the diffusion is largely affected by the molecular size of the gas and the microstructure of the membrane. This permeability-molecular size relationship is generally observed with nonpolar gases. The separation factor based on Knudsen diffusion for the hydrogen sulfide/methane pair would be less than unity (i.e., the separation would favor methane over hydrogen sulfide). It appears that the unusually great solubility of an acid gas such as hydrogen sulfide in polyphosphazene is largely the reason for the very high separation factor of greater than 20 for H_2S over CH_4.

7.2.12 Dehydration of Air and Organic Solvents

Production practice of using porous alumina membranes for dehydrating air and various alcohols has been reported in recent years [Abe, 1986; Mitsubishi Heavy Industries, 1986]. These applications utilize the principle of capillary condensation of water vapor in small pores of a membrane, thus blocking transport of other gases. The condensed water migrates through the membranes and evaporates on the permeate side of the membranes (Figure 7.4).

GA 35289

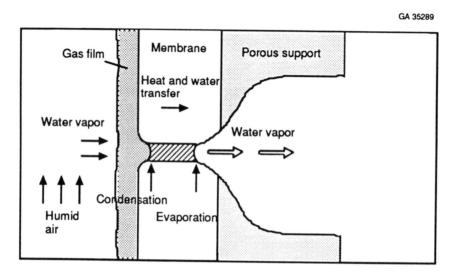

Figure 7.4 Schematic diagram of capillary condensation and transport of water vapor through a microporous membrane [Asaeda et al., 1984]

Separation of water/organic solvent mixtures. Most of the water/organic solvent mixtures have azeotropic points. Membrane separation of these mixtures is attractive in view of the potential energy savings. Several dense or nonporous organic polymer membranes are used for this purpose and enjoy high separation factors. Their permeabilities, however, are usually low. Some investigations have shown the promises of a number of porous inorganic membranes.

Given in Table 7.11 is the separation performance data of gamma-alumina membranes with pore diameters in the 1 to 3 nm range for removing water vapor from gas mixtures containing various alcohols. Exceptional binary separation factors exceeding 10 have been obtained in all cases at a temperature ranging from 65 to 89°C. Some are as high as 220. For comparison, the separation factors achievable with ideal Knudsen diffusion for those cases (water separation from methanol, ethanol or isopropanol) are only 1.3 to 1.8. It is possible that the extremely high separation factors are attributable to some coatings of sodium silicate to reduce the pore size [Asaeda et al., 1986b].

Leakage of the solvent through the membrane pinholes is one the major technical issues. To make these porous membranes viable for this application, even small pinholes in the membranes are not acceptable as they can have deleterious effects on the separation factor. One of the methods for ensuring high separation factors is to plug the pinholes with a small amount of an organic material having a relatively large molecular size.

Polyvinyl alcohol (PVA) and cellulose acetate (CA) have been found to be among these effective "sealants."

TABLE 7.11
Separation of water and hydrocarbons by inorganic membranes

Membrane material (pore dia. in nm)	Temp (°C)	S.F.	Reference	Note
γ-Al$_2$O$_3$ (~1)	65-70	60-120	Asaeda et al. [1986]	Water/methanol; coated with sodium silicate (S.F.)$_{KD}$ = 1.3; $J = 6.2 \times 10^{-2}$
γ-Al$_2$O$_3$ (~1)	78-84	90-220	Asaeda et al. [1986]	Water/ethanol; coated with sodium silicate; (S.F.)$_{KD}$ = 1.6; $J = 5.0 \times 10^{-2}$
γ-Al$_2$O$_3$ (~1)	82-88	70-140	Asaeda et al. [1986]	Water/n-propanol; coated with sodium silicate; (S.F.)$_{KD}$ = 1.8; $J = 5.0 \times 10^{-2}$
γ-Al$_2$O$_3$ (~3)	65-70	5-10	Asaeda & Du [1986]	Water/methanol; $J = 1.1 \times 10^{-1}$
γ-Al$_2$O$_3$ (~3)	78-89	8-17	Asaeda & Du [1986]	Water/ethanol; $J = 1.2 \times 10^{-1}$
γ-Al$_2$O$_3$ (~3)	82-88	11-24	Asaeda & Du [1986]	Water/n-propanol; $J = 1.2 \times 10^{-1}$
Silica-alumina (~1)	50	~600	Sakohara et al. [1989]	Water/ethanol; $J = 1.62 \times 10^{-1}$; membrane modified with PVA or CA
Silica-alumina (~1)	50	~1,200	Sakohara et al. [1989]	Water/n-propanol; $J = 1.34 \times 10^{-1}$; membrane modified by PVA
Silica-alumina (~1)	50	~1,750	Sakohara et al. [1989]	Water/acetone; $J = 1.84 \times 10^{-1}$; (S.F.)$_{KD}$ = 1.8; membrane modified by PVA
Silica-alumina (~1)	50	~1,750	Sakohara et al. [1989]	Water/THF; $J = 1.24 \times 10^{-1}$; (S.F.)$_{KD}$ = 2.0; membrane modified by PVA
Silica	60	400	van Gemert and Cuperus [1995]	Water/methanol
Silica	70	160	van Gemert and Cuperus [1995]	Water/ethanol

TABLE 7.11 – CONTINUED
Separation of water and hydrocarbons by inorganic membranes

Membrane material (pore dia. in nm)	Temp (°C)	S.F.	Reference	Note
Silica	70	550	van Gemert and Cuperus [1995]	Water/2-propanol

Note: S.F. = Separation factor; KD = Knudsen diffusion;
J = Permeate flux $(cm^3(STP)/s\text{-}cm^2)$; PVA = polyvinyl alcohol;
CA = cellulose acetate; THF = tetrahydrofuran

Sakohara et al. [1989] used this method to prepare essentially "leak-proof" silica-alumina membranes modified with PVA or CA. The silica-alumina membranes were made by treating sol-gel alumina membranes with sodium silicate. The treated membranes were then soaked in a dilute solution of PVA (in water) or CA (in acetone) at room temperature with one side of the membrane under vacuum to ensure contact of PVA or CA with the pinholes. In the case of CA, exposure to hot water for a short time is necessary to stiffen CA which resides in some large pores. The above procedure involving the organic sealants may be repeated a number of times to achieve an acceptably low level of solvent leakage through the membrane.

The modified silica-alumina membranes exhibit very high separation factors exceeding 600 for alcohols except methanol (vs. less than 90 for unmodified membranes) and exceeding 1,000 for organic solvents such as acetone and tetrahydrofuran (THF) as shown in Table 7.11. The high separation factors as a result of the modifications with PVA or CA are observed in both normal gas/vapor separation and pervaporation processes. The generally high separation factor is the result of the drastically reduced leakage rate of the solvent and essentially the same water permeation rate after modification. The possible explanation for the low separation factor of methanol/water is that neither PVA nor CA can effectively plug the small pinholes large enough for methanol to leak through.

It has also been found that the water flux depends on the osmotic pressure of the water/organic solvent mixture involved. When no pinholes are present, only water molecules can pass through a constriction to form menisci behind it. Thus a high separation factor can be attained. When selecting a sealant, considerations should be given to the chemical compatibility between the solvent and the membrane material. An example is the use of CA-modified membranes for acetone/water mixtures where acetone can conceivably dissolve some of the CA that plugs the pinholes.

Also given in Table 7.11 is the data for the separation of water from various alcohols using silica membranes [van Gemert and Cuperus, 1995]. The resulting separation factor values (160 to 550) are higher than those of the alumina membranes but lower than those of the silica-alumina membranes discussed earlier.

Dehydration of Air. Asaeda et al. [1984] evaluated gamma-alumina membranes on kaolin supports with a reported pore diameter of about 1 nm for air dehydration under a temperature of 31 to 77°C and a low TMP of 0.04 to 0.39 bar. A separation factor of 7 to 460 was obtained. The corresponding permeance is about 9.9×10^{-3} cm^3 (STP)/s-cm²-cmHg.

Equivalent separation factors (reaching 300 or higher) in plant environments for these dehydration applications have been reported [Mitsubishi Heavy Industries, 1986]. The corresponding water flux value of 1.2 L/hr-m² also seems to be very encouraging.

7.2.13 Enrichment of Hydrogen Sulfide

A hydrocarbon feed (e.g., methane, light and heavy liquids or petroleum coke and coal) is converted to synthesis gas (mostly CO and H_2 with small amounts of acid gases CO_2, H_2S and COS as impurities) in the partial oxidation process by reacting with steam and oxygen at a temperature of 1,370-1,540°C. By cooling off the hot gas, CO is converted with the steam to form CO_2 and H_2.

Enrichment of acid gases such as H_2S and CO_2 is a critical step in coal gasification because hydrogen sulfide needs to reach a sufficiently high concentration to be effectively treated by the conventional Claus unit. Currently the proven process for separating the H_2S and CO_2 from H_2 is the Rectisol process which is extremely expensive but it produces a H_2S-rich acid gas mixture that can be further handled by the inexpensive, established Claus process to recover elemental sulfur and reduce SO_x emissions.

Other cheaper processes such as amine absorption or hot potassium carbonate Hot Pot processes produce only 2-8% by volume of H_2S in the total H_2S/CO_2 stream. It is hoped that some inorganic membranes can be used to enrich the diluted H_2S stream from the aforementioned cheaper processes to a level of 15% or higher where the H_2S can be processed in a normal Claus plant and make the overall process more economical than the Rectisol process. By assuming the cost of an effective inorganic membrane to be twice that of an organic membrane for hydrogen recovery, Johnson and Schulman [1993] estimated the total savings that could be realized by employing the inorganic membrane. The savings depend to a great extent on the cost of compressing the H_2S/CO_2 mixture as the feed to the membrane system and could reach $0.53 per 1,000 standard cubic feet of hydrogen produced. It is noted that no organic membranes have been used for this application.

Separation of hydrogen from carbon dioxide has been discussed above. The separation factor appears to be reasonably good and is slightly better than what Knudsen diffusion predicts. And further improvement is considered to be quite likely. Removal of hydrogen sulfide from hydrogen has also been studied as indicated in Table 7.12. The investigations so far have not produced evidences of a high separation factor (less than 4.1 as predicted by Knudsen diffusion). With finer pore membranes emerging, this situation may be changed.

Both hydrogen sulfide and carbon dioxide are acid gases and their molecular masses are not far from each other. Thus their separation (or enrichment of hydrogen sulfide in the H_2S/CO_2 stream) poses a challenge. While no experimental study has been made on their separation with inorganic membranes, the application provides a significant incentive for inorganic membranes in the refining industry [Johnson and Schulman, 1993].

TABLE 7.12
Separation of hydrogen and hydrogen sulfide by inorganic membranes

Membrane material (pore dia. in nm)	Temp (°C)	TMP (bar)	S.F.	P (barrer)	Reference	Note
Al_2O_3 (102)	800	3.4	1.3		Kameyama et al. [1981b]	0.8 mm thick homogeneous membranes; $P/\Delta l$ = 4.0x10^{-4}
Al_2O_3 (20-40)	10	1.5-2.5	1.7		Kameyama et al. [1981b]	coating on porous Ni; $P/\Delta l$ = 1.2x10^{-3}
Glass	10		3		Kameyama et al. [1979]	
Glass (4.1)	10-800	3.2-3.5	1.9	1.7x10^4	Kameyama et al. [1981a]	homogeneous membranes;
Glass (4.5)	10-800	3.2-3.5	1.9		Kameyama et al. [1983]	
Ni (58)	10	0.5-2.9	1.3	2.0x10^6	Kameyama et al. [1983]	0.8 mm thick homogeneous membranes

Note: S.F. = Separation factor; KD = Knudsen diffusion; P = Permeability (barrer or 10^{10} x (cm^3(STP)-cm/s-cm^2-cm Hg)); $P/\Delta l$ = Permeance (cm^3(STP)/s-cm^2-cm Hg); * indicates ideal separation factor; (S.F.)$_{KD}$ = 4.1

7.2.14 Other Gas Separation Applications

Potentials for inorganic membranes in other gas separation applications are enormous in varieties. Significant technological breakthroughs and development time, however, are required to realize those potential benefits.

Hydrocarbon vapor recovery from air. Many petrochemical processes generate waste air streams that contain hydrocarbon vapors. The emission of these untreated air streams not only represents the loss of valuable hydrocarbons but also becomes a health, safety and environmental hazard. Although no clear cost advantage has been demonstrated, organic membranes have begun to be marketed for this application. It is also possible to separate air from the hydrocarbon vapor by an inorganic membrane. No data has been obtained for the separation of air and hydrocarbon vapors. Nevertheless, there is data on the separation of nitrogen from hydrocarbons as given in Table 7.13. Separation of nitrogen from hydrocarbons containing C_3 or higher carbons appears to be encouraging and worth pursuing more. Some zeolite silicalite membranes offer an ideal separation factor of as high as 78 for hydrogen/C_3+ hydrocarbons. These values are significantly higher than the Knudsen diffusion predictions.

TABLE 7.13
Separation of nitrogen and hydrocarbons by inorganic membranes

Membrane material (pore dia. in nm)	Temp(°C)	TMP (bar)	S.F.	Reference	Note
Zeolite silicalite	20	0.1-0.5	1.2-1.5*	Jia et al. [1993]	N_2/CH_4; (S.F.)$_{KD}$ = 0.76; $P/\Delta l$ =(3.7 to 14.0)x10^{-4}
Zeolite silicalite	20	0.1-0.5	1.3-1.4*	Jia et al. [1993]	N_2/C_2H_6; (S.F.)$_{KD}$ = 1.04; $P/\Delta l$ =(3.7 to 14.0)x10^{-4}
Zeolite silicalite	20	0.1-0.5	3.3-9.9*	Jia et al. [1993]	N_2/C_3H_8; (S.F.)$_{KD}$ = 1.3; $P/\Delta l$ =(3.7 to 14.0)x10^{-4}
Zeolite silicalite	20	0.1-0.5	4.6-17*	Jia et al. [1993]	$N_2/n\text{-}C_4H_{10}$; (S.F.)$_{KD}$ = 1.4; $P/\Delta l$ =(3.7 to 14.0)x10^{-4}
Zeolite silicalite	20	0.1-0.5	3.0-78*	Jia et al. [1993]	$N_2/i\text{-}C_4H_{10}$; (S.F.)$_{KD}$ = 1.4; $P/\Delta l$ =(3.7 to 14.0)x10^{-4}
Zeolite silicalite	20	0.1-0.5	32*	Jia et al. [1993]	$N_2/neo\text{-}C_5H_{12}$; (S.F.)$_{KD}$ = 1.6; $P/\Delta l$ =(3.7 to 14.0)x10^{-4}

TABLE 7.13 – CONTINUED
Separation of nitrogen and hydrocarbons by inorganic membranes

Membrane material (pore dia. in nm)	Temp(°C)	TMP (bar)	S.F.	Reference	Note
Zeolite silicalite	20	0.1- 0.5	2.6- 5.5[*]	Jia et al. [1993]	N_2/cyclohexane; $(S.F.)_{KD} = 1.7$; $P/\Delta l = (3.7$ to $14.0) \times 10^{-4}$
Zeolite silicalite	20	0.1- 0.5	45[*]	Jia et al. [1993]	N_2/benzene; $(S.F.)_{KD} = 1.7$; $P/\Delta l = (3.7$ to $14.0) \times 10^{-4}$
Glass	-193 to 327		0.5- 1.1	Tock & Kammermeyer [1969]	N_2/C_2H_4; $(S.F.)_{KD} = 1.0$

Note: S.F. = Separation factor; KD = Knudsen diffusion;
$P/\Delta l$ = Permeance $(cm^3(STP)/s$-cm^2-cm Hg); [*] indicates ideal separation factor

Separation between Hydrocarbons. Zhu et al. [1994] prepared a new class of inorganic membranes with a composite microstructure consisting of pillared clay and carbon. Deposited on the bulk support (e.g., porous glass) is an intermediate layer made from alumina pillared montmorillonite clay which is a microporous material. Another deposition with an aqueous suspension of polyvinyl alcohol (PVA) and montmorillonite clay is made followed by drying and calcining at 450-500°C in vacuum. PVA is carbonized to form a layer with fine pores during the calcination step. Further modification can be made by hydrolysis of $Zr(C_3H_7OH)_4$.

The resulting membranes show distinctly different permeation rates of several pure organic vapors. For example, the permeation rate at 60°C of benzene is ten times that of chlorobenzene which, in turn, is about ten times that of 1,3-dichlorobenzene. Moreover, the permeation rates of the two isomers, 1,2-dichlorobenzene and 1,3-dichlorobenzene, are significantly different at various temperatures. It was suggested that the controlling transport mechanism is diffusion rather than molecular sieving [Zhu et al., 1994], although no conclusive evidence is available. These order-of-magnitude differences in the permeation flux suggests potential uses of these membranes for separating selected organic mixtures.

Separation of methanol and methyl tert-butyl ether (MTBE) is an important step in the production of MTBE which is a fast growing chemical. The separation by pervaporation using silica membranes has been modestly successful [van Gemert and Cuperus, 1995]. A maximum separation factor of about 19 can be obtained.

Helium recovery. There are potentials for inorganic membranes in some very low flow-rate applications involving helium. In the reclamation of spent helium gases from airships or deep-sea diving systems, membranes can be used to separate helium from nitrogen and oxygen for recycle.

Table 7.14 provides some available data on the separation of helium from nitrogen by inorganic membranes. Way and Roberts [1992] obtained an ideal separation factor of hollow fiber silica membranes (possibly dense) in the 96 - 325 range and Koresh and Soffer [1983a; 1983b] found that the use of molecular sieve carbon membranes offers corresponding values of 20 to 40. In contrast, zeolite silicalite membranes provide an ideal separation factor of only 1.6 to 2.8 [Jia et al. [1993]. Tests by other investigators using porous alumina, glass and silica membranes show that the ideal separation factor is not far from that calculated based on Knudsen diffusion consideration alone. Clearly more definitive data is needed from separation measurements using actual gaseous mixtures.

TABLE 7.14
Separation of helium and nitrogen by inorganic membranes

Membrane material (pore dia. in nm)	Temp(°C)	TMP (bar)	S.F.	P (barrer)	Reference	Note
Al_2O_3 (~10)	25-75	0.03-0.16	2.7*	4.9×10^7	Havredaki & Petropoulos [1983]	2.35 cm thick homogeneous membrane
Al_2O_3 (10-20)	-196 to 97	0.23-1.05	3.2*	3.2×10^5	Itaya et al. [1984]	anodic oxide
Molecular sieve carbon (0.3-0.5)	25	<10	20-40*	1.2×10^3	Koresh & Soffer [1983a,b]	
Glass	-193 to 327		0.7-1.4		Tock & Kammermeyer [1969]	
Silica (~8)	25-75	0.03-0.16	2.8*	4.8×10^7	Havredaki & Petropoulos [1983]	1.51 cm thick homogeneous membrane
Silica	25-70	21	96-325*	0.14-5.7×10^2	Way & Roberts [1992]	hollow fiber
Zeolite silicalite	20	0.1-0.5	1.6-2.8*		Jia et al. [1993]	$P/\Delta l = (3.7$ to $14) \times 10^{-4}$

Note: S.F. = Separation factor; KD = Knudsen diffusion; **P** = Permeability (barrer or 10^{10} x (cm^3(STP)-cm/s-cm^2-cm Hg)); **P/Δ*l*** D= Permeance (cm^3(STP)/s-cm^2-cm Hg); * indicates ideal separation factor; (S.F.)$_{KD}$ = 2.6

On the other hand, actual binary mixture tests using porous alumina and glass membranes show separation factor values for helium recovery from oxygen that are lower than what Knudsen diffusion provides, as indicated in Table 7.15. Only Koresh and Soffer [1983a; 1983b] show an ideal separation factor of 20 to 40 with a low permeability of 1.2×10^3 barrers when molecular sieve membranes with a reported pore diameter of 0.3 to 0.5 nm are used.

Membranes have been suggested for use in the separation of helium from natural gas [Spillman, 1989] which typically contains 85% methane and 10% ethane. While no test data using real or simulated natural gas is available, there is some information on the separation of helium and ethane using alumina and silica membranes [Havredaki and Petropoulos, 1983], Table 7.16. Clearly, Knudsen diffusion is dominant in the these limited tests. Thus no promising separation performance has been demonstrated.

Other promising separation data. Porous $BaTiO_3$ membranes have been prepared on alpha-alumina supports. A CO_2 to N_2 separation factor of 1.2 at 500°C which is higher than the Knudsen diffusion prediction of 0.8 has been obtained. If there had not been pinholes on the order of 100 nm in the membranes, the separation factor would have been higher [Kusakabe and Morooka, 1994]. The maximum CO_2 permeance through the $BaTiO_3$ membranes is very high at 1.1×10^{-2} cm^3(STP)/s-cm^2-cm Hg.

TABLE 7.15
Separation of helium and oxygen by inorganic membranes

Membrane material (pore dia. in nm)	Temp(°C)	TMP (bar)	S.F.	P (barrer)	Reference	Note
Alumina (4)	538-815	1.0-3.3	1.4-1.6	(3.9 to 4.2)x10^5	Wu et al. [1993]	
Glass (4)	134	0.7-1.1	1.4-1.5	1.5x10^3	Shindo et al. [1985]	1.4 mm thick homogeneous membranes
Molecular sieve C (0.3-0.5)	25	<10	20-40*	1.2x10^3	Koresh & Soffer [1983a and 1983b]	

Note: S.F. = Separation factor; KD = Knudsen diffusion; P = Permeability (barrer or 10^{10} x (cm^3(STP)-cm/s-cm^2-cm Hg)); * indicates ideal separation factor; S.F.)$_{KD}$ = 2.8

7.3 STATUS SUMMARY OF GAS SEPARATION BY INORGANIC MEMBRANES

Research and development activities associated with inorganic membranes as gas separation media have flourished in recent years. Both material and engineering fronts have advanced. There are, however, several technical issues that need to be addressed before the full potentials of some inorganic membranes can be realized. Clearly, a more systematic and coordinated approach will facilitate the material and application development. This will be discussed below.

TABLE 7.16
Separation of helium and hydrocarbons by inorganic membranes

Membrane material (pore dia. in nm)	Temp(°C)	TMP (bar)	S.F.	P (barrer)	Reference	Note
Al_2O_3 (~10)	25-75	0.03-0.16	2.6*	4.9×10^7	Havredaki & Petropoulos [1983]	He/C_2H_6; 2.35 cm thick homogeneous membrane; (S.F.)$_{KD}$ = 2.7
SiO_2 (~8)	25-75	0.03-0.16	2.8*	4.8×10^7	Havredaki & Petropoulos [1983]	He/C_2H_6; 1.51 cm thick homogeneous membrane; (S.F.)$_{KD}$ = 2.7

Note: S.F. = Separation factor; KD = Knudsen diffusion; P = Permeability (barrer or 10^{10} x (cm^3(STP)-cm/s-cm^2-cm Hg)); *indicates ideal separation factor

7.3.1 Needs for More Binary or Multiple Gas Mixture Separation Data at High Temperatures

As have been reviewed in this chapter, many of the literature studies on gas separation by inorganic membranes are preliminary in nature, primarily dealing with the technical feasibility issue. Much data was obtained at room or low temperatures which are far from the temperature range of many potential applications which best utilize the unique characteristics of inorganic membranes. Furthermore, much of the separation factor data is expressed in terms of the ideal separation factor which is determined from single gas permeability or permeation flux data. Actual binary or multiple gas mixture data is much needed to provide more realistic projection of what actual separation performance can be expected. The ideal separation factor can only offer some hints on the potential of the membrane for separating the given pair of gases in a mixture.

The gap between the ideal and actual binary separation factors greatly depends on the membrane and gas properties as well as the operating conditions. Through an example of $He-CO_2$ separation, Fain and Roettger [1993] have demonstrated how significantly the adsorption of one gas on the walls of a membrane affects the transport of the other gas in a binary mixture. Similarly other transport mechanisms such as capillary condensation and the interaction and momentum exchange between gases can also make actual binary or multiple gas mixture measurements necessary. Clearly, more binary or actual multicomponent gas separation tests under typical application conditions need to be performed to provide definitive permeability and separation factor data to guide any necessary additional materials developments or engineering analyses for preliminary design purposes.

Numerous studies on inorganic membranes have shown that the separation factor is limited and not far from that predicted by the Knudsen diffusion. This primarily reflects the current status of material developments of inorganic membranes. The majority of commercial and developmental inorganic membranes contain macropores or mesopores. These pore sizes fall within the dominant regime of Knudsen diffusion which is of limited use for many gas separation applications from the standpoint of process economics. To break this barrier, finer pore sizes or transport mechanisms more effective than Knudsen diffusion for gas separation is essential.

7.3.2 Material and Engineering Considerations at Application Temperatures

The majority of gas separation applications use pressure difference as the driving force for the membrane separation. As such, the issues of sealing the ends of membrane elements and connecting the elements and the module or process piping are critical in providing gas-tight or essentially leakproof conditions. The seals and connections are necessary to prevent remixing of the permeate and the retentate streams.

In most of the industrially important gas separation applications, the feed streams to be processed occur at high temperatures. It is very desirable not to ramp down the stream temperature and then ramp up again after the treatment. It is exactly this reason that inorganic membranes are attractive due to their inherent thermal stabilities. Operation at high temperatures, however, not only confounds the above issues but also can affect the phases and microstructures of the membrane materials. All these factors have implications on the permeabilities and permselectivities.

As will be introduced in the next chapter, a new class of reactors called membrane reactors combines two unit operations (membrane separation and catalytic reaction) into one compact operation. Many of the membrane reactors of potential interest use inorganic membranes and are operated in gas/vapor phases at high temperatures involving gas separation. Thus membrane separation of gases is often a critical part of a membrane reactor and the aforementioned material and engineering considerations apply. Therefore, discussions on these and other additional considerations will be treated in detail later in Chapters 9 through 11.

7.3.3 Model-Based Development Strategy

A vastly growing number of investigations on inorganic membranes is witnessed today. Central to these activities is the development of membrane materials which under the applications conditions can offer high permselectivities with acceptably high permeabilities. The need for separation factors beyond those limited by Knudsen diffusion which is based on molecular mass is obvious. It appears that other transport mechanisms will be required to effect higher permselectivities.

Molecular sieving and the interactions of gas molecules with the membrane are possible alternatives. As discussed in Chapter 4, if surface diffusion is operative on a gas but not the other, it can enhance the separation factor. Although surface diffusion contribution decreases with increasing temperature, it becomes more important as the pore diameter becomes smaller. Therefore, it is possible that as inorganic membranes with smaller pore sizes become available their separation performance may increase not only due to molecular sieving effects but also surface diffusion or other transport mechanisms.

Some quantitative assessments on how these factors determine the separation factors will be very desirable to guide systematic material development efforts for designing the next generation of inorganic membranes. Plausible theoretical models based on sound physical and chemical reasoning with justified or verified assumptions have been attempted for this purpose.

For gas transport in small pores (say, less than the 10 nm range) the sizes of which are no longer much larger than those of the gas molecules, the contribution of viscous flow can be neglected and other considerations need to be factored in the model. First, the gas molecules are considered to be hard sphere with a finite size and the gas diffusion process is assumed to proceed in the membrane pores by random walk. The membrane pores are assumed to consist of smooth-wall circular capillaries. In addition to gas molecules colliding with the membrane pore walls, adsorption on the pore wall and the associated surface flow or diffusion are considered. Adsorption also effectively reduces the pore size for diffusion.

By using statistical arguments and assuming contributions due to surface diffusion and momentum transfer to be additive to that of the gas phase diffusion, Fain and Roettger [1993] have derived the following expression for the flow rate of each gas, F_i:

$$F_i = \{Cv_i \Delta P_i / Tl\} \, (r - \delta_i - \Sigma t_i)^3 \, [\, (1/f_i) + A \, (1 - \delta_i / r)^{-2}] \qquad (7\text{-}8)$$

where v_i, ΔP_i, δ_i, t_i , f_i are the molecular velocity, partial pressure drop, molecular radius, adsorbed film thickness and momentum accommodation coefficient of the i-th gas molecule. T is the temperature, l the capillary length, Σt_i the sum of the adsorbed layer thicknesses for all gases in the mixture including the i-th gas and A and C are some temperature-dependent coefficients that can be determined from such measurements as adsorption isotherms and single gas permeability measurements. The separation factor for helium and carbon dioxide based on the above equation is plotted in Figure 7.5.

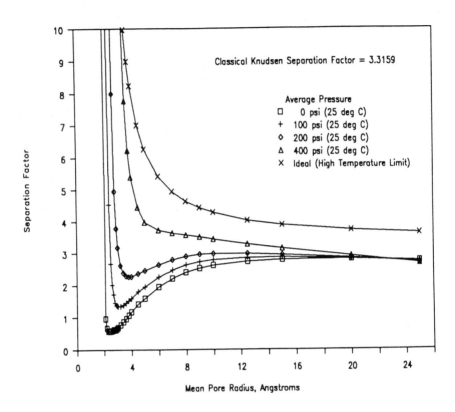

Figure 7.5 Prediction of separation factor of helium and carbon dioxide based on Eq. (7-8) [Fain and Roettger, 1993]

Based on pure gas-phase diffusion consideration, the separation factor is expected to increase monotonically as the pore size decreases. But the figure actually reveals some unexpected minimum separation factor as the pore size decreases at room temperature, particularly under lower pressures. This is attributed to surface flow of carbon dioxide which enhances its permeation relative to that of helium within that range of the pore size. This "dip" in the separation factor according to the transport model of Eq. (7-8) is in good agreement with the data calculated from actual He and CO_2 pure gas flow measurements using alumina membranes with a pore diameter as small as 1 nm [Fain and Roettger, 1993]. At high temperatures, the surface flow contribution is generally expected to be small. As the pore radius approaches 0.2 nm, the size of the larger molecule (CO_2 in this case), the separation factor sharply increases to very large numbers. The above prediction of a minimum of the ideal separation factor has been verified experimentally for other systems as well. Lin et al. [1994] prepared ultramicroporous silica membranes (down to a pore diameter close to 0.4 nm) from commercial alumina membranes with a pore diameter of 4 nm by chemical vapor deposition of silica. The silica deposition apparently takes place within the existing alumina porous structure, thus effectively reducing the final pore diameter. When the separation factor of these silica membranes was plotted against the pore diameter, the "dip" predicted by Fain and Roettger [1993] was observed, as shown in Figure 7.6. Similarly, Fain and Roettger [1994] prepared 15 ceramic membranes of varying pore sizes (some down to a pore diameter of 0.5 nm) and observed that both the room-temperature and the high-temperature separation factors for the helium/carbon tetrafluoride system display a minimum as a function of the pore size. The separation factor minima discussed above were all attributed to surface diffusion.

The above model and other similar ones indicate the need to produce membranes with pore sizes significantly smaller than what have been attainable to date. Thus sound physics-based models with verified assumptions like the one presented above not only provides the quantitative information on the approximate pore size required to achieve a given separation factor for a certain gas pair such as He-CO_2 or He-N_2 but also shows that one should expect, in some cases, to see a lower separation factor as the pore size decreases before a sharply increasing separation factor in the pore size approaching the size of the gas molecules.

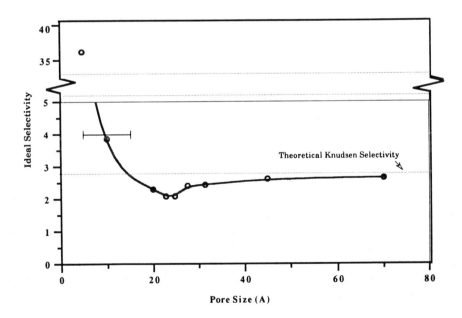

Figure 7.6 Ideal separation factor of helium/nitrogen with a series of silica-modified membranes at 25°C [Lin et al., 1994]

7.3.4 New Developments for Higher Separation Performance

New developments revealing some promises have started to emerge. One such area is zeolite membranes. It is well known that single crystal zeolites have the unusually high selectivity based on their pore sizes relative to the molecular sizes of the gases to be separated. The great challenge, however, lies in preparing polycrystal zeolites which are free from defects or large pores between crystals. A recent study [Jia et al., 1993] has shown that it is possible to prepare thin dense layers of polycrystal zeolite which is intergrown in three dimensions from individual zeolite crystallites. The separation factors obtained from these zeolite membranes appear to be encouraging. The ideal separation factor of n-butane over i-butane is as high as 6.2. The problem of intercrystal defects, however, is not totally resolved as molecules larger than the zeolite pore diameter still have limited permeabilities through the membranes.

Other molecular sieve membranes are prime candidates as well. Molecular sieve carbon membranes exhibits very high separation factors in the laboratory. Their microstructures can be tailored by adjusting the synthesis and calcination conditions; however, the issues of their mechanical, chemical and thermal stabilities under various potential application environments have not been addressed.

Other approaches to reducing the membrane pore size are being investigated. Many of them are based on the sol-gel process or chemical vapor deposition as discussed in Chapter 3. An example is the preparation of small-pore silica membranes. Amorphous silica membranes have been prepared from solutions of silicate-based polymers. More specifically, some strategies are employed: aggregation of fractal polymeric clusters, variation of sol composition, the use of organic molecular templates and modification of pore surface chemistry [Wallace and Brinker, 1993].

Yet another unique class of inorganic membrane materials called pillared clay and carbon composite membranes has been studied for gas separation [Zhu et al., 1994]. The permeation rates of benzene, chlorobenzene and 1,3-dichlorobenzene vapors through these membranes can be different by orders of magnitude as indicated earlier. This may open the door for these types of membranes for separating organic mixtures.

7.4 SENSORS FOR GASES AND CHEMICALS

Several inorganic materials, especially the ceramic type, exhibit selective transport of species in the form of gas, ions or atoms through them and can be utilized as sensors. Sensors are those devices that convert nonelectrical signals into electrical ones. Although the applications are not directly related to separation processes, their underlying principles are the same as those governing the membranes used in traditional separation applications. Their uses as gas or chemical sensors deserve brief discussions here.

7.4.1 Gas Sensors

Inorganic membranes can be used either as a sensor or as an additional layer to the sensor to increase its selectivity, and therefore its sensing sensitivity. Both types of applications will be treated separately.

<u>Ceramic oxygen sensors</u>. Ceramic sensors vary in their mode of operating mechanisms. The electrochemical type is based on physical properties of the grain itself (bulk properties) and involves transfer of chemical species. An example is yttria-, magnesia- or calcia-doped zirconia solid electrolytes as oxygen-gas sensors, particularly the first type. In these materials, a high concentration of oxygen vacancies develops which causes oxygen ions (O^{2-}) to move across the sensor thickness. Zirconia sensors generally come in a tube shape and have platinum electrodes deposited on both sides. When there is a difference in the oxygen partial pressures on the two sides of the sensor, an electromotive force develops across the electrodes which enables the calculation of an unknown pressure if the pressure on the other side is known. For higher sensitivity oxygen-gas sensors, the oxygen pump principle can be utilized [Ketron, 1989].

Zirconia sensors have been used primarily in the exhaust system of automobiles to control the air-to-fuel ratio for meeting the federal requirements on such noxious gases as carbon monoxide, methane and nitrogen oxides. The applicability of zirconia sensors for this particular application is based on the assumption that, under thermodynamic equilibrium, the partial pressure of oxygen in the exhaust gas depends primarily on the air-to-fuel ratio. To compensate for the fact that in reality equilibrium is not reached, catalytic platinum electrodes are incorporated in the zirconia sensor design [Stevens, 1986]. In the zirconia sensor, the outside of the zirconia tube is exposed to the exhaust gas while the inside is exposed to the ambient air as a reference atmosphere.

<u>Measurement of gas content of molten metal</u>. Gas dispersed in the molten metal may be released during solidification to form gas bubbles in ingots and pigs. These bubbles can cause pores and cracks during further treatment of the metal. Thus the quality of many metals is often affected by the gas content present in the molten metal and it is very desirable to determine the amount of a certain gas such as hydrogen in the molten metal bath.

The idea of a probe utilizing a porous ceramic membrane that is permeable to the gas to be measured but not to the molten nonferrous metal has been proposed [De Schutter et al., 1991]. The probe contains a tube which is closed off at its immersion extremity by the membrane. A cover is fitted to the end of the tube which melts when immersed in the molten metal. A vacuum is created inside the cover. By measuring the pressure in the tube after the membrane is immersed and the vacuum applied, the hydrogen content having diffused through the membrane is thus determined. The preferred ceramic membrane is made of alumina.

Ceramic coating to enhance gas sensor selectivity. Some inorganic materials, especially metal oxides, are known for their ability to detect organic contaminants in air such as combustible gases (methane or propane) and toxic gases (hydrogen sulfide or nitrogen oxides). In Japan, the largest application of this type of gas sensors is in the household alarms for combustible gases. Titania and tin oxides (SnO_2), for example, have been utilized in commercial conductivity detectors as a gas sensor which can simultaneously detect several gases. A wide variety of applications are possible: pollution monitoring, automotive ventilation, etc.

An important issue associated with metal oxide conductivity detectors is their relatively low selectivity over different gases. The issue is how to design and produce a gas sensor that will respond primarily to a specific gas but not to other gases that may be present. Imprecise selectivity can result in false alarm problems and reduce the accuracy of gas sensors. To help overcome this drawback, a modified two-stage sensor design was therefore proposed in which a very thin dense ceramic coating serves as a selective membrane underlying the metal oxide thin film detector [Althainz et al., 1994]. The effective transport mechanism for the selective nonporous ceramic membrane is believed to occur by the solution-diffusion mechanism as briefly described in Chapter 4. The size selective mechanism has been found to be dominant. Silica and alumina membranes can be deposited on SnO_2 by chemical vapor techniques and have been evaluated at a high temperature (350°C) for their sensitivities toward water vapor and hydrocarbon gases such as methane, propane and toluene in air. Although the additional layer of membrane prolongs the response time of the detector, the increase is not significant.

7.4.2 Ion-selective Electrodes

Membrane electrodes are used in the potentiometric methods of analysis where the membrane allows certain kinds of ions to penetrate it while rejecting others.

pH electrode. The most well known example of a membrane electrode is the glass electrode which is widely used today for measuring the pH of aqueous solutions. A long history of development has led to their high level of reliability. However, a few shortcomings of the glass electrodes still remain. They include the minimum size and the limited operation temperature (approximately 100-120°C) mainly due to the chemical instability of the glass membrane with aqueous media and to the precipitation of AgCl on the glass membrane from the reference buffer.

Attempts have been made to substitute the aqueous buffer with solid contacts such as those based on the ternary system Li-Ag-I [Kreuer, 1990]. The pH electrodes with solid contacts extend the operable temperature range to about 150°C. This allows pH measurements at a higher temperature or for those applications requiring sterilization as in fermentors. The issue of hydrolysis of the glass surface is more difficult to solve as the glass itself is a thermodynamic non-equilibrium material and consequently the glass surface structure will change with time.

For pH determination at an even higher temperature, other membrane electrode materials are being investigated. For example, an in-situ pH measurement system capable of operating at 275°C has been developed using yttria-stabilized zirconia membrane tubes [Tachibana et al., 1991]. Air is filled inside the membrane tube to balance the inside pressure with that on the high temperature aqueous side by a pressure regulator system. The inside surface of the tube is coated with Pt and functions as an oxygen electrode. The pH measurement by this new system has been found to be not affected by other redox couples contained in the solutions.

Ion electrodes. The technology developed for pH electrodes has led to the development of various glasses which respond selectively to cations other than hydrogen. Examples of these ion-selective electrodes are glass electrodes for sodium, lithium, potassium, calcium and silver ions. For example, a glass consisting of 13% BaO, 6% Na_2O, 6% CsO, 25% Al_2O_3 and 50% SiO_2 is selective to calcium ions [Truesdell and Christ, 1966] while a glass composed of 15% Li_2O, 25% Al_2O_3 and 60% SiO_2 can be used to determine lithium ions selectively [Eisenman, 1965]. These responses by the glass electrodes are related to the ion-exchange properties of their surface. It is noted that all glass electrodes are responsive to some degree to hydrogen ions and therefore should be used at sufficiently high pH conditions. Other membrane materials have also been developed and some are commercialized for analytical determination of a number of ions.

In the presence of strong oxidizing agents, glass electrodes may not be stable. For example, composites of crystalline lead and silver sulfide are used as the membrane materials for lead-selective electrodes. A major drawback of these electrodes is the ease of oxidation of the crystalline PbS. Other materials such as chalcogenide glass electrodes, particularly of the PbS-AgI-As_2S_3 system, have been found to be much more stable toward the action of oxidizing agents than the crystalline electrodes [Vlasov et al., 1989].

7.5 SUPPORTS FOR FACILITATED TRANSPORT

Facilitated transport has been briefly described in Chapter 1. In facilitated transport, the selective transport medium is a liquid or molten salt contained or immobilized in a porous support. The liquid membrane is held tightly in the support pores by capillary forces. The liquid or molten salt selectively reacts with a gas or vapor species and the reacting species diffuses across the liquid or salt and desorbed on the other side of the facilitated transport membrane. The major advantage of the facilitated transport is that diffusion is generally several orders of magnitude faster than diffusion through solid membranes. The support is, therefore, not a membrane by definition. Comprehensive review on facilitated transport are available [Goddard, 1977; Cussler and Evans, 1980; Uragami, 1990].

Inorganic supports are particularly attractive mainly for high temperature applications due to their thermal stability and chemical compatibility with many liquid membranes. Some examples of high-temperature reactive materials used in many facilitated transport membranes are molten mixtures of alkali and alkaline earth carbonate salts for hydrogen sulfide [Weiner et al., 1990], tetramethyl-ammonium fluoride tetrahydrate (a salt hydrate) for carbon dioxide [Laciak et al., 1990] and hydrogen sulfide [Appleby et al., 1993] and polyvinyl-ammonium thiocyanate for ammonia [Laciak et al., 1990]. Many ceramic materials that are themselves used as membranes in other applications are also good candidates as the support for facilitated transport. As the subject of facilitated transport can be extensive, it is not the purpose here to review the technology but instead just to demonstrate some unique uses of inorganic membrane materials in a non-separation role in otherwise separation applications.

A potentially important example where porous inorganic membranes are used as the supports for facilitated transport is the removal of gaseous contaminants such as hydrogen sulfide from coal-derived gas streams. For example, some molten carbonate salts with high solubilities of hydrogen sulfide have been demonstrated to be effective "membranes" for removing hydrogen sulfide. A usual problem, however, associated with facilitated transport is the stability of the liquid "membrane" (e.g., due to pressure upsets, wicking and evaporation). This problem is made worse by the preference that the liquid membrane stays as a very thin film to reduce the transport resistance. The major technical challenge, then, is to find a microporous support material that is compatible with the molten alkali carbonate salts or other liquid membrane materials often under high temperatures and/or high pressures. The maximum operable pressure difference is largely determined by the potential loss of the liquid membrane from the support matrix.

A number of porous ceramic materials are candidates. They include $MgAl_2O_4$, $CaAl_2O_4$, TiB_2, WC, AlN, and $LiAlO_2$, particularly the last two ceramic supports [Weiner et al., 1990; Minford et al., 1992]. Besides chemical compatibility between the liquid membrane and the support, wetting of the liquid in the support matrix is also an important consideration that affects the liquid containment. In addition, it is critical that all the pores in the support are uniformly filled with the liquid membrane or the selectivity of the facilitated transport membrane will suffer significantly. For example, Damle et al. [1994] observed that facilitated transport membranes of molten salts using disk and tubular porous supports for the removal of hydrogen sulfide from hydrogen at 560-650°C and 3.3-13 bars do not show a high selectivity for hydrogen sulfide relative to other gases. The unsatisfactory performance was attributed to leakage of the species through unfilled pores in the matrix of the ceramic support and through the seals [Damle et al., 1994].

7.6 SUMMARY

Other than isotopes separation for uranium enrichment described in Chapter 2, inorganic membranes are commercially used for particulate filtration of air or other gases in clean room applications, airborne contaminant analysis and high-purity hydrogen production. In addition, some inorganic membranes are used in pH and ion selective electrodes.

Yet greater potentials of inorganic membranes are in gas separation especially at high temperatures. This field of promising applications relies on the inherent thermal and chemical stabilities of many inorganic membranes and the possibilities of designing the desirable pore sizes and pore surfaces. Material and application development efforts in this area have escalated in recent years with promising results.

While dense inorganic membranes such as palladium-based or zirconia membranes provide extremely high-purity gases, their permeabilities are usually low, thus making the process economics unfavorable. Therefore, most of the recent investigations focus on porous inorganic membranes.

A number of potentially important gas separation applications using inorganic membranes are surveyed. These include separation of the following gas components: oxygen/nitrogen, hydrogen/hydrocarbons, hydrogen/carbon monoxide, hydrogen/nitrogen, hydrogen/carbon dioxide, hydrogen/coal gasification products, carbon dioxide/hydrocarbons, hydrogen sulfide/hydrocarbons, water/air, water/natural gas, and water/organic solvents. In addition, hydrocarbon vapor recovery from air, separation of different hydrocarbons and separation of helium from nitrogen or oxygen have also been investigated.

In many studies the separation factor, which is indicative of the membrane's ability to separate two gases in a mixture, is predominantly governed by Knudsen diffusion. Knudsen diffusion is useful in gas separation mostly when two gases are significantly different in their molecular weights. In other cases, more effective transport mechanisms are required. The pore size of the membrane needs to be smaller so that molecular sieving effects become operative. Some new membrane materials such as zeolites and other molecular sieve materials and membrane modifications by the sol-gel and chemical vapor deposition techniques are all in the horizon. Alternatively, it is desirable to tailor the gas-membrane interaction for promoting such transport mechanisms as surface diffusion or capillary condensation.

A systematic and logical approach to developing inorganic membranes that are effective for those challenging gas separation applications appears to be a development strategy based on a model incorporating plausible transport mechanisms (see section 7.3.3). The model is capable of predicting some unexpected separation performance which has been verified experimentally.

Finally, some unconventional uses of inorganic membranes as gas sensors and as the immobilizing support matrices for facilitated transport are discussed.

REFERENCES

Abe, F., 1986, European Patent Appl. 0,195,549.
Agency of Ind. Sci. Technol., 1986, Japanese Patent Appl. 61,004,513.
Althainz, P., A. Dahlke, M. Frietsch-Klarhof, J. Goschnick and H.J. Ache, 1994, Physica Status Solidi (A) **145**, 611.
Anonymous, 1995, Chem. Eng. Progr. (December), 18.
Appleby, J.B., R. Quinn and G.P. Pez, 1993, New facilitated transport membranes and absorbents for the separtion of hydrogen sulfide, presented at 205th Am. Chem. Soc. National Meeting, Denver, CO, USA.
Asaeda, M., and L. D. Du, 1986, J. Chem. Eng. Japan **19**, 72.
Asaeda, M., L. D. Du and M. Ushijima, 1984, Feasibility study on dehumidification of air by thin porous alumina gel membrane, in: Proc. 4th Int'l Drying Symp. Vol. 2 (R. Toei and A. S. Mujumdar, Eds.), p. 472.
Asaeda, M., L. D. Du and M. Fuji, 1986, J. Chem. Eng. Japan **19**, 84.
Ash, R., R. M. Barrer and T. Foley, 1976, J. Membr. Sci. **1**, 355.
Barrer, R. M., 1965, AIChE Symp. Ser. 1, Am. Inst. Chem. Eng., p. 112.
Bishop, B., R. Higgins, R. Abrams and R. Goldsmith, 1994, Compact ceramic membrane filters for advanced air pollution control, presented at 3rd Int. Conf. Inorg. Membr., Worcester, MA, USA.
Bumsted, H.E., 1973, Am. Ind. Hyg. Assoc. J. **34**, 150.
Connor, H., 1962, Platinum Metals Rev. **6**, 130.
Cussler, E.L., and D.F. Evans, 1980, J. Membr. Sci. **6**, 113.
Damle, A.S., S.K. Gangwal, R. Quinn, E. Minford, F. Herman and T. Dorchak, 1994, Facilitated transport ceramic membranes for high temperature H2S removal, presented at 3rd Int. Conf. Inorg. Membr., Worcester, MA, USA.
Dellefield, R.J., High-temperature applications of inorganic membranes, 1988, presented at the Am. Inst. Chem. Eng. National Meeting, Denver, CO, USA.
De Schutter, F.J.E., J.C. Dekeyser, S.P.M.T. De Burbure de Wesembeek and J.R. Luyten, 1991, European Patent Appl. 435,365-A1.
Eggerstedt, P.M., J.F. Zievers and E.C. Zievers, 1993, Chem. Eng. Progr. (January), 62.
Eisenman, G., 1965, Advances in Analytical Chemistry and Instrumentation, Vol. 4, John Wiley & Sons, New York, p. 213.
Fain, D.E., and G.E. Roettger, 1993, J. Eng. Gas Turbines and Power **115**, 628.
Fain, D.E., and G.E. Roettger, 1994, Hydrogen production using inorganic membranes, in: Proc. 8th Ann. Conf. Fossil Energy Mater., Oak Ridge, TN, USA, p. 51.
Gavalas, G.R., 1991, Hydrogen separation by ceramic membranes in coal gasification, in: Proc. 11th Ann. Gas. and Gas Stream Cleanup System Contractors Rev. Meeting, Vol. 2, Morgantown, WV, USA, p.583.
Gavalas, G.R., and C.E. Megiris, 1990, U.S. Patent 4,902,307.
Goddard, J.D., 1977, Chem. Eng. Sci. **32**, 795.

Grosser, K.R., 1989, Ultra clean gas delivery system, presented at Semicon East, Boston, MA, USA.

Hammel, J.J., W.P. Marshall, W.J. Robertson and H.W. Barch, 1987, European Patent Appl. 0,248,392-A3.

Havredaki, V., and J. H. Petropoulos, 1983, J. Membr. Sci. **12**, 303.

Hughes, D.T., and I.R. Harris, 1978, J. Less-Common Met. **61**, 9.

Isomura, S., T. Watanabe, R. Nakane and S. Kikuchi, 1969, Nippon Genshiryoku Gakkaishi **11**, 417.

Itaya, K., S. Sugawara, K. Arai and S. Saito, 1984, J. Chem. Eng. Japan **17**, 514.

Jensen, D., and S. Goldsmith, 1987, J. Environ. Sci. (Nov./Dec.), 39.

Jia, M.D., K.V. Peinemann and R.D. Behling, 1993, J. Membr. Sci. **82**, 15.

Johnson, H.E., and B.L. Schulman, 1993, Assessment of the potential for refinery applications of inorganic membrane technology -- an identification and screening analysis, U.S. Department of Energy Final Report under Contract No. DE-ACO1-88FE61680 (Task 23).

Kameyama, T., M. Dokiya, K. Fukuda and Y. Kotera, 1979, Sep. Sci. Technol. **14**, 953.

Kameyama, T., M. Dokiya, M. Fujishige, H. Yokokawa and K. Fukuda, 1981a, I&EC Fund. **20**, 97.

Kameyama, T., K. Fukuda, M. Fujishige, H. Yokokawa and M. Dokiya, 1981b, Hydrogen Energy Prog. **2**, 569.

Kameyama, T., M. Dokiya, M. Fujishige, H. Yokokawa and K. Fukuda, 1983, Int. J. Hydrogen Energy **8**, 5.

Ketron, L., 1989, Ceramic Bull. **68**, 860.

Koresh, J. E., and A. Soffer, 1983a, A molecular sieve carbon membrane for continuous process gas separation, in: Proc. 16th Bienn. Conf. Carbon, p. 367.

Koresh, J. E., and A. Soffer, 1983b, Sep. Sci. Technol. **18**, 723.

Koros, W.J., and G.K. Fleming, 1993, J. Membr. Sci. **83**, 1.

Kreuer, K.D., 1990, Sensors and Actuators **B1**, 286.

Kusakabe, K., and S. Morooka, 1994, Preparation of perovskite-type oxide membrane for separation of carbon dioxide at high temperature, presented at 3rd Int. Conf. Inorg. Membr., Worcester, MA, USA.

Laciak, D.V., R. Quinn, G.P. Pez, J.B. Appleby and P.S. Puri, 1990, Sep. Sci. Technol. **25**, 1295.

Lee, S.J., S.M. Yang and S.B. Park, 1994, J. Membr. Sci. **96**, 223.

Leeper, S.A., D.H. Stevenson, P.Y.C. Chiu, S.J. Priebe, H.F. Sanchez and P.M. Wikoff, 1984, Membrane technology and applications: an assessment, U.S. Department of Energy Report under Contract No. ACO7-76ID01570.

Leroux, J., and C.A. Powers, 1970, Occup. Health Rev. **21**, 26.

Lin, C.L., D.L. Flowers and P.K.T. Liu, 1994, J. Membr. Sci. **92**, 45.

Maier, W.F., 1993, U.S. Patent 5,250,184.

Matsushita Electric Ind., 1985, Japanese Patent Appl. 60,129,119.

McMahon, T.J., L. Gasper-Galvin and P. Hsu, 1990, High-temperature inorganic membrane separations, in: Proc. 10th Annual Gasification Gas Stream Cleanup Systems Contractors Review Meeting (Vo. 1, V.P. Kothari and J.L. Beeson, Eds.), Morgantown, WV, U.S.A., p. 279.

Megiris, C.E., and J.H.E. Glezer, 1992, Ind. Eng. Chem. Res. **31**, 1293.

Minford, E., F.L. Herman and R. Quinn, 1992, Facilitated transport ceramic membranes for high temperature gas cleanup, in: Proc. U.S. Dept. of Energy Contractors Rev. Meeting on Gas. Gas Stream Cleanup Syst., Morgantown, WV, USA.

Minneci, P.A., and D.J. Paulson, 1988, J. Membr. Sci. **39**, 273.

Mitsubishi Heavy Ind. and NGK Insulators, 1986, Ceramic membrane for condensable vapour separation, Product Brochure 8609-3000.

MPS Review, 1988, Modern Power Systems (August), 19.

Ohya, H., H. Nakajima, N. Togami, M. Aihara and Y. Negishi, 1994, J. Membr. Sci. **97**, 91.

Okubo, T., and H. Inoue, 1989, AIChE J. **35**, 845.

Okubo, T., K. Haruta, K. Kusakabe, S. Morooka, H. Anzai and S. Akiyama, 1991, J. Membr. Sci. **59**, 73.

Pellegrino, J.J., R. Nassimbene, A. Kirkkopru, R.D. Noble and J.D. Way, 1988, Gas s eparation using ion exchange membranes for producing hydrogen from synthesis gas, in: Proc. U.S. Dept. of Energy Contractors Rev. Meeting on Gas. Gas Stream Cleanup Syst., Morgantown, WV, USA, May.

Peterson, E.S., M.L. Stone, R.R. McCaffrey and D.G. Cummings, 1993, Sep. Sci. Technol. **28**, 423.

Philpott, J.E., 1985, Platinum Metals Rev. **29**, 12.

Philpott, J., and D.R. Coupland, 1988, Chem. Ind. **31**, 679.

Poku, J.A., and J.E. Plunkett, 1989, Assessment of membrane gas separation applications to METC supported technologies, U.S. Department of Energy Final Report under Contract No. DE-AC21-85MC21353.

Rony, P.R., 1968, Sep. Sci. **3**, 239.

Sakohara, S., S. Kitao, M. Ishizaki and M. Asaeda, 1989, in: Proc. 1st Int. Conf. Inorg. Membr., Montpellier, France, p.231.

Sano, Y., F. Mizukami, H. Yagishita, M. Kitamoto and Y. Kyozumi, 1994, Japanese Patent 06,099,044.

Shindo, Y., K. Obata, T. Hakuta, H. Yoshitome, N. Todo and J. Kato, 1981, Adv. Hydrogen Energy **2**, 325.

Shindo, Y., T. Hakuta, H. Yoshitome and H. Inoue, 1985, J. Chem. Eng. Japan **18**, 485.

Spillman, R.W., 1989, Chem. Eng. Progr. **85**, 41.

Stevens, R., 1986, Zirconia and Zirconia Ceramics, Magnesium Elektron Ltd., U.K.

Suzuki, N., and S. Tsuruta, 1986, U.S. Patent 4,599,157.

Tachibana, K., A. Hattori, M. Yokoi, T. Kodachi and N. Asaoka, 1991, Zairyo-to-Kankyo **40**, 308.

Tock, R. W., and K. Kammermeyer, 1969, AIChE J. **15**, 715.

Toyo Soda Manufacturing, 1985, Japanese Patent Appl. 60,187,320.

Truesdell, A.H., and C.L. Christ, 1966, Glass electrodes for calcium and other divalent cations, in: Glass Electrodes for Hydrogen and Other Cations (G. Eisenman, ed.), Marcel Dekker, New York, p. 293.

Tsapatsis, M., S. Kim, S.W. Nam and G. Gavalas, 1991, Ind. Eng. Chem. Res. **30**, 2152.

Uragami, T., 1990, St. Surf. Sci. Catal. **54**, 284.

Van Gemert, R.W., and F. P. Cuperus, 1995, J. Membr. Sci. **105**, 287.

Van Veen, H.M., R.A. Terpstra, J.P.B.M. Tol and H.J. Veringa, 1989, Three-layer ceramic alumina membranes for high temperature gas separation applications, in Proc. Int. Conf. Inorg. Membr., Montpellier, France.

Van Vuren, R. J., B. C. Bonekamp, K. Keizer, R. J. R. Uhlhorn, H. J. Veringa and A. J. Burggraaf, 1987, Formation of ceramic alumina membranes for gas separation, in: High Tech Ceramics (Part C), Mat. Sci. Monogr. , Vol. 38C, p. 2235.

Vlasov, Yu.G., E.A. Bychkov and A.V. Legin, 1989, Sov. Electrochem. **24**, 1087.

Wallace, S., and C.J. Brinker, 1993, Inorganic polymer-derived gas separation ceramic membranes, Quarterly report for Gas Research Institute under contract GRI 5091-222-2306.

Way, J.D., and D.L. Roberts, 1992, Sep. Sci. Technol. **27**, 29.

Weiner, S.C., E. Minford and R. Quinn, 1990, Facilitated transport ceramic membranes for high temperature gas cleanup, in: Proc. 10th Ann. Gas. Gas Stream Cleanup Syst. Contractors Rev. Meeting (V.P. Kothari and J.L. Beeson, eds.), Morgantown, WV, USA, Aug. 28-30, p. 270.

Wu, J.C.S., D.F. Flowers and P.K.T. Liu, 1993, J. Membr. Sci. **77**, 85.

Wu, J.C.S., H. Sabol, G.W. Smith, D.L. Flowers and P.K.T. Liu, 1994, J. Membr. Sci. **96**, 275.

Ye, J., W. Yuan and Q. Yuan, 1994, The preparation of ultrathin palladium-silver alloy membrane and the application of it to ultrapure hydrogen purifier, presented at 3rd Int. Conf. Inorg. Membr., Worcester, MA, USA.

Zhu, H.Y., W.H. Gao and E.F. Vansant, 1994, The pillared clay-carbon composite membrane: the preparation, vapour permeation and separation of gas mixtures, presented at 3rd Int. Conf. Inorg. Membr., Worcester, MA, USA.

CHAPTER 8

INORGANIC MEMBRANE REACTOR -- CONCEPTS AND APPLICATIONS

Membrane separation has been practiced for a few decades now, but new applications are being explored on a broad horizon. One such frontier technology area is membrane reactor (or sometimes referred to as membrane catalyst or catalyst-membrane system). It represents a variety of reactor configurations which include a membrane as part of the system for separation or purification. The technology combines the permselectivity of a membrane with a reaction, particularly a heterogeneous catalytic reaction. The integration of the two unit operations offers advantages not only in terms of system simplification, and most likely lower capital costs, but also yield improvement and selectivity enhancement. This new technology in its ultimate form uses the membrane as a catalyst or catalyst support and, at the same time, as a physical means for separating reactants and products, controlled addition of a very active reactant or selectively removing undesirable intermediate reaction products to effect the increased yield and selectivity.

Organic polymer membranes can only withstand relatively mild conditions such as those prevailing in biocatalysis involving enzymes and microorganisms and in some areas of homogeneous catalysis. Thus early studies of membrane reactors mostly dealt with low-temperature reactions or enzyme-catalyzed biotechnology applications such as saccharification of celluloses and hydrolysis of proteins. In one of the earliest membrane reactor studies, Jennings and Binning [1960] used a permselective organic membrane to remove water from other reaction mixtures during the reaction of n-butanol and acetic acid. This way the reaction can be driven to completion. Organic polymer membranes used for biocatalytic applications have been reviewed comprehensively [Flashel and Wandrey, 1979; Cheryan and Mehaia, 1986; Matson and Quinn, 1986; Chang, 1987; Belfort, 1989].

Starting in the late 1960s [Pfefferle, 1966; Gryaznov, 1969], this growing interest now is not limited to academic studies. The industry also begins to share the same vision [Roth, 1988 and 1990]. The concept of a membrane reactor/separator has been reduced to practice commercially for enzyme-catalyzed reactions with a production capacity exceeding 1,000,000 kg/year [Bratzler et al., 1986]. Hollow fiber membrane bioreactors, for example, have been available for commercial-scale production of optically-active compounds of high enantiomeric excess. An ambitious implementation plan is progressing in Japan to totally renovate and convert municipal wastewater treatment systems to one that is based on membrane bioreactors for acid and methane fermentation [Kimura, 1985]. Scaling up to such a large capacity can be relatively easily accomplished by integrating several bioreactors/separators in an appropriate process piping configuration due to the inherent modular nature of membrane units.

Many catalytic processes of industrial importance, however, involve the combination of high temperature and chemically harsh environments, a factor that strongly favors inorganic membranes. So with the introduction of commercially available glass, ceramic and metal membranes, there has been a dramatic surge of interest in the field of membrane reactor or membrane catalysis.

Some promising applications using inorganic membranes include certain dehydrogenation, hydrogenation and oxidation reactions such as formation of butadiene from butane by dehydrogenation, styrene production from dehydrogenation of ethylbenzene, dehydrogenation of ethane to ethene, water-gas shift reaction, ammonia synthesis and oxidative coupling of methane, to name just a few.

The inorganic membrane reactor technology is in a state characterized by very few in practice but many of promise. Since the potential payoff of this technology is enormous, it deserves a close-up look. This and the following three chapters are, therefore, devoted to the review and summary of the various aspects of inorganic membrane reactors: applications, material, catalytic and engineering issues.

8.1 BASIC CONCEPT AND ADVANTAGES

In many conventional reactors, the reactants are typically premixed and the products are brought in contact with the reactants until equilibrium is approached. Thus the reaction conversion is limited by the reaction equilibrium. Furthermore, uncontrolled contact between the reaction components may lead to certain reaction paths for some undesirable intermediate or side products. Finally, the valuable catalysts need to be immobilized or encapsulated in some carriers and their loss should be minimized to make the catalytic reaction processes economically viable.

Introduction of a permselective membrane to a reactor zone opens up opportunities for resolving the above important reaction performance issues. An overview of this concept will be given next followed by some examples portraying the benefits of a membrane reactor.

8.1.1 Removal of Product(s)

It is instructive to illustrate one of the major advantages of a membrane reactor by examining a generic reversible reaction:

$$A + B \leftrightarrow C + D \tag{8-1}$$

with the help of a permselective membrane. The reaction often has a limited conversion or yield dictated by the reaction equilibrium according to thermodynamics. This barrier, however, can be removed by displacing the equilibrium toward more product formation.

This can be accomplished, for example, by selectively removing either or both products through the membrane; thus, the reaction proceeds in the direction of generating more products. Therefore, proper product removal from the reaction zone shifts the equilibrium reaction to a higher conversion. For example, the equilibrium conversion of dehydrogenation of cyclohexane is only 18.7% at 200°C. However, by employing a palladium membrane to continuously remove hydrogen from the equilibrium-limited reaction products, the conversion can reach a maximum of 99.5% [Itoh, 1987b].

The implications of being able to increase the conversion of an equilibrium reaction by using a permselective membrane are several. First, a given reaction conversion may be attained at a lower operating temperature or with a lower mean residence time in a membrane reactor. This could also prolong the service life of the reactor system materials or catalysts. Second, a thermodynamically unfavorable reaction could be driven closer to completion. Thus, the consumption of the feedstock can be reduced. A further potential advantage is that, by being able to conduct the reaction at a lower temperature due to the use of a membrane reactor system, some temperature-sensitive catalysts may find new applications [Matsuda et al., 1993].

In addition to increase in reaction conversion, selectivity can also be enhanced by applying the concept of a membrane reactor. In a reaction system where some undesirable side reactions, for example, due to the high concentration of a product, may lead to a lower yield or selectivity, the dual use of a membrane as a separator and a well controlled reactor can suppress the side reactions and reduce or eliminate the formation of by-product(s). Let us assume that one of the products, C, given sufficient time or contact with some reaction components, can form some undesirable side product(s). By quickly separating the product C from the reactor, the opportunity for uncontrolled reactions is thus suppressed. A case in point is the dehydrocyclodimerization of propane to aromatics. For a given conversion of propane, the selectivity improves from 53% with a packed bed reactor to 64% with a Pd-Ag membrane reactor [Clayson and Howard, 1987]. A further example is dehydrogenation of isopentenes to produce isoprene. By removing the hydrogen evolved from the reaction zone at 410°C through a Pd-Ni alloy (5.5% Ni by weight) membrane, the formation of undesirable products such as 1,3-pentadiene or cyclopentadiene is suppressed [Gryaznov et al., 1973]. Likewise, the removal of hydrogen from the reaction zone by a palladium membrane reactor considerably reduces the undesirable reactions of hydroisomerization and hydrogenolysis competing with the isobutane dehydrogenation to form isobutene [Matsuda et al., 1993] or the undesirable hydrodealkylation of aromatics in favor of the target reaction of propane aromatization [Uemiya et al., 1990].

8.1.2 Controlled Addition of Reactant(s)

The above potential benefits of increasing conversions and/or selectivities involve preferentially removing one or more products from the reaction mixture. Similar benefits have also been shown by controlling the addition of a reactant into the reactor through a selective membrane. When a reactant is introduced to the reaction zone by the use of a permselective inorganic membrane, its concentration inside the reactor can be maintained at a low level. This can result in two immediate benefits: one is to decrease potential side reactions due to the high concentration of some reactant(s) and the other is to reduce the subsequent need of separating unconverted reactants.

The use of Pd-based membrane reactors can increase the hydrogenation rates of several olefins by more than 10 times higher than those in conventional premixed fixed-bed reactors. Furthermore, it has been pointed out that the type and state of the oxygen used to carry out partial oxidation of methane can significantly affect the conversion and selectivity of the reaction. The use of a solid oxide membrane (e.g., a yttria-stabilized zirconia membrane) not only can achieve an industrially acceptable C_2 hydrocarbon yield but also may eliminate undesirable gas-phase reactions of oxygen with methane or its intermediates because oxygen first reaches the catalyst through the solid oxide wall [Eng and Stoukides, 1991].

As will be illustrated later, one of the reactants can be introduced to the reaction zone at an adjustable rate by permeating through a permselective membrane. This mode of operation can be advantageous in selectivity over a system where premixed reactants are used. An example of this is hydrodealkylation of 1,4- and 1,5-dimethylnaphthalenes. One of their products is alpha-methylnaphthalene which can undergo isomerization into thermodynamically more stable beta-methylnaphthalene. However, when a Pd-Ni foil membrane is used to control the addition of hydrogen, beta-methylnaphthalene is essentially not present and the desired products of alpha–methylnaphthalene and naphthalene are predominant [Gryaznov et al., 1973]. This can be attributed to the lowered hydrogen pressure in the reaction mixture as a result of the controlled addition of hydrogen through the membrane. Several other hydrogenation reactions such as hydrogenation of cyclopentadiene, naphthalene and furan have been demonstrated to benefit from these hydrogen-selective membranes in their reaction selectivity.

Similarly, oxygen-selective membranes made of stabilized zirconia, perovskites or other oxides can be utilized to produce oxygen from air as a reactant for partial oxidation or oxidative coupling reactions of methane. This will eliminate the need for an expensive, dedicated oxygen purification plant by other processes. Omata et al. [1989] added air instead of pure oxygen into the reactor zone through a dense PbO membrane for the oxidative coupling of methane. Thus, nitrogen is essentially not introduced to the reaction mixture, thus avoiding the subsequent need to separate unconverted methane from nitrogen which can be very expensive. Controlled addition of reactants, as in the case of oxygen from air, also reduces the total volume of the off-gases generated.

In addition to the above potential process economic benefits, inorganic membrane reactors can also result in a safer operating environment. Some combustion reactions involve rapid release of energy when the reactants are mixed in a batch or bulk mode. Carefully controlled addition of a critical reactant (e.g., oxygen) to the reaction system (e.g., fuel) from opposite sides of the membrane reactor can minimize the potentials for uncontrolled, massive reactions leading to safety hazards. This way the fuel can be used undiluted. For example, the combustion of carbon monoxide has been carried out in a porous alumina membrane combustor [Veldsink et al., 1992]. By controlling the rates of introducing carbon monoxide and oxygen from opposite sides of the membrane, the risk of overrun reactions can be significantly lowered. The same principle can be applied to a dense membrane reactor.

Gas solubility (and thus permeation rate) in dense metal membranes typically decreases with increasing temperatures. Therefore, dense metal membrane reactors have the inherent advantage of avoiding runaway reactions.

8.1.3 Recovery of Catalysts and Engineered Catalytic Reaction Zone

In Sections 8.1.1 and 8.1.2 the permselective nature of inorganic membranes is utilized to separate reaction species resulting in favorable effects on reaction conversion, selectivity or yield. There are, however, other potential uses of these inorganic membranes that do not involve separation of reaction components.

Inorganic membranes have been tested to recover the catalysts in the reactor system. This is largely done by the appreciable size difference between the catalyst and the reaction components in relation to the membrane pore diameter. Thus the catalyst loss can be minimized. There are many studies treating some porous inorganic membranes purely as a convenient means of immobilizing catalysts in their porous structures to provide well-engineered catalytic reaction zones, especially in biocatalysis where the biocatalysts, enzymes or cells, are immobilized (e.g., [Nakajima et al., 1988]). Furthermore, these catalyst-impregnated porous matrix can provide some interesting reactor configuration. One such configuration that has been studied recently is the so called "opposing reactants geometry." This will be discussed in some detail later in this chapter.

Although the vast majority of industrial catalytic processes involve heterogeneous catalysis, homogeneous catalysts have also been used in some important industrial processes. While homogeneous catalysis enjoys the advantages of high selectivity, milder reaction conditions and better control of the nature of catalytically active compounds, it suffers from a major problem: potential catalyst loss due to the molecular dispersion of the catalyst. It is required that the catalyst be separated from the product and recycled back to the process. This is best done by an in-situ separation process. The majority of the catalysts for homogeneous catalysis are liquid. Organic polymer membranes have been used in a number of studies to retain and separate homogeneous liquid catalysts from reactants and products [Jennings and Binning, 1960; Gosser et al., 1977; Chen et al., 1992]. However, due to the typical aggressive nature of the reaction conditions involved, homogeneous catalysis appears to be a promising application area of inorganic membranes.

An innovative way of immobilizing a catalyst solution for a homogeneous catalytic reaction while simultaneously separating the product(s) and reactant(s) was demonstrated by Kim and Datta [1991] who called it supported liquid-phase catalytic membrane reactor-separator. The basic concept involves a membrane-catalyst-membrane composite as depicted in Figure 8.1 for a simple reaction:

$$A \rightarrow B \tag{8-2}$$

GA 34568.7

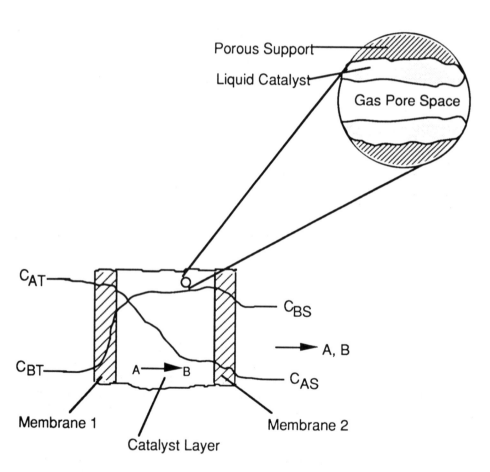

Figure 8.1 Supported liquid-phase catalytic membrane reactor-separator [Kim and Datta, 1991]

The catalyst layer consists of a porous matrix that immobilizes the catalyst solution. For liquid reactant and/or product, the catalyst solution fully occupies the pores. On the other hand, if both the reactant and the product are in a gaseous or vaporous form, it is advantageous to only partially impregnate the pores with the catalyst solution [Rony, 1968]. The catalyst-containing porous matrix is sandwiched between two membranes having different separation properties. The feed stream containing reactant A enters the reactor system from the left side and permeates through membrane 1 which is essentially a barrier to the product B. The reactant undergoes the reaction in the catalyst layer to form product B. Membrane 2 is chosen to be permeable to both the reactant and the product but not the catalyst. Thus, its primary function is to encapsulate the catalyst solution. Since the permeability of the product B through membrane 2 is significantly greater than that through membrane 1, the product diffuses to the right in the catalyst layer, permeates through membrane 2 and then, along with some unreacted A, migrates to a product collection chamber with or without the aid of a carrier or sweep fluid. It is noted that due to the permselectivity of membrane 1 toward product B, effectively the product is separated and, in cases where the product flow rate is smaller than the reactant flow rate, enriched.

Kim and Datta [1991] tested the above concept with the homogeneously catalyzed ethylene hydroformylation by hydridocarbonyltris (triphenyl phosphine) Rh (I) catalyst dissolved in dioctyl phthalate solvent. They concluded that for effective separation of the product, the transport resistance of the catalyst layer should be less than that for the membranes by controlling the liquid loading of the catalyst layer. They also pointed out that the organic membranes used can not withstand the aggressive reaction conditions and suggested that ceramic membranes appear to hold promises for practical applications.

8.1.4 Potential Advantages and Disadvantages

Although less developed than conventional reactors, membrane reactors offer potential advantages over fixed, fluidized or trickle bed reactors in several respects in addition to the integration of separation and reaction in a compact configuration: transport of reactants and products by convection rather than by diffusion alone as in many other separation processes (this increase in process rates has been estimated to result in reduction in operating costs by as much as 25% [Chan and Brownstein, 1991]), controlled pore size and distribution, controlled pore orientation, controlled and short contact time (which can provide more desirable reaction selectivity), high surface/volume ratio, less mass transfer resistance, fewer stagnant zones and less short circuiting (and thus less fouling to catalysts) and being easier to reactivate catalysts.

There are, however, some potential disadvantages due to the inherent compactness of membrane reactors. First, a membrane reactor typically provides limited contact time between reactants in the reaction zone. This may require in some cases longer membrane elements. Second, the membrane has relatively small area available for catalyst dispersion. This may limit the total catalytic surface area available and the catalyst loading.

8.2 MEMBRANE REACTOR CONFIGURATIONS

The concept of combining membranes and reactors for effectively increasing the reaction rate, selectivity or yield is intriguing and is being explored in various configurations. The difference in the configuration determines the function of the membrane reactor. In many cases as will be shown later in this chapter, the reaction equilibrium limitations are removed and the equilibrium displacement by the membrane reactor results in a conversion higher than the equilibrium value. A well studied case is the dehydrogenation of cyclohexane. If hydrogen formed is continuously removed from the reaction zone through a membrane selective to hydrogen, the conversion is much higher.

In some other cases the membrane reactor controls the fashion and the degree by which a reactant comes into contact with other reactant(s) or reaction product(s) to reduce or eliminate undesirable reactions for higher conversion or selectivities. A simple example involves the hydrogenation of phenol on a Pd-Ru foil membrane which acts as both a catalyst and a permselective transport medium of hydrogen. The yield of the reaction product, cyclohexanone, decreases rapidly with time when hydrogen and phenol vapor are premixed. However, if hydrogen permeates through the Pd-Ru membrane and reacts with phenol at the membrane surface, the yield is much higher [Gryaznov, 1992]. Still in some other cases, membrane reactors can be utilized to suppress competing adsorption which is common in conventional catalysts and is deleterious to the yield.

Various membrane reactor configurations are classified according to the relative placement of the two most important elements of this technology: membrane and catalyst. In a simpler configuration, the membrane and the reactor are connected in series (the membrane is either upstream or downstream of the reactor). The idea of combining the two unit operations, membrane and reactor, into a single one has spun a few interesting possibilities. Possible configurations encompass cases where the catalyst is physically separated from the membrane, the catalyst is attached to the membrane or the membrane is inherently catalytic and their modifications or combinations. The key features of each configuration will be highlighted here while some examples of them will appear throughout the remaining chapters of this book. It can be appreciated that the interaction between the membrane material and the permeate or reaction components is very important in determining the selectivity and, in some cases, the reaction rate or yield.

In all configurations where the membrane serves as a separator, the side of the membrane where the feed stream is introduced is called the feed side and the other side is often referred to as the permeate side. Very often to increase the transport rate of the permeate which can be a reactant(s) or product(s), a vacuum or a sweep gas is employed on the permeate side. The direction of the feed stream relative to that of the permeate-sweep gas depends on the process. Some are co-current while others are counter-current. Additionally other operational modes are possible such as intermediate ports for adding or removing reaction components (reactants, intermediate or final products), recycling of unconverted reactant(s) or permeate(s), etc.

8.2.1 Membrane Located Upstream of Reactor

When it is advantageous or necessary for a reactant stream to be concentrated or purified before reaction is initiated from either technical or economical standpoints, a membrane can be a strong candidate for the pretreatment of the reactant. However, this need can often be taken care of by directly introducing the reactant to be purified to the other side of a membrane reactor, as will be discussed later.

8.2.2 Membrane Located Downstream of Reactor

Frequently the process stream exiting a reactor requires post-processing for a variety of reasons: enriching the desired product from the product mixture to improve product yield, separating the product(s) from the reactant(s) in the reaction mixture to obtain acceptable product concentration, preventing possible side reactions as a result of prolonged contact of some components in the reaction mixture, just to name a few.

An example of this configuration is filtration of the methane fermentation broth from a sewage sludge liquor [Kayawake et al., 1991]. The liquor is treated anaerobically in a fermentor. The broth is pumped to a ceramic membrane module which is contained in the fermentor. The retentate is returned to the fermentor while the permeate is discharged to the environment. This is schematically shown in Figure 8.2. Although the membrane module is enclosed in the bioreactor for compactness and process simplification, the membrane step in essence follows the fermentation step.

8.2.3 Membrane Holding but Detached from Catalyst Pellets

In this case the membrane compartmentalizes a catalytic reactor and is not directly involved in the catalytic activity as the catalyst is not attached to the membrane. The catalyst pellets are often packed or fluidized on one (reaction) side of the membrane. The membrane functions strictly as a separator for fractionating product(s) or recovering catalyst or as a metering device for one of the reactants to provide controlled addition of that reactant to minimize undesirable side reactions.

Examples utilizing this configuration are most abundant for the packed bed membrane reactors. Reactors using a dense Pd alloy membrane as the separator and a packed bed of catalyst pellets have been demonstrated for a number of dehydrogenation reactions [Uemiya et al., 1991a; Gryaznov, 1992; Itoh, 1987a; Zhao et al., 1990b]. Porous alumina or glass membrane reactors containing a packed bed of catalyst pellets are a proven concept for increasing reaction conversion on a bench scale [Kameyama et al., 1981; Shinji et al., 1982; Itoh, 1987b].

GA 35288.2

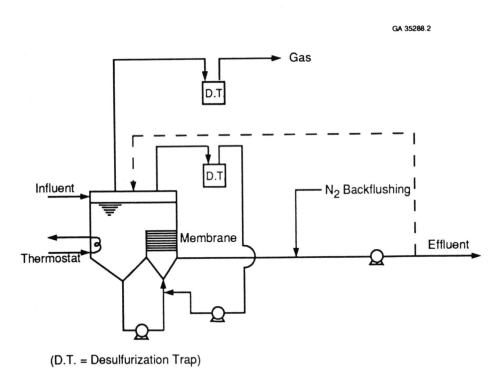

(D.T. = Desulfurization Trap)

Figure 8.2 Schematic diagram of an anaerobic fermentation system with a membrane module [Kayawake et al., 1991]

8.2.4 Membrane/Pore Surface Attached with Catalyst

When catalysts are attached to the membrane surface or, in the case of a porous membrane, on the pore surface, the catalysis and separation functions are engineered to a very compact fashion. Although very challenging, the advantages over all the previously described configurations can be significant, particularly when the membrane pores provide the efficient path for reaction (i.e., high surface area while not limited by diffusion) and directed transfer of reaction components and energy.

In a column (such as a packed or fluidized bed) reactor, the reaction conversion is often limited by the diffusion of reactant(s) into the pores of the catalyst or catalyst carrier pellets or beads. On the other hand, when the catalyst is impregnated or immobilized within membrane pores, the combination of the open pore path and the applied pressure gradient across the membrane thickness provides easier access of the reactants to the catalyst. The reaction conversion thus can be significantly improved over that for a conventional packed or moving bed reactor or a packed bed membrane reactor. This is feasible only if the product(s) can permeate through the membrane pores. It is noted that the impregnated catalyst can be liquid as well as solid. Sun and Khang [1988] compared the two modes of operation for the dehydrogenation of cyclohexane to form benzene and found that impregnating the catalyst in the membrane gives a higher conversion than packing the catalyst in the bore of the membrane element for high space time.

In this configuration, the membrane is selective to some reactants or products and the catalyst may be placed on the membrane surface or near the pore entrance such as catalytic electrodes attached to solid electrolytes or a platinum layer on the surface of a vanadium membrane as the catalyst for thermal decomposition of hydrogen sulfide [Edlund and Pledger, 1993]. Solid or liquid catalysts have been attached to membranes or their pores. It has been mentioned previously that this configuration can also be used to control the addition of a reactant through the membrane pores from the other side of the membrane where the second reactant is introduced.

A novel application of a symmetric porous membrane as a catalyst carrier but not as a permselective barrier is to use the membrane itself as the reaction zone for precise control of the stoichiometric ratio [Sloot et al., 1990]. In this case, the reactants are fed to the different sides of the membrane which is impregnated with a catalyst for a heterogeneous reaction. The products diffuse out of the membrane to its both sides. If the reaction rate is faster than the diffusion rate of the reactant in the membrane, a small reaction zone or theoretically a reaction plane will exist in the membrane. An interesting and important consequence of this type of membrane reactor is that within the reaction zone the molar fluxes of the reactants are always in stoichiometric ratio and the presence of one reactant in the opposing side of the membrane is avoided. The reaction zone can be maintained inside the membrane as long as the membrane is symmetric and not ultrathin. Therefore, membrane reactors of this fashion are particularly suited for those processes which require strict stoichiometric feed rates of premixed reactants. A symmetric porous α-alumina membrane of 4.5 mm thick was successfully tested to demonstrate the concept [Sloot et al., 1990].

8.2.5 Membrane Inherently Catalytic

Many inorganic membrane materials are also known for their intrinsic catalytic activities for certain hydrocarbon and other chemical reactions. Examples include alumina, titania and activated nickel for hydrocarbon reactions, silver for partial oxidation, and palladium and its alloys for hydrogenation and dehydrogenation. They are used or developed as conventional heterogeneous catalysts or catalyst supports. These intriguing and promising materials when made into membranes are often called "catalytic membranes" which serve as both a separator and a catalyst. When feasible, these catalytic membranes are among the most promising of the membrane reactor technology. A potential example is gamma-alumina for the well known Claus reaction where hydrogen sulfide reacts with sulfur dioxide to form elemental sulfur and water.

In the case of catalytic dense membranes such as palladium alloy sheets or tubes, a smooth membrane surface suffers from a small active surface area per unit volume of catalyst . This drawback can be remedied to some extent by adopting some conventional catalyst preparation methods to roughen the membrane surface(s) to ensure that only the region near the surface is affected unlike the Raney metal catalysts where the entire matrix is leached. For example, Gryaznov [1992] suggested the use of thermal diffusion of a chemically active metal into a Pd alloy sheet followed by acid treatment to remove this metal.

Zeolitic membranes are inherently catalytic to some hydrocarbon conversion processes. Dealkylation of ethylbenzene to benzene and ethylene is just one of many examples. The catalytic activity is associated with the acid activity of the zeolite. The permselectivity of the membrane toward ethylene alone or ethylene and benzene combined causes an increasing dealkylation and greater reaction selectivity [Haag and Tsikoyiannis, 1991]. Other potential applications of zeolitic membranes have also been suggested as given in Table 8.1. The general membrane reactor conditions range in temperature from 100 to 760°C, in pressure from 0.1 to 200 bars, in weight hourly space velocity from 0.08 to 2,000 hr^{-1} and finally in hydrogen/organic (hydrocarbons) from 0 to 100.

8.2.6 Catalytic Membrane Enhanced with Detached Catalyst Pellets

Either inherently catalytic membranes or membranes tightly bound with catalysts are conveniently referred to as catalytic membranes or catalytically active membranes. If desired or necessary, the catalytic function of a catalytic membrane can be strengthened with a second catalyst in the form of pellets.

8.2.7 Membrane as a Well-engineered Catalyst Support but Not as a Separator

All the above membrane reactor configurations utilize the permselective property of the membrane for enhancing reaction conversion, selectivity or yield, or recovery of the catalysts. There are other situations where the primary function of an inorganic membrane material is not to distinguish the permeation rates of the reaction components as they are transported through the membrane. Rather the porous structure of the membrane element serves as a well-engineered and efficient porous carrier for a catalyst or enzyme impregnated or immobilized within the membrane pores. There are many such studies falling into two major flow configurations: forced-flow and opposing-reactants modes.

TABLE 8.1
Hydrocarbon conversion processes as potential candidates for using zeolite membranes

Process	Temp. (°C)	Pressure (bar)	WHSV (hr^{-1})	Note
Cracking hydrocarbons	300-700	0.1-30	0.1-20	
Dehydrogenating hydrocarbons	300-700	0.1-10	0.1-20	
Converting paraffins to aromatics	100-700	0.1-60	0.5-400	$H_2/C_xH_y = 0$-20
Converting olefins to aromatics (e.g., benzene, toluene & xylene)	100-700	0.1-60	0.5-400	$H_2/C_xH_y = 0$-20
Converting alcohols (methanol) or ethers (dimethylether) to hydrocarbons (olefins & aromatics)	275-600	0.5-50	0.5-100	
Isomerizing xylene feedstock components	230-510	3-35	0.1-200	$H_2/C_xH_y = 0$-100
Disproportionating toluene	200-760	1-60	0.08-20	
Alkylating aromatics(benzene & alkylbenzenes) with alkylating agent (olefins, formaldehyde, alcohols)	250-500	1-200	2-2000	C_xH_y/alkylating agent = 1-20
Transalkylating aromatics with polyalkylaromatic	340-500	1-200	10-1000	aromatic/polyalkyl-aromatic = 0-100

Note: WHSV=weight hourly space velocity; H_2/C_xH_y = hydrogen/hydrocarbons mole ratio
[Adapted from Haag and Tsikoyiannis, 1991]

Forced-flow mode. Invertase, an enzyme, can be chemically immobilized to the surfaces of ceramic membrane pores by the technique of covalent bonding of silane-glutaraldehyde [Nakajima et al., 1989b]. The substrate (reactant), during the sucrose conversion process, enters the membrane reactor in a crossflow mode. Under suction from the other side of the membrane, the substrate flows into the enzyme-immobilized membrane pores where the bioconversion takes place. Both the product and the unreacted substrate indiscriminately pass through the membrane pores. Thus, no permselective properties are utilized in this case. The primary purpose of the membrane is to provide a well-engineered catalytic path for the reactant, sucrose.

As will be shown later, some ceramic membranes have been used to immobilize some biocatalysts such as enzymes for increasing the reaction rate of bioreactions. Membrane pores when mostly used as catalyst carriers are advantageous over the conventional catalyst carriers in the pellet or bead form in having less mass transfer resistance and more efficient contact of the reactant(s) with the catalyst. In a strict sense, the membrane material when used in this mode is not a membrane which is defined as a permselective medium.

A major area of applications of this type is enzyme-catalyzed processes which will be discussed in detail in Section 8.7.1.

Opposing-reactants mode. When immobilized with a catalyst or enzyme, the interconnected tortuous pores or the nearly straight pores of a symmetric inorganic membrane provides a relatively well controlled catalytic zone or path for the reactants in comparison with the pellets or beads in a fixed or fluidized bed of catalyst particles. This unique characteristic of a symmetric membrane, in principle, allows a novel reactor to be realized provided the reaction is sufficiently fast. The concept applies to both equilibrium and irreversible reactions and does not utilize the membrane as a separator. Consider a reaction involving two reactants, A and B:

$$A + B \rightarrow C \tag{8-3}$$

Shown in Figure 8.3 is the opposing reactants geometry in which the two reactants enter the membrane from its opposite sides. A reaction plane is formed inside the catalytically active membrane. This implies that the flow front of either reactant is fairly uniform due to the well-engineered microstructure of inorganic membranes. The reactants arrive at the reaction plane in a stoichiometric ratio. Thus undesirable side reactions are reduced. It is noted that as any of the reactant flow rate or concentration changes, the reaction plane will migrate to a new position inside the membrane so that mass balance is maintained.

To make the membrane reactor practical, there are two opposing considerations. On one hand, it is desirable to have the pore size of the symmetric membrane as large as possible to provide relatively high reactant fluxes. On the other hand, the pore size should not be so large that the assumption of a fast kinetics (compared to transport resistance) can no longer hold. By applying a relatively small pressure gradient across the membrane, the reaction product, C, can be directed to one side of the membrane while keeping the reaction plane inside the membrane. The consequence of this is that one of the two reactants can be kept pure enough to be recycled [Saracco and Specchia, 1994].

GA 34568.4

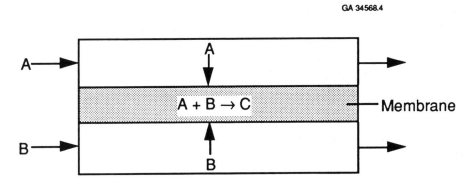

Figure 8.3 Schematic diagram of an opposing-reactants mode of a membrane used as a catalyst support but not as a separator

8.2.8 Membrane Coupling Two Reactors

When a reaction involves the evolution of a species while another reaction consumes the same species and a membrane can be found that is permselective to this species, it may be feasible to essentially combine the two membrane reactors involved into a compact single reactor configuration (see Figure 8.4). In this arrangement, a single membrane compartmentalizes the two reactor zones and keeps other reaction components separated.

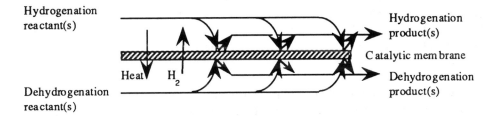

Figure 8.4 Coupling of hydrogenation and dehydrogenation reactions with a catalytic membrane

The first proof of concept involved two coupled reactions: endothermic dehydrogenation of cyclohexane and exothermic hydrogenation of pentadiene [Pfefferle, 1966; Gryaznov, 1969]. Since then most of the studies have focused on coupling dehydrogenation and hydrogenation of hydrocarbons by dense palladium-based membrane reactors. Some other possibilities include thermal decomposition of hydrogen sulfide coupled with catalytic hydrogen combustion using air [Edlund and Pledger, 1993].

There may be cases where one of the two reactions involved is coupled primarily for the purpose of consuming the permeating species so as to increase the driving force for permeation and thus the conversion. A classic group of these cases is oxidative dehydrogenation where the permeating hydrogen from a dehydrogenation reaction zone is oxidized to form water in another reaction zone separated by a membrane [Cales and Baumard, 1982; Gryaznov et al., 1986].

To conjugate two reactions with an inorganic membrane requires many considerations: the type of membrane and catalyst, material and heat balances, and operating conditions. These factors will be discussed in Chapter 11.

8.3 POTENTIALS OF INORGANIC MEMBRANE REACTORS

The integration of two unit operations, reaction and separation (by a membrane), into a single one has the great advantage of potentially reducing capital investments and operating costs. However, there are several significant challenges that need to be addressed before the technology becomes both technically feasible and economically viable on a production scale.

There have been numerous studies exploring the concept of membrane reactors. Many of them, however, are related to biotechnological applications where enzymes are used as catalysts in such reactions as saccharification of celluloses and hydrolysis of proteins at relatively low temperatures. Some applications such as production of monoclonal antibodies in a hollow fiber membrane bioreactor have just begun to be commercialized.

The greatest potentials of inorganic membranes are their uses in or as reactors for catalytic reactions. Many industrially significant reactions occur at high temperatures and often under harsh chemical environments where inorganic membranes generally are the only choice of materials due to their inherent thermal and chemical stabilities. Commercialization of inorganic membranes, especially ceramic ones, in recent years has spurred widespread interest not only in separation applications but also more importantly in membrane reactor applications and has generated a surge in the studies of inorganic membrane reactors.

8.3.1 Classification of Applications of Inorganic Membrane Reactors

For the convenience of discussion, the ever increasing volume of the literature in this field will be reviewed in the following four groups according to the type of membrane (dense or porous) and the reaction phase (liquid or gas). In those cases involving hybrid membranes where a dense membrane is deposited onto a porous support, the discussion will fall under the category of dense membranes. As will become evident, hydrogenation and dehydrogenation reactions have received most attention for three major reasons: (1) their strategic importance in the refining of petroleum and production of petrochemicals;

(2) the ease of separation of hydrogen from hydrocarbons and other gases because of its low molecular weight particularly when Knudsen diffusion is the dominant transport mechanism; and (3) the unique property of several membrane materials being capable of transporting hydrogen at an extremely high permselectivity.

As evident from the literature, the use of membrane reactors shows significant promises even in the case of porous membranes where Knudsen diffusion predominates and the separation factors are only marginal. Even a moderate separation performance can lead to a substantial improvement in reaction performance by continuously displacing the reaction.

8.3.2 Palladium and Palladium-Alloy Membrane Reactors

Not surprisingly, Pd-based membranes have been studied extensively due to their extremely high hydrogen permselectivity and well-known catalytic activity for many reactions. Palladium and other noble metal membranes also retain their thermal and corrosive resistances even at elevated temperatures. Since Pfefferle [1966] patented a hydrocarbons dehydrogenation process by the use of palladium membrane reactors, the body of literature on the subject has expanded rapidly. This field has been explored extensively by V.M. Gryaznov and his coworkers in Russia since the early 1960s. Studies have begun to emerge in Japan and the USA as well in recent years. The volume of research on this family of membrane materials far exceeds that for other types of membrane materials combined.

In some cases, the systematic research efforts in Russia have led to large pilot-plant or small-scale production of special high-value chemicals in the pharmaceutical and perfume industries. For example, vitamin K is produced by liquid-phase hydrogenation of quinone and acetic anhydride in one instead of four steps through the use of Pd-alloy membrane reactors [Gryaznov, 1986]. Intermediate products for the synthesis of other vitamins have also been obtained by hydrogenating 2-butynediol in a Pd-alloy membrane reactor [Gryaznov and Slin'ko, 1982 and Gryaznov et al., 1983a]. Linalool is a perfuming and flavoring agent. It is produced by the hydrogenation of dehydrolinalool [Karavanov and Gryaznov, 1984]. In view of their relatively low permeabilities compared to those attainable through the porous inorganic membranes, dense inorganic membrane reactors made of Pd and its alloys or other metals are more likely to be economically viable for the production of low-volume high-value products such as those mentioned above.

Pd alloys are preferred over pure Pd as the membrane materials because of several considerations. First of all, pure palladium can become embrittled after repeated cycles of hydrogen sorption and desorption. Second, the hydrogen permeabilities of certain Pd-alloys are higher than those of pure palladium. Third, the catalytic activities of the alloy membranes, in many cases, exceed that of palladium alone. Finally, palladium is very expensive. Alloying with other metals makes it more economically attractive for membrane reactor consideration. Along this line of reasoning, it is desirable to prepare as thin a Pd-based membrane as possible. A feasible alternative is to deposit a thin dense Pd-based membrane on a thicker porous inorganic support which provides the necessary mechanical strength and at the same time higher permeate flux through the composite membrane structure.

As fabricated, a Pd or its alloy membrane suffers from the relatively low catalytic activity of the attached catalyst due to the typical low surface area to volume ratio of the membrane geometry. The catalyst in a dense Pd-based membrane can be the Pd itself or its alloy or some other materials attached to the membrane. Pretreatments to the Pd-based membranes can help alleviate this problem. This and other membrane material and catalysis issues will be further covered in Chapter 9.

8.4 APPLICATIONS OF DENSE MEMBRANES IN GAS/VAPOR PHASE REACTIONS

Many large-volume industrially important reactions take place in gas or vapor phase typically at high temperatures where organic membranes lose their structural integrity and many dense inorganic membranes still function properly. Most of the aforementioned extensive studies on Pd-based membrane reactors have been focused on gas/vapor phase reactions at high temperatures and, in many cases, under harsh chemical environments. In those reaction conditions, organic membranes are essentially ruled out. To a much lesser degree, silver membranes have also been studied for their potential utilization in enhancing chemical reactions due to their oxygen permselectivity.

In addition, solid electrolyte membranes, notably metal oxide types, have been evaluated for improving industrial reactions and fuel cell applications.

8.4.1 Hydrogen-consuming Reactions

Due to the inherent catalytic activity of palladium for many hydrogenation reactions, Pd-based membranes have been examined for their potentials as membrane reactors for those reactions. Highlighted in Table 8.2 are some examples of studies using dense membranes, all of which are Pd-based, in reactions involving consumption of hydrogen such as hydrogenation, hydrodealkylation or tritium recovery. Not surprisingly, the literature in this field is dominated by investigations of hydrogenation of hydrocarbons. Today there are a variety of hydrogenation reactions practiced industrially ranging from high-volume petroleum refining and chemicals production to processes making high-value, low-volume pharmaceutical products. Except petroleum refining operations, only a few industrial hydrogenation processes are operated on a large scale. Table 8.2 includes some of these bulk processes such as selective hydrogenation of acetylenes, manufacture of cyclohexane by hydrogenating benzene and hydrodealkylation of alkyl naphthalenes. Selectivity is important in all these reactions and the use of Pd-based membranes has produced some encouraging results. It is, however, doubtful that these bulk processes employing dense Pd-based membranes will ever be widely implemented commercially in view of the low permeabilities of the dense membranes compared to the demand for throughput rate typically in bulk processes.

TABLE 8.2
Gas/vapor phase hydrogen-consuming reactions using dense membrane reactors

Reaction	Catalyst	Membrane	Conclusions	Reference
Tritium recovery $H_2 + 1/2 O_2 \rightarrow H_2O$ (T=300°C)		Pd on Al_2O_3 (disk)	Oxidative hydrogen pump for tritium recovery from breeder blankets	Drioli et al., 1990
Tritium recovery $H_2 + 1/2 O_2 \rightarrow H_2O$ (T=450°C)		Pd-V (tube)	Tritium recovery from Li-based breeder blankets	Hsu and Bauxbaum, 1986
Hydrogenation of 1,3-butadiene to butene (T=100°C)	Catalytic membrane	Pd (foil)	$Y=98\%$; Reaction rate can be over 10 times that for premixed reactants	Nagamoto and Inoue, 1986
Hydrogenation of 1,3-butadiene to butene (T=100°C)	Catalytic membrane	Pd-Ru(foil)	$C=100\%$; $S=61\%$	Armor and Farris, 1993
Hydrogenation of carbon monoxide to hydrocarbons (T=250–410°C)		Pd-Ru & Pd-Ni (foil & tube)	Membrane reactor produces C_2H_4 in favor of CH_4 & C_2H_6	Gur'yanova et al., 1988
Hydrogenation of carbon monoxide to hydrocarbons (T=290-390°C)	Ni or Cu coating	Pd-Ni (foil)	Ni favors C_3 & C_4 formation; Cu favors CH_3OH	Gur'yanova et al., 1989
Hydrogenation of carbon monoxide to hydrocarbons (T=100–400°C)		Pd-Ag or Pd-Y (disk)	Higher yield than fixed-bed reactor	Caga et al., 1989

Chapter 8

TABLE 8.2 – Continued
Gas/vapor phase hydrogen-consuming reactions using dense membrane reactors

Reaction	Catalyst	Membrane	Conclusions	Reference
Hydrogenation of carbon monoxide to hydrocarbons (T=250°C)		Pd-Ru (foil)	C=7%; C_2-C_3 are formed in favor of C_4-C_5	Gryaznov et al., 1993
Hydrogenation of carbon dioxide to hydrocarbons (T=280-380°C)	Ni or Ru coating	Pd-Ru (foil)	CH_4 is the main product	Gryaznov et al., 1981
Hydrogenation of acetylene to ethylene (T=100-180°C)		Pd-Ni (tube)	High selectivity of C_2H_4 when adsorbed & subsurface hydrogen not in equilibrium	Gryaznov and Slin'ko, 1982
Hydrogenation of ethylene to ethane (T=50-100°C)	Catalytic membrane	Pd (foil)	Reaction rate is not enhanced by Pd membrane	Nagamoto and Inoue, 1981
Hydrogenation of ethylene to ethane (T=100-450°C)		Pd-Ag or Pd-Y (disk)	C_m=60%; mainly C_1-C_4 products; Pd-Y more active	Al-Shammary et al., 1991
Hydrogenation of ethylene to ethane (T=150-300°C)		Pd or Pd-Ru (disk)	Pd-Ru gives lower conversion but higher stability	Foley et al., 1993
Hydrogenation of ethylene to ethane (T=150°C)		Pd-Ru (foil)	C=66% compared to 39% when premixing reactants	Armor and Farris, 1993
Hydrogenation of nitroethane to ethylamine (T=120°C)		Acid-treated Pd-Zn (foil)	C_m=100%; acid treatment improves permeability	Mishchenko et al., 1987
Hydrogenation of propane to methane & ethane (T=160-260°C)		Pd-Ru (capillary tube)	Hydrogen addition through membrane gives lower C & S	Skakunova et al., 1988

TABLE 8.2 – Continued
Gas/vapor phase hydrogen-consuming reactions using dense membrane reactors

Reaction	Catalyst	Membrane	Conclusions	Reference
Hydrogenation of propylene to propane (T=100°C)	Pd	Pd (foil)	Reaction rate is only slightly enhanced by Pd membrane	Nagamoto and Inoue, 1985
Hydrogenation of 1-butene to butane (T=110°C)	Au (electroplated)	Pd-Ag (tube)	Reaction rate enhanced by hydrogen addition through membrane	Yolles et al., 1971
Hydrogenation of 1-butene to butane (T=100°C)	Pd	Pd (foil)	Reaction rate is only slightly enhanced by Pd membrane	Nagamoto and Inoue, 1985
Hydrogenation of 1,3-pentadiene to cyclopentane & cyclopentene		Pd-Ru or Pd-Rh	Formation of cyclo-pentadiene is avoided	Gryaznov et al., 1987
Hydrogenation of 1,3-pentadiene to cyclopentane & cyclopentene (T=150°C)		Pd-Ru/Co	C=99.8% (vs. 90% for conventional reactors); S=92.7%	Gryaznov et al., 1993
Hydrogenation of cyclopentadiene to cyclopentane & cyclopentene (T=151°C)		Pd-Ru supported by polymer (foil)	C=89%;Y=92%	Gryaznov & Karavanov, 1979
Hydrogenation of cyclopentadiene to cyclopentane & cyclopentene (T=102°C)		Pd-Ru	S=92% for cyclopentene	Gryaznov et al., 1979
Hydrogenation of cyclopentadiene to cyclopentane & cyclopentene (T=165-240°C)		Pd-Cu (foil)	S for cyclopentene higher than fixed bed reactor	Zhernosek et al., 1979

TABLE 8.2 – Continued
Gas/vapor phase hydrogen-consuming reactions using dense membrane reactors

Reaction	Catalyst	Membrane	Conclusions	Reference
Hydrogenation of cyclopentadiene to cyclopentane & cyclopentene (T=161°C)		Pd (foil)	C=90%; S=95% for cyclopentene	Gryaznov et al., 1983b
Hydrogenation of cyclopentadiene to pentane (T=100°C)		Pd-Zn (foil)	C=100%; S=95%; acid treatment of membrane increases C & S	Mishcehnko et al., 1987
Hydrogenation of cyclohexene to cyclohexane (T≤200°C)		Pd-Ni (tube)	H_2 addition through membrane enhances reaction rate	Aguilar et al., 1977
Hydrogenation of cyclohexene to cyclohexane (T=150–250°C)		Pd-Ru (foil)	C_m=80%; S_m=100%	Armor and Farris, 1993
Hydrogenation of 1,3-cyclooctadiene to cyclooctene (T=80-200°C)		Pd-Ru (foil)	Y=83%; S=94%	Gryaznov and Slin'ko, 1982; Ermilova et al., 1985
Hydrogenation of benzene to cyclohexane (T=390°C)		Pd (tube)	Y(=4%) > Y_e(=0.1%)	Gryaznov et al., 1973
Hydrogenation of nitrobenzene to aniline (T=30-260°C)		Pd-Ru (foil)	Y_m=100%	Mischenko et al., 1979
Hydrogenation of nitrobenzene to aniline (T=250°C)		Pd-Zn (foil) acid treated	C=100%	Mischenko et al., 1987
Hydrodealkylation of toluene to benzene & methane (T=320-680°C)		Pd-W-Ru (foil)	Y=22% for benzene & Y=41% for methane at 671°C	Smirnov et al., 1977

TABLE 8.2 – Continued
Gas/vapor phase hydrogen-consuming reactions using dense membrane reactors

Reaction	Catalyst	Membrane	Conclusions	Reference
Hydrogenation of dimethyl-naphthalenes to α-methylnaphthalene & naphthalene (T≤600°C)		Pd-Ni or Pd-Mo (foil)	Further isomerization to β-methyl-naphthalene is avoided	Gryaznov et al., 1973
Hydrogenation of naphthalene to tetralin (T=80-150°C)		Pd-Rh	S=100%	Gryaznov et al., 1979
Hydrogenation of p-carboxy-benzaldehyde to p-toluenic acid (T=254°C)		Pd-Zn (foil) acid treated	C>99%	Mishchenko et al., 1987
Hydrogenation of furan to tetrahydrofuran (T=140°C)		Pd-Ni	Y=100%	Gryaznov & Karavanov, 1979
Hydrogenation of quinone & acetic anhydride to 2-methyl-1,4-diacetoxy-naphthalene (T=132°C)		Pd-Ni (tube)	Y=95%	Gryaznov & Karavanov, 1979

Note: C=conversion, S=selectivity, Y=yield, subscript m denotes the maximum value and subscript e indicates the equilibrium value without membrane

The temperature range of those hydrogenation reactions listed in Table 8.2 exceeds the upper temperature limits of organic polymer membranes. The geometry of the membrane element can be in the form of foil, disk or tube. The alloying elements used in the membranes are varied: Ag, Co, Cu, Mo, Ni, Rh, Ru, V, W, Y and Zn, especially Ag, Ni and Ru. It is interesting to note that Ni and Pd are the most common catalysts for many hydrogenation reactions industrially. These two metals readily absorb hydrogen into the interstices among metal atoms. Therefore, palladium-based membranes, although expensive, can provide both hydrogen permselectivity and catalytic activity for many hydrogenation reactions. It should be mentioned that other metals such as nickel, titanium, tantalum and vanadium, although not as well known as palladium, are also selectively permeable to hydrogen. There is, however, no application study of these metals as membranes for hydrogen separation.

The available literature data suggests that hydrogen permeates through Pd-based membranes in the form of highly active atomic hydrogen which is quite different from the hydrogen in a bulk gas phase. The atomic hydrogen readily reacts with other reactants such as hydrocarbons adsorbed on the membrane surfaces. Hydrogen diffusion through dense Pd-based membranes is not the reaction rate-limiting step [Gryaznov et al., 1973]. Thus, continuous, controlled addition of hydrogen from one side of the membrane to the other where the other reactant(s) and reaction zone are, in many cases, can better steer the reaction path and enhance the reaction rate by adjusting the permeation rate. The permeation rate depends on, among other factors, the operating temperature and the driving force (pressure or electrical potential).

Increase of the reaction conversion as a result of introducing hydrogen through Pd or Pd-alloy membranes instead of premixing it with other reactant(s) as in a conventional reactor is evident in many cases. Pd membranes improve the yield of cyclohexane in the course of hydrogenation of benzene from the equilibrium barrier of 0.1% to an actual higher level of 4% at 390°C [Gryaznov et al., 1973]. The hydrogenation rates of several olefins (e.g., 1,3-butadiene) in Pd membrane reactors have been observed to be ten times or higher than those in conventional reactors where the reactants are premixed [Nagamoto and Inoue, 1986]. Hydrogenation of carbon monoxide to form hydrocarbons in a Pd-Ag or Pd-Y membrane reactors results in a higher yield than a fixed bed reactor [Caga et al., 1989]. Acid-treated Pd-Zn membranes make it possible to achieve essentially 100% conversion for a number of gas-phase hydrogenation reactions: nitroethane to ethylamine at 120°C, nitrobenzene to aniline at 250°C, cyclopentadiene to pentane at 100°C and p-carboxybenzaldehyde to p-toluenic acid at 254°C [Mishchenko et al., 1987]. Likewise, Pd-Ru membranes are responsible for 100% conversion of 1,3-pentadiene to cyclopentane and cyclopentene vs. 90% in a conventional reactor at 150°C [Gryaznov et al., 1993]. Unlike conventional reactors where molecular hydrogen in the gas phase participates in the hydrogenation reactions, Pd-based membrane reactors engage the more active atomic hydrogen in the above reactions.

Furthermore, this mode of bringing hydrogen into contact with other reactant(s) from the opposite sides of a membrane reactor reduces the hydrogen pressure in the reaction mixture. Consequently potential side reactions are reduced and the product selectivity can thus be improved for some hydrogenation or hydrodealkylation reactions. Examples of high hydrogenation reaction selectivity are: essentially 100% for naphthalene to tetralin [Gryaznov et al., 1979], 98%+ for 1,3-butadiene to butene [Nagamoto and Inoue, 1986], 93% for 1,3-pentadiene to cyclopentane and cyclopentene [Gryaznov et al., 1973], 95% for cyclopentadiene to cyclopentene [Gryaznov et al., 1983b], just to name a few.

The expected benefits of reaction selectivity and conversion, however, are not realized in all cases. While many studies are encouraging, several others report no noticeable advantages using dense Pd or Pd-alloy membrane reactors for hydrogenation of hydrocarbons. A case in point is hydrogenation of ethylene to make ethane. When palladium is used to construct the membrane reactor, the reaction rate is not improved over a conventional reactor [Nagamoto and Inoue, 1981]. Pd-Ru, while improving the chemical stability of the membrane over Pd, gives lower reaction conversions [Foley et al., 1993]. Skakunova et al. [1988] found that hydrogen addition through Pd-based membranes actually lowers the conversion and selectivity for hydrogenation of propane to produce methane and ethane. In addition, Nagamoto and Inoue [1985] discovered that palladium membranes only slightly enhance the hydrogenation reactions of propylene or butene. In contrast, Al-Shammary et al. [1991] found that a Pd-Y membrane favors the formation of C_1-C_4 products.

A reaction with the reactants (hydrogen and a second reactant) premixed as in a conventional reactor is often limited by hydrogen adsorption while that with the hydrogen permeating through a membrane is limited only by the permeation rate (the rate of hydrogen desorption out of the membrane is very rapid). Thus the coverage of adsorbed hydrogen is greater when a membrane is used and the hydrogen permeating out of the membrane can be consumed quickly in the reaction. A reactant such as 1,3-butadiene that strongly reduces hydrogen adsorption on the catalyst hampers its own hydrogenation when it is premixed with hydrogen. Therefore its hydrogenation can benefit from employing a membrane to control the addition of hydrogen.

The unusual interaction of hydrogen with palladium-based membrane materials opens up the possibility of oxidative hydrogen pump for tritium recovery from breeder blankets. The feasibility for this potential commercial application hinges on the hot-fusion and cold-fusion technology under development [Saracco and Specchia, 1994]. At first, Yoshida et al. [1983] suggested membrane separation of this radioactive isotope of hydrogen followed by its oxidation to form water. Subsequently, Hsu and Bauxbaum [1986] and Drioli et al. [1990] successfully tested the concept of combining the separation and reaction steps into a membrane reactor operation.

8.4.2 Hydrogen-generating Reactions

There have been many studies involving evolution of hydrogen as a reaction product. Because of the relative ease of separating hydrogen from other gases or vapors using hydrogen-permeating dense membranes, the favorable effect of increasing reaction conversion by removing hydrogen from the reaction products can be found in many reactions. Hydrogen-permselective membranes are particularly attractive to dehydrogenation reactions many of which are thermodynamically limited. Highlighted in Table 8.3 are only some representative examples of the various studies on different hydrogen-generating reactions.

Palladium and selective alloys with other metals can be fabricated into dense membrane reactors in foil or tubular form, mostly in thin layers to reduce permeation resistance. In some cases, the dense membranes are supported by a thicker porous support such as glass or alumina. The alloying metals are varied: Ag, Ni, Rh, Ru and W. Silver and ruthenium are most widely used. Besides Pd and its alloys, other materials are also permselective to hydrogen and have been evaluated for membrane reactor applications. For example, dense silica is highly permselective to hydrogen and helium. Its separation factor for H_2/N_2 can exceed 5,000. Dense silica membranes have been tested successfully for dehydrogenation of isobutane to produce isobutene [Ioannides and Gavalas, 1993] and vanadium has been used as the core material for hydrogen permeation in a composite Pd-SiO_2-V-SiO_2-Pt membrane reactor [Edlund et al., 1992; Edlund and Pledger, 1993].

TABLE 8.3
Gas/vapor phase hydrogen-generating reactions using dense membrane reactors

Reaction	Catalyst	Membrane	Conclusions	Reference
Water gas shift reaction (T=400°C)	Fe_3O_4-Cr_2O_3	Pd on porous glass	C_m=99%+; $C>C_e$	Uemiya et al., 1991a
Water gas shift reaction (T=700°C)	Pt	V with Pt & Pd coatings	C=90%; resistance to poisoning by H_2S	Edlund et al., 1992
Steam reforming of methane (T=350-500°C)	Supported Ni	Pd on porous glass	C_m=90%+; $C>C_e$; Pd more effective than porous glass individually	Uemiya et al., 1991b
Steam reforming of methane (T=727°C)	Catalytic membrane	Pd	C=94%	Nazarkina and Kirichenko, 1979
Steam reforming of methane (T=700-800°C)	Catalytic membrane	Pd	C=96%; C_e=77%	Oertel et al., 1987
Methanol-water reaction to form hydrogen & carbon dioxide		Pd-Ag	Mobile ultrapure hydrogen generation	Philpott, 1975
Decomposition of hydrogen iodide (T=500°C)	Catalytic membrane	Pd-Ag	C=4%; C_e=2%	Yehenskel et al., 1979
Thermolysis of hydrogen sulfide (T=700°C)	Pt.	V with Pt & Pd coatings	C_m>99.4%; C_e=13%	Edlund and Pledger, 1993

TABLE 8.3 – Continued
Gas/vapor phase hydrogen-generating reactions using dense membrane reactors

Reaction	Catalyst	Membrane	Conclusions	Reference
Dehydrogenation of ethane to ethylene (T=455°C)	Catalytic membrane	Pd-Ag (tube)	$C \leq 0.7\%$	Pfefferle, 1966
Dehydrogenation of butane to butene (T=400°C)	Cr_2O_3-Al_2O_3	Pd-Ag	Y=18-25% (butene); Y=0.8-1.4% (butadiene)	Gryaznov, 1992
Dehydrogenation of isobutane to isobutene (T=450-500°C)	Cr_2O_3-Al_2O_3	Dense SiO_2 on porous glass	Higher Y & S when using membrane reactor	Ioannides and Gavalas, 1993
Dehydrogenation of isobutane to isobutene (T=400-500°C)	Pt-Sn on Al_2O_3	Pd on porous Al_2O_3	Sn reduces activity but suppresses coking	Matsuda et al., 1993
Dehydrogenation of butene to butadiene (T=385°C)	Cr_2O_3-Al_2O_3	Pd	$C_m=3C_e$ when hydrogen is oxidized on permeate side	Zhao et al., 1990a
Dehydrocyclo-dimerization of propane to aromatics (T=500°C)	Ga type ZSM-5 zeolite	Pd on porous Al_2O_3	$C_m>95\%$; higher C & Y (aromatics) when using membrane reactor	Uemiya et al., 1990
Dehydrocyclo-dimerization of propane to aromatics (T=550°C)	zeolite	Pd-Ag (foil)	$S(=64\%)> S((=$ 53%) with a packed bed reactor	Clayson and Howard, 1987
Dehydrogenation of isopropanol (T=156°C)		Pd-Ni	C=83%	Mikhalenko et al., 1986
Dehydrogenation of cyclohexane to benzene (T=125°C)		Pd-Ag	C=87%	Wood, 1968
Dehydrogenation of cyclohexane to benzene (T=300-500°C)		Pd-W-Ru	Y_m=51%	Smirnov et al., 1977

TABLE 8.3 – Continued
Gas/vapor phase hydrogen-generating reactions using dense membrane reactors

Reaction	Catalyst	Membrane	Conclusions	Reference
Dehydrogenation of cyclohexane to benzene (T=330-575°C)		Pd-Ru (foil)	Y_m=91%	Gryaznov et al., 1977
Dehydrogenation of cyclohexane to benzene (T=200°C)	Pt/Al₂O₃	Pd (tube)	C_m=99.5%; C_e=18.7%	Itoh, 1987b
Dehydrogenation of n-hexane to benzene (T=530-575°C)		Pd-Ru	Y=50-58% for benzene	Smirnov et al., 1977
Dehydrogenation of cyclohexanediol to pyrocatechol (T=250°C)		Pd-Rh (foil)	Y=95% (no phenol formed); S=100%	Sarylova et al., 1977
Dehydrogenation of heptane to benzene & methane (T=450-590°C)		Pd-W-Ru (foil)	Y-55% for benzene & Y=31% for methane	Smirnov et al., 1977
Dehydrogenation of dimethyl-ethylcarbinol to dimethylvinylcar binol and ter-amylalcohol		Pd	S=88% for dimethylvinyl-carbinol & S=98% for ter-amylalcohol	Sokol'skii et al., 1988

Note: C=conversion, S=selectivity, Y=yield, subscript m denotes the maximum value and subscript e indicates the equilibrium value without membrane

Palladium and some of its alloys are catalytically active to many dehydrogenation reactions. In other cases where other catalysts are required, they are either impregnated in the porous support (for those composite membranes), deposited on the membrane surfaces as a coating or packed as pellets inside the membrane element.

In contrast to the studies on gas- and vapor-phase hydrogenation reactions utilizing dense Pd-based membrane reactors, dehydrogenation reactions have been consistently observed to benefit from the concept of a membrane reactor. In almost all cases the reaction conversion is increased. This is attributed to the well known favorable effect of equilibrium displacement applied to dehydrogenation reactions which are mostly limited by the equilibrium barrier.

Dehydrogenation reactions are typically endothermic. Their equilibrium conversions, therefore, increase with increasing temperatures. Thus it may be argued that for high-temperature dehydrogenations the advantages of a membrane reactor over the conventional one may be diminishing. However, catalyst deactivation becomes more pronounced at high temperatures particularly for dehydrogenation reactions. In this case, the possibility of running a catalytic dehydrogenation reaction at a lower temperature to reach a given conversion with the use of a membrane reactor is still quite attractive.

One of the most significant dehydrogenation reactions is the production of butadiene which is the major feedstock for making butadiene-styrene and other synthetic elastomers, acrylonitrile-butadiene-styrene (ABS) resins and many chemical intermediates. Butadiene can be produced in either a two-stage process involving two dehydrogenation reactions which are equilibrium-limited or a single-stage oxidative dehydrogenation of butenes which is not equilibrium-limited. In the two-stage process, n-butane is dehydrogenated to butene first followed by dehydrogenating butene to form butadiene in the second stage. A number of investigators have studied both steps in the two-stage process by applying dense Pd-based membrane reactors [Gryaznov and Smirnov, 1974; Zhao et al., 1990a; Gryaznov, 1992; Ioannides and Gavalas, 1993; Matsuda et al., 1993].

An interesting case of using a dense inorganic membrane as a reactor is thermal decomposition (or thermolysis) of hydrogen sulfide at a high temperature of 700°C. The combination of such a corrosive gas and a high temperature poses a very challenging containment problem. The selection of materials for the membrane and protective layers requires careful considerations. This will be treated in Chapter 9.

The selectivity of the target product(s) is also enhanced in many situations over that attainable in a conventional packed, fluidized or trickle bed reactor. Removal of hydrogen from the reaction products enhances the yield of certain desired product(s). It does so by suppressing undesirable side reactions. In their studies of dehydrogenation of propane to form aromatics, Uemiya et al. [1990] concluded that the use of a palladium membrane to continually transfer hydrogen out of the reaction zone reduces occurrence of hydrocracking and dealkylation, thus leading to a higher yield of aromatics than what a conventional reactor can achieve.

Several hydrocarbon dehydrogenation reactions cause coking at high temperatures. The presence of hydrogen is known to suppress the coke formation. Therefore, it may be expected that the lower hydrogen pressure and higher olefin pressure as a result of continuously removing hydrogen from the reactor may worsen the coking problem than observed in a conventional reactor. For example, dehydrogenation of isobutane is associated with a significant carbon deposit problem [Ioannides and Gavalas, 1993].

In addition to those applications listed in Table 8.3, dehydrogenation of reforming gases, the product gases from methane steam conversion and the purge gases of ammonia synthesis have been mentioned [Gryaznov, 1992]. A novel idea of hydrogen storage for hydrogen-powered vehicles has been suggested [Philpott, 1985]. In this scenario, benzene is hydrogenated to form cyclohexane which can be distributed in a similar fashion as gasoline. The tanks of the vehicles are filled with hydrogenated hydride. Hydrogen can be released when needed by an onboard catalytic dehydrogenation reactor for cyclohexane. The key step of this idea, namely dehydrogenation of cyclohexane, is feasible at least in theory [Touzani et al., 1984].

8.4.3 Oxidation or Oxygen-Generating Reactions

Some dense inorganic membranes made of metals and metal oxides are oxygen specific. Notable ones include silver, zirconia stabilized by yttria or calcia, lead oxide, perovskite-type oxides and some mixed oxides such as yttria stabilized titania-zirconia. Their usage as a membrane reactor is profiled in Table 8.4 for a number of reactions: decomposition of carbon dioxide to form carbon monoxide and oxygen, oxidation of ammonia to nitrogen and nitrous oxide, oxidation of methane to syngas and oxidative coupling of methane to form C_2 hydrocarbons, and oxidation of other hydrocarbons such as ethylene, methanol, ethanol, propylene and butene.

TABLE 8.4
Oxidation or oxygen-generating reactions using dense membrane reactors

Reaction	Catalyst	Membrane	Conclusions	Reference
Decomposition of CO_2 to CO & O_2 (T=1681°C)		CaO-ZrO_2 (tube)	C=22% vs. C_e=1.2%;	Nigara and Cales, 1986
Decomposition of CO_2 to CO & O_2 (T=1311-1509°C)		Y_2O_3-ZrO_2 (tube)	C_m=28%; C_m>C_e	Itoh et al., 1993
Oxidation of ammonia to N_2 & NO (T=250-382°C)		Ag (tube)	Y=40% for N_2 & Y=25% for NO compared to 50% & 15% with conventional reactors	Gryaznov et al., 1986
Oxidation of methane to syngas (T=850°C)	Precious metal reforming catalyst	Perovskite-type oxide (tube)	C>98% S=90%	Balachandran et al., 1993; Balachandran et al., 1995

TABLE 8.4 – Continued
Oxidation or oxygen-generating reactions using dense membrane reactors

Reaction	Catalyst	Membrane	Conclusions	Reference
Oxidative coupling of methane to C_2 (T=744°C)	Inherently catalytic membrane	PbO/MgO on porous Al_2O_3 (tube)	$S>97\%$; air as feedstock	Omata et al., 1989
Oxidative coupling of methane to C_2 (T=470-640°C)		Ag (disk & tube)	Ethane & ethene form at lower temp. than conventional reactors; $C=0.03\%$ with $S=92\%$ or $C=3\%$ with $S=42\%$	Anshits et al., 1989
Oxidative coupling of methane to C_2 (T=700-850°C)	LiO/MgO, $MnNa_aMg_bZr_c$ O_x	Y_2O_3-TiO_2-ZrO_2 (tube)	$C=35-45\%$; $Y=20-25\%$ for C_2	Hazbun, 1989
Oxidative coupling of methane to C_2 (T=730-850°C)	MgO or ZrO_2-Y_2O_3	PbO modified by alkali-metal or alkaline earth compounds on porous & nonporous supports (tube)	$S>90\%$; activity affected by alkali-metal compounds and membrane supports	Nozaki et al., 1993
Oxidation of ethylene to ethylene oxide (T=250-400°C)	Ag	Y_2O_3-TiO_2-ZrO_2 (tube)	$C=10\%$; $S=75\%$	Hazbun, 1989
Oxidation of methanol and ethanol to acetaldehyde & formaldehyde (T=246-378°C)		Ag (tube)	Oxygen transfer through membrane faster than diffusion in vacuum	Gryaznov et al., 1986

TABLE 8.4 – Continued
Oxidation or oxygen-generating reactions using dense membrane reactors

Reaction	Catalyst	Membrane	Conclusions	Reference
Oxidation of ethanol to acetaldehyde (T=252-346°C)		Ag (tube)	Y=83% compared to 56% for premixed reactor	Gryaznov et al., 1986
Oxidation of propylene to hexadiene & benzene (T=600°C)		$La_{0.3}Bi_{1.7}O_2$ (disk)	C=3.2%; S=53 & 25% for hexadiene & benzene, resp. (vs. C=4.3%; S=39% for hexadiene+ benzene)	DiCosimo et al., 1986
Oxidation of propylene to propylene oxide (T=300-500°C)	Ag	Y_2O_3-TiO_2-ZrO_2 (tube)	C=15%; S=30%	Hazbun, 1989
Oxidation of butene to butadiene (T=462-505°C)	$W_3Sb_2O_3$	Y_2O_3-TiO_2-ZrO_2 (tube)	C=30% & S=92% at 462°C; C=57% & S=88% at 505°C	Hazbun, 1989

Note: C=conversion, S=selectivity, Y=yield, subscript m denotes the maximum value and subscript e indicates the equilibrium value without membrane

Similar to the case of dehydrogenation or other hydrogen-generating reactions, the use of a dense membrane reactor to remove oxygen from an oxygen-generating reaction such as decomposition of carbon dioxide displaces the reaction equilibrium and increases the conversion from 1.2% (limited by the equilibrium) to 22% [Nigara and Cales, 1986]. This has been confirmed by Itoh et al. [1993].

Also analogous to hydrogenation reactions, dense oxygen-permselective membranes improve the selectivity or yield of many oxidation reactions, see Table 8.4. This is accomplished by controlled addition of oxygen through the membrane to react with other reactants in the reaction zone on the feed side of the membrane. These oxygen-specific membranes can separate oxygen from nitrogen in the air and are used to introduce pure oxygen as a feedstock to the oxidation reactors. They can eliminate the need for separate oxygen plants which are rather costly; thus they are the logical candidates as the potential materials for membrane reactors involving oxidation reactions. However, if air is used as the feedstock, the nitrogen introduced will need to be separated from the reaction mixture in subsequent downstream processing steps.

These considerations can significantly improve the economics of partial oxidation of methane which encompasses two industrially important processes: synthesis gas production and direct conversion technologies for methanol from methane. The oxygen-selective membranes can also be applied to oxidative coupling of methane to form water and a methyl radical which dimerizes to produce ethane. Secondary reactions of ethane can yield ethylene which is used for fuel production or in petrochemical production through existing commercial technologies. Oxidative coupling of methane has been studied by a number of investigators using several types of dense membranes. For example, Omata et al. [1989] applied a dense PbO/MgO membrane reactor at 750°C to the oxidative coupling of methane to form ethane and ethylene with oxygen extracted from air flowing inside the membrane tube. Oxygen ion diffuses from the inside of the membrane to the outer surface and reacts with methane to produce the C_2 hydrocarbons. The selectivity to ethane and ethylene is higher than 97% in all cases which is much higher than that of the mixed gas reaction in a conventional reactor. Anshits et al. [1989] and Hazbun [1989] have also made similar studies employing silver and yttria-stabilized titania-zirconia mixed oxide membranes, respectively.

Dense perovskite-type oxides that contain transition metals (La-Sr-Fe-Co-O systems) show both ionic and electronic conductivities. These perovskites of the ABO_3 type have dopants on the A and/or B sites. These mixed-conductivity oxide membranes not only selectively separate oxygen but also transport electrons from the catalytic side of the reactor to the oxygen-reduction interface. Unlike many other solid electrolyte membranes such as stabilized zirconia membranes which are electronic insulators, they are capable of transporting oxygen without external electrical circuits or electrodes. Dense perovskite tubular and monolithic honeycomb membranes made by extrusion have been successfully used for converting methane, the major component of natural gas, to syngas (CO and H_2) at 850°C using air as the oxidant [Balachandran et al., 1993; Balachandran et al., 1995]. It is found that the physical integrity, mechanical properties and the oxygen flux of the membrane at the high application temperatures are determined by the stoichiometry of the perovskite compound. Another advantage of using these oxygen-selective membranes is that the oxidation kinetics can be regulated by the controlled addition of oxygen to the reactor. The syngas can be converted into upgraded products such as paraffins, olefins, waxes, etc.

Although involving oxidation reactions, solid oxide fuel cells will be treated in a separate section because of the unique characteristics of their applications.

8.4.4 Coupled Reactions

As mentioned earlier, two compatible reactions may be coupled or conjugated properly by a shared membrane through which the species (as a product on one side of the membrane and a reactant on the other) common to both reactions selectively passes. Summarized in Table 8.5 are some documented studies of reaction coupling using dense palladium-based membranes with the alloying component ranging from nickel, ruthenium, rhodium to silver.

TABLE 8.5
Reaction coupling using dense membrane reactors

Reactions	Catalysts	Membrane	Conclusions	Reference
Borneol dehydrogenation (A) & Cyclopentadiene hydrogenation (B)	Cu or CuO (A)	Pd-Ni, Pd-Ru or Pd-Rh	C for borneol and cyclopentadiene and S for cyclopentene are 100, 5 and 100%, respectively, and 57, 93 and 92%, respectively, under different conditions	Smirnov et al., 1983
Butane dehydrogenation (A) & Hydrogen oxidation (B)	Cr_2O_3-Al_2O_3 (A) Pd-Ru (B)	Pd-Ru (foil)	Y and S of butadiene increase	Orekhova and Makhota, 1985
1-butene dehydrogenation (A) & Hydrogen oxidation (B)	Cr_2O_3-Al_2O_3 (A) Pd (B)	Pd (foil)	$C=1.8C_e$ at 447°C	Zaho and Govind, 1990
2-butene dehydrogenation (A) & Hydrogenation of benzene (B)		Pd-Ni (foil)	C (=6%) > C_e (2%) for 2-butene; C (=4%) > C_e (=0.1%)	Gryaznov et al., 1973
Cyclohexane dehydrogenation (A) & 1,3-pentadiene hydrogenation (B)	Pt-Ru on Al_2O_3 (A) Pd-Ru (B)	Pd-Ru (foil)	C_m=99% S_m=98%	Gryaznov, 1992

TABLE 8.5 - Continued
Reaction coupling using dense membrane reactors

Reactions	Catalysts	Membrane	Conclusions	Reference
Cyclohexanol dehydrogenation (A) & Hydrogenation of phenol (B)		Pd-Ru	C=100% for cyclohexanol at 282°C; C=39% for phenol & S=95% for cyclohexanone; C for phenol increases when hydrogen/phenol ratio is increased	Basov and Gryaznov, 1985
Ethane dehydrogenation (A) & Hydrogen oxidation (B)		Pd-Ag (tube)	C (=0.7%) > C_e (0%)	Pfefferle, 1966
Methane dehydrogenation (A) & Hydrogen oxidation (B)	Pt-Sn on Al_2O_3 (A) Rh-SiO$_2$ (B) pellets	Pd (tube)	Low Y for ethane; coking occurs	Andersen et al., 1989

Note: C=conversion, S=selectivity, Y=yield, subscript m denotes the maximum value and subscript e indicates the equilibrium value without membrane

Common to those coupled reactions are a dehydrogenation reaction as the first reaction and a hydrogenation reaction as the second reaction, with the two reaction zones separated by the shared membrane. As mentioned previously, the concept was proposed by Pfefferle [1966] and Gryaznov [1969] in the 1960s. But only few significant successes have been obtained [Gryaznov, 1992]. The conversion and selectivity, especially the former, are in several cases relatively low or not sufficiently high to encourage more research activities in this field. This may not be too surprising in view of the complexity of the various issues involved in the technology which will be treated in some details later in Chapter 11.

It has been demonstrated that the conversion of a given reaction can be higher in a coupled reactor configuration than in a single membrane reactor. Given in Figure 8.5 [Gryaznov, 1992] are the conversion data of cyclohexane dehydrogenation using a Pd-Ru foil membrane with a catalyst containing 0.4% Pt and 0.4% Ru supported on alumina under two different conditions. In the first case, Ar is used as the sweep gas on the other side of the membrane for the hydrogen formed and in the second case hydrogenation of 1,3-pentadiene to form pentenes is coupled to consume the hydrogen on the other side of the membrane. It is seen that the conversion of cyclohexane is higher when reaction coupling occurs.

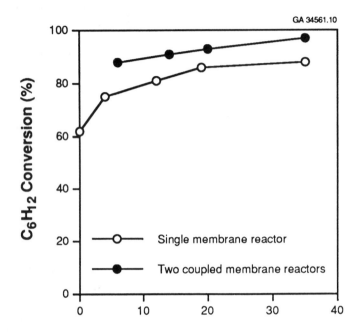

Figure 8.5 Effect of reaction coupling on conversion of cyclohexane dehydrogenation [Gryaznov, 1992]

8.5 APPLICATIONS OF DENSE MEMBRANES IN LIQUID- AND MULTI-PHASE REACTIONS

Shown in Table 8.6 are some literature data on the use of dense membrane reactors for liquid- or multi-phase catalytic reactions. Compared to gas/vapor phase application studies, these investigations are relatively few in number. Most of them involve hydrogenation reactions of various chemicals such as acetylenic or ethylenic alcohols, acetone, butynediol, cyclohexane, dehydrolinalool, phenylacetylene and quinone. As expected, the majority of the materials adopted as membrane reactors are palladium alloy membranes. High selectivities or yields are observed in many cases. A higher conversion than that in a conventional reactor is found in a few cases.

TABLE 8.6
Liquid- and multi- phase reactions using dense membrane reactors

Reaction	Catalyst	Membrane	Conclusions	Reference
Dehydrogenation of methanol, formic acid or formaldehyde in eletrolyte ($T=25^{\circ}C$)	Pt coating	Pd or Pd-alloy	Study of electrocatalysis of CHO-compounds	Guther and Vielstich, 1982
Hydrogenation of acetylenic/ethylenic alcohols ($T=100\text{-}130^{\circ}C$)		Pd-Ru	100-fold activity increase than conventional Raney catalysts	Gryaznov et al., 1982
Hydrogenation of acetone ($T=47\text{-}227^{\circ}C$)	embedded Al_2O_3	Pd-Ru-In	Alumina increases catalytic activity	Gryaznov et al., 1993
Hydrogenation of 2-butyne-1,4-diol to cis/trans-butenediol ($T=60^{\circ}C$)		Pd-Ru	$S=95\text{-}96.5\%$ for cis-butenediol	Gryaznov et al., 1982 & 1983a
Hydrogenation of cyclohexane ($T=70^{\circ}C$)		Pd-Ru	Low conversions: $C \leq 4.4\%$	Farris and Armor, 1993
Hydrogenation of dehydrolinalool to linalool ($T=170^{\circ}C$)		Pd-Ru	$Y_m=95\%$	Karavanov and Gryaznov, 1984
Hydrogenation of phenylacetylene	Pd alloy	Pd alloy spiral tube	$S_m=92\%$; C is 10 times that of premixed system	Gryaznov, 1992
Hydrogenation of quinone & acetic anhydride to vitamin K ($T=132^{\circ}C$)		Pd-Ni	$Y=95\%$ (vs. 80% for traditional process)	Gryaznov, 1986

TABLE 8.6 – Continued
Liquid- and multi- phase reactions using dense membrane reactors

Reaction	Catalyst	Membrane	Conclusions	Reference
Hydrogenation of sodium *p*-nitrophenolate to sodium *p*-aminophenolate (T=20-60°C)		Polarized Ni	Yield of Na p-aminophenolate reaches a max. at a membrane potential of -0.2 V	Nogerbekov et al., 1984

Note: *C*=conversion, *S*=selectivity, *Y*=yield, subscript *m* denotes the maximum value and subscript *e* indicates the equilibrium value without membrane

Dense inorganic membranes (primarily Pd-based) have been used in relatively small-volume high-value commercial applications. An example of this is the manufacture of vitamins. Contrary to the conventional process for producing vitamin K in several steps, a new single-step process involving a Pd-Ni membrane reactor has been developed [Gryaznov and Karavanov, 1979; Gryaznov, 1986]. In that process, the vitamin can be made by hydrogenating a mixture of quinone and acetic anhydride in a Pd-Ni membrane reactor. Besides reducing the number of preparation steps required in the conventional process, the yield is increased from 80 to 95%. Similarly, 2-butynediol can be converted to some intermediate products, cis- and trans-butenediols, for the synthesis of other vitamin products [Gryaznov and Slin'ko, 1982; Gryaznov et al., 1983a]. Dense Pd-Ru membrane reactors at 60°C give a high cis-butenediol at 95-96%. A further example of liquid-phase consumer-product application of dense inorganic membrane reactors is in the perfume industry. Specifically, dehydrolinalool is hydrogenated to produce linalool which is found in many oils such as bergamot and rosewood [Karavanov and Gryaznov, 1984]. At 170°C, the yield of linalool reaches 95% in an Pd-Ru membrane reactor.

8.6 APPLICATIONS OF POROUS MEMBRANES IN GAS/VAPOR PHASE REACTIONS

Dense inorganic membranes typically exhibit very high permselectivities. On the other hand, their permeabilities are mostly low compared to those of porous membranes. Since throughput is a major consideration for bulk processing, a high membrane permeability is in many cases desirable at the expense of the permselectivity provided the latter is at an economically acceptable level. The relatively low permselectivity of porous membranes can be compensated to some extent by recycling a portion of the retentate stream back to the feed stream.

The advent of reliable quality ceramic membranes entering the industrial market has heightened the interest for porous inorganic membrane reactors at high temperatures, typically associated with gas or vapor phase reactions. As will become evident, most of the porous inorganic membranes that are used in membrane reactor studies are made of ceramic materials. This is mostly driven by the high temperature requirements of many industrially important catalytic reactions.

8.6.1 Hydrogenation

Due to its small molecular weight, hydrogen permeates through microporous inorganic membranes at a faster rate than the majority of other gases. Therefore, like the case of dense palladium-based membranes, hydrogenation and dehydrogenation reactions are more likely to be the candidates for the applications of these microporous membranes. Table 8.7 provides a partial representation of documented studies of gas or vapor phase hydrogenations using porous inorganic membranes.

TABLE 8.7
Gas/vapor phase hydrogen-consuming reactions using porous membrane reactors

Reaction	Catalyst	Membrane	Conclusions	Reference
Hydrogenation of ethene to ethane (T=200°C)	Pt or Os; impregnated in membrane	Al_2O_3 (disk)	Conversion to ethane detected; no data provided	Furneaux et al., 1987
Hydrogenation of 2-butene to butane	Pt or CaA-zeolite layer	Stainless steel (disk)	Only trans-2 butene is selectively separated & hydrogenated	Suzuki, 1987
Hydrogenation of n-hexane & methylcyclo-hexane to ethene & propene (T=270°C)	Pt or CaY-zeolite layer	Ni-Al alloy (disk)	Only hexane is selectively separated & hydrogenated	Suzuki, 1987

Note: C=conversion, S=selectivity, Y=yield, subscript *m* denotes the maximum value and subscript *e* indicates the equilibrium value without membrane

Furneaux et al. [1987] used porous alumina membrane reactors to hydrogenate ethene to form ethane at 200°C with Pt or Os as the catalyst impregnated in the alumina membranes. Conversion to ethane was detected but no data was provided. Suzuki [1987] tested porous stainless steel and nickel-aluminum alloys as membrane reactors for hydrogenation reactions. Hydrogenation of 2-butene with stainless steel as the membrane material and Pt or CaA-zeolite as the catalyst results in selective permeation of trans-2 butene and subsequent hydrogenation to butane. Similarly, when n-hexane and methylcyclohexane are fed to one side of a Ni-Al alloy membrane reactor, only hexane selectively permeates through the membrane. Thus hexane is added to the reaction zone on the other side of the membrane in a controlled dosage and, with Pt or CaY-zeolite, selective hydrogenation is carried out.

8.6.2 Dehydrogenation

Given in Table 8.8 is a summary of experimental data available on dehydrogenation reactions using porous inorganic membrane reactors with a variety of catalysts. Some of those reactions are industrially important and will be discussed separately as follows.

TABLE 8.8
Gas/vapor phase hydrogen-generating reactions using porous membrane reactors

Reaction	Catalyst	Membrane	Conclusions	Reference
Decomposition of H_2S (T=800°C)	MoS_2 or WS_2 (packed bed)	Glass (tube)	$C=2C_e$	Kameyama et al., 1981
Decomposition of H_2S (T=800°C)	MoS_2 (packed bed)	Al_2O_3 (tube)	$C=2C_e$	Kameyama et al., 1983
Decomposition of H_2S (T=800°C)	MoS_2; impregnated in membrane	Al_2O_3 (tube)	$C=14\%$ $>C_e=3.5\%$	Abe, 1987
Water-gas shift reaction to H_2 & CO_2 (T=130-165°C)	$RuCl_3 \cdot 3H_2O$ (film)	Glass (tube)	C up to 85% at lower T than in industrial processes (200-400°C)	Seok and Hwang, 1990
Steam reforming of methane to CO, CO_2 & H_2 (T=400-625°C)	NiO	Al_2O_3 (tube)	C up to 1.1 times C_e	Tsotsis et al., 1992
Dehydrogenation of methanol to HCHO & H_2 (T=300-400°C)	Ag	Glass (tube)	$C_m=91\%$ at 300°C	Song and Hwang, 1991
Dehydrogenation of methanol to HCHO & H_2 (T=500°C)	ZnO deposition	Al_2O_3 (disk)	Y is enhanced 1.3 times by membrane	Zaspalis et al., 1991a

TABLE 8.8 – Continued
Gas/vapor phase hydrogen-generating reactions using porous membrane reactors

Reaction	Catalyst	Membrane	Conclusions	Reference
Dehydrogenation of methanol to HCHO & H_2 (T=500°C)	ZnO (packed bed)	Al_2O_3 (disk)	C increased from 15 to 25%; S decreased from 80 to 75%; Y increased from 14 to 19% when membrane used	Zaspalis et al., 1991b
Dehydrogenation of ethane to ethene (T=200°C)	Pt, Os or Cr_2O_3 impregnated in membrane	Al_2O_3 (disk)	Conversion to ethene; no data provided	Furneaux et al., 1987
Dehydrogenation of ethane to ethene (T=450-600°C)	Pt impregnated in membrane	Al_2O_3 (tube)	$C_m = 6C_e$	Champagnie et al., 1990
Dehydrogenation of ethane to ethene (T=550-675+°C)	Pt impregnated in membrane	Al_2O_3 (tube)	$C=35\%$ and $C_e=15\%$ at $T = 600°C$	Champagnie et al., 1992
Dehydrogenation of propane to propene (T=575°C)	$Cr_2O_3/$ Al_2O_3 (packed bed)	Al_2O_3 (tube)	C increased from 40 to 59%; $S=90\%$ when membrane reactor with pre- & post- dehydrogena- tion zones used	Bitter, 1988
Dehydrogenation of propane to propene (T=520-600°C)	Pt-Mg-Al_2O_3	Al_2O_3 (tube)	C & Y higher than those of conventional packed bed reactors	Ziaka et al., 1993
Dehydrogenation of n-butane to butene (T=300-500°C)	Pt on SiO_2 (packed bed)	Al_2O_3 (disk)	C increased by 1.5 & S increased by 1.6 when membrane used	Zaspalis et al., 1991b

TABLE 8.8 – Continued
Gas/vapor phase hydrogen-generating reactions using porous membrane reactors

Reaction	Catalyst	Membrane	Conclusions	Reference
Dehydrogenation of isobutene & propene to propane	Pt or CaA-zeolite layer	Stainless steel (disk)	Only H_2 & C_3H_8 present in permeate	Suzuki, 1987
Dehydrogenation of 2-methylbutenes to isoprene (T=550°C)	Pt or Sn-promoted zinc aluminate	Al_2O_3 (tube)	C increased from 27 to 37%; S=85% when membrane reactor with pre- & post-dehydrogenation zones used	Bitter, 1988
Dehydrogenation of cyclohexane to benzene (T=170-360°C)	Pt on Al_2O_3 support (packed bed)	Glass (tube)	C increased from 33 to 80% (C_e=35%) when membrane used	Shinji et al., 1982
Dehydrogenation of cyclohexane to benzene (T=180-220°C)		Glass (tube)	C=24% > C_e=10% at 180°C	Itoh et al., 1985
Dehydrogenation of cyclohexane to benzene (T=200-350°C)	Pd impregnated in membrane	C (tube)	C=80-100% > C_e=20%	Fleming, 1990
Dehydrogenation of cyclohexane to benzene (T=180-220°C)	Pt on Al_2O_3 support (packed bed)	Glass (tube)	C increased from 18 to 45% when membrane used	Itoh, 1987b
Dehydrogenation of cyclohexane to benzene (T=180-220°C)	Pt on Al_2O_3 support (packed bed)	Glass (tube)	C=45% much higher than C_e=19%	Itoh et al., 1988
Dehydrogenation of cyclohexane to benzene (T=220-325°C)	Pt impregnated in membrane	Glass (tube)	For high space time, C_{CMR} > C_{IMRCF} > C_e	Sun and Khang, 1988

TABLE 8.8 – Continued
Gas/vapor phase hydrogen-generating reactions using porous membrane reactors

Reaction	Catalyst	Membrane	Conclusions	Reference
Dehydrogenation of cyclohexane to benzene (T=197°C)	Pt on Al_2O_3 support (packed bed)	Al_2O_3 (hollow fiber)	$C=3C_e$	Okubo et al., 1991
Dehydrogenation of cyclohexane to benzene (T=187°C)	Pd	Glass (tube)	$C_m=75\%$ when Pd deposited next to membrane surfaces	Cannon and Hacskaylo, 1992
Dehydrogenation of cyclohexane to benzene (T=200-220°C)	Pt on SiO_2 pellet	Al_2O_3 (tube)	C up to twice that in packed bed reactor	Tiscareno-Lechuga et al., 1993
Dehydrogenation of ethylbenzene to styrene (T=625°C)	$Li_{0.5}$-$Fe_{2.4}$-$Cr_{0.1}$-O_4 with K_2O & V_2O_5 (packed bed)	Al_2O_3 (tube)	Y increased from 51 to 65%; $S=94\%$ when membrane reactor with pre- & post-dehydrogenation zones used	Bitter, 1988
Dehydrogenation of ethylbenzene to styrene (T=600-640°C)	Fe_2O_3 on Al_2O_3 support (packed bed)	Al_2O_3 (tube)	C increased by 15% and Y increased by 2-5% when membrane used	Wu et al., 1990
Dehydrogenation of ethylbenzene to styrene (T=500-600°C)	Fe_2O_3	Al_2O_3 (tube)	C is 20-23% higher than C_e	Moser et al., 1992
Dehydrogenation of ethylbenzene to styrene (T=580-610°C)	Cr/K-Fe_2O_3	Al_2O_3 (tube)	Y is increased when a packed bed reactor is replaced by a membrane reactor	Gallaher et al., 1993

TABLE 8.8 – Continued
Gas/vapor phase hydrogen-generating reactions using porous membrane reactors

Reaction	Catalyst	Membrane	Conclusions	Reference
Dehydrogenation of ethylbenzene to styrene (T=555-602°C)	K-Fe$_2$O$_3$	Al$_2$O$_3$ (tube)	$C > C_e$; steam reduces coking problem	Tiscareno-Lechuga et al., 1993

Note: C=conversion, S=selectivity, Y=yield, subscript m denotes the maximum value, subscript e indicates the equilibrium value without membrane, CMR=catalytic membrane reactor, $IMRCF$=catalytically inert membrane with catalyst pellets on the feed side

Styrene production from ethylbenzene. One of the very important industrial catalytic processes is dehydrogenation of ethylbenzene to produce styrene. The reaction is thermodynamically limited which forces the commercial production conditions toward high temperatures and relatively low pressures. This causes reduction in the selectivity of styrene due to the formation of other undesirable byproducts. Superheated steam is introduced for three major benefits: providing some of the necessary heat of reaction, oxidizing the coke that forms on the catalyst surface and diluting the reaction mixture to favor the product formation. With hydrogen as a product, this reaction appears to be amenable to porous inorganic membrane reactors for improving the conversion by continuously removing hydrogen to displace the equilibrium. This has been the rationale for the great interest in studying this process in recent years [Bitter, 1988; Wu et al., 1990; Liu et al., 1990; Tiscareno-Lechuga et al., 1993]. The prevailing operating temperature ranges from 500 to 650°C. The temperature in combination with the steam used constitutes one of the most harsh environments among various dehydrogenation processes. Many ceramic membranes are capable of withstanding the conditions without serious structural deformation or chemical transformation.

Studies using both laboratory-prepared and commercial inorganic membranes packed with formed catalysts have shown a conversion of as high as 70% which represents an improvement of about 15% over the case without the use of a membrane [Bitter, 1988; Wu et al., 1990]. This potentially offers significant energy savings in reaction and product separation. The well accepted standard alumina-supported iron oxide catalyst doped with potassium carbonate is used. The styrene selectivity is also increased by 2 to 5% to the 90 - 95% level as a result of using the membrane reactor approach. The simultaneous improvements in both the conversion and the selectivity are in striking contrast to the general observation with conventional reactors where typically selectivity is inversely proportional to conversion.

In many dehydrogenation reactions, carbon deposition can become a significant issue. This has been observed for dehydrogenation of ethylbenzene. Introduction of steam reduces the coking problem [Tiscareno-Lechuga et al., 1993]. More discussions will be made later in Chapter 11 on the general issues related to coking.

Dehydrogenation of cyclohexane to benzene. Another well studied reaction using porous inorganic membrane reactors is dehydrogenation of cyclohexane to make benzene. The operable temperature range is from about 170 to 360°C using a precious metal (e.g., Pt or Pd) as the catalyst either impregnated in the membrane pores or on a carrier such as alumina or silica.

Microporous alumina membrane tubes have been tested as the membrane reactors for this reaction with a maximum conversion about twice or three times that of the equilibrium conversion when the Pt catalyst is introduced as pellets in a packed bed [Okubo et al., 1991; Tiscareno-Lechuga et al., 1993]. Porous tubular Vycor glass membrane reactors have been utilized as well [Shinji et al., 1982; Itoh et al., 1985 and 1988]. The resulting conversion is at least twice the limit imposed by the equilibrium and, depending on the temperature and other conditions, can appreciably exceed that limit [Itoh, 1987b; Cannon and Hacskaylo, 1992]. Sun and Khang [1988] have also found that a catalyst-impregnated membrane reactor results in a higher conversion than a membrane reactor containing pellets of the catalyst with both types of reactors having a conversion greater than the equilibrium value.

Dehydrogenation of aliphatic hydrocarbons. A number of aliphatic hydrocarbons experience enhanced dehydrogenation conversions by carrying out the reactions in porous inorganic membranes. Most of the studies use porous alumina membrane tubes as the reactors.

Conversion of ethane to ethene in an alumina membrane reactor with Pt as the catalyst impregnated in the membrane pores can be as high as 6 times that of the equilibrium value in the 450-600°C range [Champagnie et al., 1990]. A higher selectivity can reach 98% with a lower conversion at a temperature greater than 675°C [Champagnie et al., 1992]. Propane was found to be dehydrogenated to propene at a higher degree in an alumina membrane reactor than in a conventional packed bed reactor. Bitter [1988] used Cr_2O_3 on Al_2O_3 as the catalyst bed at 575°C to achieve a conversion of 59% and a selectivity of 90% with pre- and post-dehydrogenation zones installed to convert extra residual propane to propene. Ziaka et al. [1993] obtained similar results with a Pt-Mg-Al_2O_3 catalyst for the temperature range of 520-600°C.

Zaspalis et al. [1991b] and Bitter [1988] utilized alumina membrane reactors containing Pt catalysts to examine dehydrogenation of n-butane to butene and 2-methylbutenes to isoprene, respectively. Both the conversion and selectivity improved by using the membrane reactors. The increase of conversion is about 50% in both cases. Moreover, Suzuki [1987] used stainless steel membranes and Pt or CaA-zeolite layer catalysts to perform dehydrogenation of isobutene and propene to produce propane.

Decomposition of hydrogen sulfide. Dissociation of hydrogen sulfide to elemental sulfur is presently carried out by a combination of cooling, condensing and leaching steps which is not energy efficient. The procedures potentially can be replaced by a membrane separation/catalysis process at high temperatures (e.g., 800°C). Inorganic membranes are suitable for this application.

Porous alumina and glass membranes have been found to be effective for decomposing hydrogen sulfide at 800°C by using packed beds of MoS_2 or WS_2 catalysts or impregnating the membrane tubes with the catalysts [Kameyama et al., 1981 and 1983; Abe, 1987]. When a packed bed of catalyst is employed, the reaction conversion due to equilibrium displacement reaches a level twice the equilibrium conversion in a conventional packed bed reactor. If, on the other hand, the catalyst is impregnated in the porous membrane matrix, the realized conversion is even higher [Abe, 1987]. Viscous flow through the membrane is believed to be a limiting factor for increase in the conversion. Kameyama et al. [1983] found that a larger-pore alumina membrane with a 30-fold increase in the permeability but a lower permselectivity than a porous glass membrane can reach essentially the same conversion. This will the an important implication when a large throughput of a membrane reactor is economically vital.

The chemical environment of hydrogen sulfide is quite corrosive especially at high temperature. This poses a potential material challenge as far as the membrane and reactor vessel are concerned. This will be discussed in Chapter 9.

Other hydrogen-generating reactions. An important implication of conversion enhancement due to the use of inorganic membrane reactors is that, for the same conversion, the reaction may be performed at a lower temperature. This was clearly demonstrated, for example, in the water-gas shift reaction to hydrogen and carbon dioxide which was conducted in a porous glass membrane reactor and achieved a conversion of up to 85% at a lower temperature (130-165°C) than in industrial processes (typically 200-400°C) [Seok and Hwang, 1990]. The catalyst employed is a hydrated $RuCl_3$. Methane can be steam reformed to produce hydrogen, carbon monoxide and carbon dioxide using a porous alumina membrane reactor and NiO as the catalyst with a relatively small improvement in conversion [Tsotsis et al., 1992].

Furthermore, Song and Hwang [1991] and Zaspalis et al. [1991a and 1991b] have investigated dehydrogenation of methanol. Formaldehyde (HCHO) is formed in addition to hydrogen. When silver is used as the catalyst contained inside a porous glass membrane reactor, a maximum conversion of 91% can be achieved at 300°C. On the other hand, the conversion increases only from 15 to 25% when ZnO deposition or pellets are used as the catalyst contained in a porous alumina membrane reactor at 500°C. Correspondingly the yield increases from 14 to 19% in the latter case [Zaspalis et al., 1991b].

8.6.3 Other Applications

Listed in Table 8.9 are other gas/vapor phase reactions taking advantage of the general concept of an inorganic membrane reactor.

TABLE 8.9
Other gas/vapor phase reactions using porous membrane reactors

Reaction	Catalyst	Membrane	Conclusions	Reference
Reduction of nitric oxide with NH_3 (T=300-350°C)	V_2O_5	γ-Al_2O_3 & TiO_2 (disk)	C=100%; S=75-80%	Zaspalis et al., 1991d
Decomposition of RuO_4 to RuO_2 & O_2 (T=500°C)	Fe_2O_3 impregnated	Fe_2O_3- molecular sieve-clay	C_m=99.4%	Peng et al., 1983
Oxidative dehydrogenation of methanol to formaldehyde (T=20-500°C)	Inherently catalytic membrane	Al_2O_3 (disk)	Reaction rate 10 times that of fixed bed reactors; S_m=85%	Zaspalis et al., 1988
Oxidative dehydrogenation of methanol to formaldehyde (T=200-500°C)	Inherently catalytic membrane	Al_2O_3 (disk)	C_m=90%; S_m=17%; reaction rate 10 times that of fixed bed reactors	Zaspalis et al., 1991a
Oxidative dehydrogenation of methanol to formaldehyde (T=180-500°C)	Ag on inherently catalytic membrane	Al_2O_3 (disk)	C_m=50%; S_m=60%	Zaspalis et al., 1991c
Conversion of propene to ethene & butene (T=20-22°C)	Rhenium oxide coating	Glass (tube)	C_m=40% > C_e=34%	Seok and Hwang, 1990
Oxidation of volatile organic carbon	Precious metals	Ceramic (honeycomb monolith)	C_m>99%	Bishop et al., 1994

Note: C=conversion, S=selectivity, Y=yield, subscript m denotes the maximum value, and subscript e indicates the equilibrium value without membrane

A number of oxide compounds can be decomposed or reduced to other more desired forms. Examples are decomposition of carbon dioxide and RuO_4 and reduction of nitric oxide with ammonia. A maximum conversion of 28% in the CO_2 decomposition was obtained in a porous yttria-stabilized zirconia membrane reactor at 1509°C [Itoh et al., 1993]. This level of conversion is higher than what an equilibrium calculation will predict. RuO_4 was successfully transformed into RuO_2 and oxygen through the use of a porous membrane reactor made from iron oxide-molecular sieve-clay composites with iron oxide impregnated in the membrane pores. The conversion can be as high as 99.4% at 500°C [Peng et al., 1983]. Additionally, it has been demonstrated that the reduction of nitric oxide can be essentially 100% complete with ammonia at 300 to 350°C using gamma-alumina or titania membrane reactors with V_2O_5 as the catalyst [Zaspalis et al., 1991d]. The corresponding selectivity is 75-80%.

Using nearly straight-pore alumina membrane plates as both a separator and a catalyst, Furneaux et al. [1987] demonstrated the dehydration of isopropanol to form propene and hydrogenolysis of ethane to make methane. No quantitative information, however, was provided for the conversion or yield of the associated reactions.

One of the potentially wide-spread applications under development is catalytic filters for air pollution control. This combines separation and catalytic oxidation into one unit operation. One possibility is the oxidation of volatile organic carbon (VOC) by employing a porous honeycomb monolithic ceramic membrane filter. Inside the pores are deposited an oxidation catalyst such as precious metals. The resulting VOC removal efficiency can exceed 99% [Bishop et al., 1994].

8.6.4 Nonseparative Catalytic Reaction Applications

As briefly mentioned earlier in this chapter, the porous matrix of an inorganic membrane can be applied as a well-engineered catalytic reaction zone where the catalyst is immobilized on the pore surface or the membrane is inherently catalytic to the reaction involved.

The opposing-reactant geometry has been studied recently by several investigators for a number of catalytic reactions including destruction of NO_x [Zaspalis et al., 1991d], conversion of hydrogen sulfide to sulfur [Sloot et al., 1990 and 1992] and oxidation of carbon monoxide [Veldsink et al., 1992].

Apparently at a temperature above 300°C, the oxidation kinetics of NO_x and ammonia gas is so fast that slip of the reactants, when fed from the opposite sides of an alumina or alumina-titania membrane, can be avoided. Vanadium oxide, used as the catalyst for the reaction, is impregnated onto the membrane pore surface. The conversion of NO_x can reach 70% with the selectivity for nitrogen up to 75% in the temperature range of 300 to 350°C [Zaspalis et al., 1991d].

Sloot et al. [1990 and 1992] successfully investigated the reaction between hydrogen sulfide and sulfur dioxide at a temperature ranging from 200 to 270°C using the opposing-reactant mode. Alpha-alumina membrane was employed as the catalytic reaction zone which was impregnated with transition-phase alumina as the catalyst.

Similarly, the fast oxidation reaction of carbon monoxide proves to be amenable to the concept of opposing-reactant geometry [Veldsink et al., 1992]. In this case, alpha-alumina membrane pores are deposited with platinum as the catalyst for the reaction to proceed at about 250°C.

8.7 APPLICATIONS OF POROUS MEMBRANES IN LIQUID- AND MULTI-PHASE REACTIONS

Typically liquid-phase reactions do not require high temperatures, and as such organic membranes may be suitable for the membrane reactor applications. Justification of using inorganic membranes for these applications comes from such factors as better chemical stability and better control and containment of the catalysts.

8.7.1 Food and Biotechnology Processing

Enzymatic reactions are commonly observed or practiced in various kinds of food and biotechnology products. With the goals of reducing operating costs and improving product quality, a number of enzyme immobilization techniques have been developed in recent decades [Woodward, 1985]. The availability of robust membranes, particularly porous inorganic membranes, has improved the enzyme immobilization technology. One type of membrane bioreactors immobilizes enzyme in the membrane pores by dead-end filtration of the enzyme solution.

Enzyme immobilization. It has been known for some time that many enzyme-catalyzed processes such as hydrolysis and isomerization of low-molecular-weight substrates exhibit significantly higher reaction rates when the enzymes are immobilized in a membrane. Enzymes covalently bound to organic polymer membranes have been studied and successfully demonstrated for mostly integrated bioconversion and separation processes. Some of those membrane reactors are commercially practiced. A prototype of a forced-flow membrane bioreactor using an organic membrane has been commercially available and applied to the production of cyclo-dextrin from starch [Okada et al., 1987] and the hydrolysis of starch by glucoamylase [Sahashi et al., 1987]. Ceramic membranes in many cases are preferred over organic membranes due to their higher structural stability and better thermal and chemical resistances. These are important considerations for biocatalysts particularly in view of the requirement of membrane sterilization. It is expected that inorganic membranes will be applied to a large extent in this area of food and biotechnology in the future.

Two different biocatalytic reactions for low-molecular-weight substrates using ceramic membranes as the carrier to immobilize enzymes were studied by Nakajima et al. [1989a]: inversion of sucrose to form glucose and fructose using invertase as the enzyme and isomerization of glucose to produce fructose using glucose isomerase as the enzyme. Various ceramic membranes, many of which have asymmetric alumina microstructures and a pore diameter ranging from 0.04 to 10 μm, are first imparted with functional amino groups on their surfaces by treating with silanes followed by soaking in a 5% glutaraldehyde solution. The solutions of the enzymes, after pH adjustment, were then supplied to the membrane pores from the support layer side by applying a transmembrane pressure difference in a dead-end filtration mode. It was found that the above chemical immobilization procedures produced much higher concentrations of immobilized enzymes than physical immobilization schemes. Amino groups are thus bound to the treated surface of the membrane pores. A schematic diagram conceptually showing the difference in enzyme immobilization between a conventional column reactor and a membrane reactor is give in Figure 8.6 [Nakajima et al., 1988].

The substrate (sucrose or glucose) enters the membrane reactor in a crossflow filtration mode. It reacts while being forced through the membrane pores under an applied pressure difference from the side of the finer-pore layer. Conversion of the sucrose is essentially 100% and that of the glucose reaches beyond 35%. A ten-fold reaction rate was observed in a ceramic membrane reactor in contrast to a conventional column reactor where the enzymes were immobilized in porous beads [Nakajima et al., 1989b]. This significant increase in productivity is probably due to the high concentration of the enzyme in the porous ceramic membrane, which results in more effective contact between the substrate and the enzymes inside the membrane reactor. The typical residence time through the membrane is a few seconds or minutes compared to a few hours for the case of a column reactor. This characteristic makes the membrane bioreactors suitable for sanitary conditions. Diffusion of the substrate within the bead in a column reactor is rather limited. The reaction rate responds quickly to the applied pressure difference in a membrane reactor. Therefore, the applied pressure can be used to control the reaction which is different than the conventional enzyme reactor where temperature is often employed to control the reactor. Membranes with a mean pore diameter of about 0.5 μm are found to be most suitable for achieving high productivity of sucrose inversion as too small pores significantly reduce the permeate flux of the substrate solution.

By using the membrane bioreactors, even the impure substrate such as 50% molasses (containing particles and high-molecular-weight substances) produced successful results leading to 100% sucrose conversion [Nakajima et al., 1993]. The enzyme becomes deactivated after about 100 hours in the presence of impurities in the molasses solution. This deactivation time is compared to about 200 hours when a pure sucrose solution is used. The conversion of sucrose to ructooligosaccharides is also effectively improved with enzyme immobilized membrane reactors [Nakajima et al., 1990]. High potentials also exist for the enzymatic reactions of low-molecular-weight substrates using hydrolase or transferase as the enzyme.

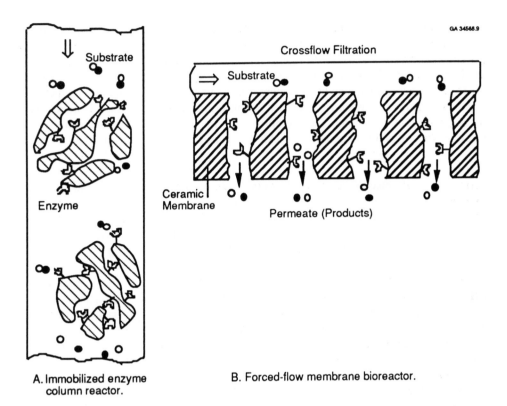

A. Immobilized enzyme column reactor.

B. Forced-flow membrane bioreactor.

Figure 8.6 Schematic difference between an enzyme-immobilized column reactor and a forced-flow membrane bioreactor [Nakajima et al., 1988]

8.7.2 Hydrocarbon Processing

Given in Table 8.10 are examples of the studies on the use of porous inorganic membranes in liquid-phase processes involving hydrocarbons.

Hydrogenation of dehydrolinalool via a porous stainless steel membrane reactor was found to offer high conversions and high selectivities [Gryaznov, 1992].

Many chemicals are produced from catalytic reactions between volatile and nonvolatile reactants and this type of multiphase reactions can benefit from the use of a membrane reactor. A class of reactions that fit this category is liquid phase oxidations and hydrogenations. For example, the hydrogenation of α-methylstyrene to cumene has been investigated in a tube consisting of a γ-alumina membrane layer inside an α-alumina macroporous support [Harold et al., 1989; Cini and Harold, 1991]. The membrane tube serves as a controlled gas-liquid contacting device where the gas stream flows through the tube core and the liquid stream containing the nonvolatile and dissolved volatile reactants flows on the shell side. Thus the volatile and nonvolatile reactants are well segregated. It has the apparent advantage when the feed liquid is not saturated with the gas reactant. The γ-alumina membrane is impregnated with a noble metal catalyst such as Pd. The liquid reactant fills the membrane pores so that the supply of either gas or liquid reactant to the reaction zone is more direct and efficient than in other multiphase reactor geometry such as slurry, fixed-bed bubble or trickle-bed reactors. It is found that, for the same catalyst volume, the overall reaction rate in a membrane tube can exceed that of a string of pellets (a prototype trickle-bed reactor) by up to a factor of 20.

TABLE 8.10
Liquid- and multi-phasehydrocarbon reactions using porous membrane reactors

Reaction	Catalyst	Membrane	Conclusions	Reference
Reaction of CH_4, C_2H_6 & C_3H_6 with H_2O_2 (T=120°C)	Nafion layer	C (disk)	Reaction promoted by Nafion	Parmaliana et al., 1992
Conversion of α– methylstyrene to cumene (T=40°C)	Pd	Al_2O_3 (tube)	Reaction rate increased up to 20 times that of trickle bed reactor	Cini and Harold, 1991
Hydrogenation of dehydrolinalool (T=173°C)	Pd powder	Stainless steel	C is 25 times that of a dense Pd-Ru membrane; S_m=95%	Gryaznov, 1992
Hydrogenation of nitrobenzoic acid (T=35-96°C)	Pd	Calcium-aluminum silicate (disk)	Gas-liquid reactor	De Vos and Hamrin, 1982; De Vos et al., 1982

Note: C=conversion, S=selectivity, Y=yield, subscript m denotes the maximum value and subscript e indicates the equilibrium value without membrane

8.7.3 Other Applications

Sewage wastes containing a large amount of fermentable organic components can be treated biologically under anaerobic conditions. In the anaerobic digestion process for sewage sludge, short-chain fatty and volatile acids are derived from soluble organic materials by acid-producing bacteria. A further step of gasification by methane-producing bacteria converts volatile acids to methane and carbon dioxide as the off-gas of the digester. Methane constitutes the major component of the off-gas with the remaining being carbon dioxide. This anaerobic digestion process generates methane as a fuel which, although having a lower heating value than natural gas, can be used to partially offset the energy costs required for the waste treatment.

A methane fermentation bioreactor in which a ceramic membrane separator is enclosed (Figure 8.2) was used for anaerobic digestion of a liquor obtained from a heat-treated sewage sludge [Kayawake et al., 1991]. The installation of the membrane module inside the fermentor is designed to offer a compact system and reduce the permeability power required. In essence, the fermentation step is followed by membrane separation and the retentate is recycled back to the fermentation tank. The liquor, with coarse solid particles removed, is fed to the digester at 35°C as a substrate. The fermentation broth in the digester approaches perpendicularly a bundle of ceramic membrane tubes having a mean pore diameter of 0.1 μm from the outer surfaces and is filtered by applying suction to the inside of the tubes. The properly filtered permeate liquor collected inside the membrane tubes can then be discharged to the environment or recycled to the digester. The produced gas, after desulfurization, is stored in a gas holder and then burned with a gas burner.

The outer surfaces of the membrane tubes are subject to the growth or microbial compaction when the crossflow velocity is low. This affects the permeate flux. To maintain the permeate flux at a reasonably high level by reducing the effects of these deposits, nitrogen gas is regularly utilized to provide backflushing. Some details of a typical backflushing system are given in Chapter 5. The permeate flux can also be increased by circulating the gas produced by the fermentation. This semi-continuous digester-separator setup was demonstrated to generate transparent liquor containing essentially no suspended solids and drastic decreases in COD, BOD and total organic carbon (TOC) by 80, 95 and 75%, respectively, compared to the influent to the fermentor. Although the final quality of the treated liquor did not reach the environmentally acceptable levels, further improvements are considered to be likely [Kayawake et al., 1991] and continuous methane fermentation by the combination of fermentor-membrane appears to have a higher treatment efficiency than other methods.

8.7.4 Phase-transfer Catalysis

A membrane can be used in the so-called phase transfer catalysis as a separator between two immiscible liquids or a liquid and a gas. It serves as a well controlled contact surface. An interesting type of membrane reactor has been suggested in which a ceramic membrane is applied to regulate the contact between a gas and a liquid stream on the opposites of the membrane [De Vos, 1982; De Vos et al., 1982]. Hydrogenation of nitrobenzoic acid can be effectively performed with a porous calcium-aluminum silicate membrane reactor which essentially becomes a gas-liquid reactor.

8.8 APPLICATIONS OF ELECTROCATALYTIC MEMBRANE REACTORS

The applications discussed so far all utilize pressure difference as the driving force to transport species across a membrane and no electrical effects occur. There is another class of membrane reactors that involve electrocatalytic effects through the use of two electrodes, generally made of metals, which are attached to a solid electrolyte membrane. In a closed circuit configuration, either electric current is generated as a result of a reaction taking place at one of the electrodes or it is applied to force the migration of a permeating ionic species through the membrane to effect a reaction at one of the electrodes. The former mode is called a fuel cell and the latter is called electrochemical pumping (EOP) (Figures 8.7a and 8.7b). The major difference between the two configurations lies in that EOP uses an electric current generator to effect reactions while a fuel cell converts fuel-initiated reactions to electric current.

These applications will be briefly treated in this section. As will become evident, the solid electrolyte membrane materials are either stabilized oxides or mixed oxides. Further details of science and technology of electrocatalytic membrane reactors beyond the scope of this chapter can be found in a number of excellent reviews [Gellings et al., 1988; Stoukides, 1988].

8.8.1 Electrochemical Oxygen Pumping (EOP)

When an electric current is imposed across a solid electrolyte membrane by the action of a current generator, the permeating species (often oxygen ions) migrates or "jumps" through the membrane toward or away from the catalyst which is either the cathode or anode. This is generally called electrochemical oxygen pumping (EOP). Thus a chemical reaction occurs on the catalyst electrode as a consequence of the addition or removal of the permeating reactant relative to the catalytic zone or the catalyst activity is modified due to the migration of the permeating species. The reaction conversion is proportional to the amount of the permeate molecules in the former case but not in the latter case.

Catalytic application studies employing EOP via solid electrolyte membranes are summarized in Table 8.11. EOP has been applied to decomposition reactions of NO, NO_x and SO_2 and oxidation reactions of a number of inorganic as well as organic compounds. The increase of reaction rates due to EOP (compared to the case when no EOP is applied) can be drastic. In some extreme cases, the rate increase is as high as six orders of magnitude. The conversion, however, is low in most cases. The effect of EOP on selectivity is not consistent: higher in some cases but lower in other. The solid electrolyte membranes used are all stabilized zirconia (mostly by yttria and in limited cases by calcium oxide or scandium oxide). The more commonly used catalysts include silver and platinum. It appears that EOP by solid oxide electrolyte membranes are still far from being practical for industrial applications until the low conversion issue is solved.

GA 35288.1

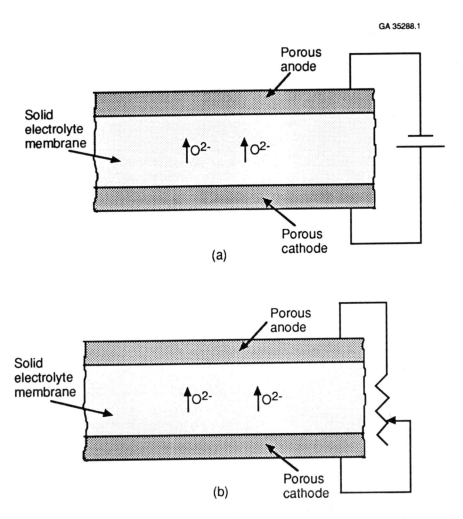

Figure 8.7 Schematic diagrams showing two electrocatalytic membrane reactor configurations: (a) electrochemical oxygen pumping and (b) solid oxide fuel cell operation.

8.8.2 Fuel Cells

Fuel cells are desirable for power generation and the development of solid oxide fuel cells (SOFC) has received considerable attention in recent years. The basic principle of SOFC is schematically shown in Figure 8.7b. SOFC is an electrochemical device for converting fuel into electrical energy. The membrane is sandwiched between two porous electrodes. An oxidant such as oxygen or air is fed to a porous cathode and the oxygen migrates through the solid oxide electrolyte membrane to the anode side as oxygen ions.

After releasing on the anode side, the oxygen reacts with the fuel (e.g., hydrogen, carbon monoxide or various gas mixtures containing methane, carbon dioxide, carbon monoxide, hydrogen, or hydrocarbons, etc.) to form combustion products.

TABLE 8.11
Catalytic applications of solid electrolyte membrane reactors using electrochemical oxygen pumping (EOP)

Reaction	Catalyst	Membrane	Conclusions	Reference
Decomposition of NO to N_2 & O_2 (T=600-800°C)	Pt	Sc_2O_3-ZrO_2 (disk)	Reaction rate up to 3 orders of magnitude higher by EOP	Pancharatnam et al., 1975
Decomposition of NO to N_2 & O_2 (T=900°C)	ZrO_2 catalytic	Sc_2O_3-ZrO_2 (disk)	Reaction rate up to 6 orders of magnitude higher by EOP	Gur and Huggins, 1979
Decomposition of NO_x to N_2 & O_2 (T=650-1050°C)	Transition metal oxide	Y_2O_3-ZrO_2 (honeycomb)	C=91%	Cicero and Jarr, 1990
Decomposition of SO_2 to S_8 & O_2 (T=650-1050°C)	Transition metal oxide	Y_2O_3-ZrO_2 (honeycomb)	C=38%	Cicero and Jarr, 1990
Methanation (T=500-1090°C)	Ni, Fe, Co & Pt	Y_2O_3-ZrO_2 (tube)	Higher S by EOP; too low CH_4 conc.	Gur and Huggins, 1986
Oxidation of CO to CO_2 (T=250-600°C)	Pt	Y_2O_3-ZrO_2 (disk)	Reaction rate up to 500 times higher by EOP	Yentekakis and Vayenas, 1988
Oxidation of CO, CH_4 & C_2H_4 (T=600-1000°C)	Ag, Pt, Ni, In_2O_3-SnO_2	CaO-ZrO_2 or Y_2O_3-ZrO_2 (tube)	Reaction rate increases exponentially with applied voltage	Otsuka et al., 1984
Oxidation of SO_2 to SO_3 (T=450-550°C)	Pt, V_2O_5	Y_2O_3-ZrO_2 (tube)	EOP enhances catalytic activity	Mari et al., 1979
Oxidation of CH_4 to C_2 hydrocarbons (T=700°C)	Ag-Bi_2O_3	Y_2O_3-ZrO_2 (tube)	C_m=2%;Y_m=1.8%; S_m≥90%; C higher, S lower & Y unchanged by EOP	Otsuka et al., 1985

TABLE 8.11 – Continued
Catalytic applications of solid electrolyte membrane reactors using electrochemical oxygen pumping (EOP)

Reaction	Catalyst	Membrane	Conclusions	Reference
Oxidation of CH_4 to C_2 hydrocarbons (T=700°C)	Ag-Li/MgO	Y_2O_3-ZrO_2 (disk)	EOP to membrane favors CO_2; EOP from membrane favors CO	Seimanides et al., 1986
Oxidation of CH_4 to C_2 hydrocarbons (T=750-900°C)	Ag, Ag-Bi_2O_3, Cu or Fe	Y_2O_3-ZrO_2 (tube)	High S (>90%) only at low C; Fe catalyzes syngas	Eng and Stoukides, 1988
Oxidation of CH_4 to C_2 hydrocarbons (T=700-900°C)	Ag-Bi_2O_3, Ag-Li/MgO, Ag-Sn_2O_3, Cu, Mo-Ag, Pt, Pt-Bi_2O_3	Y_2O_3-ZrO_2 (disk)	C_m=12%; Y_m=3.1%	Mazanec, 1988
Oxidation of CH_4 to C_2 hydrocarbons (T=700°C)	Al-Li/Cl/NiO	Y_2O_3-ZrO_2 (disk)	C_m=5%; S_m=92%	Otsuka et al., 1988
Oxidation of CH_4 to C_2 hydrocarbons (T=700°C)	Bi_2O_3-Pr_6O_{11}	Y_2O_3-ZrO_2 (tube)	C=4%; Y=1.6%	Steele et al., 1988
Oxidation of CH_4 to C_2 hydrocarbons (T=735-750°C)	Ag	Y_2O_3-ZrO_2 (tube)	C<0.1%; EOP decreases S	Nagamoto et al., 1990
Oxidation of CH_3OH to HCHO, CO & CH_4 (T=550-700°C)	Ag	Y_2O_3-ZrO_2 (disk)	C up to 6 times higher by EOP to anode	Neophytides and Vayenas, 1989
Oxidation of CH_3OH to HCHO, CO & CH_4 (T=327-627°C)	Pt	Y_2O_3-ZrO_2 (disk)	Reaction rate up to 15 times higher by EOP	Vayenas and Neophytides, 1991
Oxidation of C_2H_4 to CH_3CHO (T=400°C)	Pt	Y_2O_3-ZrO_2 (disk)	Y=4.2% & S=0.67% by EOP (vs. Y=2% & S=0.59% without EOP)	Stoukides and Vayenas, 1981

TABLE 8.11 – Continued
Catalytic applications of solid electrolyte membrane reactors using electrochemical
oxygen pumping (EOP)

Reaction	Catalyst	Membrane	Conclusions	Reference
Oxidation of C_2H_4 to CH_3CHO (T=277-452°C)	Pt	Y_2O_3-ZrO_2 (tube)	Reaction rate up to 50 times higher by EOP	Bebelis and Vayenas, 1989
Oxidation of C_2H_4 to CH_3CHO (T=368-470°C)	Ag	Y_2O_3-ZrO_2 (disk)	S for C_2H_4O lower by EOP to catalyst	Bebelis and Vayenas, 1992
Oxidation of C_3H_6 to C_2H_5CHO (T=390-400°C)	Ag	Y_2O_3-ZrO_2 (tube)	S increases by EOP	Stoukides and Vayenas, 1984
Oxidation of C_3H_6 to CH_2CHCHO (T=450°C)	MoO_3-Bi_2O_3 on Ag or Au	Y_2O_3-ZrO_2 (disk)	S=60, 30 & 10% for CH_2CHCHO, CO_2 & CO, repectively	Hayakawa et al., 1987
Oxidation of ethylbenzene to styrene (T=575-600°C)	Pt	Y_2O_3-ZrO_2 (tube)	Reaction rate up to 600 times higher but S lower by EOP	Michaels and Vayenas, 1984

Note: *C*=conversion, *S*=selectivity, *Y*=yield, subscript *m* denotes the maximum value and subscript *e* indicates the equilibrium value without membrane

The chemical potentials difference between the two electrodes caused by the reaction generates an electromotive force. Through an external load, a circuit is complete. The chemical energy from the oxidation of the fuel is thus transformed into electrical energy.

The membrane should possess high anionic conductivity at high temperatures under highly oxidizing and reducing atmosphere and should be essentially gas impermeable. Stabilized zirconia systems are the best known materials and have been extensively studied because of their anionic conductivity over a wide range of oxygen partial pressures. Other candidates include CaO-ZrO_2, Sc_2O_3-ZrO_2, ThO_2-Y_2O_3, ThO_2-CeO, ThO_2-La_2O_3, Bi_2O_3-LaO, some of their ternary oxide mixtures and substituted perovskites.

The potential of fuel cells for simultaneously generating power and performing a chemical reaction such as partial oxidation is considered to be one of the most promising technologies. The reaction can produce some valuable intermediate chemicals or remove certain pollutants. An example of the latter is a perovskite fuel cell for hydrogen sulfide oxidation and electricity generation. The chemically aggressive nature of hydrogen sulfide poses a problem of material selection for the anode and membrane and will be discussed in Chapter 9.

One of the most advanced processes for deriving energy from coal is fuel cells. Solid oxide membranes are the core to this direct energy conversion device. Hydrogen (e.g., from a coal gasifier) as a fuel is fed to the anode side while oxygen or air is introduced at the cathode. Oxygen passes through the oxygen ion-conducting membrane and reacts with hydrogen to form water and release the chemical energy in the form of electrical energy through an external circuit connected to the two electrodes. In the fuel cell configuration, the oxygen-selective membrane serves as a species separator and a partition for the reactor.

Summarized in Table 8.12 are the literature data on fuel cell applications using solid oxide electrolyte membranes. In addition to the most studied yttria-stabilized zirconia, calcium oxide- and scandium oxide-stabilized zirconia have also been investigated with limited success. The temperature range within which the stabilized zirconia electrolyte membranes are operative is about 700 to 1,000°C. The high temperatures are required to achieve a power density even reasonably high enough to be considered for potential applications. This is a consequence of the relatively low electrical conductivity of typical solid electrolytes [Gur and Huggins, 1983]. In fact, one of the critical factors in determining the future success of SOFC is the oxygen ion resistance of the dense solid oxide membrane. Therefore, the thinner the membrane is, the higher the membrane permeability (and thus the efficiency of SOFC). Higher selectivities are also observed with thinner membranes. During oxidation of ammonia, the selectivity for NO improves as the yttria-stabilized zirconia membrane becomes thinner and can be as high as 97% using a 200 μm thick membrane [Farr and Vayenas, 1980; Vayenas and Ortman, 1981].

The catalyst employed depends on the reaction involved although Pt, Ag and Au are frequently used. Different catalysts for otherwise the same fuel cell and reaction can lead to rather different product distributions. Ethane and ethylene are the products from partial oxidation of methane using Bi_2O_3-LaO as the catalyst in a stabilized zirconia membrane fuel cell. It has been shown that when iron catalyst is used instead as the anode, the product contains synthesis gas (CO and hydrogen) as well as ethane and ethylene [Eng and Stoukides, 1988]. Some solid electrolytes can also function as catalysts for certain reactions. For example, Di Cosimo et al. [1986] discovered that Bi_2O_3-LaO, being a solid electrolyte itself, plays a central role in the catalysis of selective oxidation of propylene due to the lattice oxygen in the electrolyte. The catalysts can be deposited on the solid electrolyte membranes by applying slurries containing the materials followed by drying and calcining in air at high temperature.

In many oxidation reactions, the conversion and selectivity from a fuel cell operation tend to compromise each other. At a high selectivity, the conversion is often low and vice versa. Another important issue related to solid oxide fuel cells is that heat from many exothermic partial oxidation reactions does not provide sufficient energy to justify the process. The resultant power density is usually low. The above two issues can be appreciated by the results from a study of partial oxidation of methane [Eng and Stoukides, 1988].

TABLE 8.12
Fuel cell applications using dense membrane reactors

Reaction	Catalyst	Membrane	Conclusions	Reference
Oxidation of H_2 to H_2O (T=810-1094°C)	Pt	CaO-ZrO$_2$ (tube)	Productivity limited by electrical resistance of membrane	Weissbart and Ruka, 1962
Oxidation of H_2 to H_2O (T=708-840°C)	Pt	Y$_2$O$_3$-ZrO$_2$ (monolith)	Higher power density with a crossflow fuel cell	Michaels et al., 1986
Oxidation of H_2 & CO (T=1000°C)	Ni-Zr cermet	Y$_2$O$_3$-ZrO$_2$ (tube)	Industrial-scale fuel cell	Isenberg, 1981
Oxidation of NH$_3$ to N$_2$ & NO (T=627-927°C)	Pt	Y$_2$O$_3$-ZrO$_2$ (tube)	S =60-70% for NO	Farr and Vayenas, 1980
Oxidation of NH$_3$ to N$_2$ & NO (T=727-927°C)	Pt or Pt-Rh	Y$_2$O$_3$-ZrO$_2$ (disk)	C_m=39%; S_m=97%; poer density=0.6 mW/cm^2	Vayenas and Ortman, 1981
Oxidation of CH$_4$ (T=750-900°C)	Ag, Ag-Bi$_2$O$_3$ or Fe	Y$_2$O$_3$-ZrO$_2$	S_m>90%; C_m=<2%); syngas produced when Fe is used	Eng and Stoukides, 1988
Oxidation of CH$_4$ (T=760°C)	Pt-Sm$_2$O$_3$-La$_{0.83}$Sr$_{0.1}$MnO$_3$	Y$_2$O$_3$-ZrO$_2$ (disk)	Low C; S>90%	Pujare and Sammells, 1988
Oxidation of CH$_4$ (T=840°C)	Ag-Pd	Y$_2$O$_3$-ZrO$_2$	C=24%; Y=10%	Belyaev et al., 1989
Oxidation of CH$_4$ (T=800°C)	KF, BaCO$_3$, NaCl, Bi$_2$O$_3$ or Sm$_2$O$_3$ on Au	Y$_2$O$_3$-ZrO$_2$	C=1.9%; Y=1.6%	Otsuka et al., 1990
Oxidation of CH$_4$ (T=750-840°C)	Ag	Y$_2$O$_3$-ZrO$_2$ (disk)	C=4%; Y=2.5%	Belyaev et al., 1990
Oxidation of H_2, CO & CH$_4$ (T=700°C)	Pt or Au	Sc$_2$O$_3$-ZrO$_2$ (disk)	Reactivity:H_2> CO>CH$_4$	Goffe and Mason, 1981

TABLE 8.12 - Continued
Fuel cell applications using dense membrane reactors

Reaction	Catalyst	Membrane	Conclusions	Reference
Oxidation of H_2, CO, CH_4, CH_3OH & C_2H_5OH (T=700-850°C)	Pt or Au	Sc_2O_3-ZrO_2 (disk)	"Blackened" electrolyte is more effective	Nguyen et al., 1986
Oxidation of CH_4 & NH_3 to HCN (T=800-1000°C)	Pt-Rh	Y_2O_3-ZrO_2 (tube)	S>75% for HCN; power density=10 mW/cm^2	Kiratzis and Stoukides, 1987
Steam reforming (T=800-900°C)	Ni-cermet anode	Y_2O_3-ZrO_2	C increases with increasing catalyst electrode potential	Vayenas et al., 1992

Note: C=conversion, S=selectivity, Y=yield, subscript m denotes the maximum value and subscript e indicates the equilibrium value without membrane

Methane can be catalytically oxidized in the fuel cell mode to simultaneously generate electricity and C_2 hydrocarbons by dimerization of methane using a yttria-stabilized zirconia membrane. A catalyst, used as the anode, is deposited on the side of the membrane that is exposed to methane and the cathode is coated on the other side of the membrane. When the catalyst Ag-Bi_2O_3 is used as the anode for the reaction at 750-900°C and atmospheric total pressure, the selectivity to ethane and ethylene exceeds 90%. But this high selectivity is at the expense of low power output and low overall methane conversion (less than about 2%).

An extensive review has been made on catalytic and electrocatalytic methane oxidation with solid oxide membranes including the fuel cell mode [Eng and Stoukides, 1991].

There is a class of nonporous materials called proton conductors which are made from mixed oxides and do not involve transport of molecular or ionic species (other than proton) through the membrane. Conduction of protons can enhance the reaction rate and selectivity of the reaction involved. Unlike oxygen conductors, proton conductors used in a fuel cell configuration have the advantage of avoiding dilution of the fuel with the reaction products [Iwahara et al., 1986]. Furthermore, by eliminating direct contact of fuel with oxygen, safety concern is reduced and selectivity of the chemical products can be improved. The subject, however, will not be covered in this book.

8.9 CURRENT SATE OF TECHNOLOGY AND CHALLENGES

Despite their potential advantages over organic membranes because of their inherent stabilities, inorganic membranes as separators and reactors at the current stage of development are subject to several material, catalysis and engineering limitations. Those limitations which will be treated in more detail in subsequent chapters are more a reflection of the early stage of developments than practical or theoretical limits. It is expected that more interest and demonstrated feasibilities will motivate more focused development efforts to address those issues.

8.10 SUMMARY

The core to the concept of an inorganic membrane reactor is to steer a membrane-selective species of a reaction mixture in or out of the membrane, which also serves as part of the reactor boundaries, to improve either the reaction conversion or product distribution. This is accomplished primarily by equilibrium displacement due to the permselective removal of a product or by avoiding or reducing undesirable side reactions (between certain chemical species including intermediate products) via controlled addition of some key reactant to the reaction zone through the membrane.

Enhancement in conversion by the usage of a membrane reactor has been demonstrated for many dehydrogenation reactions. Product selectivity of some hydrogenation and other reactions are found to improve with a permselective membrane as part of the reactor. Several dense metal as well as solid electrolyte membranes and porous metal as well as various oxide membranes have been discovered to be effective for the reaction performance.

In addition, solid electrolyte membranes potentially can be used for electrochemical oxygen pumping and as fuel cells which can produce chemicals while generating electricity.

While the aforementioned and other novel membrane reactors hold great promises, many material, catalysis and engineering issues need to be fully addressed before the inorganic membrane reactor technology can be implemented in an industrial scale. This is particularly true for many bulk-processing applications at high temperatures and often harsh chemical environments. Those issues will be treated in the subsequent chapters.

REFERENCES

Abe, F., 1987, European Pat. Appl. 228,885.
Aquilar, G., V.M. Gryaznov, L.F. Pavlova and V.D. Yagodovski, 1977, React. Kinet. Catal. Lett. **7**, 181.
Al-Shammary, A.F.Y., I.T. Caga, J.M. Winterbottom, A.Y. Tata and I.R. Harris, 1991, J. Chem. Tech. Biotechnol. **52**, 571.

Andersen, A., I.M. Dahl, K.J. Jens, E. Rytter, A. Slagtern and A. Solbakken, 1989, Catal. Today **4**, 389.

Anshits, A.G., A.N. Shigapov, S.N. Vereshchagin and V.N. Shevnin, 1989, Kinet. Catal. **30**, 1103.

Armor, J.N., and T.S. Farris, 1993, Membrane catalysis over palladium and its alloys, in: Proc. 10th Int. Congr. Catal., Budapest, Hungary (1992), p1363.

Balachandran, U., S.L. Morissette, J.T. Dusek, R.L. Mieville, R.B. Poeppel, M.S. Kleefisch, S. Pei, T.P. Kobylinski and C.A. Udovich, 1993, Development of ceramic membranes for partial oxygenation of hydrocarbon fuels to high-value-added products, in: Proc. Coal Liquefaction and Gas Conversion Contractors' Review Conf., Pittsburgh, 1993 (U.S. Dept. of Energy).

Balachandran, U., J.T. Dusek, S.M. Sweeney, R.P. Poeppel, R.L. Mieville, P.S. Maiya, M.S. Kleefisch, S. Pei, T.P. Kobylinski, C.A. Udovich and A.C. Bose, 1995, Am. Ceram. Soc. Bull. **74**, 71.

Basov, N.L., and V.M. Gryaznov, 1985, Dehydrogenation of cyclohexanol and hydrogenation of phenol into cyclohexanone on a membrane catalyst produced from a palladium-ruthenium alloy, in: Membrane Catalyst Permeable to Hydrogen or Oxygen (V.M. Gryaznov, Ed.), Akad. Nauk SSSR, Inst. Neftekhim. Sint., Moscow, USSR, p. 117.

Bebelis, S., and C.G. Vayenas, 1989, J. Catal. **118**, 125.

Bebelis, S., and C.G. Vayenas, 1992, J. Catal. **138**, 588.

Belfort, G., 1989, Biotechnol. Bioeng. **33**, 1047.

Belyaev, V.D., V.A. Sobyanin, V.A. Arzhannikov and A.D. Neuimin, 1989, Dokl. Acad. Nauk. SSSR **305**, 1389.

Belyaev, V.D., O.V. Bazman, V.A. Sobyanin and V.N. Parmon, 1990, St. Surf. Sci. Catal. **55**, 469.

Bishop, B., R. Higgins, R. Abrams and R. Goldsmith, 1994, Compact ceramic membrane filters for advanced air pollution control, in: Proc. 3rd Int. Conf. Inorg. Membr., Worcester, MA, USA.

Bitter, J.G.A., 1988, U.K. Pat. Appl. 2,201,159A.

Bratzler, R.L., S.L. Matson, J.L. Lopez and S.A. Wald, 1986, New reactor systems for the large scale stereoselective synthesis and separation of optically active fine chemicals, World Biotech. Rep. **2**, 89.

Caga, I.T., J.M. Winterbottom and I.R. Harris, 1989, Catal. Lett. **3**, 309.

Cales, B., and J.F. Baumard, 1982, High Tem.-High Press. **14**, 681.

Cannon, K.C., and J.J. Hacskaylo, 1992, J. Membr. Sci. **65**, 259.

Champagnie, A.M., T.T. Tsotsis, R.G. Minet and I.A. Webster, 1990, Chem. Eng. Sci. **45**, 2423.

Champagnie, A.M., T.T. Tsotsis, R.G. Minet and E. Wagner, 1992, J. Catal. **134**, 713.

Chan, K.K., and A.M. Brownstein, 1991, Ceram. Bull. **70**, 703.

Chang, H.N., 1987, Biotechnol. Adv. **5**, 129.

Chen, S., H. Fan and Y.K. Kao, 1992, Chem. Eng. J. **49**, 35.

Cheryan, M., and M.A. Mehaia, 1986, Chemtech (November), 676.

Cicero, D.C., and L.A. Jarr, 1990, Sep. Sci. Technol. **25**, 1455.

Cini, P., and M.P. Harold, 1991, AIChE J. **37**, 997.

Clayson, D.M., and P. Howard, 1987, U.K. Pat. Appl. 2,190,397A.

De Vos, R., and C.E. Hamrin, Jr., 1982, Chem. Eng. Sci. **37**, 1711.

De Vos, R., V. Haziantoniou and N.H. Schoon, 1982, Chem. Eng. Sci. **37**, 1719.

Di Cosimo, R., J.D. Burrington and R.K. Grasselli, 1986, U.S. Patent 4,571,443.

Drioli, E., V. Violante and A. Basile, 1990, in: Proc. 5th World Congress on Filtration, Nice, France, p. 449.

Edlund, D.J., W.A. Pledger, B.M. Johnson and D.T. Friesen, 1992, Toward economical and energy-efficient production of hydrogen from coal using metal-membrane reactors (paper 11F), presented at 5th NAMS Annual Meeting, Lexington, Kentucky, USA.

Edlund, D.J., and W.A. Pledger, 1993, J. Membr. Sci. **77**, 255.

Eng, D., and M. Stoukides, 1988, Partial oxidation of methane in a solid electrolyte cell, in: Proc. 9th Int. Congr. Catal., Calgary, Canada (Chem. Soc. Canada), p. 974.

Eng, D., and M. Stoukides, 1991, Catal. Rev. - Sci. Eng. **33**, 375.

Ermilova, M.M., N.V. Orekhova, L.S. Morozova and E.V. Skakunova, 1985, Membr. Katal. 70.

Farr, R.D., and C.G. Vayenas, 1980, J. Electrochem. Soc. **127**, 1478.

Farris, T.S., and J.N. Armor, 1993, Appl. Catal. **96**, 25.

Flaschel, E., and C. Wandrey, 1979, DECHEMA-Monogr. **84**, 337.

Fleming, H.L., 1990, Dehydration and recovery of organic solvents by pervaporation, in: Proc. 8th Annual BBC Membr. Planning Conf., Cambridge, MA, USA, p. 293.

Foley, H.C., A.W. Wang, B. Johnson and J.N. Armor, 1993, Effect of a model hydrogenation on a catalytic palladium membrane, in: Proc. ACS Symp. Selectivity in Catal., **517**, 168.

Furneaux, R.C., A.P. Davidson and M.D. Ball, 1987, European Pat. Appl. 244,970.

Gallaher, G.R., T.E. Gerdes and P.K.T. Liu, 1993, Sep. Sci. Technol. **28**, 309.

Gellings, P.J., H.J.A. Koopmans and A.J. Burggraaf, 1988, Appl. Catal. **39**, 1.

Goffe, R.A., and D.M. Mason, 1981, J. Appl. Electrochem., **11**, 447.

Gosser, L.W., W.H. Knoth and G.W. Parshall, 1977, J. Mol. Catal. **2**, 253.

Gryaznov, V.M., 1969, U.S.S.R. Authors Certif. No. 27,4092.

Gryaznov, V.M., 1986, Plat. Met. Rev. **30**, 68.

Gryaznov, V.M., 1992, Plat. Met. Rev. **36**, 70.

Gryaznov, V.M., and A.N. Karavanov, 1979, Khim.-Farm. Zh. **13**, 74.

Gryaznov, V.M., and M.G. Slin'ko, 1982, Faraday Disc. Chem. Soc. **72**, 73.

Gryaznov, V.M., and V.S. Smirnov, 1974, Russ. Chem. Rev. **43**, 821.

Gryaznov, V.M., V.S. Smirnov and M.G. Slin'ko, 1973, Heterogeneous catalysis with reagent transfer through the selectively permeable catalysts, in: Proc. 5th Int. Congr. Catal. **2**, 1139.

Gryaznov, V.M., V.P. Polyakova, E.M. Savitsky and E.V. Khrapova, 1977, U.S. Patent 4,026,958.

Gryaznov, V.M., V.S. Smirnov, V.M. Vdovin, M.M. Ermilova, L.D. Gogua, N.A. Pritula and I.A. Litvinov, 1979, U.S. Patent 4,132,668.

Gryaznov, V.M., V.S. Smirnov and M.G. Slin'ko, 1981, Stud. Surf. Sci. Catal. **7** (pt. A), 224.

Gryaznov, V.M., T.M. Belosljudova, A.P. Maganjuk, A.N. Karavanov, A.M. Ermolaev, and I.K. Sarycheva, 1982, U.K. Pat. Appl. 2,096,595.

Gryaznov, V.M., A.N. Karavanov, T.M. Belosljudova, A.M. Ermolaev, A.P. Maganjuk and I.K. Sarycheva, 1983a, U.S. Patent 4,388,479.

Gryaznov, V.M., V.S. Smirnov, V.M. Vdovin, M.M. Ermilova, L.D. Gogua, N.A. Pritula and G.K. Fedorova, 1983b, U.S. Patent 4,394,294.

Gryaznov, V.M., V.I. Vedernikov and S.G. Gul'yanova, 1986, Kinet. Catal. **26**, 129.

Gryaznov, V.M., A.P. Mischenko and M.E. Sarylova, 1987, U.K. Pat. Appl. 2,187,758A.

Gryaznov, V.M., O.S. Serebryannikova, Yu.M. Serov, M.M. Ermilova, A.N. Karavanov, A.P. Mischenko and N.P. Orekhova, 1993, Appl. Catal. **96**, 15.

Gur, T.M., and R.A. Huggins, 1979, J. Electrochem. Soc. **126**, 1067.

Gur, T.M., and R.A. Huggins, 1983, Science **219**, 967.

Gur, T.M., and R.A. Huggins, 1986, J. Catal. **102**, 443.

Gur'yanova, O.S., Yu.M. Serov, S.G. Gul'yanova and V.M. Gryaznov, 1988, Kinet. Catal. **28**, 728.

Gur'yanova, O.S., Yu.M. Serov, S.G. Gul'yanova and V.M. Gryaznov, 1989, Kinet. Catal. **30**, 406.

Guther, W., and W. Vielstich, 1982, Electrochemica Acta **27**, 811.

Haag, W.O., and J.G. Tsikoyiannis, 1991, U.S. Patent 5,019,263.

Harold, M.P., P. Cini, B. Patenaude and K. Venkataraman, 1989, AIChE Symp. Ser. **268**, 26.

Hayakawa, T., T. Tsunoda, H. Orita, T. Kameyama, H. Takahashi, K. Fukuda and K. Takehira, 1987, J. Chem. Soc. Chem. Commun. , 780.

Hazbun, E.A., 1989, U.S. Patent 4,827,071.

Hsu, C., and R.E. Bauxbaum, 1986, J. Nucl. Mater. **141/143**, 238.

Ioannides, T., and G.R. Gavalas, 1993, J. Membr. Sci. **77**, 207.

Isenberg, A.O., 1981, Sol. St. Ionics **3/4**, 431.

Itoh, N., Y. Shindo, K. Haraya, K. Obata, T. Hakuta and H. Yoshitome, 1985, Int. Chem. Eng. **25**, 138.

Itoh, N., 1987a, AIChE J. **33**, 1576.

Itoh, N., 1987b, Properties of a composite charged reverse osmosis membrane, presented at Int. Congr. Membr. Membr. Processes, Tokyo, Japan, p. 352.

Itoh, N., Y. Shindo, K. Haraya and T. Hakuta, 1988, J. Chem. Eng. Japan **21**, 399.

Itoh, N., M.A. Sanchez C., W.C. Xu, K. Haraya and M. Hongo, 1993, J. Membr. Sci. **77**, 245.

Iwahara, H., T. Esaka, H. Uchida, T. Yamauchi and K. Ogaki, 1986, Sol. St. Ionics **18&19**, 1003.

Jennings, J.F., and R.C. Binning, 1960, U.S. Patent 2,956,070.

Kameyama, T., M. Dokiya, M. Fujishige, H. Yokokawa and K. Fukuda, 1981, Ind. Eng. Chem. Fundam. **20**, 97.

Kameyama, T., M. Dokiya, M. Fujishige, H. Yokokawa and K. Fukuda, 1983, Int. J. Hydrogen Energy **8**, 5.

Karavanov, A.N., and V.M. Gryaznov, 1984, Kinet. Catal. **25**, 60.

Kayawake, E., Y. Narukami and M. Yamagata, 1991, J. Ferment. Bioeng. **71**, 122.

Kim, J.S., and R. Datta, 1991, AIChE J. **37**, 1657.

Kimura, S., 1985, Desalination **53**, 279.

Kiratzis, N., and M. Stoukides, 1987, J. Electrochem. Soc. **134**, 1925.

Liu, Y., A.G. Dixon, Y.H. Ma and W.R. Moser, 1990, Sep. Sci. Technol. **25**, 1511.

Mari, C.M., A. Molteni and S. Pizzini, 1979, J. Electrochem. Acta **24**, 745.

Matson, S.L., and J.A. Quinn, 1986, Ann. NY Acad. Sci. **469** (Biochem. Eng. 4), 152.

Matsuda, T., I. Koike, N. Kubo and E. Kikuchi, 1993, Appl. Catal. **96**, 3.

Mazanec, T., 1988, U.S. Patent 4,793,904.

Michaels, J.N., and C.G. Vayenas, 1984, J. Electrochem. Soc. **131**, 2544.

Michaels, J.N., C.G. Vayenas and L.L. Hegedus, 1986, J. Electrochem. Soc. **133**, 522.

Mikhalenko, N.M., E.V. Khrapova and V.M. Gryaznov, 1986, Kinet. Katal. **27**, 138.

Mischenko, A.P., V.M. Gryaznov, V.S. Smirnov, E.D. Senina, I.L. Parbuzina, N.R. Roshan, V.P. Polyakova and E.M. Savitsky, 1979, U.S. Patent 4,179,470.

Mischenko, A.P., V.M. Gryaznov, M.E. Sarylova and V.A. Bednyakova, 1987, U.K. Pat. Appl. 2,187,759A.

Moser, W.R., Y. Becker, A.G. Dixon and Y.H. Ma, 1992, Catalytic inorganic membranes and their modifications for use in catalyzed chemical processes, presented at 5th North American Membrane Society Annual Meeting, Lexington, KY, paper 11E.

Nagamoto, H., and H. Inoue, 1981, J. Chem. Eng. Japan **14**, 377.

Nagamoto, H., and H. Inoue, 1985, Chem. Eng. Commun. **34**, 315.

Nagamoto, H., and H. Inoue, 1986, Bull. Chem. Soc. Japan **59**, 3935.

Nagamoto, H., K. Hayashi and H. Inoue, 1990, J. Catal. **126**, 671.

Nakajima, M., N. Jimbo, K. Nishizawa, H. Nabetani and A. Watanabe, 1988, Process Biochem. **23**, 32.

Nakajima, M., N. Jimbo, H. Nabetani and A. Watanabe, 1989a, Use of ceramic membrane for enzyme reactors, in: Proc. 1st Int. Conf. Inorg. Membr., Montpellier, France (Trans Tech Publ., Zürich) p. 257.

Nakajima, M., A. Watanabe, N. Jimbo, K. Nishizawa and S. Nakao, 1989b, Biotechnol. Bioeng. **33**, 856.

Nakajima, M., K. Nishizawa, H. Nabetani and A. Watanabe, 1990, Food Biotechnol. **4**, 365.

Nakajima, M., K. Nishizawa and H. Nabetani, 1993, Bioproc. Eng. **9**, 31.

Nazarkina, E.B., and N.A. Kirichenko, 1979, Khim. Technol. Topl. Masel. **3**, 5.

Neophytides, S., and C.G. Vayenas, 1989, J. Catal. **118**, 147.

Nguyen, B.C., T.A. Lin and D.M. Mason, 1986, J. Electrochem. Soc. **133**, 1807.

Nigara, Y., and B. Cales, 1986, Bull. Chem. Soc. Jpn. **59**, 1997.

Nogerbekov, B.Yu., N.N. Gudeleva and R.G. Mustafina, 1984, Tr. Inst. Org. Katal. Elektrokhim. Akad. Nauk Kaz. SSR **23**, 214.

Nozaki, T., S. Hashimoto, K. Omata and K. Fujimoto, 1993, Ind. Eng. Chem. Res. **32**, 1174.

Oertel, M., J. Schmitz, W. Weirich, D. Jendryssek-Neumann and R. Schulten, 1987, Chem. Eng. Technol. **10**, 248.

Okada, T., M. Ito, H. Ishizuka and K. Hibino, 1987, in: Proc. Agricultural Chemical Soc. Japan, Tokyo, Japan, p. 379.

Okubo, T., K. Haruta, K. Kusakabe, S. Morooka, H. Anzai and S. Akiyama, 1991, Ind. Eng. Chem. Res. **30**, 614.

Omata, K., S. Hashimoto, H. Tominaga and K. Fujimoto, 1989, Appl. Catal. **52**, L1.

Orekhova, N.V., and N.A. Makhota, 1985, Effect of hydrogen removal through membrane of palladium alloys on dehydrogenation of butane (in Russian), in: Membrane Catalysts Permeable to Hydrogen or Oxygen (V.M. Gryaznov, Ed.) Akad. Nauk SSSR, Inst. Neftekhim. Sint., Moscow, USSR, p. 49.

Otsuka, K., S. Yokoyama and A. Morikawa, 1984, Bull. Chem. Soc. Japan **57**, 3286.

Otsuka, K., S. Yokoyama and A. Morikawa, 1985, Chem. Lett. , 319

Otsuka, K., K. Suga and I. Yamanaka, 1988, Catal. Lett. **1**, 423.

Otsuka, K., K. Suga and I. Yamanaka, 1990, Catal. Today **6**, 587.

Pancharatnam, S., R.A. Huggins and D.M. Mason, 1975, J. Electrochem. Soc. **127**, 869.

Parmaliana, A., F. Frusteri, F. Arena and N. Giordano, 1992, Catal. Lett. **12**, 353.

Peng, N., F. Wang and D. Zhou, 1983, Fushe Fanghu {Radiation Protection} **3**, 154.

Pfefferle, W.C., 1966, U.S. Patent 3,290,406.

Philpott, J.E., 1975, Plat. Met. Rev. **20**, 110.

Philpott, J.E., 1985, Plat. Met. Rev. **29**, 12.

Pujare, N.U., and A.F. Sammells, 1988, J. Electrochem. Soc. **135**, 2544.

Rony, P.R., 1968, Sep. Sci. **3**, 239.

Roth, J.F., 1988, Future catalysis for the production of chemicals, in: Catalysis 1987 (Elsevier, Amsterdam) p. 925.

Roth, J.F., 1990, St. Surf. Sci. Catal. **54**, 3.

Sahashi, H., T. Okada and K. Hibino, 1987, in: Proc. Agricul. Chem. Soc. Japan, Tokyo, Japan, p. 380.

Saracco, G., and V. Specchia, 1994, Catal. Rev.-Sci. Eng. **36**, 305.

Sarylova, M.E., A.P. Mishchenko, V.M. Gryaznov and V.S. Smirnov, 1977, Izv. Akad. Nauk SSSR Ser. Khim., No. 2, 430.

Seimanides, S., M. Stoukides and A. Robbat, 1986, J. Electrochem. Soc. **133**, 1535.

Seok, D.R., and S.T. Hwang, 1990, St. Surf. Sci. Catal. **54**, 248.

Shinji, O., M. Misono and Y. Yoneda, 1982, Bull. Chem. Soc. Japan **55**, 2760.

Skakunova, E.V., M.M. Ermilova and V.M. Gryaznov, 1988, Bull. Acad. Sci. USSR Chem. Ser. **5**, 858.

Sloot, H.J., G.F. Versteeg and W.P.M. van Swaaij, 1990, Chem. Eng. Sci. **45**, 2415.

Sloot, H.J., C.A. Smolders, W.P.M. van Swaaji and G.F. Versteeg, 1992, AIChE J. **38**, 887.

Smirnov, V.S., V.M. Gryaznov, V.I. Lebedeva, A.P. Mischenko, V.P. Polyakova and E.M. Savitsky, 1977, U.S. Patent 4,064,188.

Smirnov, V.S., M.M. Ermilova, V.M. Gryaznov, N.V. Orekhova, V.P. Polyakova, N.R. Roshan and E.M. Savitsky, 1983, U.K. Patent 2,068,938.

Sokol'skii, D.V., B. Yu. Nogerbekov and L.A. Fogel, 1988, Zh. Fiz. Khim. **61**, 677.

Song, J.Y., and S.T. Hwang, 1991, J. Membr. Sci. **57**, 95..

Steele, B.H.C., I. Kelly, H. Middleton and R. Rudkin, 1988, Sol. St. Ionics **28/30**, 1547.

Stoukides, M., 1988, Ind. Eng. Chem. Res. **27**, 1745.

Stoukides, M., and C.G. Vayenas, 1981, J. Catal. **70**, 137.

Stoukides, M., and C.G. Vayenas, 1984, J. Electrochem. Soc. **131**, 839.

Sun, Y.M., and S.J. Khang, 1988, Ind. Eng. Chem. Res. **27**, 1136.

Suzuki, H., 1987, U.S. Patent 4,699,892.

Tiscareno-Lechuga, F., G.C. Hill, Jr. and M.A. Anderson, 1993, Appl. Catal. **96**, 33.

Touzani, A., D. Klvana and G. Bélanger, 1984, Int. J. Hydrogen Energy **9**, 929.

Tsotsis, T.T., A.M. Champagnie, S.P. Vasileiadis, E.D. Ziaka and R.G. Minet, 1992, Chem. Eng. Sci. **47**, 2903.

Uemiya, S., T. Matsuda and E. Kikuchi, 1990, Chem. Lett. 1335.

Uemiya, S., N. Sato, H. Ando and E. Kikuchi, 1991a, Ind. Eng. Chem. Res. **30**, 585.

Uemiya, S., N. Sato, H. Ando, T. Matsuda and E. Kikuchi, 1991b, Appl. Catal. **67**, 223.

Vayenas, C.G., and D. Ortman, 1981, U.S. Patent 4,272,336.

Vayenas, C.G., and S. Neophytides, 1991, J. Catal. **127**, 645.

Vayenas, C.G., S. Bebelis, I.V. Yentekakis and H.G. Linz, 1992, Catal. Today **11**, 303.

Veldsink, J.W., R.M.J. van Damme, G.F. Versteeg and W.P.M. van Swaaij, 1992, Chem. Eng. Sci. **47**, 2939.

Weissbart, J., and R. Ruka, 1962, J. Electrochem. Soc. **109**, 723.

Wood, B.J., 1968, J. Catal. **11**, 30.

Woodward, J. (Ed.), 1985, Immobilized Cells and Enzymes - A Practical Approach, IRL Press, Oxford, UK.

Wu, J.C.S., T.E. Gerdes, J.L. Pszczolkowski, R.R. Bhave, P.K.T. Liu and E.S. Martin, 1990, Sep. Sci. Technol. **25**, 1489.

Yehenskel, J., D. Leger and F. Courvoisier, 1979, Adv. Hydrogen Energy Progr. **2**, 569.

Yentekakis, I.V., and C.G. Vayenas, 1988, J. Catal. **111**, 170.

Yolles, R.S., B.J. Wood and H. Wise, 1971, J. Catal. **21**, 66.

Yoshida, H., S. Konishi and Y. Naruse, 1983, Nucl. Technol. Fusion **3**, 471.

Zaho, R., and R. Govind, 1990, Sep. Sci. Technol. **25**, 1473.

Zaspalis, V.T., K. Keizer, J.G. van Ommen, W. van Praag, J.R.H. Ross and A.J. Burggraaf, 1988, presented at AIChE National Meeting, Denver, CO.

Zaspalis, V.T., W. van Praag, K. Keizer, J.G. van Ommen, J.R.H. Ross and A.J. Burggraaf, 1991a, Appl. Catal. **74**, 205.

Zaspalis, V.T., W. van Praag, K. Keizer, J.G. van Ommen, J.R.H. Ross and A.J. Burggraaf, 1991b, Appl. Catal. **74**, 223.

Zaspalis, V.T., W. van Praag, K. Keizer, J.G. van Ommen, J.R.H. Ross and A.J. Burggraaf, 1991c, Appl. Catal. **74**, 235.

Zaspalis, V.T., W. van Praag, K. Keizer, J.G. van Ommen, J.R.H. Ross and A.J. Burggraaf, 1991d, Appl. Catal. **74**, 249.

Zhao, R., N. Itoh and R. Govind, 1990a, Am. Chem. Soc. Sym. Ser. **437**, 216.

Zhao, R., R. Govind and N. Itoh, 1990b, Sep. Sci. Technol. **25**, 1473.

Zhernosek, V.M., N.N. Mikhalenko, S. Iokhannes, E.V. Khrapova and V.M. Gryaznov, 1979, Kinet. Catal. **20**, 753.

Ziaka, Z.D., R.G. Minet and T.T. Tsotsis, 1993, J. Membr. Sci. **77**, 221.

CHAPTER 9

INORGANIC MEMBRANE REACTORS -- MATERIAL AND CATALYSIS CONSIDERATIONS

One of the most important reasons for considering inorganic membranes for either separation or reaction, particularly the latter, is their generally high thermal, chemical and structural resistances. The material aspects of their use are paramount: how they interact with the process streams and respond to the operating conditions and environments. Catalysis considerations are equally important as they directly influence the reaction conversion and selectivity. The interplay between catalysis and separation brings new opportunities as well as challenges. The design, installation and operation of inorganic membrane reactors without considering both types of issues are doomed to fail.

9.1 MATERIAL CONSIDERATIONS IN APPLICATION ENVIRONMENT

The issues related to containment of reaction components are sometimes underestimated in their importance. Their consequences, however, will surface quickly early in full implementation. Some general material characteristics have been discussed in Chapter 4, but they are mostly limited to the temperature and chemical environment typical of traditional separation applications. The material aspects of the membrane related to high-temperature reactive and permeating conditions will be reviewed in this section.

9.1.1 Permeability and Permselectivity

Permselectivity is crucial to the utility of any types of membranes. If the permselectivity toward a particular reaction species is high, the separation is quite clean and the need for further separation processing downstream of the membrane reactor is reduced. When a permeate of very high purity is required in some cases, dense membranes are preferred. While a high permselectivity is generally desirable, there may be situations where a high permeate flux in combination with a moderate permselectivity is a better alternative to a high permselectivity with a low permeability, particularly when recycle streams are used.

Permeability vs. permselectivity dilemma. Like its organic counterpart, an inorganic membrane generally does not exhibit a high permeability and a high permselectivity at the same time. Typically dense membranes are associated with relatively high permselectivities but relatively low permeabilities and porous membranes are just the opposite. This type of permeability versus permselectivity dilemma of organic membranes for gas separation has been solved in some cases by molecularly engineering the chemical nature of the membrane material through functional group substitutions or modifications.

This can be illustrated by the performance of "typical" commercial organic polymer membranes in separating the CO_2/CH_4 gas mixture. Shown in Figure 9.1 is the so-called trade-off curve between the ideal separation factor (an indicator of permselectivity) of CO_2/CH_4 and the CO_2 permeability [Koros and Fleming, 1993]. The curve represents a trend line to be expected of the separation characteristics of typical commercially available membranes including polycarbonate. By tailoring the chemical structure of the polycarbonate (as indicated by the three points above the curve), the limits set by the conventional membrane fabrication chemistry can be removed. Thus membranes with both high permselectivity and high permeability are technically feasible. It may be expected that a similar type of material design will break the barrier of the compromise between the two important separation properties.

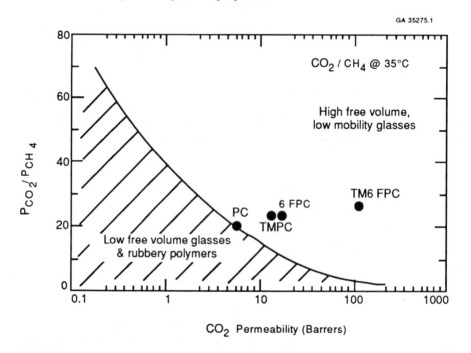

Figure 9.1 "Trade-off" curve between the ideal separation factor of CO_2/CH_4 and the CO_2 permeability [Koros and Fleming, 1993]

Material design and selection. The permeate flux of a membrane has an enormous impact on the process economics. This can be particularly essential when dense membranes are used due to their extremely high permselectivity. When choosing a dense membrane among several candidate materials with comparable permselectivities, the one with a higher flux is preferred.

For example, in selecting a solid state oxide membrane for oxygen transport, the oxygen permeability of a perovskite-type oxide membrane can be two orders of magnitude higher than that of stabilized zirconia [Balachandran et al., 1993]. Moreover, the oxygen permeance of a yttria-stabilized zirconia (YSZ) membrane can be substantially increased by doping the YSZ with selected metals. When titanium is introduced into the YSZ membranes, the oxygen permeance can be increased about 100 times over the undoped YSZ membranes [Arashi and Naito, 1992].

A further example is the traditionally favored dense Pd membrane. Palladium has limited catalytic activity for a variety of hydrogenation reactions and can suffer from hydrogen embrittlement. And it is quite expensive and not easy to fabricate compared to other metals. Possible alternative hydrogen-permeable metals are vanadium, tantalum and niobium. Recognizing this, Edlund and Pledger [1993] prepared a composite membrane with vanadium as the base metal because of its high permeability to hydrogen. The high hydrogen permeability is primarily due to the high diffusivity of hydrogen in vanadium, particularly at relatively low temperatures [Takata and Suzuki, 1993]. In addition, compared to palladium, vanadium is less expensive and not prone to hydrogen embrittlement under certain conditions. Beside vanadium, other selected metals may be used to substitute Pd. Tantalum has been suggested to be used as the core material for highly hydrogen-selective composite membranes. It is much less expensive than Pd and apparently diffuses hydrogen rapidly [Anonymous, 1995].

It is well known that a pure palladium membrane suffers from cracking, rupture or other mechanical failures and a proper second metal to be alloyed with Pd enhances its mechanical strength. The secondary and, when present, the ternary metal can have a significant influence over the resulting hydrogen permeability. It appear that the effects can be either enhancement or retardation. Given in Table 9.1 are the hydrogen permeabilities of various Pd alloys relative to that of pure palladium at 350°C and under 20 bars [Itoh, 1990]. While certain metals at a small to large dosage (e.g., 10% Y or 23% Ag) can increase the hydrogen permeability, other metals (e.g., Ru at 5% or 10% Ni) create the opposite effects.

When designing palladium alloy membranes, the composition of the alloying elements is optimized by balancing various membrane performance considerations. For the dehydrogenation of borneols and the hydrogenation of cyclopentadiene, Smirnov et al. [1983] determined the optimal composition of the secondary metal (Ni, Rh or Ru) for palladium-based membranes by considering mechanical properties and hydrogen permeability of the membranes. On one hand, the quantity of Ni, Rh or Ru needs to be greater than 5% to impart satisfactory mechanical properties in the presence of hydrogen. On the other hand, the Ni, Rh or Ru content in the alloy should be less than 10% by weight to avoid an undesirable reduction in the hydrogen permeability.

Chapter 9

Membrane treatment. When using dense palladium-based membranes, activation pretreatment at a high temperature (such as 265°C) in air followed by hydrogen (for about an hour or so in each step) yields a higher permeability than no activation, particularly at a lower temperature. The activation may need to be repeated a few times to be effective. Shown in Tables 9.2 and 9.3 are hydrogen permeation data of activated and non-activated pure Pd and Pd-Ru(5%) and Pd-Ru(5%)-Ag(30%) membranes as a function of the application temperature [Armor and Farris, 1993]. It is evident that within the temperature range studied the permeability increases with increasing temperature. A Pd-based membrane with activation treatment can offer a 30 to 50% greater hydrogen permeability compared to that without activation.

TABLE 9.1
Effects of alloying metal on the hydrogen permeability of a palladium-based membrane

Alloying metal content (%)[a]	Relative hydrogen permeability[b]
Y (10%)	3.76
Ag (23%)	1.73
Ce (7.7%)	1.57
Au (5%)	1.06
Cu (40%)	1.06
B (0.5%)	0.94
Ru (5%)	0.33
Ni (10%)	0.19

Note: a: Remaining content being Pd; b: Relative permeability = hydrogen permeability of Pd alloy membrane/hydrogen permeability of pure Pd
[Itoh, 1990]

TABLE 9.2
Hydrogen permeation flux data of activated and non-activated pure palladium foil membranes (100 μm thick)

Temperature (°C)	Hydrogen permeation flux (μmoles/cm^2-min)	
	Non-activated	Activated
107	8.6	4.7
157	17	17
206	21	22
257	24	25

Note: Activation in air then in 50% H_2/N_2 both at 265°C for 1 hour; feed rate=64 ml/min of 50% H_2/N_2; permeate sweep rate=20 ml/min
[Armor and Farris, 1993]

Membrane thickness. As expected, as the membrane thickness increases, the permeability decreases. At a temperature of about 100°C, a four-fold reduction of the Pd-membrane thickness results in a corresponding four-fold increase in the hydrogen permeability. As the temperature increases, however, this difference decreases with increasing temperature.

There has been a large volume of data showing the benefit of having thin dense membranes (mostly Pd-based) on the hydrogen permeation rate and therefore the reaction conversion. An example is catalytic dehydrogenation of propane using a ZSM-5 based zeolite as the catalyst and a Pd-based membrane. Clayson et al. [1987] selected a membrane thickness of 100 μm and achieved a yield of aromatics of 38% compared to approximately 80% when a 8.6 μm thick membrane is used [Uemiya et al., 1990].

TABLE 9.3
Hydrogen permeation flux data (μmoles/cm^2-min) of activated and non-activated palladium alloy membranes

Temp. (°C)	100 μm thick Pd-Ru(5%)	25 μm thick Pd-Ru(5%)		25 μm thick Pd-Ru(5%)-Ag(30%)	
		Non-activated	Activated	Non-activated	Activated
101	3.9	3.4	15	2.0	13
152	8.6	16	22	14	34
205	13	25	28	33	40
255	17	29	33	36	42

Note: Activation in air then in 50% H_2/N_2 both at 265°C for 1 hour; feed rate=61 ml/min; permeate sweep rate=20 ml/min
[Armor and Farris, 1993]

The effects of the membrane thickness on the reactor performance, however, are not as straightforward as they first appear to be. The reaction conversion and selectivity in a membrane reactor largely depend on the permeation rate of a reaction component relative to the reaction rate of that component. The permeation rate of a gas is influenced by the porosity, pore diameter and thickness of the membrane, particularly the last two variables. How the permeated species are removed from the permeate side in the case of the reaction taking place on the feed side also affects the reactor performance.

When a sweep gas is used, a typical family of conversion curves for a membrane reactor where a product is selectively removed to effect equilibrium displacement to increase reaction conversion are shown in Figure 9.2 for dehydrogenation of cyclohexane with Ar as the sweep gas [Itoh et al., 1985]. When no membrane is employed in the reactor, the maximum achievable conversion is that of the equilibrium value. When a permselective membrane is used, the conversion exceeds the equilibrium limit. As the membrane thickness decreases, the hydrogen permeation rate increases thus leading to a conversion increase. However, as a thinner membrane is used, not only the hydrogen permeation rate becomes higher but also the permeation rate of the reactant cyclohexane. As the loss of unconverted cyclohexane to the permeate side increases, the total conversion of cyclohexane on both the feed and the permeate sides begins to decrease. Therefore, an optimal membrane thickness corresponding to a maximum conversion exists for a constant feed rate of cyclohexane. This maximum conversion is equivalent to the minimum combined molar flow rate of cyclohexane in the tube and shell regions. Thus, in principle, an optimal membrane thickness can be fabricated for a particular application during membrane preparation.

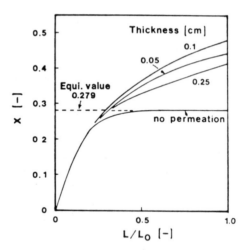

Figure 9.2 Calculated total conversion profile of cyclohexane to benzene in a porous shell-and-tube Vycor glass membrane reactor with membrane thickness as a parameter [Itoh et al., 1985]

In Figure 9.3, the conversion of thermal decomposition of carbon dioxide in a dense yttria-stabilized zirconia membrane reactor is shown. Similarly to the above case, when Ar is utilized to sweep the permeate, a membrane thickness exists that gives a maximum conversion for a given sweep gas flow rate and carbon dioxide feed rate. This is attributed to back permeation of oxygen to the reaction side downstream from the reactor entrance due to a higher oxygen partial pressure on the permeate side than that on the reaction side. The oxygen permeated back to the reaction side causes the reverse reaction to occur, thus reducing the conversion of carbon dioxide. If, on the other hand, vacuum is applied to the permeate side, the permeated species are quickly removed which leaves no opportunity for some of the permeated species to back permeate into the feed side. In this case, the reaction conversion consistently increases as the membrane thickness decreases.

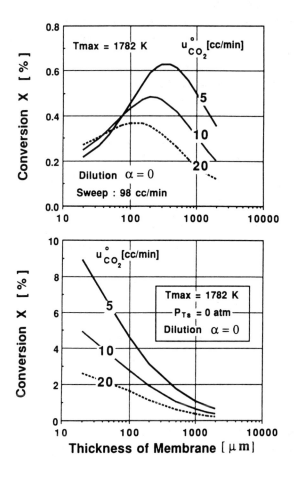

Figure 9.3 Conversion of thermal decomposition of carbon dioxide in a dense yttria-stabilized zirconia membrane reactor as a function of membrane thickness when a sweep gas is used (top) and when vacuum is applied (bottom) [Itoh et al., 1993]

Composite membrane structure. In the area of inorganic membranes, it has long been recognized that some dense membranes such as palladium-based membranes and zirconia membranes display very high selectivity for hydrogen and oxygen, respectively. Their permeability, however, is much lower than those of porous membranes. In contrast, porous inorganic membranes, with some exceptions, generally do not have as high a permselectivity as dense membranes. A potential solution to the aforementioned permeability versus permselectivity dilemma facing inorganic membranes is to employ a hybrid membrane in which a dense membrane forms a composite with a dense or porous support which have higher permeabilities. In this case the dense membrane dissociates the hydrogen molecule on one surface of the composite membrane and recombines hydrogen atoms on the other side. The permselective membrane layers should be as thin as possible for higher hydrogen permeation provided physical integrity is maintained to avoid pinhole defects.

Dense metals other than Pd can transport hydrogen more rapidly and thus can be used as the support for Pd membrane. An example is tantalum. Composite dense membranes have been made by coating both sides of a 13-μm thick tantalum foil with a 1-μm layer of pure Pd with a specific crystallographic orientation. The coating can be applied by physical vapor deposition in a vacuum of 10^{-4} torr. The resulting permeability is an order of magnitude higher and the cost is two orders of magnitude lower than those of a pure Pd [Anonymous, 1995]. Along the line of employing a porous support, thin dense palladium films have been deposited on porous glass and ceramic supports based on this concept [Uemiya et al., 1991a and 1991b; Collins and Way, 1993].

Furthermore, many inorganic membranes utilize the concept of an asymmetric or composite structure where one or more porous support layers provide the necessary mechanical strength while a defect-free separating membrane layer, as thin as possible, is responsible for the permselectivity. A mechanically strong support is especially critical when dealing with high-pressure high-temperature gas-phase membrane reactor applications. The thin layer and the defect-free integrity of the membrane as required by the permeability and permselectivity considerations, respectively, are only possible with a bulk support that possesses the following characteristics: being smooth, defect-free, highly porous (having a lower flow resistance compared to the membrane layer), strong bonding with the membrane and mechanically strong under pressure (no compaction). A severe consequence of using rough-surface support is the local distortion of the deposited membrane which is deleterious to permselectivity [Saracco and Specchia, 1994].

Membrane surface contamination. Although not as hydrogen selective as Pd and its alloys, other metals such as niobium and vanadium in dense form also have moderate to high hydrogen permselectivity and potentially can be considered as membrane materials. Inevitably the membrane surface is contaminated with non-metal impurities prior to or during separation or membrane reactor applications.

How severely the contaminants affect the hydrogen permeation flux depends on the nature of the contaminant and the membrane material. Surface contamination has been suggested to increase the rate of hydrogen permeation through metal membranes for exhaust gas pumping and purification for fusion reactors [Livshits et al., 1990]. It has been observed that, for a copper membrane, less than one monolayer of sulfur on the membrane surface increases the hydrogen permeation flux considerably [Yamawaki et al., 1994].

The presence of non-metal contaminants, however, may actually reduce hydrogen permeation rate. Using niobium membranes, Yamawaki et al. [1994] found that continuous exposure of the membrane surface to oxygen gas results in reduction of the hydrogen permeation flux. Moreover, when the same niobium membranes are in contact with hydrocarbons, the resulting surface concentration of carbon and the corresponding effects on hydrogen permeation vary with the type of hydrocarbon. For example, acetylene causes a substantial increase of carbon concentration on the membrane surface which leads to a decrease of hydrogen permeation flux through the dense Nb membrane. Contrary to acetylene, methane does not result in any appreciable increase of surface concentration of carbon or change of hydrogen permeation rate. The effects of carbon contaminants are very relevant to membrane reactor applications which involve many hydrocarbon reactions.

Similarly, it is difficult to prepare vanadium surface free of oxygen, sulfur and carbon which are considered to be barriers to hydrogen absorption [Takata and Suzuki, 1993].

It has been proposed that the surface impurity composition on a metal membrane can be controlled by the combination of thermal annealing and ion bombardment [Yamawaki et al., 1987].

9.1.2 High Temperature Effects

When using commercial inorganic membranes for separation or membrane reactor applications at relatively high temperatures, say, greater than 400°C or so, care should be taken to assess their thermal or hydrothermal stability under the application conditions. This is particularly relevant for small pore membranes because they are often made at a temperature not far from 400°C. Even if such an exposure does not yield any phase changes, there may be particle coarsening (and consequently pore growth) involved.

Thermal stability. Thermal stability of several common ceramic and metallic membrane materials has been briefly reviewed in Chapter 4. The materials include alumina, glass, silica, zirconia, titania and palladium. As the reactor temperature increases, phase transition of the membrane material may occur. Even if the temperature has not reached but is approaching the phase transition temperature, the membrane may still undergo some structural change which could result in corresponding permeability and permselectivity changes. These issues for the more common ceramic membranes will be further discussed here.

Alumina membranes have been the most common among commercial inorganic membranes. General chemistry and phase transformation of alumina have also been extensively studied. Many phases of alumina exist and the temperature ranges of their occurrence are determined by their thermal as well as chemical history, as illustrated in Figure 9.4 [Wefers and Misra, 1987]. The alpha phase is the only thermodynamically stable oxide of aluminum. Therefore, alpha-alumina membranes are thermally very stable for practically any gas separation or membrane reactor applications. Other metastable phases, however, are present in some commercial membranes. Gamma-alumina membranes and their modifications are such an example. Upon heating, gamma alumina, one of the poorly crystalline transition aluminas, undergoes transition to delta (between 700 and 800°C) to theta (between 900 and 1,000°C) and finally to alpha phase (between 1,000 and 1,100°C). The gamma phase is tetragonal in crystal structure, delta phase is orthorhomic or tetragonal, theta phase is monoclinic phase and alpha phase is hexagonal [Wefers and Misra, 1987].

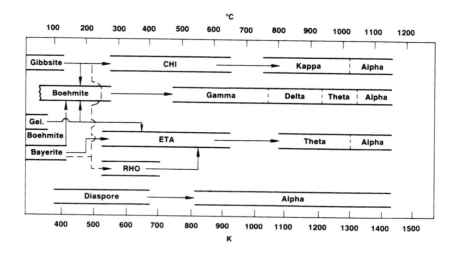

Figure 9.4 Phase transition of alumina [Wefers and Misra, 1987]

The actual phase transition temperatures of alumina membranes have been studied by various investigators and are found to semi-quantitatively agree with the above ranges [Leenaars et al., 1984; Larbot et al., 1988; Burggraaf et al., 1989; Chai et al., 1994]. The X-ray diffraction patterns of a sol-gel derived alumina membrane as a function of the calcination temperature are presented in Figure 9.5 [Chai et al., 1994]. The poorly crystalline alumina at 500°C goes through the gamma-alumina stage before transformation to the crystalline alpha phase around 1,000°C. The membrane pore size becomes significantly larger as alumina is transformed to the alpha phase. This is shown in Table 9.4 for a sol-gel derived alumina membrane heat treated at different temperatures for 34 hours [Burggraaf et al., 1989]. The pore diameter slowly increases from 2.5 nm as boehmite to 4.8 nm as gamma-alumina at 800°C to 5.4 nm as theta-alumina at 900°C. Then the membrane pore diameter changes abruptly to about 78 nm as alpha-alumina at 1,000°C. The corresponding changes in porosity are not as drastic.

Titania membranes prepared at a temperature lower than about 350°C are essentially amorphous. At 350°C or so, phase transition to a crystalline phase of anatase begins to occur. The transformation to anatase (tetragonal in crystallinity) is complete and the new phase of rutile (also tetragonal) begins at a temperature close to 450°C. Transformation to rutile is complete at about 600°C [Xu and Anderson, 1989]. Thus, at a temperature between 450 and 600°C, both anatase and rutile phases are present. It has been suggested that this temperature range may be lower at 350-550°C [Larbot et al., 1986; Burggraaf et al., 1989].

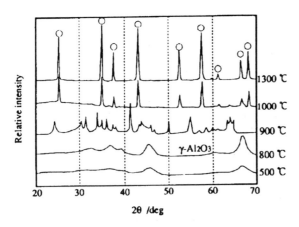

Figure 9.5 X-ray diffraction patterns of alumina membrane after calcination at different temperatures (open circle "o" indicating alpha-alumina) [Chai et al., 1994]

A sharp change in porosity of the titania membranes occurs at 350°C. This may be attributed to the crystallization of the constituent titania particles [Xu and Anderson, 1994]. Thus crystallization is likely to be a key factor affecting thermal stability of sol-gel derived titania membranes.

Silica membranes have also been studied by several investigators for use in gas separation and membrane reactors. They are thermally very stable up to about 500°C. Sintering and densification temperatures of silica membranes depend on the water/alkoxide ratio in the sol-gel process for making the membranes [Langlet et al., 1992]. Crystallization of amorphous silica particles in the membranes takes place at temperatures around 1,000°C [Larbot et al., 1989]. However, pore growth can gradually increase the resulting membrane permeability in the temperature range of 500 to 900°C [Hyun and Kang, 1994].

TABLE 9.4
Changes of pore diameter and porosity of alumina membranes heat treated at different temperatures

Temperature (°C)	Phase	Pore diameter (nm)	Porosity (%)
200	boehmite	2.5	41
400	γ-Al$_2$O$_3$	2.7	53
500	γ-Al$_2$O$_3$	3.2	54
800	γ-Al$_2$O$_3$	4.8	55
900	θ-Al$_2$O$_3$	5.4	48
1,000	α-Al$_2$O$_3$	78	41

Note: Heat treatment time = 34 hours
[Adapted from Burggraaf et al, 1989]

Finally, interest in zirconia membranes has increased in recent years. Zirconia exhibits three well-defined phases in the order of increasing temperature: the monoclinic, tetragonal and cubic phases. However, it has been suggested that a low-temperature metastable tetragonal phase exists in contrast to the high-temperature tetragonal phase [Cot, 1991; Colomban and Bruneton, 1992]. For pure zirconia membranes, it appears that the following phase transitions occur [Stevens, 1986; Colomban and Bruneton, 1992]:

Amorphous to metastable tetragonal (low-temperature) - about 430°C
Metastable tetragonal (low-temperature) to monoclinic - about 600-800°C
Monoclinic to tetragonal (high-temperature) - about 950-1230°C
Tetragonal (high-temperature) to cubic - about 2370°C

The transition temperature from the low-temperature tetragonal phase to the monoclinic phase occurs at a lower temperature (600-700°C) for the more acidic gels (pH = 1.5 and 6.0) than the more basic gels (pH = 12) which transforms at 800°C [Colomban and Bruneton, 1992]. Like in other ceramic systems, a small addition of other oxide can change the nature of the phase(s) formed at a given temperature. For example, with 10% molar GeO$_2$ present, the zirconia cubic phase can exist with tetragonal which is stable up to 1,300°C [Colomban and Bruneton, 1992].

An important implication of the phase transformation and crystallization discussed above is that they are often accompanied by a volume change which can cause shrinkage or expansion and consequently cracking. This is attributed to the different theoretical densities of different phases. A notable example is the transition from the high-temperature tetragonal to the monoclinic phase upon cooling. The transition is associated with a large volume expansion, approximately 3 to 5%. This can exceed elastic and fracture limits of the membranes, thus leading to cracks [Stevens, 1986]. The cracks can be avoided by adding the so-called stabilizing agents, typically such oxides as yttria, ceria, calcia or magnesia.

There have been many evidences indicating that the thermal stability of a ceramic material depends not only on its precursor, but also any dopants, the size of the phase transforming crystals and the applied stress [Colomban and Bruneton, 1992]. For example, a pure alumina membrane, after exposed to 1,000°C, experiences significant shrinkage and grain growth, thus leading to crack formation in the membrane. If, however, the membrane is doped with BaO or La_2O_3, very little phase transition occurs and the membrane microstructure is not significantly altered at the same temperature [Chai et al., 1994]. Apparently, the gamma to alpha phase transition is based on a nucleation and growth mechanism and some alkaline earth or rare earth oxides suppress the phase transformation by inhibiting the nucleation of α-phase. Thus, the doped alumina membranes have higher thermal stability than undoped membranes.

The membrane should be operated far from the limits set by the thermal stability consideration. Moreover, the maximum operable temperature of an inorganic membrane reactor may very well be dictated by the consideration of other materials in direct or indirect contact with the process stream. Some of these materials are used for the end seal and joining of the membrane and the module housing.

Examining the response of the membrane to a temperature range defined by the operating window of a membrane reactor should be one of the minimum and earliest characterization steps.

Hydrothermal stability. Steam or water vapor is sometimes present in the reaction mixture not necessarily as a reactant or product. It is a convenient form of heat source for endothermic reactions such as dehydrogenation of hydrocarbons. It can also oxidize carbon deposits that deactivate many catalysts. Steam is regularly used in the dehydrogenation of ethylbenzene to produce styrene for the above reasons.

Hydrothermal stability of the membrane in a membrane reactor under an environment like this is a critical factor in its service life as well as performance. Water is known to promote phase transition of metal oxides at a lower temperature.

Gamma alumina membranes have been commercially available for some time. Compared to alpha-alumina membranes, these membranes have a much more limited thermal and hydrothermal resistance. For this reason, Wu et al. [1990b] exposed the membranes to the application temperature of 600 to 640°C in the presence of water vapor for 2 hours prior to applying commercial gamma-alumina membranes with an approximate pore diameter of 4 nm in dehydrogenation of ethylbenzene. It is found that the modified membrane has a higher nitrogen permeability after the exposure. Once modified, the membrane shows the same nitrogen permeability before and after each subsequent membrane reactor test. The estimated pore diameter increases from 4 nm to 6-9 nm, resulting from the modification. The membrane layer thickness and its overall microstructure do not show any noticeable changes.

Porous glass membranes are known to suffer from partial silica dissolution upon long exposure to water especially at high temperatures. This can result in undesirable structural changes which, in turn, affect the separation properties of the membranes. Chemical treatments can be applied to inhibit the silica dissolution process. These treatments include surface modification of the glass membranes [Schnabel and Vaulont, 1978], pretreating the feed stream with a chemical such as aluminum chloride [Ballou et al., 1971] and the introduction of zirconia to the glass composition [Yazawa et al., 1991].

Exposure to moisture at a temperature in the 200-300°C range tends to enhance the tetragonal to monoclinic transition which is detrimental to the mechanical strength of a zirconia membrane. This effect of water promoting phase transition at a lower temperature is common among metal oxides. Therefore, such long-term exposure tends to cause a drastic reduction in fracture strength [Stevens, 1986].

Silica structure normally densifies above 800°C. The presence of water, however, may have accelerated the densification process. Silica-containing membranes are not thermally stable if water is present even at 600°C. They are subject to permanent structural densification. This structural change causes reduction in the membrane permeability.

Matching of thermal expansion coefficients. In addition to possible phase transition or structural change problems described above, there is also the issue of matching the thermal expansion coefficients of the components in a composite membrane which is composed of layers of different materials. If the thermal expansion coefficients of two adjacent layers of materials as part of the membrane element differ appreciably, any thermal cycles in normal operation or erratic temperature excursions can induce significant stresses on the components and jeopardize the physical integrity of the membrane.

The issue of mismatch of thermal expansion coefficients similar to that for a composite membrane is also very critical for fuel cells. In the fuel cells, electrodes are attached to solid electrolyte membranes. Significant temperature variations during applications, pretreatments or regeneration of the membranes (e.g., decoking) can cause serious mechanical problems associated with incompatible thermal expansions of different components. A possible partial solution to the above problem is to use partially stabilized instead of fully stabilized zirconia. The former has a significantly lower thermal expansion coefficient than the latter.

Furthermore, at high temperatures, the effect of thermal expansion on the overall dimensions of the membrane element and module must be taken into considerations. This is particularly true for those endothermic reactions which require that the membrane modules be heated, for example, in a furnace. Given in Table 9.5 are the mean thermal expansion coefficient data from various literature sources for a wide variety of membrane, sealing and piping materials within the general temperature range of 25 to 1,000°C. Most of the data are applicable only to dense materials. While the data may be used as a guide when designing the membrane reactor, it is noted that the coefficient data vary not only with the temperature but also with the literature source [mostly from Perry and Chilton, 1973; Lay, 1983]. It is obvious from Table 9.5 that to connect a ceramic membrane element to a stainless steel, for example, the significant difference in the thermal expansion coefficients needs to be addressed. Some of the solutions are discussed later in this chapter.

TABLE 9.5
Thermal expansion coefficient and thermal shock resistance data for various materials in an inorganic membrane reactor

Material	Mean thermal expansion coeff. ($\times 10^{-6}/°C$)	Thermal shock resistance
Alumina (dense)	8-8.5	Good
Carbon (graphite)	2-5	Excellent
Carbon-48% porosity (graphite)[a]	3.6	
Carbon	2.3-3.6	Excellent
Carbon-48% porosity[a]	4.7	
Cordierite	2.6	Excellent
Chromium oxide	7.8	Poor
Magnesia	13.9	Poor
Mullite	5.0	Good
Silica	0.5-0.8	Excellent
Silicon carbide	4.8	Excellent
Silicon nitride	3.0	Excellent
Spinel	9.1	Moderate
Stabilized zirconia (dense)	10	Moderate
Thoria	9.4	Poor
Tin oxide	4.2	Good
Titania (rutile)	9.1	Poor
Buna-N[a]	80-100	
Teflon[a]	100-190	
Glass[a]	3.2	Poor
Palladium	12	
Silver[c]	19	
Carbon steel	12-14	
Stainless steel 304[a]	17	

TABLE 9.5 – Continued
Thermal expansion coefficient and thermal shock resistance data for various materials
in an inorganic membrane reactor

Material	Mean thermal expansion coeff. (x10^{-6}/°C)	Thermal shock resistance
Stainless steel 316[a,b]	14-16	

Note: All data taken from Lay [1983] except [a] from Perry and Chilton [1973], [b] from
Saracco & Specchia [1994] and [c] from Jordan and Greenspan [1967]

Thermal shock resistance. Temperature swing as part of the normal cycles of operation
or regeneration of the membranes or membrane reactors can lead to deleterious thermal
shock. The materials for the various components in a membrane reactor should be
carefully selected to impart good thermal shock resistance. This is particularly important
for high temperature reactions. Also listed in Table 9.5 is a summary of various
membrane materials along with qualitative description of their resistance to thermal
shock. Again, the available data apply to dense materials. While various metal oxides
have been made into commercial inorganic membranes, they tend to be affected by
thermal shock much more than other ceramic materials.

For those dense solid electrolyte membranes using metal oxides, the degree of
stabilization can make a difference in the resulting thermal shock resistance. For
example, the fully stabilized zirconia has poor thermal shock resistance compared to the
partially stabilized zirconia.

9.1.3 High-Temperature Structural Degradation of Chemical Nature

The combination of high temperature and chemical exposure poses a very challenging
material problem that is quite common in high-temperature membrane reactor
applications. The consequence of structural degradation as a result of such a
combination not only affects the permeability and permselectivity but also leads to
physical integrity or mechanical properties. These issues apply to both metal and
ceramic membranes.

Membrane embrittlement. Structural changes or degradations in mechanical properties
of the membranes in response to the application environments can be a significant issue.
Pd alloys are preferred over pure Pd as the material of choice for dense membranes in
that pure Pd can suffer from embrittlement after repeated cycles of hydrogen sorption
and desorption particularly at a temperature higher than approximately 200°C. It is
believed that there are two hydride phases of pure palladium: α phase at lower hydrogen
partial pressures and β phase at higher pressures. The two phases can coexist under

certain conditions. Pure palladium is subject to the $\alpha \rightarrow \beta$ phase transformation due to the adsorption of hydrogen which changes the atom spacing in the metal lattice which in turn causes structural distortion even only after 30 hydrogenation-dehydrogenation cycles at about 300°C or below upon exposure to a hydrogen-rich environment [Philpott, 1985]. This has seriously inhibited the commercial use of pure Pd for hydrogen purification despite of its excellent hydrogen selectivity. The problem has been solved, to some extent, by the addition of a second metal such as silver in the alloy [Hunter, 1960; Uemiya et al., 1991b]. The Pd-Ag alloy has been found to be not only stable through the aforementioned phase transformation, but also easier and less expensive to be fabricated.

There has been some suggestion on one possible mechanism by which hydrogen embrittlement occurs for a wide variety of materials under hydrogen bearing environments. It is called "interfacial weakening" which manifests itself in the form of interfacial separation. Based on some metallographic evidences for the participation of inclusions in hydrogen embrittlement processes, Sudarshan et al. [1988] proposed that inclusion-matrix interfaces serve as initiation sites for cracks under exposure to hydrogen.

The composition and the associated phases of a membrane can significantly affect its structural stability. It has been found that the physical integrity and mechanical properties of a perovskite-type membrane at a high application temperature are determined by the stoichiometry of the membrane compound [Balachandran et al., 1993]. Under the application conditions, certain stoichiometry of the membrane compound responds to the oxygen gradient at a high temperature with different phases formed. The perovskite $La_{0.2} Sr_{0.8} Fe_{0.6} Co_{0.4} O_x$ has a cubic structure in an oxygen-rich (e.g., 20% oxygen gas) atmosphere. However, when the oxygen partial pressure is below 5%, the cubic phase is transformed to an oxygen-vacancy-ordered phase. The phase transition is accompanied by a substantial volume expansion [Balachandran et al., 1995]. This leads to the lattice mismatch and consequently nonuniform volumetric expansions between materials in different parts (e.g., the inner and the outer walls) of the membrane. With the brittle nature of these oxides, this can cause fracture.

Corrosive reaction streams. In some application environments, the reactive or corrosive nature of one or more of the reaction components in a membrane reactor can pose a great technical challenge to the selection as well as the design of the membrane element. Feed streams often contain some impurities that may significantly affect the performance of the membrane. Therefore, attention should also be paid to the response of the selected membrane material to certain impurities in the reactant or product streams. Care should be taken to pretreat the feed streams to remove the key contaminants as far as the membrane is concerned in these cases. For example, palladium alloy membranes can not withstand sulfur- or carbon-containing compounds at a temperature higher than, say, 500°C [Kameyama et al., 1981]. Even at 100°C, the rate of hydrogen absorption (and, therefore, permeation) in a pure palladium disk is reduced by the presence of sulfur at a surface coverage of 40 to 100% monolayer [Peden et al, 1986].

Often the above challenge warrants major modifications to the membrane and the membrane reactor as a whole. This consideration can be well illustrated with the design of an inorganic membrane reactor for the thermolysis of hydrogen sulfide [Edlund and Pledger, 1993]. To achieve an almost complete conversion of the reaction, the membrane needs to have a very high selectivity to hydrogen which is one of the reaction products along with sulfur. This essentially precludes the use of ceramic membranes and narrows down the material choice to some metals. A further important factor is that hydrogen sulfide undergoes irreversible chemical reactions with many metals including stainless steel and palladium at a high temperature such as 700°C. For economic and hydrogen embrittlement reasons, vanadium is selected as the base membrane material over the popular choice of palladium or its alloys.

The decision to use vanadium as the membrane base material is accompanied by two issues related to chemical degradation. First of all, like most of the metals, vanadium is subject to chemical reaction by hydrogen sulfide. An effective solution to this is the use of a platinum coating on the feed side where hydrogen sulfide is. The platinum used also serves as an efficient catalyst for the thermolysis of hydrogen sulfide. A thin intermetallic-diffusion barrier in the form of a silica layer is sandwiched between platinum and vanadium to prevent contact between the two metals. Furthermore, vanadium can suffer from oxidation, the conditions of which could occur during start-up and shut-down operations on the permeate side of the membrane reactor. Therefore, a noble metal such as palladium or platinum can be used to prevent the oxidation of vanadium. A palladium coating is preferred due to its higher hydrogen permeability and lower cost than a platinum layer. Again, a silica layer is used to separate palladium and vanadium to avoid direct contact of the two metals.

Exposures of some metal oxide membranes, both dense and porous, to extreme pH conditions (e.g., pH less than 2 or greater than 12) can cause structural degradations, particularly with extended contact time. The extent of degradation depends on the specific phase of the material, porosity, and temperature. Steam can also be deleterious to some metal oxide and Vycor glass membranes. For example, as mentioned earlier, porous glass membranes undergo slow structural changes upon exposure to water due to partial dissolution of silica.

Non-oxide ceramic materials such as silicon carbide has been used commercially as a membrane support material and studied as a potential membrane material. Silicon nitride has also the potential of being a ceramic membrane material. In fact, both materials have been used in other high-temperature structural ceramic applications. Oxidation resistance of these non-oxide ceramics as membrane materials for membrane reactor applications is obviously very important. The oxidation rate is related to the reactive surface area; thus oxidation of porous non-oxide ceramics depends on their open porosity. The generally accepted oxidation mechanism of porous silicon nitride materials consists of two concurrent processes: the diffusive Knudsen flow of oxygen and the reaction between oxygen and silicon nitride at the pore walls to form silica. At high temperature, the diffusion step is rate limiting and the oxidation reaction is localized at the outer region of the pores [Desmaison, 1990]. The silica formed has a larger volume than silicon nitride. This causes the closure of the pores. Therefore, an essential requirement for minimizing internal oxidation of the non-oxide ceramic membrane material is to produce membranes with very small pores, to oxidize above a threshold temperature (e.g., 1,100°C for silicon nitride) and to minimize impurities.

Hot gas corrosion behavior of various silicon-based non-oxide ceramic materials (e.g., silicon carbide, silicon nitride, etc.) can vary widely depending on the stoichiometry, structure and sintering aids. Silicon carbide materials exhibit excellent corrosion resistance towards sulfur-containing atmospheres even at a high temperature around 1,400°C [Förthmann and Naoumidis, 1990].

9.1.4 Sealing of End Surfaces of Elements

Keeping the permeate and the retentate or reaction products divided after membrane separation is essential from the viewpoints of process efficiency as well as safety in some cases (e.g., when hydrogen is involved). As mentioned in Chapter 5, various methods are used to seal the ends of the membrane elements and make the connection between the elements and the module housing or headers fluid-tight. This type of problem becomes particularly challenging when dealing with compounding effects of high temperature, high pressure and harsh chemical environments.

The porosity at both ends of a tubular or monolithic honeycomb membrane element can be a potential source of leakage. These extremeties need to be made impervious to all reaction components, either permeates or retentates so the two do not remix. Typically the end surfaces and the outer surfaces near the ends of a membrane element are coated with some impervious enamel or ceramic materials which can withstand relatively high temperatures.

Many enamel or ceramic end seal materials are alumina-based. Some of them are readily available commercially. High-temperature grade ceramic end seal materials have been used in cartridge heaters, thermocouples, power resistors, etc. Some of them can be used in environments exceeding 1,500°C in temperature. They can be applied to the area requiring sealing by dipping, brushing or injection.

Commonly used enamels contain multiple ingredients. Typically silica, alumina and other metal oxides such as calcia are the major ingredients. In addition, some organic additives such as dispersants and viscosity modifiers (e.g., polyvinyl alcohol) are used to ensure that the starting slip consists of finely divided and dispersed particles in the submicron range in order to seal the pores on the end surfaces. The pores can be filled with very fine ceramic particles [Garcera and Gillot, 1986] or calcined colloidal silica

[Omata et al., 1989]. The slip should also have the proper consistency (viscosity) for processing purposes. The composition of the slip depends on the application and the environment involved and is not well published in the literature. The slip can be applied to the end surfaces by the slip casting techniques described in Chapter 3, followed by drying and melting. After application to the surface to be sealed, the slip is heated above the melting point of the enamel composition. The selection of the slip composition should be such that the resulting melting point of the enamel is not too high to alter the physical or chemical characteristics of the membrane element.

Other ceramic sealing materials containing calcined colloidal particles are being pursued by the membrane manufacturers as the issue of sealing extremities of a membrane element continues to be one of the important materials selection and engineering challenges for inorganic membranes.

When selecting the sealing material, caution should be exercised to ensure chemical compatibility with the application environment and, equally important, should possess a thermal expansion coefficient closely matching that of the bulk of the membrane matrix material so that cracking does not occur. Moreover, compression end seal is likely to be used at high temperature. Therefore, the sealing material needs to have a sufficiently high compressive strength to survive the application environment.

9.1.5 Element and Module Assembling (Joining)

In addition to the end seal issue just described, there is another critical material engineering issue facing membrane reactors. It concerns the connection between the membrane element and module housing or piping. In fact, this is considered to be one of the most critical issues to be addressed to make inorganic membrane reactors technically feasible and economically viable.

Element-to-module connection. As mentioned previously, it is typical that one or more membrane elements are placed in parallel within a metal casing or housing to form a module which in turn is combined with other modules to become a system. Inevitably these membrane elements are linked to metallic or plastic piping or vessels through some connectors or housings, such as a header. To render the membrane module packing leakproof, especially at high temperatures, high pressures and harsh chemical environments, is one of the critical issues to be resolved before many industrial gas phase reactions can utilize porous inorganic membrane reactors. The acceptable gas-tightness of the connections between membrane elements and modules is often determined relative to the membrane permeation rate. In many cases as long as the leakage rate is significantly smaller than the permeate rate, the sealing is considered satisfactory. The leakage issue is not just process economics related, but may also be a safety concern which often dictates if a process is acceptable or not.

Most of the commercial ceramic membranes today come equipped with gaskets or O-rings made of silicon, fluorinated polymers, PTFE (polytetrafluoroethylene) or EPDM (a terpolymer elastomer made from ethylene-propylene diene monomer). Sometimes high vacuum grease is applied to the O-rings to ensure the seal. These elastomeric or polymeric gaskets are particularly suitable for sealing the gap between ceramic membrane elements and metal housing when the application requirements call for an ultraclean (e.g., microelectronics and food and beverage uses) and easy to change system to be operated at a low temperature (<100°C). They can not, however, survive the prevailing conditions existing in many high temperature (say, >200-300°C) membrane reactor applications. Other methods are required.

There are several possible ways of accomplishing this. These include localized cooling at the joining area, mechanical seal (such as Swagelok), ceramic-to-ceramic, ceramic-to-metal seals (cermets), gland packing, and special types of seals.

A relatively simple method is to raise the reaction zone temperature to the required high level but maintain the element-module connection at the ends of the module at a temperature low enough to use the elastomeric or polymeric gaskets mentioned earlier. An example of this practice is to place the reaction zone in the middle of the membrane reactor length in a tube furnace while the areas near the gaskets at both ends are far away from the heated region or are cooled (e.g., with cooling water) to a temperature that the gaskets can withstand [Capannelli et al., 1993]. The drawback of this method is that it is not practical for large-scale industrial applications because of the extensive membrane length and space required.

Nourbakhsh et al. [1989] employed a compressible Knudsen flow-type seal for flat membrane reactors. To avoid membrane cracking due to the difference in the thermal expansion coefficients of the membrane and the reactor wall, the membrane reactor is equipped with a pneumatic system which compresses the membrane after the desired operating temperature has been established and all reactor parts have reached steady state temperatures. This type of sealing is also limited to laboratory applications and not practical for production environments.

Depending on the application conditions, pure graphite as the gasket materials for the element-to-module connection can hold its integrity at different temperatures. In an oxidation atmosphere, the graphite gaskets are usable up to approximately 450°C. In contrast, in a reducing environment as prevailing in dehydrogenation reactions, graphite strings can survive higher temperatures. To ensure gas-tightness at high temperatures, the graphite strings may need to be wrapped around the membrane element several times followed by tightening with compression fittings (such as Swagelok fittings). [Champaignie et al., 1990a and 1990b].

Similar to the end sealing issues mentioned above, some module packing considerations are mismatching of the thermal expansion coefficients and chemical compatibility. Shown in Table 9.5 is the comparison of the values of the thermal expansion coefficient for various ceramic materials (including glass) used in commercial membranes and various metals which can be used for the module housing. In general, the ceramic materials exhibit lower thermal expansion coefficients than the metals. The thermal expansion coefficient consideration is crucial whenever dissimilar materials are used for the membrane and the reactor housing which are to be joined.

Ceramic-to-ceramic joining. Ceramic-to-ceramic joining technology has been in existence for a long time. Extensive research has been conducted on joining ceramics to ceramics by a brazing technique with either a metallic or glassy interlayer [Bates et al., 1990]. Bird and Trimm [1983] used Versilic tubing to seal carbon membranes into a glass cell and found that the seal was stable up to approximately 200°C but showed some leakage after prolonged temperature cycling. Carolan et al. [1994] provides a method for connecting a magnesia-supported mixed oxide membrane tube to an alumina tube. A glass ring with a softening temperature of 810°C was placed between the two tubes and, upon heating the assembly to a temperature higher than the softening point, the glass melted to form a seal for the joining area between the two ceramic tubes.

A porous alumina membrane tube is joined with nonporous alpha-alumina tubes for the non-reaction zones by the adhesive-welding technique [He et al., 1994]. First, the joining ends of the membrane and the nonporous tubes are cut and ground in a V-shape. Then the two tubes are connected together with a ceramic adhesive (such as Rebond Ceramic Adhesive 904) on a mold plate, cured at room temperature for 24 hours and calcined at 850°C for about half an hour. Finally the joint area is coated twice with a glaze material (such as NJ-1 glaze) and dried under an infra-red lamp before sintering at 920°C for an hour. The mechanical strength of the joint is found to be higher than that of the membrane tube.

Ceramic-to-metal joining. The needs to join a ceramic membrane element to a metal pipe, module housing or reactor vessel arise from time to time as the ceramic membranes are increasingly considered for high temperature membrane reactor applications. Since ceramic-to-metal joining is becoming one of the critical technical areas in inorganic membrane reactors, it will be discussed in greater detail here. Conventionally ceramic components (e.g., ceramic membranes) are joined to metal structures (e.g., reactor or module housing) by a class of processes consisting of two steps: metalization of the ceramics and brazing. The best known process is the moly-manganese process. Typically metalization takes place at a high temperature around 1,500°C.

A simpler (one-step) and more economical joining process is direct brazing in a furnace under a vacuum or inert gas atmosphere through the use of active filler metals [Mizuhara et al., 1989]. An active element such as the commonly used Ti in the filler metal forms a true alloy with the base metal. The difference in the thermal expansion coefficients between the ceramic membrane and the metal housing can lead to high stress at the brazed area during cooling from the brazing temperature to the ambient temperature. Therefore, the cooling rate from the brazing temperature should be low (e.g., 5°C/min or lower). The joint properties by the ceramic-metal brazing are influenced by the variables related to those three materials involved. The ceramic surface should be free of defects. This can be accomplished, for example, by a sintering or lapping process. It is desirable that the metal has a low yield strength or shows low blushing (or surface flow) of the filler metal. Extensive blushing effectively depletes the amount of the active element available for wetting the ceramics. The filler metal should be ductile in order to reduce the thermal stress due to mismatch of the thermal expansion coefficients. A thicker filler metal tends to generate joints with higher shear strength.

The mismatch in the thermal expansion coefficients (TEC) between a porous ceramic element and a metal component of a module or reactor is even more than that between a dense ceramic component and the same metal. This mismatch can be lessened by employing several brazing interlayers between the two base materials to be joined. The choices of the interlayer materials are such that the TEC mismatches are progressively smaller and the materials are ductile. The ductility can be further enhanced by using thin metallic connectors [Velterop and Keizer, 1991]. A process has been patented for coating the porous ceramic membrane element at its connecting ends with a brazing material which reacts with its base material at a high temperature around 1,500°C. The reaction can yield a structure progressively changing from being porous to dense [Saracco and Specchia, 1994]. The dense portion can then be brazed using the aforementioned brazing process.

The brazing techniques in several variations including the ones described above have been used by several industrial as well as academic groups to provide the ceramic-to-metal joining. They are currently labor intensive but their potentials for large scale applications are great. Other techniques such as vacuum soldering and micro-arc welding are also some possibilities for making the required ceramic-to-metal joints [Velterop and Keizer, 1991].

A new technique is under development for depositing metal onto a ceramic substrate (e.g., ceramic membrane) at room temperature. This ion vapor deposition process combines ion implementation and electron beam deposition [Anonymous, 1988]. The ceramic substrate is irradiated with argon or nitrogen ions while the metal is vaporized by an electron beam and deposited onto the ceramic surface. The ion implementation directs the metal to the surface to be joined.

Ceramic materials typically have low tensile strengths compared to compressive strengths which are five to ten times higher. Therefore, the ceramic-to-metal joint construction should only impose compressive stress on the ceramic-metal interface. This, for example, can be achieved through the use of a metallic bellow. By joining the bellow directly to the porous ceramic element and welding the bellow under compressive conditions to the module or reactor housing, the joint will stay at all times under a compressive force [Velterop and Keizer, 1991]. To reinforce the joint, a dense ceramic ring can be used between the membrane tube and the bellow. Furthermore, the joining of the ring to the bellow can be strengthened by applying a multiple-layer brazing technique. In this technique, various brazing materials having a fracture property progressively ranging from being brittle (ceramic) to ductile (metal) are used, thus making the joint capable of withstanding thermal stresses at high temperatures due to the TEC mismatch [Saracco and Specchia, 1994].

Additionally, the packing media need to be able to withstand thermal shock or fatigue due to, for example, reactor start-up and shut-down cycles, and, in those cases involving ceramic membranes, absorb mechanical shock from the operation of the unit. Due to their relatively brittle nature, for example, commercial ceramic membranes currently often use some flexible materials, such as polymers, for the connectors to adsorb mechanical vibration.

9.1.6 Effects of Membrane Structure on Reaction Performance

In porous composite membranes, the support layer(s) can play an important role in the reactor performance. This, for example, is the case with consecutive reactions such as partial oxidations where intermediate products are desirable. Harold et al. [1992] presented a concept in which two reactants are introduced to a two-layer membrane system from opposite sides: ethylene on the membrane side while oxygen on the support side. The mass transfer resistance of the support layer lowers the oxygen concentration in the catalytic zone and directs the preferred intermediate product, acetaldehyde, toward the membrane side. Thus the support layer structure enhances the yield of acetaldehyde.

More discussions on the membrane microstructure affecting the membrane reactor performance will be made in Chapter 10 and 11.

9. 2 CATALYSIS CONSIDERATIONS

When combining the separator and the reactor functions into one compact physical unit, factors related to catalysis need to be considered in addition to those related to selective separation discussed in previous chapters. The selection of catalyst material, dispersion and heat treatment and the strategic placement of catalyst in the membrane reactor all can have profound impacts on the reactor performance. The choice of membrane material and its microstructure may also affect the catalytic aspects of the membrane reactor. Furthermore, when imparting catalytic activity to inorganic membranes, it is important to understand any effects the underlying treatments may have on the permeability and permselectivity of the membranes.

When used as monolithic, porous catalytic membranes, the platinum group metals provide higher mechanical stability and heat conductivity than conventional supported metal catalysts. The cost, however, can be an issue. An economically feasible solution to this problem is to introduce the metal catalyst particles into a narrow region just below the surface of a less expensive porous membrane.

9.2.1 Placement of Catalyst Relative to Membrane

Where and how the catalyst is placed in the membrane reactor can have significant impact not only on the reaction conversion but also, in some cases, the yield or selectivity. There are three primary modes of placing the catalyst: (1) A bed of catalyst particles or pellets in a packed or fluidized state is physically separated but confined by the membrane as part of the reactor wall; (2) The catalyst in the form of particles or monolithic layers is attached to the membrane surface or inside the membrane pores; and (3) The membrane is inherently catalytic. Membranes operated in the first mode are sometimes referred to as the (catalytically) passive membranes. The other two modes of operation are associated with the so called (catalytically) active membranes. In most of the inorganic membrane reactor studies, it is assumed that the catalyst is distributed uniformly inside the catalyst pellets or membrane pores. As will be pointed out later, this assumption may lead to erroneous results.

The above issues will be described individually as follows.

9.2.2 Catalyst Bed Separated from Membrane

Many of the earlier studies on inorganic membrane reactors use a bed of catalyst particles contained inside the membrane element. Transport of the reaction components into and out of the catalyst particles mostly rely on diffusion. In addition, transport of the reaction product(s) out of the catalyst bed may not be as efficient as the other modes of catalyst placement relative to the membrane.

The distribution of the catalyst in a catalyst pellet and its effects on the effectiveness factor, selectivity or yield have been studied extensively. The results depend on the catalyst loading and have been reviewed by Gavriilidis et al. [1993]. When the catalyst loading is not high, the catalyst surface area is a linear increasing function of the loading. The optimal catalyst distribution inside the pellet is a Dirac delta function in general [Wu et al., 1990a]. As the loading becomes high, the catalyst surface area increases nonlinearly with the loading and the optimal catalyst distribution generally can not be predicted analytically [Baratti et al., 1993].

Yeung et al. [1994] extended the studies to a general case of a bed of catalyst pellets on the feed side of a membrane reactor where the membrane is catalytically inert for an arbitrary number of reactions with arbitrary kinetics under nonisothermal conditions. Their conclusions are similar to those for the case of pellets in a fixed bed reactor [Baratti et al., 1993]. It appears that the presence of a catalytically inert membrane and a permeate stream do not affect the nature of the optimal catalyst distribution but may change the specific locations and number of steps. The optimal distribution of catalyst within the pellets is a multiple step function.

For a more specific case of an isothermal first-order reversible reaction of the type

$$A \leftrightarrow B \tag{9-1}$$

under well-mixed fluid phase conditions and no interphase mass transfer resistance, Yeung et al [1994] solved the governing transport and reaction equations analytically for both uniform and Dirac delta distributions of catalyst activity. They concluded that, by placing the catalyst near the surface instead of the center of the pellets, the total conversion (on both the feed and the permeate sides) increases to higher values as compared to the uniformly distributed case. This is illustrated in Figure 9.6 where, for an inert membrane reactor with catalyst on the feed side (IMRCF), the cases $\bar{s}=1$ and $a(s)=1$ represent the Dirac delta distribution located at the feed side (surface step distribution) and the uniform distribution, respectively. The Dirac delta distribution near

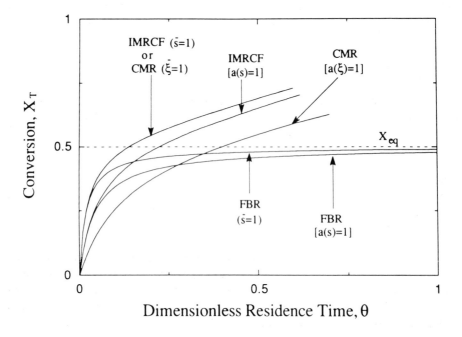

Figure 9.6 Total conversion for inert membrane reactor with catalyst on the feed side (IMRCF), catalytic membrane reactor (CMR) and conventional fixed-bed reactor (FBR) with uniform and Dirac delta catalyst activity distributions as a function of the dimensionless residence time [Yeung et al., 1994]

the external surface of the pellet, which is superior to a uniform activity profile in total conversion, maximizes the reactant concentration at the catalyst site. For both a Dirac delta distribution on the external surface of a pellet and a uniformly distributed catalyst activity profile in a pellet, the IMRCF performs better than the corresponding fixed-bed reactor (FBR) without a membrane. As expected, the conversions in FBR are bound by the equilibrium values On the contrary, given sufficient residence time in the IMRCF, the total conversion can exceed the equilibrium value as a result of continuous selective removal of the product.

It is noted that, if the Dirac delta distribution is placed away from the external surface of the pellet (i.e., $\bar{s}<1$), the conversion can be lower than that for the case of a uniformly distributed catalyst. This is shown in Figure 9.7(a) [Yeung et al., 1994].

Besides total conversion, other reaction performance index may benefit from optimizing the catalyst distribution and location. Examples are product purity on the feed or permeate side and product molar flow rate on the feed or permeate side. Yeung et al. [1994] have also investigated these aspects and provided comparisons among IMRCF, FBR and catalytic membrane reactor (CMR) in Figure 9.8. It is apparent that the various reaction performance indices call for different optimal catalyst distributions.

9.2.3 Catalyst Attached to Membrane

In the second mode of incorporating a catalyst into a membrane reactor, the catalyst is attached to the membrane surface on the feed or permeate side or to the surfaces of the membrane pores. This case and the third mode where the membrane is inherently catalytic are often called catalytic membrane reactor (CMR).

Catalysts such as the platinum group metals can be used in dispersed or monolithic solid form. The catalyst can be deposited on the surface of a membrane (dense or porous) or, in the cases of catalyst particles, dispersed in the sub-surface layer or throughout the matrix of a porous membrane.

There is evidence that, in some cases of gas-phase or multiphase (e.g., gas-liquid) catalytic reactions, attachment of catalyst to the membrane in some form is more desirable than having the catalyst bed separated from the membrane in their catalytic performance [Sun and Khang, 1988; Harold et al., 1989; Zaspalis et al., 1991a]. Zaspalis et al. [1991a] estimated that, when impregnated in membrane pores, a catalyst could be ten times more active than in the form of pellets. As long as the pore size is sufficiently large, the reactant(s) and product(s) can be transported to and from the catalytic sites by convection which is more efficient than diffusion.

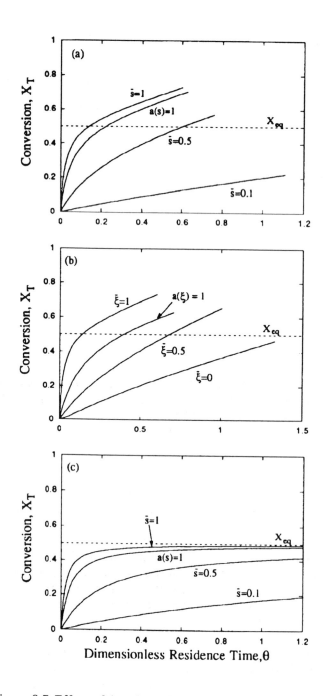

Figure 9.7 Effect of location of Dirac delta catalyst activity distribution on total conversion as a function of dimensionless residence time for (a) IMRCF, (b) CMR and (c) FBR [Yeung et al., 1994]

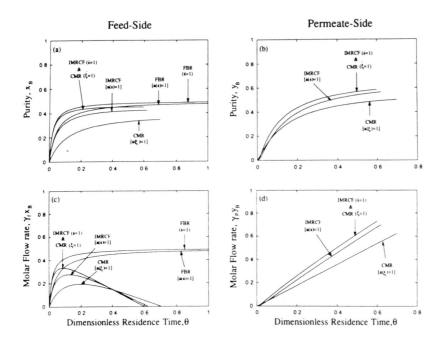

Figure 9.8 Comparison of IMRCF, CMR and FBR with uniform and Dirac delta catalyst activity distributions for (a) product purity on the feed side, (b) product purity on the permeate side, (c) product molar flow rate on the feed side, and (d) product molar flow rate on the permeate side as a function of the dimensionless residence time [Yeung et al., 1994]

Beside the benefit of having less mass transfer resistance from the catalyst particles to the membrane, attachment of catalyst to the membrane is also preferred over a bed of catalyst in certain reactions where the contact between the reactant(s) and the catalyst is controlled. An example is the partial oxidation or oxidative coupling of methane [Eng and Stoukides, 1991]. In this case, the reaction conversion and selectivity have been observed to depend on the chemical form of the reactant oxygen and its rate of delivery to the reaction zone through the membrane. This approach has also been demonstrated through an analysis for utilizing a catalytically impregnated porous membrane tube as a catalyst for gas-liquid reactions [Harold et al. 1989].

Catalysts can be contained in the membrane pores or on the membrane surface by several impregnation and adsorption techniques commonly used for conventional catalyst preparation. To introduce catalyst inside the pores, for example, Sun and Khang [1988] use chloroplatinic acid solution to impregnate Pt catalyst in porous glass membrane pores. Ni, Fe Ag, and Pd catalysts of varying particle sizes can be deposited in porous alumina membranes by immersing the membrane in an electrolysis bath of a salt solution of the metal to be introduced [Moskovits, 1984]. Pd particles can be dispersed in porous stainless steel membranes by the condensation of Pd and toluene vapors followed by melting the resultant glassy solid [Gryaznov, 1992].

Catalyst attached to membrane surface. When depositing catalyst particles on the surface of a catalytically inert, dense membrane, the membrane surface layer should be porous in nature to provide a high surface area catalyst support for strong adhesion of the catalyst particles. A layer or multi-layers of catalyst particles can be coated on inorganic membrane surfaces by several methods: Pd by vapor deposition [Gryaznov et al, 1979], Pd and Pt by solution deposition [Gryaznov et al., 1983; Guther and Vielstich, 1982].

Catalyst attached to membrane pore surface. The final distribution of the catalyst in the membrane pores can significantly impact the reactor performance. The optimal form of the catalyst distribution for maximizing conversion was studied mathematically by Keller et al. [1984]. They determined that the optimal distribution of the catalyst concentration is of the Dirac delta function.

The concept of optimizing the catalyst distribution and location in a pellet for maximizing the total conversion, selectivity or product purity (i.e., product to reactant ratio) has been extended to the case of a single isothermal first-order reversible reaction in a porous membranes whose pore surfaces are deposited with the catalyst [Yeung et al., 1994]. Similar to pellets in an IMRCF, CMR with the catalyst attached to pore surfaces benefits from having a surface step distribution instead of a uniform distribution (i.e., $a(\xi)=1$). The surface step distribution is a Dirac delta distribution on the feed side of the membrane (i.e., $\bar{\xi}=1$). In fact, as shown in Figure 9.6, the conversion of the CMR and the IMRCF is identical when the Dirac delta is located on the feed side of the membrane (for the CMR) or on the external surface of the pellet (for the IMRCF). This conversion value is also larger than the conversion in a FBR. However, with a uniform catalyst distribution, the CMR conversion may be lower than the FBR conversion at a low residence time [Yeung et al., 1994].

When the Dirac delta distribution is placed closer to the permeate side (i.e., a subsurface step distribution) of an CMR, the total conversion is actually lower than that with a uniform catalyst distribution (Figure 9.7). For a performance index other than the total conversion (such as product purity or product molar flow rate), the optimal distribution of the catalyst concentration can be rather complex even for reversible first-order reactions as displayed in Figure 9.8.

It is expected that, for more complex reactions and conditions (e.g., multiple reactions, nonisothermal conditions, etc.), step distributions of a relatively narrow width (approaching a Dirac delta type) placed at a specific location within the membrane pore structure will be optimal for catalytic membrane reactors [Yeung et al., 1994]. This type of subsurface step distribution of catalyst activity has been recognized.

In some cases it is actually advantageous to deposit the catalyst (e.g., Pt) near the ends of the (alumina) membrane pores on the low pressure (permeate) side for the oxidation of carbon monoxide to form carbon dioxide [Furneaux, 1987].

The optimal position to place the catalysts inside a porous inorganic membrane structure have also been studied by Cannon and Hacskaylo [1992]. In their study, palladium is impregnated at different sites depth-wise on the pore walls of porous Vycor glass membranes by an adsorption and a reaction method. The adsorption method results in Pd deposition primarily within the first several microns of the inner and outer surfaces of the membrane tube. In contrast, membranes impregnated by the reaction of palladium allyl chloride dimer show a rather uniform deposition of Pd inside the membrane pores. When the membrane reactors impregnated by the above methods are used for the dehydrogenation of cyclohexane to benzene, the resulting conversions are quite different. While the non-uniformly impregnated membrane reactor reaches a conversion of over 75%, the uniformly impregnated one exhibits only 39%.

Preparation of membranes with pores having non-uniform catalyst distributions. It has been indicated that special, non-uniform catalyst distributions inside the membrane pores offer optimal reaction performance indices such as the total conversion, product purity and product molar flow rate. Specifically, surface step distributions near the pore mouths or subsurface step distributions inside the membrane pore channels are preferred.

To prepare this type of catalyst distribution and positioning, Yeung et al. [1994] suggested a sequential slip-casting technique for ceramic membranes made by the sol-gel process described in Chapter 3. In this new strategy for making non-uniform catalytic membrane reactors, the membrane is built layer-by-layer using two types of sols: a catalytically inert sol and a catalytically active sol containing the desired catalyst. The nature, number and thickness of each layer is determined by the desired catalyst distribution and its location. By manipulating the number and the thickness of layers for each type of sol, one can control the thickness and position of the active layer within the membrane. For example, to have a platinum-impregnated γ-Al$_2$O$_3$ membrane with a step distribution of Pt concentration in the pores along the membrane thickness, a catalytically active sol can be prepared by using a sol impregnated with chloroplatinic acid which is slip-cast to form a layer of the membrane that is catalytically active. For the portions of the membrane that are desired to be catalytically inert, the regular sol without catalyst impregnation should be used instead.

To better control the thickness and densification of each layer, a technique utilizing the osmotic pressure can be applied. Here, while the membrane layer is being deposited onto the inside surface of a porous support tube, the support tube is immersed in a concentrated sucrose solution. By the principle of osmotic pressure, water permeates through the support layer from the deposition solution side to the sucrose solution side, thus densifying the membrane layer formed [Yeung et al., 1994].

Reasonably precise placement of the catalyst in a catalytic membrane reactor appears to be feasible with other techniques [Lin et al., 1989; van Praag et al., 1989].

9.2.4 Inherently Catalytic Membranes

As indicated in Chapter 8, the most compact configuration of a membrane reactor is where the membrane (selective layer) is also catalytic for the reaction involved.

Many of the metal oxide materials used for making ceramic membranes, particularly the porous type, have also been used or studied as catalysts or catalyst supports. Thus, they are naturally suitable to be the membrane as well as the catalyst. For example, alumina surface is known to contain acidic sites which can catalyze some reactions. Alumina is inherently catalytic to the Claus reaction and the dehydration reaction for amine production. Silica is used for nitration of benzene and production of carbon bisulfide from methanol and sulfur. These and other examples are highlighted in Table 9.6.

Zeolitic membranes possess catalytic activity, either acid activity, or metal activity, or both. The catalytic activity can be modified according to the need of the reaction involved [Haag and Tsikoyiannis, 1991]. For example, the acid activity of siliceous zeolites can be adjusted by the amount of trivalent elements (especially aluminum) by the degree of cation exchange from salt form to hydrogen form or by thermal or steam treatment. Substitution by a trivalent element such as aluminum introduces a negative charge which must be balanced. If proton is used, the membrane is acidic. The charge may also be balanced by cation exchange with alkali or alkaline earth metal cations. The acid activity of aluminum phosphate-type zeolites can be increased by incorporation of activating agents such as silica. Zeolite has been known to be a catalyst for catalytic cracking.

On the metallic membrane side, a well known type of material with this characteristic is Pd and certain Pd alloys. Palladium is known to be catalytic to many reactions including oxidation, hydrogenation and hydrocracking. It has been found that the catalytic activity of selected binary Pd alloys is higher than that of pure Pd. Silver catalyzes a number of oxidation reactions such as oxidation of ethylene and methanol. In addition, nickel is catalytic to many industrially important reactions.

TABLE 9.6
Examples of common inorganic membrane materials which are inherently catalytic to selected reactions

Membrane material	Catalyst form	Reaction
Alumina	Al_2O_3	Claus process
Alumina	Al_2O_3	Dehydration for amine production
Alumina	Al_2O_3-Cr_2O_3	Dehydrogenation of butane
Alumina	Al_2O_3-SiO_2	Catalytic cracking for gasoline production
Silica	SiO_2	Nitration of benzene
Silica	SiO_2	Carbon bisulfide from methane and sulfur
Silica	SiO_2-Al_2O_3	Methylamine from methanol and hydrogen sulfide
Silica	SiO_2-Al_2O_3	Alkylation for ethylbenzene production
Silica	SiO_2-K_2SO_4-V_2O_5	Naphthalene oxidation
Silica	SiO_2-Al_2O_3 and Pt on SiO_2	Hexane isomerization
Zeolite	Zeolite on aluminum silicate	Catalytic cracking
Palladium	Supported Pd	Oxidation of acetylene and acetic acid
Palladium	Pd on Al_2O_3	Hydrogenation of acetylene
Palladium	Pd on zeolite	Hydrocracking
Palladium	Pt-Pd	Oxidation of carbon monoxide
Silver	Supported Ag	Oxidation of ethylene
Silver	Ag	Oxidation of methanol
Nickel	Supported Ni	Steam reforming
Nickel	Supported Ni	Hydrogenation of benzene
Nickel	Supported Ni	Methanation
Nickel	Supported Ni	Hydrogenation of vegetable oils

9.2.5 Supported Liquid-phase Catalyst Sandwiched between Two Different Membranes

As mentioned in Chapter 8, many catalysts in homogeneous catalysis are liquid which are acids, bases or other ion-generating compounds. The corrosive nature of these catalysts and the reaction mixtures suggests the utility of inorganic membranes. Two major technical problems need to be solved. One is the immobilization or encapsulation of the liquid catalyst. The other is separation of the catalyst from the reactants and products. Separation between the reactants and products is also desirable.

A feasible method of solving the above problems has been proposed by Kim and Datta [1991]. It is illustrated in Figure 8.1. The catalyst is immobilized in a porous matrix support which is sandwiched between two different membranes. The first membrane is predominantly permeable to the reactant and the second membrane to both the reactant and the product. Neither of the two membranes is permselective toward the liquid catalyst. This arrangement of catalyst relative to the membranes has been demonstrated using organic membranes which, although conceptually feasible, may degrade under the reaction conditions. Provided that the required unique permselectivities can be tailor-made, inorganic membranes are ideal for this type of applications. Many ceramic membranes, for example, have composite structures and, therefore, an additional membrane layer can in principle be deposited on the other side of the porous support to make the membrane-catalyst-membrane composites. The greatest challenge, however, is to impregnate the porous support with the liquid catalyst prior to the deposition of the second membrane layer. An alternative is to produce the second membrane layer as a stand-alone thin membrane which is then brought into contact with the catalyst-containing support.

9.2.6 Selection of Catalyst Placement

The choice of the above three modes of catalyst placement relative to the membrane can significantly affect the reactor performance. From the analysis of catalytically active and passive (inert) membrane reactors [Sun and Khang, 1988], it appears that the critical parameter determining the choice is the reaction residence time. At low residence times, the difference between a catalytically active and a catalytically passive membrane is not significant. However, as the reaction residence time becomes high, the catalytically active membrane shows a higher reaction conversion.

The selection of where the catalyst is best placed also depends on the nature of the permeating reactant(s). Take the example of a dense hydrogen-permeable Pd-based membrane. Apparently the hydrogen permeating through the membrane is delivered to the reaction zone in an active monoatomic form which should be in direct contact with the catalyst surface. Thus, the catalyst in these cases is best situated right at the membrane surface. Wood and Wise [1966], for example, dispersed gold catalyst on a palladium membrane for the hydrogenation of cyclohexane and some alkenes based on this reasoning. Gaseous hydrogen molecules are not as active as those desorbed out of the Pd-based membranes. The enhanced catalytic property of this so called "subsurface" hydrogen has shown benefits to some partial hydrogenation reactions. Similar results have been observed with the so called "subsurface" oxygen through a dense silver membrane in some partial oxidation reactions.

Although technically more superior to the catalytically passive membranes in reaction performance under certain conditions, the catalyst-modified membranes pose a few technical challenges. The first issue is the control of the resulting membrane pore sizes and their distribution. There has been no indication of a robust procedure for incorporating catalysts either only on the membrane surface or in the membrane pores such that the final pore size distribution is uniform or can be predicted. The second concern is the physical and chemical stabilities of the catalyst. After an extended period of operation, the catalyst may be physically detached from the membrane or eroded, or catalytically poisoned. It is generally easier to replace or regenerate the catalyst in the catalytically passive mode.

9.2.7 Preparation of Membrane Catalyst

For those cases where the catalyst is not attached to the membrane, catalyst pellets are typically used. The preparation of catalyst pellets has been extensively documented and will not be reviewed here. Only the preparation of the catalyst inside the membrane porous structure is discussed in this chapter.

Given the relatively limited surface area available for catalyst on the surface of the membrane or its pores, a high catalyst surface area can be achieved with minimum catalyst loading by applying a very fine dispersion of catalyst particles. Dispersion of the catalyst particles affects the reaction kinetics and selectivity.

Two methods have been adopted for preparing catalysts for the membrane reactors: impregnation and ion exchange. The former method has been predominantly used when attaching catalysts to membranes.

Wet impregnation. The first method is the commonly used wet impregnation of catalysts on their carriers which has been adapted to prepared various catalytic membranes, particularly ceramic membranes. Usually one end of the membrane element is plugged during catalyst impregnation. A batch solution containing the desirable catalyst in an inorganic or organic solvent is used to fill the inside of the membrane element. Alternatively the membrane element is dip coated with the solution to impregnate the catalyst from the outer surface of the element. Typically, the treated membrane is air calcined or heated under a reducing atmosphere (e.g., in hydrogen) or subjected to both calcination and reduction.

Shown in Table 9.7 are some examples of incorporating catalysts into porous ceramic membranes. Both metal and oxide catalysts have been introduced to a variety of ceramic membranes (e.g., alumina, silica, Vycor glass and titania) to make them catalytically active. The impregnation/heat treatment procedures do not appear to show a consistent cause-and-effect relationship with the resulting membrane permeability. For example, no noticeable change is observed when platinum is impregnated into porous Vycor glass membranes using chloroplatinic acid [Champagnie et al., 1990a]. On the contrary, when porous alumina membranes are treated by the incipient wetness technique with iron, potassium, aluminum and copper nitrate solutions, the permeability is significantly lowered probably due to pore blockage. Similarly, wet impregnation of rhodium acetylacetonate on silica membranes causes considerable permeability reduction [Raman et al., 1993]. More definitive studies in this area are needed.

TABLE 9.7
Wet impregnation of catalysts in ceramic membranes

Catalyst	Impregnation solution	Membrane material	Reference
Fe, K, Al, Cu	Nitrate solution	Porous alumina	Liu et al. [1990]
Pt	Chloroplatinic acid solution	Porous Vycor glass	Champagnie et al. [1990a]
Pt	Chloroplatinic acid solution	Porous alumina	Uzio et al. [1991]; Yeung et al. [1994]
Pd	Nitrate solution	Porous Vycor glass	Cannon and Hacskaylo [1992]
Ag	Nitrate solution	Porous alumina	Zaspalis et al. [1991a]
Ag	Electroless plating solution	Porous alumina	Yeung et al. [1994]
V_2O_3	Toluene solution of vanadium acetylacetonate	Porous titania & alumina	Zaspalis et al. [1991b]
Rh	Tetrahydrofuran solution of rhodium acetylacetonate	Porous silica	Raman et al. [1993]

Monolayer metal complexation. A special impregnation method requires that the membrane or pore surface contain surface functional groups which interact with certain metal precursor molecules to form metal complexes. Upon reduction, the complexes yield monolayer dispersion of metal catalysts. This approach provides better control of metal loading and dispersion at the molecular level. It offers high activity per unit of catalyst loading. Vanadium acetylacetonate has been used as a precursor to disperse vanadia in titania and gamma-alumina membranes to prepare monolayers of catalyst particles [Zaspalis et al., 1991b]. A well mixed solution of vanadium acetylacetonate in toluene is batch filtered through the titania and alumina membranes. The initial amount of metal complex present in the toluene solution is equivalent to six monolayer coverages of the membrane surface. The resulting catalyst loading is equivalent to about

70% monolayer coverage on the porous titania membrane, as determined by X-ray fluorescence.

Raman et al. [1993] applied the same concept to uniformly disperse ca. 6 nm rhodium metal particles onto the surface of an amine-derivatized silica membrane. First the silica membrane is exposed to a solution of an amine-derivatized silylating agent, N--(2-aminoethyl)-3-aminopropyltrimethoxysilane, in methanol for 20 hours at room temperature. The silanol groups on the silica surface react with the silylating agent to provide the metal complexation sites. The amount of silylating agent used is in large excess of the stoichiometric amount needed for the complete reaction of the hydroxyl groups. The silylated membrane tube is then filled with a solution of tetrahydrofuran containing the metal-organic compound $[(1,5$-cyclooctadiene)RhCl$]_2$ at 30°C for 2 hours followed by a hydrogen reduction at 200-225°C also for 2 hours to form uniformly dispersed rhodium crystallites. The metal-organic molecules are used as the reagents for site specific surface reactions leading to dispersion of metal complexes at the molecular level.

The dispersion described above is determined by the number and distribution of the amine complexing sites which in turn depend on the number and distribution of surface hydroxyl groups and steric constraints imposed by the pore size [Raman et al., 1993]. Modest reduction in helium and nitrogen permeabilities as a result of the dispersion procedure seems to indicate that there is no significant particle penetration into the membrane pores and most of the catalyst deposition occurs on the membrane surface. Apparently, the reduced membrane pore size due to silylation prevents significant penetration of the large metal organic precursor molecules.

Ion exchange. Alumina-supported iron oxides have been prepared as the catalyst for dehydrogenation of ethylbenzene by ion exchanging with an iron precursor solution [Wu et al., 1990b]. An iron oxalate solution is circulated through a bed of granular activated alumina under a controlled pH of 3.5 for 2 hours. The ion exchanged catalyst is then air calcined at 700 to 800°C. If needed, additional ion exchange can be applied in a similar fashion after calcination.

Heat treatment. A catalyst is activated by converting it into an active form typically by heat treatment steps that involve calcination, decomposition, reduction or their combinations. During the preparation of the catalyst on the membrane surface or pore wall, the condition for any required heat treatments to disperse and activate the catalyst should be carefully selected so as not to change the microstructural and chemical characteristics of the membrane. Often the air calcination temperatures are higher than those used for reduction. A low treatment temperature is preferred because it will minimize catalyst particle coarsening and maintain adequate total catalyst surface area. The relatively low hydrogen reduction temperature (for example, 200-225°C when preparing rhodium particles [Raman et al., 1993] or 350°C when converting platinum catalyst [Champagnie et al., 1990a]) is advantageous in that it will not affect the microstructural or chemical stabilities of most ceramic membranes as some oxidative heat treatments do.

Surface treatment of dense membranes for catalytic activity. For an inherently catalytic or catalyst-impregnated dense membrane reactor, the typical surface area to volume ratio available to the catalyst is low due to the geometry of the membrane element. Although the hydrogen permeating through Pd- or Pd-alloy membranes exists as atomic hydrogen which is more active catalytically than molecular hydrogen from the bulk phase, the low surface area to volume ratio for the catalyst imposes as a serious drawback of the membrane catalyst compared to the conventional pellet catalyst.

The aforementioned low ratio problem can be at least partially compensated by treating one or both of the membrane surfaces. One solution to increase the surface area per unit volume is to render the membrane surface sufficiently rough by local thermal diffusion of a chemically active metal followed by an acid treatment. One or both sides of a catalytically active Pd-based membrane are first deposited with a thin layer of catalytically inactive metal (e.g., copper or mercury) by such techniques as vacuum atomization, electroplating, diffusion welding, casting and dipping. The ratio of the coating to the membrane thickness lies between 0.01 to 0.1. The coated Pd-based membrane is then subjected to heat treatment at 300 to 800°C for copper coating or at -10 to 150°C for mercury coating to ensure mutual diffusion between the coating metal and the membrane. After the coating material is chemically removed (e.g., with trichloroacetic acid for copper coating or with a 40-60% aqueous solution of iron chloride or with a 20-30% nitric acid [Gryaznov et al., 1987]. After drying, one or two porous layers are formed on the Pd-based membrane which is catalytically more active than a non-treated dense membrane.

The roughened surfaces prepared by the above method provide the higher specific surface areas for catalyst dispersion to be effective [Gryaznov, 1992]. Acid-treated Pd-Zn membranes appear to be very effective in improving both conversion and selectivity of a number of hydrocarbon hydrogenation reactions, resulting in essentially complete conversions in several cases [Mishchenko et al., 1987].

9.2.8 Catalytic Design of Palladium-Based Membrane Reactors

An inherent advantage of palladium-based membranes is that the hydrogen transported through the membrane is in the active monoatomic form. When a catalyst such as gold is finely dispersed on a palladium membrane, the monoatomic hydrogen which comes into contact with the catalyst surface becomes much more active than gaseous hydrogen molecules. In this manner, cyclohexene is hydrogenated to cyclohexane with an improved reaction rate on a dense Pd-Ag membrane electroplated with gold particles [Wood and Wise, 1966]. Similarly, hydrogenation of 1-butene to butane can be enhanced by employing a Pd-Ag membrane which is gold-electroplated [Yolles et al., 1971].

Not only the permeability, permselectivity and mechanical properties, but also the catalytic properties are affected by the two hydride forms that can exist in palladium. The α phase corresponds to solid solutions with a H/Pd ratio of about 0.1 and the β phase with a H/Pd ratio of about 0.6. The phase change is associated with a large change in lattice constant that often leads to microcracks and distortion in the palladium membrane. As a result, the mechanical properties are reduced. The α/β transformation depends on the operating conditions such as temperature and hydrogen partial pressure. Repeated thermal cycles, for example, between 100 and 250°C under 1 atm of hydrogen pressure can make a 0.1 mm thick Pd foil expand to become 30 times thicker [Armor, 1992].

Various alloying elements for palladium have been attempted in order to improve catalytic activity, durability or other considerations. It has been suggested that the preferred alloying metal(s) to be those elements of the Group VI to VIII in the periodic table [Gryaznov et al., 1977]. Some metals effective for this purpose are Ni, Ru, Rh, Cu and Ag at a dosage of 2 to 15% of the alloys by weight. The experimental results, however, have been at times confusing as has been pointed out to some degree in Chapter 8. A further example of this lack of design rules governing the selection of optimal membrane materials (e.g., the nature and amount of alloying metal) for achieving maximum conversion or selectivity has been observed by Armor [1992] and others. For the same dehydrogenation reaction of cyclohexane, Pd-Rh (7-20%) alloy membranes are catalytically effective but Pd-Ru membranes are not. And, for the same Pd-Ru (5%) membranes, catalytic hydrogenation of cyclohexene takes place while similar reaction of toluene does not [Armor and Farris, 1993].

It appears that whether the concept of applying Pd-based membrane reactors can be reliably applied depends to a great extent on the nature and amount of the alloying element(s) for the membrane material, amounts of the two hydride phases (α vs. β) present in the membrane, activation temperature and procedure, the reactant(s) other than hydrogen (especially, hydrocarbon), impurities such as sulfur compounds, carbon monoxide and soot and the underlying operating conditions. A more systematic understanding of the nature of these interactions is essential before Pd-membrane reactor technology can be implemented for production where robustness of the operating window is often critical.

9.2.9 Catalytic Effects of Membrane Reactor Components

Given a catalytic reaction, one also needs to consider any unintended catalytic effects of the materials of construction including the membrane and processing equipment.

Relevant to this issue is dehydrogenation of ethylbenzene for the manufacture of styrene which uses alumina supported iron oxide as the preferred catalyst in most cases. Therefore, when an alumina membrane is used in conjunction with stainless steel piping or vessels as the membrane reactor, caution should be exercised. An estimate of the effects of their exposure to the reaction mixture at the application temperature of 600 to 640°C is desirable. Wu et al. [1990b] estimated that the alumina membrane contributes to less than 5% conversion of ethylbenzene and the stainless steel tubing or piping could account for as much as 20% conversion. The high activity of the stainless steel is attributed to iron and chromium oxide layers that may form on the wetted surface.

9.3 SUMMARY

Material and catalysis aspects are the two foremost important factors when considering inorganic membrane reactors. They need to be evaluated early in the development stage.

Besides the critical issue of containment and sealing, the choice of the materials for the membrane and other membrane reactor components affects the permeability and permselectivity, operable temperature, pressure and chemical environments and reaction performance. Important material parameters include the particular chemical phase, thickness, thermal properties and surface contamination of the membrane, membrane/support microstructure, and sealing of the end surfaces of the membrane elements and of the joining areas between elements and module components. The conventional permeability versus permselectivity dilema associated with membranes needs to be addressed before inorganic membrane reactors are used in bulk processing.

The placement of the catalyst in the reactor relative to the membrane has significant impacts on the reaction performance. Three commonly used configurations are: the catalyst bed separated from the membrane, the catalyst is attached to the membrane surface or pore surface and the membrane being inherently catalytic. In the case of impregnating the catalyst inside a porous inorganic membrane, the optimal distribution and location of the catalyst can be different for different performance indices (e.g., total conversion, product purity and product molar flow rate, etc.). A uniform distribution of catalyst concentration is, in almost all cases, not the optimal distribution. Instead, step distributions are preferred. Some catalyst impregnation techniques enable reasonably precise placement of the catalyst in the membrane. Finally some guidelines for the selection of catalyst placement are given.

REFERENCES

Anonymous, 1988, Ceram. Bull. **67**, 1646.
Anonymous, 1995, Chem. Eng. (August), 19.
Arashi, H., and H. Naito, 1992, Solid State Ionics **53**, 431.
Armor, J.N., 1992, Chemtech **22**, 557.

Armor, J.N., and T.S. Farris, 1993, Membrane catalysis over palladium and its alloys, in: Proc. 10th Int. Congr. Catal., Budapest, Hungary (1992), p. 1363.

Balachandran, U., S.L. Morissette, J.T. Dusek, R.L. Mieville, R.B. Poeppel, M.S. Kleefisch, S. Pei, T.P. Kobylinski and C.A. Udovich, 1993, Development of ceramic membranes for partial oxygenation of hydrocarbon fuels to high-value-added products, in: Proc. Coal Liquefaction and Gas Conversion Contractors' Review Conf., Pittsburgh (U.S. Dept. of Energy).

Balachandran, V., J.T. Dusek, S.M. Sweeney, R.B. Poeppel, R.L. Mieville, P.S. Maiya, M.S. Kleefisch, S. Pei, T.P. Kobylinski, C.A. Udovich and A.C. Bose, 1995, Am. Ceram. Soc. Bull. **74**, 71.

Ballou, E.V., T. Wydeven and M.I. Leban, 1971, Environm. Sci. Technol. **5**, 1032.

Baratti, R., H. Wu, M. Morbidelli and A. Varma, 1993, Chem. Eng. Sci. **48**, 1869.

Bates, C.H., M.R. Foley, G.A. Rossi, G.J. Sundberg and F.J. Wu, 1990, Ceram. Bull. **69**, 350.

Bird, A.J., and D.L. Trimm, 1983, Carbon **21**, 177.

Burggraaf, A.J., K. Keizer and B.A. van Hassel, 1989, Solid State Ionics **32/33**, 771.

Cannon, K.C., and J.J. Hacskaylo, 1992, J. Membr. Sci. **65**, 259.

Capannelli, G., A. Bottino, G. Gao, A. Grosso, A. Servida, G. Vitulli, A. Mastantuono, G. Lazzaroni and P. Salvadori, 1993, Catal. Lett. **20**, 287.

Carolan M.F., P.N. Dyer, S.M. Fine, A. Makitka, III, R.E. Richards and L.E. Schaffer, 1994, U.S. Patent 5,332,597.

Chai, M., M. Machida, K. Eguchi and H. Arai, 1994, J. Membr. Sci. **96**, 205.

Champagnie, A.M., T.T. Tsotsis, R.G. Minet and I.A. Webster, 1990a, Chem. Eng. Sci. **45**, 2423.

Champagnie, A.M., T.T. Tsotsis, R.G. Minet and I.A. Webster, 1990b, Studies of ethane dehydrogenation in a ceramic membrane reactor, presented at Int. Congr. Membr. Membr. Proc., Chicago, IL, USA.

Clayson, D.M., and P. Howard, 1987, UK Pat. Appl. GB-2190397A.

Collins, J.P., and J.D. Way, 1993, Ind. Eng. Chem. Res. **32**, 306.

Colomban, Ph., and E. Bruneton, 1992, J. Non-Crystalline Solids **147&148**, 201,

Cot, L., 1991, J. Chim. Phys. **88**, 2083.

Desmaison, J., 1990, High temperature oxidation of nonoxide structural ceramics: use of advanced protective coatings, in: High Temperature Corrosion of Technical Ceramics (R.J. Fordham, Ed.), Elsevier Applied Science, London, U.K., p. 93.

Edlund, D.J., and W.A. Pledger, 1993, J. Membr. Sci. **77**, 255.

Eng, D., and M. Stoukides, 1991, Catal. Rev. - Sci. Eng. **33**, 375.

Förthmann, R., and A. Naoumidis, 1990, Influence of hot gas corrosion on the bending strength of silicon carbide materials, in: High Temperature Corrosion of Technical Ceramics (R.J. Fordham, Ed.), Elsevier Applied Science, London, U.K., p. 217.

Furneaux, R.C., A.P. Davidson and M.D. Ball, 1987, European Pat. Appl. 244,970.

Garcera, D., and J. Gillot, 1986, European Patent 170,025.

Gavriilidis, A., A. Varma and M. Morbidelli, 1993, Catal. Rev. - Sci. Eng. **35**, 399.

Gryaznov, V.M., 1992, Plat. Met. Rev. **36**, 70.

Gryaznov, V.M., V.S. Smirnov and M.G. Slin'ko, 1977, Binary palladium alloys as selective membrane catalysts., in: Proc. 6th Int. Congr. Catal., Vol. 2, p. 894.

Gryaznov, V.M., V.S. Smirnov, V.M. Vdovin, M.M. Ermilova, L.D. Gogua, N.A. Pritula and I.A. Litvinov, 1979, U.S. Patent 4,132,668.

Gryaznov, V.M., V.S. Smirnov, V.M. Vdovin, M.M. Ermilova, L.D. Gogua, N.A. Pritula and G.K. Fedorova, 1983, U.S. Patent 4,394,294.

Gryaznov, V.M., A.P. Mischenko, V.S. Smirnov, M.E. Sarylova and A.B. Fasman, 1987, U.K. Pat. Appl. 2,187,756A.

Guther, W., and W. Vielstich, 1982, Electrochimica Acta **27**, 811.

Haag, W.O., and J.G. Tsikoyiannis, 1991, U.S. Patent 5,019,263.

Harold, M.P., P. Cini, B. Patenaude and K. Venkataraman, 1989, AIChE Symp. Ser., **85** (268), 26.

Harold, M.P., V.T. Zaspalis, K. Keizer and A.J. Burggraaf, 1992, Improving partial oxidation product yield with a catalytic inorganic membrane, presented at 5th NAMS Annual Meeting, Lexington, KY, paper 11B.

He, R., Y. Lu and Y.H. Ma, 1994, Preparation of reactor tube by welding a porous membrane with a non-porous ceramic tube, presented at 3rd Int. Conf. Inorg. Membr., Worcester, MA, USA.

Hunter, J.B., 1960, Platinum Metals Rev. **4**, 130.

Hyun, S.H., and B.S. Kang, 1994, J. Am. Ceram. Soc. **77**, 3093.

Itoh, N., 1990, Sekiyu Gakkai-Shi (J. Japan Petro. Inst.) **33**, 136.

Itoh, N., Y. Shindo, K. Haraya, K. Obata, T. Hakuta and H. Yoshitome, 1985, Int. Chem. Eng. **25**, 138.

Itoh, N., M.A. Sanchez C., W.C. Xu, K. Haraya and M. Hongo, 1993, J. Membr. Sci. **77**, 245.

Jordan, G., and R.P. Greenspan, 1967, Tech. Quart. (Master Brewers Assoc. Am.) **4**, 114.

Kameyama, T., K. Fukuda, M. Fujishige, H. Yokokawa and M. Dokiya, 1981, Hydrogen Energy Progr. **2**, 569.

Keller, J.B., M.S. Falkovitz and H. Frisch, 1984, Chem. Eng. Sci. **39**, 601.

Kim, J.S., and R. Datta, 1991, AIChE J. **37**, 1657.

Koros, W.J., and G.K. Fleming, 1993, J. Membr. Sci. **83**, 1.

Langlet, M., D. Walz, P. Marage and J.C. Joubert, 1992, J. Non-Cryst. Solids **147&148**, 488.

Larbot, A., J.A. Alary, J.P. Fabre, C. Guizard and L. Cot, 1986, Mater. Res. Soc. Symp. Proc. (Better Ceramics through Chemistry II) **73**, 659.

Larbot, A., J.A. Alary, C. Guizard, J.P. Fabre, N. Idrissi and L. Cot, 1988, Ceramurgia **18**, 216.

Larbot, A., A. Julbe, C. Guizard and L. Cot, 1989, J. Membr. Sci. **44**, 289.

Lay, L.A., 1983, Corrosion Resistance of Technical Ceramics, National Physical Laboratory, U.K.

Leenaars, A.F.M., K. Keizer and A.J. Burggraaf, 1984, J. Mater. Sci. **19**, 1077.

Lin, Y.S., L.G.J. de Haart, K.J. de Vries and A.J. Burggraaf, 1989, EUROCERAMICS **3**, 590.

Liu, Y., A.G. Dixon, Y. H. Ma and W.R. Moser, 1990, Sep. Sci. Technol. **25**, 1511.

Livshits, A.I., M.E. Notkin and A.A. Samartsev, 1990, J. Nucl. Mater. **170**, 79.

Mischenko, A.P., V.M. Gryaznov, M.E. Sarylova and V.A. Bednyakova, 1987, U.K. Pat. Appl. 2,187,759A.

Mizuhara, H., E. Huebel and T. Oyama, 1989, Ceram. Bull. **68**, 1591.

Moskovits, M., 1984, U.S. Patent 4,472,533.

Nourbakhsh, N., A. Champagnie, T.T. Tsotsis and I.A. Webster, 1989, AIChE Symp. Ser. **85**, 75.

Omata, K., S. Hashimoto, H. Tominaga and K. Fujimoto, 1989, Appl. Catal. **52**, L1.

Peden, C.H.F., B.D. Kay and D.W. Goodman, 1986, Surf. Sci. **175**, 215.

Perry, R.H. and C.H. Chilton, eds.,1973, Chemical Engineers' Handbook, 5th ed. (McGraw-Hill. N.Y., USA)

Philpott, J.E., 1985, Platinum Metals Rev. **29**, 12.

Raman, N.K., T.L. Ward, C.J. Brinker, R. Sehgal, D.M. Smith, Z. Duan, M. Hampden-Smith, J.K. Bailey and T.J. Headley, 1993, Appl. Catal. A: General **69**, 65.

Saracco, G., and V. Specchia, 1994, Catal. Rev.-Sci. Eng. **36**, 305.

Schnabel, R., and W. Vaulont, 1978, Desalination **24**, 249.

Smirnov, V.S., M.M. Ermilova, V.M. Gryaznov, N.V. Orekova, V.P. Polyakova, N.R. Roshan and E.M. Savitsky, 1983, U.K. Patent 2,068,938.

Stevens, R., 1986, Zirconia and Zirconia Ceramics (2nd ed.), Magnesium Elektron, U.K.

Sudarshan, T.S., C.K. Waters and M.R. Louthan, Jr., 1988, J. Mater. Eng. **10**, 215.

Sun, Y.M., and S.J. Khang, 1988, I&EC Res. **27**, 1136.

Takata, K., and T. Suzuki, 1993, Mater. Sci. Eng. **A163**, 91.

Uemiya, S., T. Matsuda and E. Kikuchi, 1990, Chem. Lett. 1335.

Uemiya, S., N. Sato, H. Ando, Y. Kude, T. Matsuda and E. Kikuchi, 1991a, J. Membr. Sci. **56**, 303.

Uemiya, S., T. Matsuda and E. Kikuchi, 1991b, J. Membr. Sci. **56**, 315.

Uzio, D., A. Giroir-Fendler, J. Lieto and J.A. Dalmon, 1991, Key Eng. Mat. **61** and **62**, 111.

van Praag, W., V.T. Zaspalis, K. Keizer, J.G. van Ommen, L.R.H. Ross and A.J. Burggraaf, 1989, EUROCERAMICS 3, 605.

Velterop, F.M., and K. Keizer, 1991, Key Eng. Mat. **61 & 62**, 391.

Wefers, K., and C. Misra, 1987, Oxides and hydroxides of aluminum, Alcoa Tech. Paper 19 (rev.), Aluminum Co. of America, Pittsburgh, PA, USA.

Wood, B.J., and H. Wise, 1966, J. Catal. **5**, 135.

Wu, H., A. Brunovska, M. Morbidelli and A. Varma, 1990a, Chem. Eng. Sci. **45**, 1855.

Wu, J.C.S., T.E. Gerdes, J.L. Pszczolkowski, R.R. Bhave, P.K.T. Liu and E.S. Martin, 1990b, Sep. Sci. Technol. **25**, 1489.

Xu, Q., and M.A. Anderson, 1989, Mater. Res. Soc. Symp. Proc. (Multicomponent Ultrafine Microstructure) **132**, 41.

Xu, Q., and M.A. Anderson, 1994, J. Am. Ceram. Soc. **77**, 1939.

Yamawaki, M., T. Namba, K. Yamaguchi and T. Kiyoshi, 1987, Nucl. Inst. Meth. **B23**, 498.

Yamawaki, M., V. Bandurko, R. Satoh and K. Yamaguchi, 1994, Control of hydrogen permeation through metal membranes by varying surface contamination condition, in: Proc. 18th Symp. Fusion Technology (K. Herschbach, W. Maurer and J.E. Vetter, Eds.), Elsevier Sci., Netherlands, p. 1075.

Yazawa, T., H. Tanaka, H. Nakamichi and T. Yokoyama, 1991, J. Membr. Sci. **60**, 307.

Yeung, K.L., R. Aravind, R.J.X. Zawada, J. Szegner, G. Cao and A. Varma, 1994, Chem. Eng. Sci. **49**, 4823.

Yolles, R.S., B.J. Wood and H. Wise, 1971, J. Catal. **21**, 66.

Zaspalis, V.T., W. van Praag, K. Keizer, J.G. van Ommen, J.R.H. Ross and A.J. Burggraaf, 1991a, Appl. Catal. **74**, 205.

Zaspalis, V.T., W. van Praag, K. Keizer, J.G. van Ommen, J.R.H. Ross and A.J. Burggraaf, 1991b, Appl. Catal. **74**, 249.

CHAPTER 10

INORGANIC MEMBRANE REACTORS -- MODELING

10.1 OVERVIEW OF MEMBRANE REACTOR MODELING

When implemented in the field, inorganic membrane reactors invariably will be large in scale and complex in configuration. Their reliable design and operation rest on the foundation of good understanding and design of laboratory and pilot reactors. An important tool that helps build that foundation is validated mathematical models.

For ease of fabrication and modular construction, tubular reactors are widely used in continuous processes in the chemical processing industry. Therefore, shell-and-tube membrane reactors will be adopted as the basic model geometry in this chapter. In real production situations, however, more complex geometries and flow configurations are encountered which may require three-dimensional numerical simulation of the complicated physicochemical hydrodynamics. With the advent of more powerful computers and more efficient computational fluid dynamics (CFD) codes, the solution to these complicated problems starts to become feasible. This is particularly true in view of the ongoing intensified interest in parallel computing as applied to CFD.

10.2 GENERAL FRAMEWORK OF MODEL DEVELOPMENT

In a simple membrane reactor, basically the membrane divides the reactor into two compartments: the feed and the permeate sides. The geometries of the membrane and the reaction vessel can vary. The feed may be introduced at the entrance to the reactor or at intermediate locations and the exiting retentate stream, for process economics, may be recycled back to the reactor. Furthermore, the flow directions of the feed and the sweep (including permeate) streams can be co-current or counter-current or some combinations. It is obvious that there are numerous possible process and equipment configurations even for a geometrically simple membrane reactor.

Discussions in this chapter will focus mostly on a relatively simple geometry of a shell-and-tube membrane reactor with a feed stream and a sweep stream flowing on the opposite sides (the tubular and annular regions) of the membrane which may be dense or porous in nature. The catalyst may be located in the tubular region, the membrane matrix and/or the annular region. Normally, however, the catalysts are placed in one or two of the those three regions. In special cases such as reaction coupling through a membrane, catalyst beds may be present in both the feed and permeate sides. Where appropriate, other complexities will be discussed. However, many assumptions will be made to facilitate the development of relatively simple models of some generality.

10.3 MODELING OF SHELL-AND-TUBE MEMBRANE REACTORS

Among the various models proposed for simple shell-and-tube membrane reactors, three general categories emerge: packed-bed catalytic membrane reactors, fluidized-bed catalytic membrane reactors and catalytic nonpermselective membrane reactors. Their differences primarily are: where the catalytic zone(s) is or are and, in those cases containing catalyst beds, how the bed is sustained. Special cases of each category derived from the general models will be discussed in some detail in this chapter. Some of the membrane reactor dynamics described by these models as well experimental observations will be treated in Chapter 11 along with other engineering considerations.

10.4 PACKED-BED CATALYTIC MEMBRANE TUBULAR REACTORS

The first type of generic model for shell-and-tube membrane reactors refers to a nonisothermal packed-bed catalytic membrane tubular reactor (PBCMTR) whose cross-sectional view is shown in Figure 10.1. Mathematical models for this type of membrane reactor have been reviewed quite extensively by Tsotsis et al. [1993b].

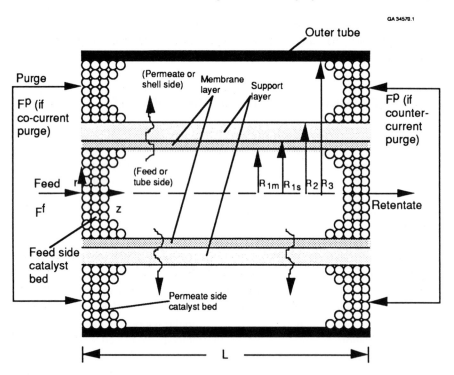

Figure 10.1 Schematic diagram of a packed-bed membrane shell-and-tube reactor

10.4.1 Problem Statement and Assumptions

Consider the PBCMTR in Figure 10.1. A concentric dense or porous membrane tube consisting of a membrane layer adjacent to a support layer is contained inside an impervious outer tube to form the membrane reactor. Assume that the reactant(s) enters the tube side at $z = 0$ and the retentate exits at $z = L$. In a co-current (or counter-current) flow configuration with respect to the feed stream, the purge (or called sweep) stream enters at $z = 0$ (or $z = L$) and carries the permeate with it on the shell side to the other end of the tube. Although Figure 10.1 assumes that the reactant is introduced to the tube side as the feed stream, the governing equations to be described later can be modified in a straightforward manner to be applied to those cases where the reactant enters the annular region on the shell side.

In a generalized case, both the tube and the shell sides are packed with catalyst beds (e.g., in a reaction coupling situation explained in Chapter 8) and the membrane layer is catalytic either inherently or through impregnation on the pore surfaces. Two commonly occurring membrane reactor configurations will be treated as special cases of this generalized model.

10.4.2 General Models Describing Transport Phenomena in Packed-Bed Catalytic Membrane Tubular Reactors

In the following, steady state mass and heat transport equations are written for species j and reaction i in each of the regions depicted in Figure 10.1. Assumptions are made that:

(1)   The diffusivities $(D_j^t, D_j^m, D_j^u \text{ and } D_j^s)$ and thermal conductivities $(\lambda^t, \lambda^m, \lambda^u, \lambda^s)$ are independent of concentrations $(C_j^t, C_j^m, C_j^u \text{ and } C_j^s)$ and temperatures $(T^t, T^m, T^u, \text{ and } T^s)$.

(2)   All axial diffusion and heat conduction terms can be ignored when compared to their radial counterparts.

(3)   The only significant mass and heat transfer mechanisms in the membrane and the support layers are radial diffusion and conduction and reaction effects. The membrane or the support matrix containing the fluid is treated as a continuum with effective diffusivities and thermal conductivities.

(4)   Radial velocities on the tube and shell sides are both negligible compared to axial velocities.

(5)   The interfacial mass- and heat-transfer resistance between any two adjacent regions is negligible compared to the corresponding internal mass- or heat-transfer resistances in each region.

(6)   Ideal gas law applies.

Finally, it is assumed that multiple species ($j = 1,...J$) react in multiple reactions ($i = 1,...I$) and the rate of the reaction i can be expressed in the following form:

$$r_i^{\Phi} = k_i^{\Phi} \mathfrak{R}_i^{\Phi}(C_j^{\Phi}, T^{\Phi}, K_{ei}^{\Phi}) \tag{10-1a}$$

$$= k_i^{\Phi} \frac{\prod\limits_{j}^{J}(C_j^{\Phi})^{\beta_{ij}^{F\Phi}} - \prod\limits_{j}^{J}(C_j^{\Phi})^{\beta_{ij}^{R\Phi}} / K_{ei}^{\Phi}}{G_i^{\Phi}(C_j^{\Phi}, T^{\Phi})} \tag{10-1b}$$

where Φ refers to a specific region of the shell-and-tube membrane reactor (t = tube side; m = membrane layer; u = support layer; and s = shell side), k_i^{Φ} the reaction rate constant for the reaction i in the region Φ, described by an Arrhenius expression $A_i^{\Phi} \exp(-E_i^{\Phi} / RT)$, $\mathfrak{R}_i^{\Phi}$ the reaction dependence form such that the reaction term is equal to $k_i^{\Phi} \mathfrak{R}_i^{\Phi}$, K_{ei}^{Φ} the equilibrium constant for the reaction i and C_j^{Φ} and T^{Φ} the concentration of species j and the temperature, respectively, in the region Φ. G_i^{Φ} represents some expressions unique to certain reactions and it is often equal to unity.

Furthermore,

$$\beta_i^{F\Phi} = \sum_j^J \beta_{ij}^{F\Phi} \tag{10-2a}$$

defines the overall reaction order of the i forward reaction and

$$\beta_i^{R\Phi} = \sum_j^J \beta_{ij}^{R\Phi} \tag{10-2b}$$

gives the overall reaction order of the i reverse reaction.

Under the above assumptions, the convective mass and heat transfer equations governing the changes of concentrations of species j and temperatures for the feed (tube) side, membrane layer, support layer and permeate (shell) side (i.e., $C_j^t, T^t, C_j^m, T^m, C_j^u, T^u, C_j^s$ and T^s, respectively) are:

On the tube (feed) side:

$$v^t \frac{\partial C_j^t}{\partial z} = D_j^t \frac{1}{r}\frac{\partial}{\partial r}\left(r\frac{\partial C_j^t}{\partial r}\right) + \sum_i^I \upsilon_{ij}^t \eta_i^t k_i^t \mathfrak{R}_i^t (C_j^t, T^t, K_{ei}^t) \quad \text{(for } j = 1 \text{ to } J)$$ (10-3)

$$\left(\sum_j^J v^t C_j^t c_{pj}\right)\frac{\partial T^t}{\partial z} = \lambda^t \frac{1}{r}\frac{\partial}{\partial r}\left(r\frac{\partial T^t}{\partial r}\right)$$

$$+ \sum_i^I \left(-\Delta H_i^t\right)\eta_i^t k_i^t \mathfrak{R}_i^t \left(C_j^t, T^t, K_{ei}^t\right)$$ (10-4)

In the permselective and catalytic membrane layer:

$$D_j^m \frac{1}{r}\frac{\partial}{\partial r}\left(r\frac{\partial C_j^m}{\partial r}\right) + \sum_i^I v_{ij}^m k_i^m \mathfrak{R}_i^m \left(C_j^m, T^m, K_{ei}^m\right) = 0 \quad \text{(for } j = 1 \text{ to } J)$$ (10-5)

$$\lambda^m \frac{1}{r}\frac{\partial}{\partial r}\left(r\frac{\partial T^m}{\partial r}\right) + \sum_i^I \left(-\Delta H_i^m\right)k_i^m \mathfrak{R}_i^m \left(C_j^m, T^m, K_{ei}^m\right) = 0$$ (10-6)

In the inert porous support layer:

$$D_j^u \frac{1}{r}\frac{\partial}{\partial r}\left(r\frac{\partial C_j^u}{\partial r}\right) = 0 \quad \text{(for } j = 1 \text{ to } J)$$ (10-7)

$$\lambda^u \frac{1}{r}\frac{\partial}{\partial r}\left(r\frac{\partial T^u}{\partial r}\right) = 0$$ (10-8)

Finally, on the shell (permeate) side:

$$v^s \frac{\partial C_j^s}{\partial z} = D_j^s \frac{1}{r}\frac{\partial}{\partial r}\left(r\frac{\partial C_j^s}{\partial r}\right) + \sum_i^I \upsilon_{ij}^s \eta_i^s k_i^s \mathfrak{R}_i^s \left(C_j^s, T^s, K_{ei}^s\right) \quad \text{(for } j = 1 \text{ to } J)$$ (10-9)

$$\left(\sum_j^J v^s C_j^s c_{pj}\right)\frac{\partial T^s}{\partial z} = \lambda^s \frac{1}{r}\frac{\partial}{\partial r}\left(r\frac{\partial T^s}{\partial r}\right)$$

$$+ \sum_i^I \left(-\Delta H_i^s\right)\eta_i^s k_i^s \mathfrak{R}_i^s \left(C_j^s, T^s, K_{ei}^s\right)$$ (10-10)

where

$D_j^t, D_j^m, D_j^u, D_j^s$ = diffusivity of species j through the respective region

v^t, v^s = velocity on tube and shell sides, respectively

$v_{ij}^t, v_{ij}^m, v_{ij}^s$ = stoichiometric coefficient for species j in reaction i in the respective region

$\eta_i^t \eta_i^s$ = effectiveness factor for reaction i on the tube and shell sides, respectively

C_{pj} = heat capacity for species j

$\lambda^t, \lambda^m, \lambda^u, \lambda^s$ = effective thermal conductivity of content in the respective region

$\left(-\Delta H_i^t\right), \left(-\Delta H_i^m\right), \left(-\Delta H_i^s\right)$ = heat of reaction for reaction i in the respective region

The effectiveness factors η_i^t and η_i^s in Equations (10-3) to (10-4) and Equations (10-9) to (10-10) account for the effects of interphase and intraparticle mass transfer resistances on the tube (feed) and shell (permeate) sides, respectively. When these resistances are not negligible, the fugacity of a reactant within a catalyst particle is lower than that in the bulk fluid and may vary radially in the particle. As a result, the actual reaction rate would be lower than that based on the bulk fluid fugacity. Thus the effectiveness factor is the average reaction rate in the catalyst particle divided by the average reaction rate based on the bulk stream conditions. The fugacity distribution inside a spherical particle can be obtained by solving differential mass balance equations. The effectiveness factors can then be calculated (see, for example, Collins et al. [1993]) along the length of the membrane reactor.

It is relevant at this point to stress the importance of preserving the nonisothermal condition of the reactor. Most of the modeling studies of membrane reactors assume an isothermal operation. However, as it has been demonstrated experimentally [Becker et al., 1993], this assumption is incorrect and, more often than not, a temperature profile exists along the membrane reactor length.

The above set of equations for the four regions (tube side, membrane layer, support layer and shell side) are to be solved with prescribed inlet and other boundary conditions. They require an inlet condition for each of the J species and the temperature on both the tube and shell sides. Additionally, the following boundary conditions in the r-direction are needed: two for each of the J species and the temperature for each of the four regions.

Depending on the composition of the incoming reactant stream on the tube side, the inlet concentration may take the form:

$$C_j^t\big|_{z=0} = \begin{cases} C_{j0}^t \ for \ j = j_f \\ 0 \ for \ j \neq j_f \end{cases} \tag{10-11}$$

where j_f's are the species contained in the feed stream and C^t_{j0} are some given values and the inlet temperature is

$$T^t\big|_{z=0} = T^t_0 \qquad\qquad (10\text{-}12)$$

Similarly, on the shell side,

$$C^s_j\big|_{z=0} = \begin{cases} C^s_{j0} \ for \ j = j_p \\ 0 \ for \ j \neq j_p \end{cases} \qquad\qquad (10\text{-}13\mathrm{a})$$

where C^s_{j0} are given and j_p's are the species comprised of the entering shell side stream. In many cases the shell side inlet stream contains only the inert sweep (or purge) gas. When the sweep gas is inert with respect to the reactions,

$$C^s_j\big|_{z=0} = 0 \quad (\text{inert sweep gas}) \qquad\qquad (10\text{-}13\mathrm{b})$$

The inlet temperature may be described by:

$$T^s\big|_{z=0} = T^s_0 \qquad\qquad (10\text{-}14)$$

In the radial direction, all the species concentrations and temperatures in each region require two boundary conditions. The symmetry condition at the tube center-line calls for

$$\frac{\partial C^t_j}{\partial r}\bigg|_{r=0} = 0 \qquad\qquad (10\text{-}15)$$

and the impervious outer tube wall leads to a no flux boundary condition

$$\frac{\partial C^s_j}{\partial r}\bigg|_{r=R_3} = 0 \qquad\qquad (10\text{-}16)$$

Furthermore, the four regions are coupled by the requirement that the concentrations and temperatures as well as their fluxes in the radial direction be continuous at the interfaces of two adjacent regions:

$$C^t_j\big|_{r=R_{1m}} = C^m_j\big|_{r=R_{1m}} \qquad\qquad (10\text{-}17)$$

$$D^t_j \frac{\partial C^t_j}{\partial r}\bigg|_{r=R_{1m}} = D^m_j \frac{\partial C^m_j}{\partial r}\bigg|_{r=R_{1m}} \qquad\qquad (10\text{-}18)$$

$$C_j^m\big|_{r=R_{1s}} = C_j^u\big|_{r=R_{1s}} \tag{10-19}$$

$$D_j^m \frac{\partial C_j^m}{\partial r}\bigg|_{r=R_{1s}} = D_j^u \frac{\partial C_j^u}{\partial r}\bigg|_{r=R_{1s}} \tag{10-20}$$

$$C_j^u\big|_{r=R_2} = C_j^s\big|_{r=R_2} \tag{10-21}$$

$$D_j^u \frac{\partial C_j^u}{\partial r}\bigg|_{r=R_2} = D_j^s \frac{\partial C_j^s}{\partial r}\bigg|_{r=R_2} \tag{10-22}$$

The various radii R_{1m}, R_{1s}, R_2 and R_3 are shown in Figure 10.1.

Similarly, the thermal boundary conditions are stated as

$$\frac{\partial T^t}{\partial r}\bigg|_{r=0} = 0 \tag{10-23}$$

$$T^t\big|_{r=R_{1m}} = T^m\big|_{r=R_{1m}} \tag{10-24}$$

$$\lambda^t \frac{\partial T^t}{\partial r}\bigg|_{r=R_{1m}} = \lambda^m \frac{\partial T^m}{\partial r}\bigg|_{r=R_{1m}} \tag{10-25}$$

$$T^m\big|_{r=R_{1s}} = T^u\big|_{r=R_{1s}} \tag{10-26}$$

$$\lambda^m \frac{\partial T^m}{\partial r}\bigg|_{r=R_{1s}} = \lambda^u \frac{\partial T^u}{\partial r}\bigg|_{r=R_{1s}} \tag{10-27}$$

$$T^u\big|_{r=R_2} = T^s\big|_{r=R_2} \tag{10-28}$$

$$\lambda^u \frac{\partial T^u}{\partial r}\bigg|_{r=R_2} = \lambda^s \frac{\partial T^s}{\partial r}\bigg|_{r=R_2} \tag{10-29}$$

$$\lambda^s \frac{\partial T^s}{\partial r}\bigg|_{r=R_3} = h^w (T^o - T^s)\big|_{r=R_3} \tag{10-30}$$

where h^w is the heat transfer coefficient of the reactor outer wall and T^o the temperature of the medium or the heating/cooling fluid surrounding the outer wall on the shell side.

At $z=0$, the continuity conditions such as Equation (10-17) are inconsistent with a uniform inlet condition such as Equation (10-11). One way to resolve this dilemma is to use an artificial function $g(z)$ defined as [Becker et al., 1993]

$$g(z) = \begin{cases} C^t_{j0}\left[e^{-z/z_c} + (z/z_c)e^{-z/z_c}\right] & for\ j = j_f \\ 0 & for\ j \neq j_f \end{cases} \tag{10-31}$$

such that Equation (10-17) is replaced by

$$C^t_j\big|_{r=R_{1m}} - g(z) = C^m_j\big|_{r=R_{1m}} \tag{10-32}$$

This results in interface concentrations on the membrane side to be different from the uniform inlet concentration, but the two concentrations are very quickly brought to equality. The rate of approaching equality depends on the choice of the parameter z_c. Thus $C^m_j\big|_{r=R_{1m}}$ is zero at $z=0$ but increases rapidly to $C^t_j\big|_{r=R_{1m}}$ as z increases.

So far the problem has been formulated to account for mass and heat transport. To complete the model, some momentum transfer submodel needs to be established. It is generally assumed that the equations describing the fluid flow in the feed and permeate regions are independent of the above mass or heat transfer equations. In reality the change in the flow rate due to the change in the number of moles after reaction is insignificant only when the concentration of the reactant(s) or permeate(s) in the carrier gas(es) is dilute. In many other situations, the coupling effects of membrane permeation and axial pressure drop on either side of the composite membrane need to be considered [Fang et al., 1975]. The coupling, of course, will significantly add to the complexity of the problem statement.

The simplest way of characterizing the fluid flows in the tube or shell regions is to assume a plug flow and negligible pressure drop. Plug flow conditions are a good approximation at a high Reynolds number, for example, in flows through a packed bed of catalyst particles where the small openings for the flow make the velocity profile inconsequential. There is some indication that the exact velocity profile used for the feed or the permeate side does not result in significant differences in the simulated reactor performance.

Negligible pressure drop may be a reasonable assumption for an unpacked tubular or annular region especially for a reasonably short reactor or for a packed region where the catalyst particle size is relatively large compared to the opening area for the fluid flow. Otherwise, pressure drop equations applicable to packed beds of catalyst particles are needed.

Pressure drop due to fluid flow through a tube or annulus with or without packed beds can be calculated according to the following equation

$$-\frac{dp}{dz} = 4f\frac{(\frac{1}{2}\rho\bar{v}^2)}{L_c}$$ (10-33)

where f is the Fanning friction factor, ρ the fluid density, $\bar{v}$ the average axial velocity (or superficial velocity in the case of fluid flow through a packed bed) and L_c is a characteristic length (equal to the inner tube or outer tube diameter when no packed bed is present or equal to the equivalent diameter of catalyst particles). The friction factor, f, is in turn given by

$$f = \frac{16}{D_1\bar{v}^t\rho^t}$$ (10-34a)

for an empty tube (on the tube side) of an inside diameter of D_1, where ρ^t is the density of the gas mixture and [Bird et al., 1960]

$$f = \frac{16}{D_2\bar{v}^s\rho^s}\left\{\frac{ln(D_1/D_2)}{ln(D_1/D_2)[1+(D_1/D_2)^2]+[1-(D_1/D_2)^2]}\right\}$$ (10-34b)

for an empty annulus (on the shell side) of an outer and an inner diameter of D_2 and D_1, respectively, and

$$f = \frac{(1-\varepsilon)}{\varepsilon^3}\left[\frac{75(1-\varepsilon)}{\rho d_p\bar{v}/\mu} + 0.875\right]$$ (10-34c)

for a packed bed with a void fraction, ε, and an equivalent particle diameter of d_p. Equation (10-34c) combined with Equation (10-33) constitutes the Ergun equation.

It is noted that Equations (10-34a) and (10-34b) can be applied to laminar flows. Other expressions are available for turbulent flows (e.g., Bird et al., [1960]). Equation (10-34c) is applicable over a wide range of flow rates.

Thus, the combination of Equation (10-33) and Equation (10-34a), (10-34b) or (10-34c) describes the pressure drop due to fluid flow in the tubular and annular regions. The choice among Equations (10-34a), (10-34b) or (10-34c) depends on whether a packed bed of catalyst particles is present in the given region.

The mass-, heat- and momentum-transfer equations and their corresponding boundary conditions discussed so far are obviously very complex and their solutions are not trivial to obtain. Moreover, the thickness, diffusivities and conductivity of each layer in the membrane element are difficult to measure. It is, therefore, convenient and reasonable to consider the permselective membrane layer and the support layer(s) as an integral region with effective thickness, diffusivities and conductivity for the composite region. And it is also desirable to search for simpler models which are capable of providing the engineering insight needed for optimal design and operation of inorganic membrane reactors. The collapse of two regions into one reduces the number and complexity of the governing equations. Many membrane reactor models in the literature, in fact, eliminate the membrane region altogether and account for the permeation effects through the use of interfacial mass transfer terms in the differential mass balance equations for the feed and permeate regions.

The majority of the membrane reactor models (except, for example, that of Becker et al. [1993]) have further simplified the governing equations in the tubular and annular regions by performing macroscopic mass and energy balances for those two regions. This is done by taking differential balances only in the axial direction by assuming plug flows. One can essentially replace the radial diffusion terms in Equations (10-3), (10-4), (10-9), and (10-10) with constant terms determined by the permeate fluxes and heat fluxes at the feed side-membrane, membrane-permeate side and, finally, permeate side-outer wall interfaces. This approach has been taken by Tsotsis et al. [1993b] for a general model describing a nonisothermal packed-bed catalytic membrane reactor consisting of three regions: the membrane (or membrane/support matrix), tube side and shell side. Their model will be presented in the following as an approximation to the more accurate, and yet complex, model discussed earlier. The molar flow rate of species j ($\dot{n}_j^\Phi$ for each region Φ) instead of concentrations will be used on the tube and shell sides. Thus

$$\dot{n}_j^\Phi = F^\Phi C_j^\Phi \tag{10-35}$$

where F^Φ represents the total volumetric flow rate in the region Φ.

Furthermore, the membrane-support composite will be treated as a single region "m" and the inner tube diameter is now designated as R_1. See Figure 10.2.

Macroscopic balance model. On the tube (feed) side:

$$\frac{d\dot{n}_j^t}{dz} = 2\Pi R_1 D_j^m \left(\frac{dC_j^m}{dr}\right)_{r=R_1} + \pi R_1^2 \sum_i^I \upsilon_{ij}^t \eta_i^t k_i^t \mathfrak{R}_i^t\left(C_j^t, T^t, K_{ei}^t\right) \tag{10-36}$$

$$\left(\sum_j^J \dot{n}_j^t C_{pj}\right)\frac{dT^t}{dz} = 2\pi R_1 \lambda^m \left(\frac{dT^m}{dr}\right)_{r=R_1} +$$

$$\pi R_1^2 \sum_i^I \left(-\Delta H_i^t\right)\eta_i^t k_i^t \mathfrak{R}_i^t\left(C_j^t, T^t, K_{ei}^t\right) \tag{10-37}$$

GA 34570.2

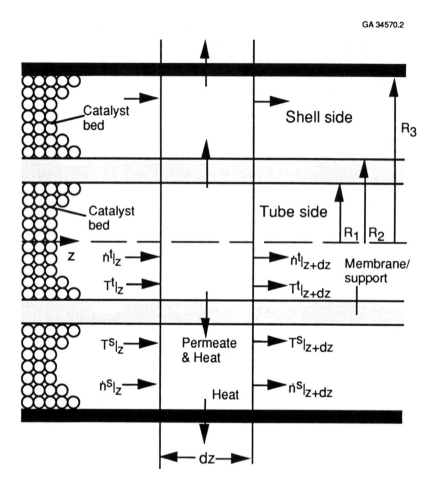

Figure 10.2 Schematic diagram of a plug-flow macroscopic model for a packed-bed membrane shell-and-tube reactor

subject to the inlet conditions

$$\left.\begin{array}{l} \dot{n}_j^t = \dot{n}_{j0}^t = F_0^t C_{j0}^t \\ T^t = T_0^t \end{array}\right\} \quad \text{at } z = 0 \tag{10-38}$$

where C_{pj} is the heat capacity of species j and $\dot{n}_{j0}^t$ and F_0^t the molar flow rate of species j and the total volumetric flow rate at the inlet of the tube (feed) side. C_{j0}^t is the inlet concentration of species j on the tube side.

The pressure drop occurring in the tube region is given by the Ergun equation, Equation (10-34c), for a packed bed:

$$-\frac{dp^t}{dz} = 4f^t\frac{\{\frac{1}{2}\rho^t(\bar{v}^t)^2\}}{d_p^t} \tag{10-39}$$

where

$$f^t = \frac{(1-\varepsilon^t)}{(\varepsilon^t)^3}\left[\frac{75(1-\varepsilon^t)}{\rho^t d_p^t \bar{v}^t/\mu^t} + 0.875\right] \tag{10-40}$$

The fluid density ρ^t is calculated from

$$\rho^t = \sum_j^J C_j^t M_j = \sum_j^J \frac{p_T^t X_j^t}{RT^t} M_j \tag{10-41}$$

where p_T^t is the total pressure on the tube side, and X_j^t and M_j the mole fraction and molecular weight of the species j. Equation (10-39) requires a boundary condition (pressure) to be specified, for example, at the reactor exit $z = L$.

In the membrane matrix:

The mass and heat transfer for this region can still be accounted for by Equations (10-5) and (10-6):

$$D_j^m\frac{1}{r}\frac{\partial}{\partial r}\left(r\frac{\partial C_j^m}{\partial r}\right) + \sum_i^I v_{ij}^m k_i^m \Re_i^m\left(C_j^m, T^m, K_{ei}^m\right) = 0 \quad \text{(for } j = 1 \text{ to } J) \tag{10-5}$$

$$\lambda^m\frac{1}{r}\frac{\partial}{\partial r}\left(r\frac{\partial T^m}{\partial r}\right) + \sum_i^I\left(-\Delta H_i^m\right)k_i^m \Re_i^m\left(C_j^m, T^m, K_{ei}^m\right) = 0 \tag{10-6}$$

The associated boundary conditions are

$$\left.\begin{array}{l} C_j^m = C_j^t = C_T^t X_j^t \\[2mm] \lambda^m\frac{dT^m}{dr} = h^t(T^m - T^t) \end{array}\right\} \quad \text{at } r = R_1 \tag{10-42}$$

and

$$
\left.\begin{aligned}
C_j^m &= C_j^s = C_T^s X_j^s \\
-\lambda^m \frac{dT^m}{dr} &= h^s (T^m - T^s)
\end{aligned}\right\} \quad \text{at } r = R_2
$$

(10-43)

where $C_T^s \left(= \sum_j^J C_j^s \right)$ is the total concentration in the shell region and h^t and h^s the heat transfer coefficients for the tube and shell sides of the membrane, respectively. It is noted that since T^t and T^s do not depend on the radial position in the model, they represent the "bulk" temperatures in their respective regions. The usual continuity boundary condition of the form $T^t = T^m$ at $r=R_1$ or $T^m = T^s$ at $r=R_2$ does not apply here because T^m is but T^t and T^s are not dependent on r.

On the shell (permeate) side:

$$
\pm \frac{d\dot{n}_j^s}{dz} = -2\pi R_2 D_j^m \left(\frac{dC_j^m}{dr} \right)_{r=R_2} +
$$

$$
\pi \left(R_3^2 - R_2^2 \right) \sum_i^I \upsilon_{ij}^s \eta_i^s k_i^s \Re_i^s \left(C_j^s, T^s, K_{ei}^s \right)
$$

(10-44)

$$
\pm \left(\sum_j^J \dot{n}_j^s C_{pj} \right) \frac{dT^s}{dz} = -2\pi R_2 \lambda^m \left(\frac{dT^m}{dr} \right)_{r=R_2}
$$

$$
+\pi \left(R_3^2 - R_2^2 \right) \sum_i^I \left(-H_i^s \right) \eta_i^s k_i^s \Re_i^s \left(C_j^s, T^s, K_{ei}^s \right)
$$

$$
+2\pi R_3 U^w (T^o - T^s)
$$

(10-45)

and

$$
-\frac{dp^s}{dz} = 4 f^s \frac{(\frac{1}{2}\rho^s \bar{v}_p^2)}{d_p^s}
$$

(10-46)

where, if a catalyst bed is present in the shell region,

$$
f^s = \frac{(1-\varepsilon^s)}{(\varepsilon^s)^3} \left[\frac{75(1-\varepsilon_2^s)}{\rho^s d_p^s \bar{v}^s / \mu^s} + 0.875 \right]
$$

(10-47)

and ρ^s, μ^s, ε^s, $\bar{v}^s$ and d_p^s are defined similarly as their counterparts in the tube region. T^o is the temperature of the surrounding or the cooling/heating fluid. The overall heat transfer coefficient through the outer wall between the two "bulk" temperatures is U^w.

The "±" sign in Equations (10-44) and (10-45) is determined by the flow direction on the permeate (shell) side relative to the feed (tube) side. "+" applies if the two streams are co-current and "−" if counter-current.

The above equations are subject to the following boundary conditions:

$$\left.\begin{array}{l} \dot{n}_j^s = \dot{n}_{j0}^s = F_0^s C_{j0}^s \\ T^s = T_0^s \end{array}\right\} \quad at\ z = 0 \tag{10-48}$$

and the pressure at the reactor exit must be specified

$$p^s = p_L^s \quad (at\ z = L) \tag{10-49}$$

For a counter-current stream in the permeate region, the boundary condition described by Equation (10-48) is specified at $z=L$ instead.

Tsotsis et al. [1993b] assume the effectiveness factors η_i^t and η_i^s to be equal to 1. This assumption can be justified for those reactions that are not diffusion limited as can be determined by experiments with different particle sizes of the catalyst. They then use the membrane radius, membrane length, feed and purge gas concentrations and temperature and other properties of membranes and gases and operating parameters to non-dimensionalize the above set of governing equations and boundary conditions derived from the macroscopic mass and heat balances. The details will not be presented here and readers are referred to the original source. The physical significance of some of those dimensionless groups or variables will be mentioned wherever they are appropriate to the discussions later in this chapter as well as later chapters.

10.4.3 Solution Methodology

As mentioned earlier, most membrane reactor models are based on isothermal macroscopic balances in the axial direction and do not solve the transport equations for the membrane/support matrix. They all account for the effects of membrane permeation through the use of some common relevant parameters (as a permeation term) in the transport equations for both the feed and permeate sides. Those parameters are to be determined experimentally. The above approach, of course, is feasible only when the membrane (or membrane/support) is not catalytic.

When only the feed side and permeate side mass balance equations are considered under the isothermal condition, the resulting equations are a set of first-order ordinary differential equations. Furthermore, a co-current purge stream renders the set of equations an initial value problem and well established procedures such as the

Runge-Kutta-Gill method, implicit Euler scheme or finite element technique [Agarwalla and Lund, 1992] can be adopted to obtain the solution. On the other hand, a counter-current purge stream creates a split boundary value problem and solution techniques such as the iterative convergence method, the adaptive random search procedure [Mohan and Govind, 1988a] and the orthogonal collocation technique [Collins et al., 1993] have been applied.

When the membrane region is considered in the governing equations, the radial diffusion term for the membrane layer makes the problem a boundary value problem. Orthogonal collocation [Becker et al., 1993] can be utilized. Alternatively, the boundary value problem could be converted to an initial value problem and solved by the shooting method [Gerald and Wheatley, 1984]. However, this approach may produce a numerically unstable solution under fast reaction conditions [Sun and Khang, 1990]. The steady state problem then needs to be solved from the transient problem by adding time derivative terms to the describing equations.

For systems involving recycle streams or intermediate feed locations, the method of successive substitution can be used [Mohan and Govind, 1988a]. Moreover, multiple reactions including side reactions and series, parallel or series-parallel reactions result in strongly coupled differential equations. They have been solved numerically using an implicit Euler method [Bernstein and Lund, 1993].

10.4.4 Special Case: Catalytic Membrane Tubular Reactor with Packed Bed on One Side

The models presented so far are quite general with respect to the catalytic activities of the various regions: tube side, membrane (and support) layer and shell side. In practice, however, not all the regions are catalytic and almost all inorganic membrane reactor case studies only involve one or two catalytic regions.

First consider the case where the membrane is catalytic and so is either the feed or the permeate side of the membrane (but not both sides as in the general model).

Tsotsis et al. [1992] considered a case where two reaction zones exist in a porous membrane reactor: one inside the membrane matrix and the other in the tubular region which is packed with catalyst particles. They presented a packed-bed catalytic membrane tubular reactor model under isothermal and co-current flow conditions. Thus, Equations (10-37), (10-6) and (10-45) all reduce to the condition

$$T^t = T^m = T^s = T^t_0 \tag{10-50}$$

Furthermore, Equations (10-36), (10-5) and (10-44) apply with $k_i^s = 0$ because no reaction is assumed to take place on the shell side. The pressure drop on the tube side is described by the Ergun equation, Equation (10-39) in combination with Equation (10-40). The pressure drop on the shell side without catalyst particles is assumed to be zero.

The macroscopic mass balance model by Tsotsis et al. [1992], when applied to the reaction of ethane dehydrogenation, compares well with experimental data and both show higher conversions than the corresponding equilibrium values based on either tube- or shell-side conditions (pressure and temperature). This is clearly a result of the equilibrium displacement due to the permselective membrane. The conversion, as expected, increases with increasing temperature.

A similar type of membrane reactor has been used and modeled for the ethylbenzene dehydrogenation reaction [Becker et al., 1993]. It is assumed that the reaction rate is the same in the membrane layer as in the catalyst particles on the tube side. Four regions are considered: the tube (feed) side which is catalytic, the catalytic membrane layer, the inert support layer and the inert shell (permeate) side.

Assuming an isothermal condition and negligible axial diffusion in all regions and axial convection in the membrane and support layers, Becker et al. [1993] presented a detailed model of the form represented by Equations (10-3), (10-5), (10-7) and (10-9) with $k_i^s = 0$. Pressure drops in both tube and shell regions are assumed to be negligible. The model prediction does not agree well with experimental results. When the membrane layer and the tubular region are both catalytic, the experimental data shows an increase in conversion of approximately 10% while the model predicts essentially no change in conversion with or without the catalytic effect of the membrane.

This discrepancy is partially attributed to the very short contact time through the membrane layer for any significant additional conversion due to the catalytic membrane layer. It may also be the result of the actual catalyst distribution in the membrane and support layers. It has been pointed out that the reaction conversion of cyclohexane to benzene in a porous catalytic membrane reactor is extremely sensitive to the distribution of the Pd catalyst through the membrane [Cannon and Hacskaylo, 1992]. The last point is supported by the observation that, when the membrane is not catalytic and the remaining conditions are the same, experimental data and model prediction agree [Becker et al., 1993].

Some other uncertainties associated with packed-bed catalytic membrane reactors are the following. The reaction rate in the membrane layer is not easy to assess and it is likely to be different from that in the catalyst bed. The extremely small thickness of the membrane layer compared to the dimensions in the tubular and the annular regions makes the direct determination of the membrane-related parameters difficult, if not impossible in some cases. Furthermore, obtaining accurate data of the effective diffusivity for the membrane, particularly in the presence of a support layer, is not straightforward and often involves a high degree of uncertainty.

10.4.5 Special Case: Packed-Bed Inert Membrane Tubular Reactor

By "inert" it means that the membrane is a separator but not a catalyst. Many membrane reactor modeling studies consider only those cases where the membrane is catalytically inert and the catalyst is packed most often in the tubular (feed) region but sometimes in the annular (permeate) region. When it is assumed that no reaction takes place in the membrane or membrane/support matrix, the governing equations for the membrane/support matrix are usually eliminated. The overall effect of membrane permeation can be accounted for by a permeation term which appears in the macroscopic balance equations for both the feed and permeate sides. Thus, the diffusional gradient term

$$2\pi R_n D_j^m \left(\frac{dC_j^m}{dr} \right)_{r=R_n} \quad (n = 1 \, or \, 2) \tag{10-51a}$$

in Equations (10-36) and (10-44) is often replaced by a permeation term of the following form

$$\beta_m \left[\left(p_j^t \right)^n - \left(p_j^s \right)^n \right] \tag{10-51b}$$

where β_m is a constant dependent on some membrane properties such as membrane thickness, mean pore size, etc. and p_j^t and p_j^s the partial pressures of species j on the tube and shell sides, respectively. The value of the exponent n is determined by the type of the membrane used and the underlying transport mechanism(s). As discussed in Chapter 4, n is often taken to be 0.5 or higher for dense metal membranes and 1 for porous membranes.

Similarly, the term representing heat transfer through the membrane in Equations (10-37) and (10-45)

$$2\pi R_n \lambda^m \left(\frac{dT^m}{dr} \right)_{r=R_n} \quad (n=1 \text{ or } 2) \tag{10-52}$$

is typically substituted by an equivalent heat exchange term involving the use of an overall heat transfer coefficient through the membrane between two "bulk" temperatures T^t and T^s:

$$2\pi R_1 U^m (T^t - T^s) \tag{10-53}$$

Perfect mixing model. It is noted that the macroscopic balance equations such as Equations (10-36) and (10-44) assume plug-flow conditions and exhibit dependency on one dimension only – the axial (z) direction. There have been the so-called single-cell or perfect mixing (continuous stirred tank) models where the concentrations of the various species are assumed to be uniform everywhere in a given region [e.g., Mohan and Govind, 1988b; Itoh et al., 1990; Sun and Khang, 1990]. In these cases, the mass and energy balance equations reduce to a set of algebraic equations. They can be derived in a fairly straightforward way. See Figure 10.3 where both the tube and shell sides are assumed to be perfectly mixed in each region. For example, the mass balance equation on the tube side under the perfect mixing condition becomes

$$n_j^t - (n_j^t)_0 = \Psi + \pi R_1^2 L \sum_i^I \upsilon_{ij}^t \eta_i^t k_i^t \mathfrak{R}_i^t (C_j^t, T^t, K_{ei}^t) \tag{10-54}$$

where the subscript 0 represents the inlet condition, L the reactor length and the permeation term, Ψ, can be expressed as

$$\Psi = 2\pi R_1 L D_j^m \left. \frac{dC_j^m}{dr} \right|_{r=R_1} \tag{10-55}$$

or

$$\Psi = -2\pi R_1 L K_j (p_j^t - p_j^s) \tag{10-56}$$

where K_j is the effective permeability of the species j through the membrane and p_j^t and p_j^s are the partial pressures of the species j on the tube and shell sides, respectively.

Similarly, the mass balance equation for the shell side and the energy balance equations for both regions can be developed.

For convenience of discussion, modeling studies of packed-bed inert membrane tubular reactors will be divided into two categories depending on the type of inorganic membranes: dense or porous.

Dense palladium-based membranes. Shown in Table 10.1 are modeling studies of packed-bed dense membrane shell-and-tube reactors. All utilized Pd or Pd-alloy membranes except one [Itoh et al., 1993] which used yttria-stabilized zirconia membranes. As mentioned earlier, the permeation term used in the governing equations for the tube and shell sides of the membrane is expressed by Equation (10-51b) with n equal to 0.5 [e.g., Itoh, 1987] or 0.76 [e.g., Uemiya et al., 1991].

Itoh and his co-investigators have developed a number of packed-bed Pd-based
membrane shell-and-tube reactors for various reactions: dehydrogenation of
cyclohexane, ethylbenzene, and 1-butene and water-gas shift reaction. Oertel et al.

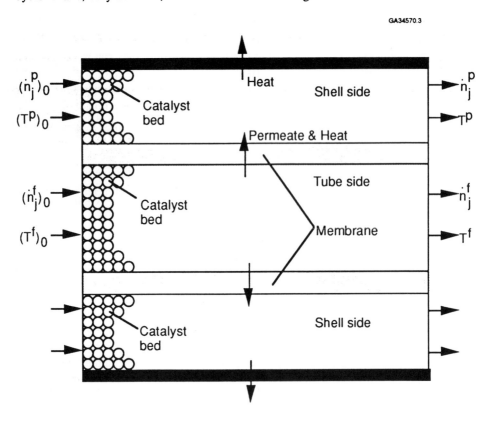

Figure 10.3 Schematic diagram of a single cell or perfect mixing model for a packed-bed
membrane shell-and-tube reactor

[1987] has modeled the methane steam reforming reactions in a packed-bed dense
membrane reactor. While most of the studies assume an isothermal condition and the
catalyst bed on the tube side, non-isothermal operations have also been modeled, so have
been those cases containing the catalyst bed on the shell side. Moreover, most of the
assumed flow conditions for the tube and shell sides are plug-flow but well-mixed
conditions have also been considered.

In all the studies listed in Table 10.1, the transport equations for the membrane layer are
avoided because the membrane is not catalytic. The presence of the membrane is
reflected in the permeation terms in both transport equations for the tube and shell sides.
Even in a case where the surface of a Pd membrane is considered to catalyze oxidation

of hydrogen [Itoh and Govind, 1989b], the membrane equations are not considered and the reaction is accounted for in the shell region (i.e., Equations (10-44) and (10-45) apply).

TABLE 10.1
Modeling studies of packed-bed dense membrane shell-and-tube reactors

Flow condition (tube/shell)	Thermal condition	Catalyst bed	Reaction (s)	Reference
PF/PF	Isothermal	Tube side	Cyclohexane dehydrogenation	Itoh, 1987
PF/PF; PF/PM; PM/PF; PM/PM	Isothermal	Tube side	Cyclohexane dehydrogenation	Itoh et al., 1990
PF/PF	Isothermal	Tube side	Ethylbenzene dehydrogenation	Itoh & Xu, 1991
PF/PF	Isothermal	Shell side	Water-gas shift	Uemiya et al., 1991
PF/PF	Adiabatic & isothermal	Tube side	1-butene dehydrogenation (tube side); oxidation of H_2 (shell side)	Itoh & Govind, 1989b
PF/PF	Isothermal & non-isothermal	Tube side	1-butene dehydrogenation (tube side); oxidation of H_2 (shell side)	Itoh & Govind, 1989a
PF/PF	Isothermal & adiabatic	Tube side	Cyclohexane dehydrogenation (tube side); oxidation of H_2 (shell side)	Itoh, 1990
PF/PF	Non-isothermal	Shell side	Methane steam reforming	Oertel et al., 1987
PF/PM	Isothermal	No catalyst required	Thermal decomposition of CO_2	Itoh et al., 1993

Note: PF = plug flow; PM = perfect mixing (continuous stirred tank)

In the isothermal models, Equations (10-37) and (10-45) are not applicable and they are replaced by

$$T^t = T^s = constant = T_0^t \qquad (10\text{-}57)$$

If the reaction does not occur on the shell side, $k_i^s = 0$. Likewise, $k_i^t = 0$ when catalyst is present only on the shell side. Moreover, plug-flow conditions call for the use of Equation (10-36) or (10-44) for the macroscopic mass balance and Equation (10-37) or (10-45) for the corresponding heat balance. On the other hand, when perfect mixing prevails in the tube or shell region, Equation (10-54) and Equation (10-55) or (10-56) are used in conjunction with associated heat balance equations.

Dehydrogenation of cyclohexane to produce benzene has been studied by Itoh [1987] and Itoh et al. [1990]. Both studies assume an isothermal condition and a catalyst bed in the tubular region.

The reaction rate for cyclohexane dehydrogenation

$$C_6H_{12} \rightleftharpoons C_6H_6 + 3H_2 \qquad (10\text{-}58)$$

is given in terms of the partial pressures of the reaction components:

$$r_c = \frac{k_c \left(K_e P_c / P_H^3 - P_B \right)}{1 + K_B K_e P_c / P_H^3} \qquad (10\text{-}58a)$$

where the subscripts C, H and B stand for cyclohexane, hydrogen and benzene, respectively, and k_c, K_e and K_B are the reaction rate constant, reaction equilibrium constant and adsorption equilibrium constant for benzene, respectively.

The hydrogen permeation rate is described by Equation (10-51b) with $n=0.5$ and β_m given by

$$\beta_m = \left[2\pi L / \ln(R_2 / R_1) \right] (C_0 / \sqrt{P_0}) D_H^m \qquad (10\text{-}59)$$

where $p_0 =$ 101.3 kPa

$C_0 =$ equilibrium hydrogen concentration in Pd at a partial pressure of p_0

$D_H^m =$ diffusivity of hydrogen through the Pd membrane

With the feed in argon (as an inert carrier gas) introduced to the tube side and argon as the purge gas to the shell side, the experimental data agrees with the model prediction [Itoh, 1987] and both yield higher conversions than the equilibrium values. It is found that, as expected, increasing the purge gas flow rate or decreasing the feed flow rate increases the conversion.

The flow conditions (e.g., co-current vs. counter-current and plug flow vs. perfect mixing) on both sides of the membrane can have significant effects on the conversion [Itoh et al., 1990] and will be discussed in Chapter 11.

A non-isothermal plug-flow membrane reactor on both sides of the membrane has been developed and applied to the methane steam reforming reaction to produce synthesis gas at high temperatures according to [Oertel et al., 1987]

$$CH_4 + H_2O \rightarrow CO + 3H_2 \tag{10-60a}$$

$$CO + H_2O \rightarrow CO_2 + H_2 \tag{10-60b}$$

The overall reaction rate is expressed as

$$r_{CH_4} = k_{CH_4} \left(P_{CH_4} - \frac{P_{CO}P_{H_2}^3}{P_{H_2O}K_{e1}} \right) \tag{10-60c}$$

where the equilibrium constant K_{e1} refers to the reaction given in Equation (10-60a). Steam reforming of desulfurized natural gas with a hydrogen selective membrane not only converts feed hydrocarbons but also directly produces high-purity hydrogen at high temperatures.

The hydrogen permeation rate through a Pd membrane is assumed to follow a Sievert law (i.e., Equation (10-51b) with $n=0.5$). The membrane reactor contains an impervious reformer tube which is packed with Ni-coated Raschig rings and among the rings a number of Pd tubular membranes are inserted. A porous tube is centrally placed inside the reformer tube to provide a path for the product gases. The reformer tube is surrounded by a heating pipe. Heated helium gas flowing in a counter-current direction inside the annular region between the reformer tube and the heating pipe exchanges the heat with the reformer tube region where the endothermic methane steam reforming reaction takes place. However, heat transfer between the reformer and product tubes is assumed to be negligible. Heat transfer between the membrane tubes and the reformer tube is assumed to involve convection only but that between the reformer tube and product tube can be ignored. In contrast, heat is transferred between the heating pipe and the reformer tube by convection as well as radiation. Thus, Equation (10-45) can still be applied by adding a radiative heat transfer term. Pressure drops on the shell side (reformer tube) and the tube side (Pd tubes) are assumed to be negligible.

Figure 10.4 shows the methane conversion as a function of the normalized reactor length according to the model [Oertel et al., 1987]. The packed-bed inert membrane reactor (solid line) is superior to a conventional reactor (dashed line) without the Pd tubes. The advantage can be 36% higher than that in the conventional reactor for a Pd thickness of 50 μm.

Dehydrogenation of ethylbenzene to form styrene

$$C_6H_5C_2H_5 \rightleftharpoons C_6H_5C_2H_3 + H_2 \tag{10-61}$$

with a rate equation (with the subscripts E, S and H representing ethylbenzene, styrene and hydrogen, respectively).

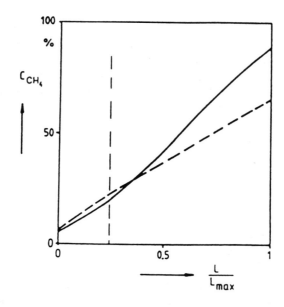

Figure 10.4 Methane conversion from steam reforming in a packed-bed Pd membrane reactor [Oertel et al., 1987]

$$r_1 = \frac{k_1(p_E - p_s p_H / K_e)}{p_T(1 + Kp_s)^2} \tag{10-61a}$$

where p_T is the total pressure can be rather complex and accompanied by many possible side reactions [e.g., Itoh and Xu, 1991; Wu and Liu, 1992]. Itoh and Xu [1991] consider the following side reactions:

$$C_6H_5C_2H_5 \rightarrow C_6H_6 + C_2H_4 \tag{10-62}$$

with a corresponding rate equation

$$r_2 = \frac{k_2 p_E}{p_T(1 + Kp_s)^2} \tag{10-62a}$$

and

$$C_6H_5C_2H_5 + H_2 \rightarrow C_6H_5CH_3 + CH_4 \tag{10-63}$$
with

$$r_3 = \frac{k_3 p_E p_H}{(p_T)^2 (1+Kp_s)^2} \tag{10-63a}$$

Applying an isothermal and plug-flow membrane reactor (on both sides of the membrane) to the above reactions, Itoh and Xu [1991] concluded that: (1) the packed-bed inert membrane reactor gives conversions higher than the equilibrium limits and also performs better than a conventional plug-flow reactor without the use of a permselective membrane; and (2) the co-current and counter-current flow configurations give essentially the same conversion.

The water-gas shift reaction

$$CO + H_2O \rightleftharpoons CO_2 + H_2 \tag{10-64}$$

having a rate expression

$$r_{wg} = \frac{k_{wg} (p_{CO} p_{H_2O} - p_{CO_2} p_{H_2} / K_e)}{1 + 4.4 \, p_{H_2O} + 13 \, p_{CO_2}} \tag{10-64a}$$

has been studied both experimentally and theoretically in an isothermal packed-bed inert membrane reactor with the Fe/Cr_2O_3 catalyst packed in the shell region [Uemiya et al., 1991]. The model results agree well with the experimental data and both give higher CO conversion than the equilibrium values.

The concept of reaction coupling where two separate reactions on the opposite sides of a membrane share the membrane as a medium to transfer heat and a permeating species that is a product from one reaction and becomes a reactant for the other reaction. Reaction coupling engaging dehydrogenation of 1-butene [Itoh, 1990] on the tube side with oxidation of hydrogen on the shell side has been modeled assuming plug-flow conditions and no pressure drops on both sides of the membrane.

Catalyst particles are packed on the tube side. Thus, Equations (10-36) and (10-37) apply. The oxidation of hydrogen is believed to be catalyzed by the surface of the Pd membrane. However, the reaction is accounted for in the reaction term on the shell side [Itoh and Govind, 1989a]. Thus, Equations (10-44) and (10-45) are used. The permeation rate of hydrogen is described by Equation (10-51b) with $n=0.5$.

Furthermore, the reaction rate for the 1-butene dehydrogenation

$$C_4H_8 \rightarrow C_4H_6 + H_2 \qquad\qquad (10\text{-}65)$$

to produce butadiene is assumed to follow

$$r_B = \frac{k_B\left[p_B^{1/2} - (p_D p_H / K_e)^{1/2}\right]}{p_T^{1/2}(1 + 1.21 p_B^{1/2} + 1.263 p_D^{1/2})^2} \qquad\qquad (10\text{-}65a)$$

where the subscripts B, D and H stand for 1-butene, butadiene and hydrogen, respectively. The coupled reaction of oxidation of hydrogen on the shell side

$$2H_2 + O_2 \rightarrow 2H_2O \qquad\qquad (10\text{-}66)$$

is assumed to be of the form

$$r_o = \frac{k_o p_H^2 p_o}{p_T^3} \qquad\qquad (10\text{-}66a)$$

where the subscripts O and H refer to oxygen and hydrogen, respectively. As mentioned previously, p_T is the total pressure. Shown in Figure 10.5 are the calculated temperature and conversion profiles as functions of the length of the adiabatic packed-bed membrane reactor with a dimensionless effective thermal conductivity of the membrane as a parameter [Itoh and Govind, 1989b]. As this parameter increases, the difference in temperature between the tube and shell sides decreases. It is seen that complete conversion is reached within a short length of the adiabatic membrane reactor which outperforms an isothermal membrane reactor.

Itoh and Govind [1989b] further analyzed an isothermal packed-bed inert membrane reactor but under a counter-current flow configuration. Under the conditions studied, the authors found that the counter-current flow configuration provides a much greater conversion than the co-current flow mode.

Using the reaction rate expression in Equation (10-58a) and the model developed earlier [Itoh and Govind, 1989a], Itoh [1990] analyzed the performance and thermal behavior of a packed-bed, plug-flow inert membrane reactor for cyclohexane dehydrogenation under adiabatic conditions. The reactor length required to attain complete conversion can be short depending on the heat transfer coefficient through the membrane. A smaller value of the heat transfer coefficient appears to accelerate the reaction and result in a shorter reactor length required to reach 100% conversion. Itoh [1990] found the model agrees well with experiments for the oxidative dehydrogenation of cyclohexane under isothermal conditions.

Dense metal oxide membranes. Itoh et al. [1993] studied thermal decomposition of carbon dioxide using yttria-stabilized zirconia membranes. The dependence of the oxygen permeation rate on the partial pressures of the oxygen on both sides of the membrane can be rather complex as pointed out in Chapter 4. Nevertheless, simplifications adopting the form in Equation (10-51b) with a constant exponent n (e.g., equal to 0.25 in [Itoh et al., 1993]) have been made.

The reactor is assumed to be isothermal with the rate for the reaction

$$2\,CO_2 \rightleftharpoons 2CO + O_2 \tag{10-67}$$

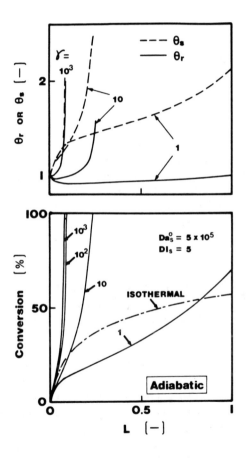

Figure 10.5 Temperature and conversion profiles in an adiabatic packed-bed membrane reactor with a dimensionless effective thermal conductivity of the membrane as a parameter [Itoh and Govind, 1989b]

to be given by

$$r_{CO_2} = k_{CO_2} (p_{CO_2} - p_{CO} \, p_{O_2}^{1/2} / K_e)$$ (10-67a)

The reaction does not require any catalyst at high temperatures (1,300-1,500°C). The feed (carbon dioxide) enters the reactor from the shell side under assumed plug flow conditions and the permeating oxygen exits on the tube side under assumed perfect mixing conditions. It is not certain if the perfect mixing conditions prevail in the experimental set-up.

Figure 10.6 compares the model and experimental results of direct thermal decomposition of CO_2 using a dense yttria-stabilized zirconia membrane shell-and-tube reactor [Itoh et. al., 1993]. The agreement for the reactor conversion is very good. At a CO_2 feed rate of less than 20 cm^3/min and with a membrane thickness of 2,000 μm, the conversion is significantly enhanced by the use of a permselective membrane for oxygen. Beyond a feed rate of 20 cm^3/min., however, the difference in conversion between a membrane and a conventional reactor.

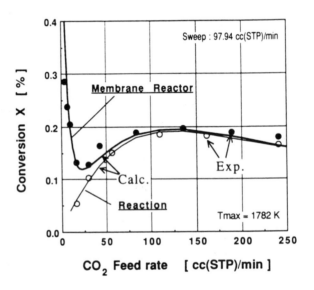

Figure 10.6 Comparison of modeling and experimental conversions of thermal decomposition of carbon dioxide in a dense zirconia membrane reactor [Itoh et al., 1993]

Microporous membranes. While dense metal or metal oxide membranes possess exceptionally good permselectivities, their permeabilities are typically lower than those of porous inorganic membranes by an order of magnitude or more. Commercial availability of porous ceramic membranes of consistent quality has encouraged an ever increasing number of experimental and theoretical studies of porous inorganic membrane reactors.

Table 10.2 lists most of the published models of packed-bed (inert) porous membrane shell-and-tube reactors. The membrane materials used to validate the models are primarily porous Vycor glass or alumina membranes.

Many models do not include mass balance equations for the membrane layer. In these situations, the permeation flux of species j through a porous membrane is commonly expressed as a special form of Equation (10-51b) with $n = 1$:

$$J_j = \frac{P_j \Delta p_j}{\Delta \ell} \tag{10-68}$$

where P_j is the membrane permeability, $\Delta \ell$ the membrane thickness and Δp_j the pressure difference between the tube and shell sides. In the case of Knudsen diffusion which plays an important role in many of the current generation of porous inorganic membranes, the Knudsen permeability is related to some membrane properties by

$$\left(P_j\right)_{Knudsen} = \frac{k_K r_p \varepsilon_m}{k_t \sqrt{TM_j}} \tag{10-69}$$

where r_p is the pore radius, ε_m the porosity and k_t the tortuosity factor of the membrane. The membrane permeability is shown to be inversely proportional to the square root of the temperature, T, and the molecular weight of species j.

Dehydrogenation of cyclohexane in packed-bed membrane reactors has been modeled extensively. Other reactions such as dehydrogenation of ethylbenzene and ethane, decomposition of hydrogen iodide and ammonia, partial oxidation of methane and propylene disproportionation have also been studied. Moreover, series and parallel reactions and their combinations have been included to partially account for the practical complexity of some industrially important reactions. The catalyst is packed either in the tubular or annular region. While most of the models assume a constant or measured temperature profile, Mohan and Govind [1988b] have taken into account heat transfer which is more realistic. Most of the model studies assume plug-flow conditions for both the tube and shell sides, but perfectly mixed conditions have also been included [Mohan and Govind, 1988b; Sun and Khang, 1988 and 1990].

In those limited cases where the diffusion equation for the membrane/support composite is solved in conjunction with the governing mass transport equations for the tube and shell sides [Sun and Khang, 1988 and 1990; Agarwalla and Lund, 1992], Equation (10-5) applies with $k_i^m = 0$ for catalytically inert membranes. In addition, either Equations (10-36) for plug flows or Equation (10-54) for perfect mixing needs to be solved for the tube and shell sides. Most models consider only the tube and shell sides and insert the membrane permeation term in the mass balance equations for the respective regions.

TABLE 10.2
Modeling studies of packed-bed porous membrane shell-and-tube reactors

Flow condition (tube/shell)	Thermal condition	Catalyst bed	Reaction (s)	Reference
PF/PF	Measured	Tube side	Cyclohexane dehydrogenation	Shinji et al., 1982
PF/PF	Isothermal	Tube side	HI decomposition	Itoh et al., 1984
PF/PF	Isothermal	Tube side	Cyclohexane dehydrogenation	Itoh et al., 1985
PF/PF	Isothermal	Tube side	Cyclohexane dehydrogenation	Mohan & Govind, 1986
PF/PF	Isothermal	Tube side	$A \leftrightarrow bB + cC$	Mohan & Govind, 1988a
PF/PF, PM/PM	Adiabatic	Shell side	$A \leftrightarrow B$ & ethylbenzene dehydrogenation	Mohan & Govind, 1988b
	Isothermal		HI decomposition, propylene disproportionation & cyclohexane dehydrogenation	Mohan & Govind, 1988c
PM/PM	Isothermal	Shell side	Cyclohexane dehydrogenation	Sun & Khang, 1988
PM/PM	Isothermal	Shell side	Cyclohexane dehydrogenation, $A+B \rightarrow C+D$ & $A+3B \rightarrow C+D$	Sun & Khang, 1990
PF/PF	Isothermal	Shell side	Partial oxidation of methanol	Song & Hwang, 1991
PF/PF	Isothermal	Tube side	$A \rightarrow B \rightarrow C$	Agarwalla & Lund, 1992
PF/PF	Isothermal	Tube side	$A \rightarrow B \rightarrow C$	Lund, 1992
PF/PF	Isothermal	Tube side	Ethylbenzene dehydrogenation & side reactions	Wu & Liu, 1992
PF/PF	Isothermal	Tube side	Series & series-parallel reactions	Bernstein & Lund, 1993
PF/PF	Isothermal	Tube side	Ammonia decomposition	Collins et al., 1993
PF/PF	Isothermal	Tube side	Ethane dehydrogenation	Tsotsis et al., 1993a

Note: PF = plug flow; PM = perfect mixing (continuous stirred tank)

Similar to corresponding cases of dense membrane reactors, the assumption of a packed bed only on the tube side calls for $k_i^s = 0$ ($i = 1$ to I). Likewise, $k_i^t = 0$ if the catalytic reaction occurs on the shell side instead.

It appears that Shinji et al. [1982] started the modeling of a packed-bed porous membrane shell-and-tube reactor for cyclohexane dehydrogenation. The same reaction has also been used as an example for several models developed by other investigators [Itoh et al., 1985; Mohan and Govind, 1986; Mohan and Govind, 1988c; Sun and Khang, 1988; Sun and Khang, 1990]. Some variations of the reaction rate for the cyclohexane dehydrogenation reaction (see Equation (10-58)) were used by different investigators. The simplest, and the most commonly used, expression is [Itoh et al., 1985]

$$r_C = k_C \left(p_C - \frac{p_B p_H^3}{K_e} \right) \tag{10-70a}$$

where C, B and H represent cyclohexane, benzene and hydrogen, respectively. Shinji et al. [1982] found an empirical rate equation

$$r_C = k_C \left\{ p_C - \frac{(p_H)_e^2}{K_e} p_B p_H \right\} \tag{10-70b}$$

which differs from Equation (10-70a) by the replacement of p_H^3 by $(p_H)_e^2 \cdot p_H$. The parameter $(p_H)_e$ is the partial pressure of hydrogen at equilibrium and k_C the rate constant which is temperature dependent according to the Arrhenius expression. Another slightly more complex reaction rate expression [Itoh et al., 1988] was given in Equation (10-58a).

Using Pt/Al$_2$O$_3$ as the catalyst bed inside a porous glass membrane to validate their isothermal packed-bed membrane tubular reactor, Shinji et al. [1982] found the agreement to be very good. They also discovered that as the cyclohexane feed rate decreases or as the sweep rate increases, the overall cyclohexane conversion increases. The highest conversion can be more than twice the equilibrium value which would be expected in the absence of the preferential separation of hydrogen. Based on a similar model, Itoh et al. [1985] also determined that the maximum conversion (which depends on the frequency factor in the rate constant, k_C) can exceed twice the equilibrium conversion. Furthermore, they have calculated the molar compositions of the gases along the reactor length in the tubular and annular regions as shown in Figure 10.7 [Itoh et al., 1985]. The concentration of cyclohexane decreases as it moves away from the entrance in the tubular region while increases in the annular region due to the incomplete rejection of cyclohexane by the porous membrane. The concentrations of benzene and hydrogen (both reaction products), as expected, are higher as the axial distance of the reactor increases. The rate of increase of hydrogen concentration in the annular region is particularly pronounced.

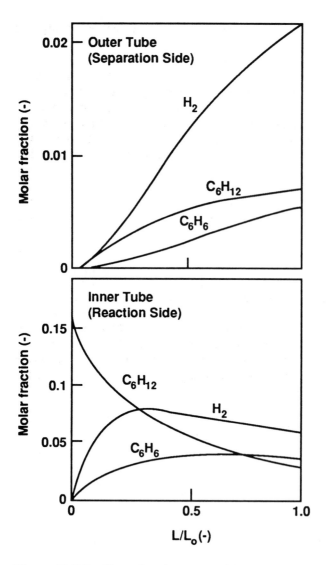

Figure 10.7 Profiles of molar compositions of gases in a packed-bed porous glass membrane reactor [Itoh et al., 1985]

In the same study [Itoh et al., 1985], the molar flow rate of cyclohexane at the reactor outlet is calculated as a function of the membrane thickness which has the most effect on the permeation rate of gases. For a given constant inlet molar flow rate, the reaction does not proceed beyond the equilibrium conversion for a conventional reactor. With membrane permeation, however, the overall conversion (i.e., the combined conversions of cyclohexane on the tube and shell sides) reaches a maximum for a certain membrane thickness. This optimum thickness depends on the reaction rate constant through the frequency factor.

Using a developed plug-flow membrane reactor model with the catalyst packed on the tube side, Mohan and Govind [1986] studied cyclohexane dehydrogenation. They concluded that, for a fixed length of the membrane reactor, the maximum conversion occurs at an optimum ratio of the permeation rate to the reaction rate. This effect will be discussed in more detail in Chapter 11. They also found that, as expected, a membrane with a highly permselective membrane for the product(s) over the reactant(s) results in a high conversion.

Sun and Khang [1988 and 1990] have also modeled packed-bed inert membrane tubular reactors for cyclohexane dehydrogenation. They constructed an isothermal model consisting of a diffusion equation for the membrane/support matrix, Equation (10-5), and perfect mixing approximations for both tube and shell sides, like Equation (10-54) with Equation (10-55). The model calculations show good agreement with experimental data using Vycor glass membranes and Pt as the catalyst and the conversions with permselective membranes are greater than those obtained under the equilibrium condition [Sun and Khang, 1988]. The authors also compared the packed-bed inert membrane reactor with other types of reactors with or without permselective membranes. The relative advantages of those reactors from cyclohexane dehydrogenation and other reactions depend on the operating conditions, membrane properties and reaction stoichiometry [Sun and Khang, 1990]. These effects of the reactor type will be dealt with later.

The decomposition of hydrogen iodide

$$2HI \rightleftharpoons H_2 + I_2 \tag{10-71}$$

can be described by [Itoh et al., 1984]

$$r_{HI} = k_{HI} \frac{X_{HI} - \left(X_{H_2} X_{I_2} / K_e \right)^{0.5}}{\left[H \left(K_{I_2} X_{I_2} \right)^{0.5} \right]^2} \tag{10-71a}$$

or by [Mohan and Govind, 1988c]

$$r_{HI} = k_{HI} \frac{X_{HI} - X_{I_2}^{0.5} / K_e}{\left[1 + K_{I_2} X_{I_2}^{0.5} \right]^3} \tag{10-71b}$$

Itoh et al. [1984] studied the decomposition of HI as a special case of a general reversible reaction

$$A \rightleftharpoons cC + dD \tag{10-72}$$

The HI reaction expressed by Equation (10-71) does not involve volume change. Therefore, the conversion is not expected to exceed the equilibrium value by lowering the reaction pressure or introducing any diluent. Furthermore, since the temperature dependency of the reaction equilibrium constant is comparatively small, the conversion can not be increased considerably by raising the reaction temperature either. Continuous removal of the products, hydrogen or iodine, or both by a permselective membrane, however, can enhance the reaction conversion by displacing the equilibrium. The resulting conversion is greater than the equilibrium level, particularly when the porous symmetric (glass) membrane thickness is small. An upper limit on the conversion exists due to permeation of the reactant. It is also noted that under otherwise identical operating conditions an optimum thickness exists corresponding to the maximum conversion. The same trend has been observed for cyclohexane dehydrogenation and discussed earlier. The authors suggest two methods of designing more effective membrane reactors. The first method is to manufacture thinner membranes on the order of 1 μm which even when deposited on a porous support appears to be unrealistically thin from the standpoint of mechanical strength. The second method is to impart a high membrane packing density (i.e., a high permeation area per unit volume of the membrane reactor). This suggestion leads to such a choice as the hollow fiber shape.

Mohan and Govind [1988c] applied their isothermal packed-bed porous membrane reactor model to the same equilibrium-limited reaction and found that the reactor conversion easily exceeds the equilibrium value. The HI conversion ratio (reactor conversion to equilibrium conversion) exhibits a maximum as a function of the ratio of the permeation rate to the reaction rate. This trend, which also occurs with other reactions such as cyclohexane dehydrogenation and propylene disproportionation, is the result of significant loss of reactant due to increased permeation rate. This loss of reactant eventually negates the equilibrium displacement and consequently the conversion enhancement effects.

Ethylbenzene dehydrogenation in a packed-bed shell-and-tube reactor using a porous membrane has been modeled by Mohan and Govind [1988b] and Wu and Liu [1992]. The kinetics for the reaction

$$C_6H_5C_2H_5 \rightleftharpoons C_6H_5C_2H_3 + H_2 \tag{10-73}$$

can be expressed as

$$r_{EB} = k_{EB}\left(p_{EB} - \frac{p_{ST}p_H}{K_e}\right) \tag{10-73a}$$

where the subscripts EB, ST and H stand for ethylbenzene ($C_6H_5C_2H_5$), styrene ($C_6H_5C_2H_3$) and hydrogen, respectively.

Mohan and Govind [1988b] modeled both mass and heat transfer in the membrane reactor and found that the use of porous permselective membrane can improve the reactor conversion above that attainable with an impermeable-wall reactor under equilibrium conditions. As pointed out earlier in the discussion of results for the reaction of HI decomposition [Mohan and Govind, 1988c], the modeling calculations for ethylbenzene dehydrogenation show a similar negative effect of reactant loss due to permeation on the conversion. Furthermore, the use of a plug-flow membrane reactor is superior to that of a continuous stirred tank membrane reactor for an endothermic reaction such as ethylbenzene dehydrogenation.

Wu and Liu [1992] attempted to approximate the reactions involved during ethylbenzene dehydrogenation under an industrial setting by considering many possible side reactions. They included in their isothermal model for plug flows on both sides of the membrane the following five main side reactions with their corresponding reaction rate expressions:

$$C_6H_5C_2H_5 + H_2 \rightarrow C_6H_5CH_3 + CH_4 \qquad \text{(10-74)}$$

$$r_{S1} = k_{S1}(P_{EB}P_H) \qquad \text{(10-74a)}$$

$$C_6H_5C_2H_5 \rightarrow C_6H_6 + C_2H_4 \qquad \text{(10-75)}$$

$$r_{S2} = k_{S2}P_{EB} \qquad \text{(10-75a)}$$

$$\tfrac{1}{2}C_2H_4 + H_2O \rightarrow CO + 2H_2 \qquad \text{(10-76)}$$

$$r_{S3} = k_{S3}P_{C_2H_4}P_{H_2O} \qquad \text{(10-76a)}$$

$$CH_4 + H_2O \rightarrow CO + 3H_2 \qquad \text{(10-77)}$$

$$r_{S4} = k_{S4}P_{CH_4}P_{H_2O} \qquad \text{(10-77a)}$$

$$CO + H_2O \rightarrow CO_2 + H_2 \qquad \text{(10-78)}$$

$$r_{S5} = k_{S5}P_T P_{CO}P_{H_2O} \qquad \text{(10-78a)}$$

where the subscripts EB and H, as previously defined, represent ethylbenzene and hydrogen, respectively, and T stands for the total pressure on the reaction side. The reverse reaction rates for the side reactions considered were assumed to be negligibly small compared to their forward reaction rates. Knudsen diffusion was assumed to be dominant which calls for the use of Equations (10-68) and (10-69). Macroscopic balance equations of the form given by Equations (10-36) and (10-44) with the permeation terms replaced by Equation (10-56) were derived for the tube and shell sides. For pressure drop calculations, Equation (10-39) with Equation (10-40) and Equation (10-46) with Equation (10-47) were used. The authors found that the overall yield to styrene increases as the Damköhler number on the feed side (as a measure of the reactor residence time) or the ratio of hydrogen permeability to reaction rate increases.

They also investigated a hybrid reactor in order to maximize hydrogen removal while minimizing ethylbenzene loss on the tube side. In this hybrid reactor system, a conventional packed-bed reactor is followed by a packed-bed inert membrane reactor (Figure 10.8). Ethylbenzene can be sufficiently converted in the first-stage packed-bed reactor. The remaining reaction components in the retentate are fed to the second stage, the packed-bed membrane reactor. Hydrogen can be selectively removed in this membrane reactor. Since part of the ethylbenzene is already consumed in the first stage, its loss in the second stage is reduced. It has been estimated that an increase of 5% or more in the overall yield of styrene over the thermodynamic limit can be realized by such a hybrid reactor system.

GA 35274.2

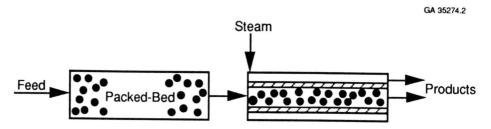

Figure 10.8 Schematic of add-on stage membrane reactors [Wu and Liu, 1992]

Mohan and Govind [1988c] have also studied propylene disproportionation to form ethylene and butene

$$2C_3H_6 \rightleftharpoons C_2H_4 + C_4H_8 \tag{10-79}$$

with a rate expression in terms of mole fractions

$$r_P = k_P \frac{X_P^2 - X_E X_B / K_e}{F(X_P, X_E, X_B)} \tag{10-79a}$$

where F is a complex empirical function and the subscripts P, E and B designate propylene, ethylene and butene, respectively. Their findings of propylene disproportionation were very similar to those of cyclohexane dehydrogenation and HI decomposition [Mohan and Govind, 1988c], as discussed previously. The negating effect of increasing permeation rate, which leads to more reactant loss, on the reactor conversion is more pronounced for propylene disproportionation than for the other two reactions (see Figure 10.9).

The process of catalytic partial oxidation of methanol with air in a fuel-rich mixture has been commercialized to produce formaldehyde since about 1890. Various catalysts have been used, but silver catalyst is by far most widely used. Song and Hwang [1991] used a packed-bed porous membrane tubular reactor with the catalyst packed on the shell side. They used a model constructed essentially from Equations (10-36) and (10-44) with the permeation terms replaced by some terms similar to Equation (10-56) and $k_i^t = 0$. The partial oxidation of methanol can be conveniently described by

$$CH_3OH + \tfrac{1}{2}O_2 \rightarrow HCHO + H_2O \tag{10-80a}$$

$$CH_3OH \rightarrow HCHO + H_2 \tag{10-80b}$$

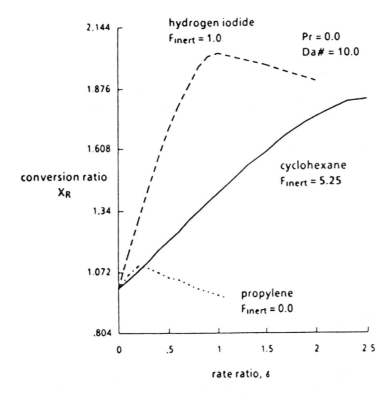

Figure 10.9 Effect of permeation to reaction rate ratio on reaction conversion in a porous membrane reactor [Mohan and Govind, 1988c]

Song and Hwang [1991] used a rate equation

$$r_M = k_M \frac{p_M}{\left(1 + K_M p_M + K_H p_H\right)} \cdot \frac{(p_O)_0^{0.5}}{1 + K_O (p_O)_0^{0.5}} \tag{10-80c}$$

in their model where the subscript M, H and O represent methanol, hydrogen and oxygen, respectively, and $(p_O)_0$ is the oxygen partial pressure at the reactor inlet. They found that, through the model and experiments using porous Vycor glass membranes, the conversion can be improved by increasing the space time and membrane permeation area (Figure 10.10). The packed-bed membrane reactor performs better than a packed-bed reactor. They concluded that further improvement in conversion is possible by utilizing a membrane with a higher permeability or permselectivity or by using recycle streams.

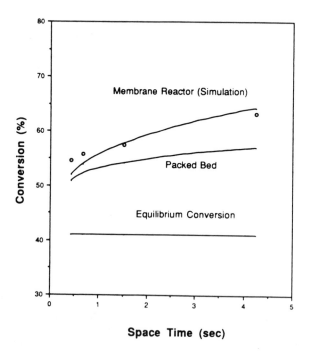

Figure 10.10 Effect of space time on conversion of partial oxidation of methanol in a porous Vycor glass membrane reactor (350°C, 60% air) [Song and Hwang, 1991]

Tsotsis et al. [1993a] presented an isothermal packed-bed porous membrane tubular reactor for ethane dehydrogenation to form ethylene

$$C_2H_6 \rightleftharpoons C_2H_4 + H_2 \tag{10-81}$$

with Pt/γ-Al_2O_3 as the catalyst packed on the tube side of an alumina membrane, they adopted the following kinetic equation

$$r_{C_2H_6} = k_{C_2H_6}\left(p_{C_2H_6} - \frac{p_{C_2H_4}p_{H_2}}{K_e}\right)$$
(10-81a)

for the reaction rate. Plug flow is assumed for both tube and shell sides. Thus Equations (10-36) and (10-44) apply with $k_i^s = 0$. The tube side pressure drop is described by the Ergun equation while the shell side pressure drop is assumed to be negligible. They found that the membrane reactor conversion is higher than equilibrium values and the conversion can be increased by increasing the sweep flow rate which causes a decreased hydrogen partial pressure on the shell side and consequently a higher permeation rate of hydrogen out of the reaction zone.

Gasification of coal containing organic nitrogen and sulfur compounds generates trace amounts of ammonia and other toxic and corrosive impurities. The ammonia thus formed must be removed prior to combustion of the synthetic gas to reduce the emission of NO_x in the gas turbine. The decomposition of ammonia

$$NH_3 \rightleftharpoons 1/2\,N_2 + 3/2\,H_2$$
(10-82)

is catalyzed by Ni/Al_2O_3 with the reaction kinetics described by (assuming that the ideal gas law applies)

$$r_A = k_A\left[\left(\frac{p_{NH_3}^2}{p_{H_2}^3}\right)^\beta - \frac{p_{N_2}}{K_e}\left(\frac{p_{H_2}^3}{p_{NH_3}^2}\right)^{1-\beta}\right]$$
(10-82a)

where β is a positive constant to be determined experimentally. The reaction is limited by equilibrium at the gasifier conditions. Therefore, the use of a permselective membrane can achieve a higher conversion and improve the efficiency of the power generation systems. Collins et al. [1993] presented a mathematical model for a plug-flow packed-bed membrane tubular reactor with a co-current or counter-current sweep flow to investigate the above reaction. Governing equations similar to Equations (10-36) and (10-44) with the permeation term replaced by Equation (10-56) were used. The authors calculated the effectiveness factor by solving the mass transfer equations for reaction components inside a catalyst particle. The effectiveness factor thus obtained was used in their model. Both co-current and counter-current flow configurations were considered. For membrane reactors having Knudsen diffusion selectivities, the co-current flow generally provides a higher conversion than the counter-current flow. If membranes with a hydrogen permselectivity greater than what Knudsen diffusion offers are used, the choice between co-current and counter-current flow configurations depends on the operating conditions. The authors further concluded that, to achieve a desired high conversion of ammonia, a porous membrane with a selectivity for hydrogen over nitrogen of greater than 50 is needed. Such a high target of separation factor has yet to be realized on a demonstration or production scale although some experimental silicate-modified porous alumina membranes have shown a promising hydrogen/nitrogen selectivity of 72 [Wu et al., 1994].

Series and series-parallel reactions in a packed-bed plug-flow membrane reactor have been analyzed by Lund and his co-investigators [Agarwalla and Lund, 1992; Lund, 1992; Bernstein and Lund, 1993]. First consider the following consecutive catalytic reactions:

$$A \xrightarrow{k_1} R \tag{10-83a}$$

with a rate equation

$$r_1 = k_1 p_A \tag{10-83b}$$

and

$$R \xrightarrow{k_2} C \tag{10-84a}$$

with

$$r_2 = k_2 p_R \tag{10-84b}$$

The intermediate product R is the desired product. The reaction rate constant k_2 is an order of magnitude greater than k_1 which makes it difficult to achieve a high selectivity to R using a conventional packed-bed, plug-flow reactor. This challenge can be met with the help of a membrane that selectively lets R pass through, thus reducing its further undesirable reaction.

Co-current and counter-current flow regimes have been modeled. It is noted that their model reaction system, Equations (10-83a) and (10-84a), is irreversible and under kinetic control, a contrast to most of other inorganic membrane reactor studies. Modeling results reveal that although both co-current and counter-current conditions are equally effective in many cases, the co-current condition appears to be preferred over the counter-current mode in terms of the selectivity to the intermediate product R. For the co-current flow configuration the selectivity varies with the conversion. At the lowest conversions, the selectivity is similar to that provided by a conventional plug-flow reactor. At intermediate conversions, the selectivity is enhanced by the use of a membrane permselective to R. At the highest conversions, the selectivity is slightly decreased.

For the above series reactions, the most critical parameters related to the yield of the intermediate product are the Damköhler-Peclet number product (rate of reaction relative to rate of membrane permeation) which should be 0.1 to 10 and the ratio of the permeability of the intermediate product to the permeability of the reactant which should be below 100. Other parameters also affect the reactor performance: the relative inert gas sweep rate, the applied transmembrane pressure difference and the rate of the desired reaction relative to that of the undesired reaction [Bernstein and Lund, 1993].

The above analyses for series reactions were expanded to the following series-parallel reactions:

$$A + B \rightarrow R + T \tag{10-85}$$

with a rate expression

$$r_3 = k_3 p_A p_B \tag{10-85a}$$

and

$$R + B \rightarrow S + T \tag{10-86}$$

with a reaction rate given by

$$r_4 = k_4 p_R p_B \tag{10-86a}$$

The intermediate product R is the desired product. In addition to the critical parameters identified for series reactions, the relative permeabilities of the reactants are also important when series-parallel reactions like those shown in Equations (10-85) and (10-86) are considered.

10.4.6 Special Case: Catalytic Membrane Tubular Reactors (CMTR)

A special situation arises when no catalyst is used on either the feed or permeate side and the membrane itself is catalytic relative to the reaction of interest.

Catalytic membrane reactors represent the most compact and yet challenging membrane reactor design. The membrane material may be inherently catalytic or rendered catalytic by impregnating a catalyst on the surface of the membrane itself or the pores inside the membrane/support matrix. When the inner tube of a shell-and-tube reactor is a permselective and also catalytic membrane, the reactor is called catalytic membrane tubular reactor. Under this special circumstance, $k_i^t = 0 = k_i^s$ for Equations (10-36) to (10-37) and (10-44) to (10-45), assuming plug flows on both the tube and shell sides. The transport equations for the membrane zone, Equations (10-5) to (10-6), hold.

Champagnie et al. [1992] adopted the aforementioned model to describe the performance of an isothermal shell-and-tube membrane reactor for ethane dehydrogenation in a co-current flow mode. Using Equation (10-81a) to represent the reaction kinetics and assuming no reactions and pressure drops on both the tube and shell sides, they were able to fit their experimental data (using Pt-impregnated alumina membranes) with the model. The reactor conversion which increases with increasing temperature is significantly greater than the equilibrium limits based on the tube or shell side conditions. In addition, the conversion improves as the flow rate of the sweep stream increases. The authors also studied the effect of the membrane layer thickness on the conversion which is found to reach a maximum for a certain membrane thickness. This finding for a catalytic membrane reactor is similar to that discussed earlier for packed-bed inert membrane reactors [Itoh et al., 1984] as illustrated in Figure 10.8.

Sun and Khang [1988 and 1990] have also modeled isothermal CMTR but assumed that the flow conditions on both sides of the membrane to be well mixed, in contrast to the general plug-flow assumptions as in Champagnie et al. [1992]. Thus, Equation (10-5) applies for the membrane region and Equations (10-54) and (10-55) and their equivalents describe the mass transfer on the tube and shell sides. In their experimental set up for cyclohexane dehydrogenation (see Equation (10-70a) for reaction rate) the reactant stream enters the shell side and the permeating species leave the reactor through the tube side. The assumption of perfect mixing on both sides of the membrane may be questionable.

Numerical simulations using the perfect mixing (or single cell) model show good agreement with experimental data obtained with a Pt-impregnated Vycor glass membrane tubular reactor [Sun and Khang, 1988]. The conversion enhancement due to the equilibrium displacement by the permselective membrane is significant, particularly for high space time cases. Pronounced internal mass transfer resistance exists in the catalytic membrane. The catalytic membrane tubular reactor (CMTR) was compared to a packed-bed (feed side) inert membrane tubular reactor (PBIMTR). The results indicated that a CMTR outperforms a PBIMTR when the space time is high, particularly for those reaction systems with an increase of the total number of moles as the reaction proceeds (Figure 10.11a). The reverse is true when the space time is low (Figure 10.11b). It is noted that at a low space time, it is possible that above a certain temperature the CMTR gives a conversion even lower then the equilibrium value based on the feed condition.

Sun and Khang [1990] extended their theoretical analysis to compare a CMTR with a PBIMTR, a conventional plug-flow reactor (PFR) and a conventional continuous stirred tank (perfect mixing) reactor (CSTR) under the same condition of a fixed amount of catalyst. Three types of general reactions were considered to account for the variation in the overall change in the number of moles due to reaction.

The first type of reactions involve an overall increase in the volume after reaction (gas-phase reactions with $\Delta n > 0$)

$$A \rightleftarrows B + 3C \tag{10-87}$$

of which cyclohexane dehydrogenation is an example. The reaction rate can be expressed as

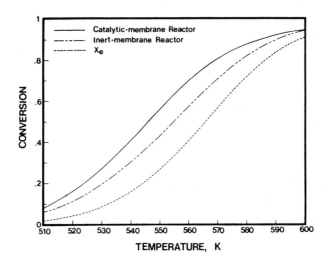

Figure 10.11a Conversion of cyclohexane dehydrogenation in a catalytic membrane reactor for high-space-time operations [Sun and Khang, 1988]

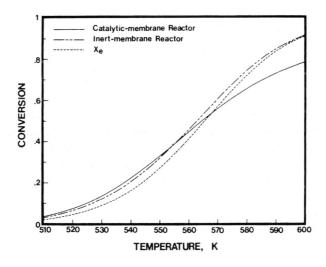

Figure 10.11b Conversion of cyclohexane dehydrogenation in a catalytic membrane reactor for low-space-time operations [Sun and Khang, 1988]

$$r = k\left(p_A - \frac{p_B p_C^3}{K_e}\right) \tag{10-87a}$$

The second type of reactions result in no net change in the volume due to reaction (most liquid-phase reactions or gas-phase-reactions with $\Delta n = 0$)

$$A + B \rightleftharpoons C + D \tag{10-88}$$

The associated reaction rate is

$$r = k\left(p_A p_B - \frac{p_C p_D}{K_e}\right) \tag{10-88a}$$

and the third category of reactions are represented by

$$A + 3B \rightleftharpoons C + D \tag{10-89}$$

which has a net decrease in the volume after reaction (gas-phase reactions with $\Delta n < 0$) and a rate equation

$$r = k\left(p_A p_B^3 - \frac{p_C p_D}{K_e}\right) \tag{10-89a}$$

A common basis was established for comparing various reactors: the same volume of catalysts and the permeability of a product (species C) being 6 times that of a reactant (species A). Sun and Khang [1990] summarized their simulation results in Table 10.3.

For the first type of reactions ($\Delta n > 0$), the PFR and the CSTR operated at the permeate-side pressure perform better than the CMR or the PBIMR. The performance of the CMR is slightly better than that of PBIMR with catalysts on the feed side when the pressure drop across the membrane is low. For the second type of reactions ($\Delta n = 0$), both CMR and PBIMR perform better than the conventional PFR and the CSTR due to the equilibrium displacement induced by selective removal of a product. The PBIMR is preferred over the CMR at a longer space time. For the third type ($\Delta n < 0$), the PBIMR outperforms any other reactors at a longer diffusional space time. The CMR in this case does not provide advantages due to the undesirable equilibrium effect induced by the pressure variation in the membrane.

In the above examples of catalytic membrane tubular reactors (CMTR), the transport equations for the catalytic membrane zone (i.e., Equations (10-5) and (10-6)) are considered and solved. A simpler and less rigorous approach to modeling a CMTR is to neglect the membrane layer(s) and account for the catalytic reactions in the tube core or annular region. This approach was adopted by Wang and Lin [1995] in their modeling of a shell-and-tube reactor with the membrane tube made of a dense oxide membrane (a doped or undoped yttria-stabilized zirconia membrane) the surface of which is assumed to be catalytic to oxidative coupling of methane similar to Li/MgO. The overall reactions are given by

$$CH_4 + 2\,O_2 \rightarrow CO_2 + 2H_2O \qquad\qquad (10\text{-}90a)$$

$$2CH_4 + 0.5\,O_2 \rightarrow C_2H_6 + H_2O \qquad\qquad (10\text{-}90b)$$

TABLE 10.3
Reactor selection guide in the order of preference

Category	Low space time	High space time
$\Delta n > 0$	1. PFR (at p_p) 2. CSTR (at p_p) 3. PFR (at p_F) 4. PBIMR 5. CSTR (at p_F) 6. CMR	1. PFR (at p_p) 2. CSTR (at p_p) 3. CMR 4. PBIMR 5. PFR (at p_F) 6. CSTR (at p_F)
$\Delta n \approx 0$	1. PFR (at p_F) 2. PBIMR 3. CSTR (at p_F) 4. CMR	1. PBIMR 2. CMR 3. PFR (at p_F) 4. CSTR (at p_F)
$\Delta n < 0$	1. PFR (at p_F) 2. PBIMR 3. CSTR (at p_F) 4. PFR (at p_p) 5. CSTR (at p_p) 6. CMR	1. PBIMR 2. PFR (at p_F) 3. CSTR (at p_F) 4. CMR 5. PFR (at p_p) 6. CSTR (at p_p)

[Sun and Khang, 1990]

Further oxidation of C_2H_6 to form CO_2 and H_2O is also considered in their analysis. Their kinetic rate expressions are complicated and involve reactions between adsorbed methane or ethane and the lattice oxygen, filling of oxygen vacancy by oxygen, coupling of the methyl radicals to form ethane, and reactions between the methyl or ethyl radicals and oxygen to form carbon dioxide. The equations can be found in their paper.

In the models presented by Wang and Lin [1995], the catalytic reaction terms appear in the transport equations for the tube core where methane is introduced. Air flows in the annular region and permeates through the dense oxide membrane to the tube side to react with methane. Transport of oxygen takes place as a result of the defects of oxygen vacancy and electron-hole in the oxide layer.

The authors considered two different flow configurations for both sides of the membrane tube: plug-flow and perfect mixing. For both bases, isothermal conditions were assumed. Thus, Equations (10-37) and (10-45) degenerate into $T^t = 0$ and $T^s = 0$, respectively. In their plug-flow CMTR model, transport equations for the various reaction components similar to Equations (10-36) and (10-44) are used for the tube and shell sides, respectively. Since no reaction occurs on the shell side, $k_i^s = 0$ for both oxygen and nitrogen. In their perfect mixing (or CSTR) model, transport equations in the form of Equation (10-54) describe the membrane reactor behavior. Instead of applying the generally used, simplified expression for the oxygen permeation rate such as Equation (10-56), they employed three oxygen flux expressions for three steps involved in oxygen permeation: surface reaction on the membrane surface exposed to air, bulk electrochemical diffusion in the membrane matrix and surface reaction on the membrane surface exposed to methane.

Based on the above two models, Wang and Lin [1995] analyzed the performances of the plug-flow and perfectly-mixed membrane reactors in which oxidative coupling of methane is carried out. For a plug-flow membrane reactor with an inner diameter of 1 cm and a length of 1 m of a dense oxide membrane, the axial profiles of the yield and selectivity of C_2 products and the oxygen partial pressure on the tube (reaction) side are shown in Figure 10.12. As the axial distance from the reactor inlet increases, both the yield and oxygen partial pressure first increase and reach a maximum before declining to an asymptotic level. Not shown in Figure 10.12, the methane conversion quickly reaches 100% at $z = 0.25$ m. The yield arrives at the maximum at $z = 0.2$ m. The selectivity decreases with z particularly after $z = 0.25$ m because of further oxidation of C_2 products. The yield and selectivity curves merge beyond $z = 0.25$ m. The plateaus exhibited by the yield, selectivity and oxygen partial pressure are explained by the fact that all reactions are essentially complete at about $z = 0.7$ m due to insufficient supply of oxygen.

In the case of a perfectly mixed membrane reactor, the issue of sufficient oxygen permeation to sustain C_2 products formation becomes critical. When a dense oxide membrane with a low oxygen permeance is used, the C_2 products yield increases rapidly with the methane flow rate but drops off quickly at an impractically low methane feed rate. The situation improves as a dense oxide membrane with a much higher oxygen permeance is utilized, as shown in Figure 10.13 for an oxide membrane with an oxygen permeance equivalent to that of a 0.5 μm Ti-doped yttria-stabilized zirconia membrane. The initial rate of increase of the yield with the methane flow rate is not as drastic as in the case of a membrane with a low oxygen permeance. However, high yields can be achieved at a reasonable flow rate of methane. For example, at a methane feed rate of 5 cm^3/s, a C_2 products yield exceeding 75% and a selectivity of more than 85% can be obtained [Wang and Lin, 1995].

The membrane reactor dynamics for oxidative coupling of methane will be further discussed in Chapter 11.

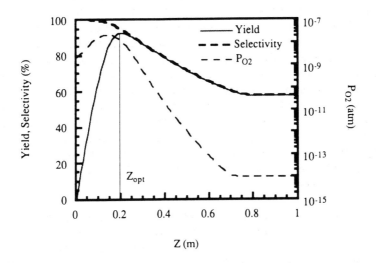

Figure 10.12 Profiles of yield and selectivity of C_2 products and oxygen partial pressure on reaction side [Wang and Lin, 1995]

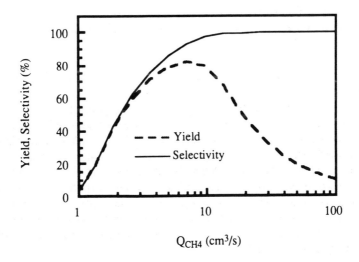

Figure 10.13 Yield and selectivity of C_2 products in a perfectly mixed membrane reactor [Wang and Lin, 1995]

10.5 FLUIDIZED-BED CATALYTIC MEMBRANE TUBULAR REACTORS

A fixed-bed reactor often suffers from a substantially small effectiveness factor (e.g., 10^{-2} to 10^{-3} for a fixed-bed steam reformer according to Soliman et al. [1988]) due to severe diffusional limitations unless very small particles are used. The associated high pressure drop with the use of small particles can be prohibitive. A feasible alternative is to employ a fluidized bed of catalyst powders. The effectiveness factor in the fluidized bed configuration approaches unity. The fluidization system also provides a thermally stable operation without localized hot spots. The large solid (catalyst) surface area for gas contact promotes effective catalytic reactions. For certain reactions such as ethylbenzene dehydrogenation, however, a fluidized bed operation may not be superior to a fixed bed operation. To further improve the efficiency and compactness of a fluidized-bed reactor, a permselective membrane has been introduced by Adris et al. [1991] for steam reforming of methane and Abdalla and Elnashaie [1995] for catalytic dehydrogenation of ethylbenzene to styrene.

Adris et al. [1991] and Abdalla and Elnashaie [1995] have developed steady-state non-isothermal fluidized-bed membrane reactor models. In their conventional two-region models [Levenspiel, 1962], the fluidized bed consists of a dense or emulsion phase which contains most of the catalyst solids and a lean or bubble phase which is essentially solids-free. Heat may be supplied by heat pipes or other means. The following assumptions were made: (1) Reaction occurs only in the well-mixed dense phase which has a uniform temperature; (2) the bubble phase is transported in plug flow in the vertical direction; (3) the gaseous flow inside the membrane tubes is plug flow; (4) the two-phase theory applies inside the reaction vessel using an "average" bubble size and the dense-phase characteristics are determined by the conditions at incipient fluidization; (5) the ideal gas law applies in the bubble and dense phases; and (6) the mass and heat transfer resistances between the particles and the gas in the dense phase are negligible.

Due to their complexity, the model equations will not be derived or presented here. Details can be found elsewhere [Adris, 1994; Abdalla and Elnashaie, 1995]. Basically mass and heat balances are performed for the dense and bubble phases. It is noted that associated reaction terms need to be included in those equations for the dense phase but not for the bubble phase. Hydrogen permeation, the rate of which follows Equation (10-51b) with $n=0.5$, is accounted for in the mass balance for the dense phase. Hydrodynamic parameters important to the fluidized bed reactor operation include minimum fluidization velocity, bed porosity at minimum fluidization, average bubble diameter, bubble rising velocity and volume fraction of bubbles in the fluidized bed. The equations used for estimating these and other hydrodynamic parameters are taken from various established sources in the fluidized bed literature and have been given by Abdalla and Elnashaie [1995].

The overall mass and heat interphase exchange coefficients can be obtained by summing the resistances from bubble to cloud and from cloud to emulsion (dense phase) in series. The overall bed to surface heat transfer coefficient is assumed to be mostly attributed to particle convection, gas convection and radiation. Heat transfer resistance on the heat pipe side, where applicable, is assumed to be negligible.

The resulting equations describing mass and heat transport are highly non-linear algebraic equations which can be solved numerically using a common procedure such as the Newton-Raphson technique.

Two design configurations for the fluidized-bed membrane reactors for steam reforming of methane are schematically shown in Figures 10.14a and 10.14b. The major difference between the two designs is in the process stream flowing inside the Pd tubes bundled across the reactor vessel. In the first design (Configuration I), some of the hydrogen produced by the reactions permeates through the Pd membranes. The remaining reaction products pass through a cyclone to remove the entrained solids and then a pressure reducing valve before being combined with the permeated hydrogen to form the final product. The reduced pressure on the tube side of the Pd membrane creates a transmembrane pressure difference to promote hydrogen permeation.

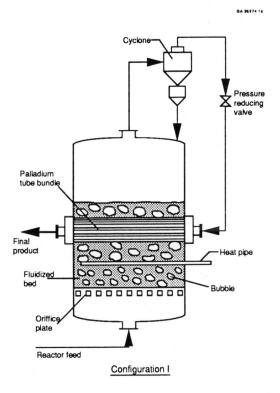

Figure 10.14a Configuration I of a fluidized-bed membrane reactor for steam reforming of methane (Adris et al., 1991]

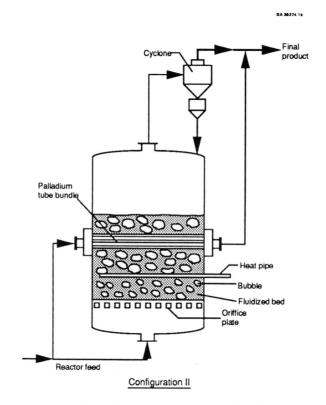

GA 36274 1b

Figure 10.14b Configuration II of a fluidized-bed membrane reactor for steam reforming of methane [Adris et al., 1991]

In the second design, part of the feed stream goes directly to the tube side of the Pd membranes while the remaining undergoes reactions in the fluidized bed of the reactor vessel. The reactor exit stream is processed by a cyclone before reporting to the final product stream which also contains the unreacted feed and the permeated hydrogen in the mixture. In this configuration, the driving force for hydrogen permeation is the hydrogen concentration difference across the membranes.

The Pd membrane tubes are positioned above a certain height from the gas distributor so that they are immersed in a gas mixture containing a considerable amount of hydrogen produced from the following reactions:

$$CH_4 + H_2O \rightleftharpoons CO + 3H_2 \tag{10-91}$$

with a rate expression

$$r_1 = k_1 \left[\left(p_{CH_4} \, p_{H_2O} / p_{H_2}^{2.5} \right) - \left(p_{H_2}^{0.5} \, p_{CO} / K_{e1} \right) \right] / E^2 \tag{10-91a}$$

$$CO + H_2O \rightleftharpoons CO_2 + H_2 \tag{10-92}$$

with

$$r_2 = k_2 \left[\left(p_{CO} \, p_{H_2O} / p_{H_2} \right) - \left(p_{CO_2} / K_{e2} \right) \right] / E^2 \tag{10-92a}$$

and

$$CH_4 + 2H_2O \rightleftharpoons CO_2 + 4H_2 \tag{10-93}$$

with

$$r_3 = k_3 \left[\left(p_{CH_4} \, p_{H_2O}^2 / p_{H_2}^{3.5} \right) - \left(p_{H_2}^{0.5} \, p_{CO_2} / K_{e1} K_{e2} \right) \right] / E^2 \tag{10-93a}$$

where

$$E = 1 + K_{CO} \, p_{CO} + K_{H_2} \, p_{H_2} + K_{CH_4} \, p_{CH_4} + K_{H_2O} \, p_{H_2O} / p_{H_2} \tag{10-94}$$

Generally, configuration I is desired for such applications as hydrogen plants in petroleum refineries and for hydrocracking and hydrotreating. Configuration II would be the choice for applications where moderate conversions are necessary to keep some hydrocarbons in the product stream for further processing such as ammonia production plants.

Using the feed conditions for a typical industrial reformer in their fluidized-bed membrane reactor models, Adris et al. [1991] calculated the reactor parameters and operating conditions required to achieve a comparable overall methane conversion for both configurations I and II and a conventional fixed-bed unit. The results given in Table 10.4 clearly illustrates the potential advantages of a fluidized-bed membrane reactor. The reaction temperatures, catalyst amounts, heat transfer areas and total reactor volumes required are significantly lower than those typical of a conventional reformer. For example, the reactor volume, heat transfer area and catalyst amount are reduced by 60, 94 and 97%, respectively, for both design options. The product gas temperatures are lowered by 37 and 24°C for configurations I and II, respectively. The resulting exit steam conversion and hydrogen yield are essentially the same as those provided by the conventional steam reformer. The overall hydrogen yield is the total amount of hydrogen produced (both from the reactor exit and through the membrane walls) divided by the amount of methane in the feed stream.

The direct benefit of having a permselective membrane in a fluidized-bed steam
reformer can be evaluated by modeling the reactor with and without the membrane for
otherwise identical reactor features and feed conditions. An example of the comparison

TABLE 10.4
Novel reformer configurations vs. conventional unit

	Fixed bed plant data	Novel design config. I	Novel design config. II
Total reaction volume (m^3)	10.44	3.89	3.89
Heat transfer area (m^2)	671.2	42.2	42.2
Catalyst mass (kg)	22540	585	585
Heating media temp. (°K)	1283	1283	1283
Exit methane conversion	0.579	0.574	0.576
Exit steam conversion	0.212	0.210	0.211
Product gas temp. (°K)	980.2	943.2	956.3
Product gas mole fractions			
Methane	0.059	0.061	0.060
Steam	0.521	0.524	0.526
Carbon monoxide	0.026	0.024	0.025
Carbon dioxide	0.065	0.064	0.063
Hydrogen	0.327	0.325	0.326
Palladium tube area (m^2)	--	61.0	61.0
Hydrogen yield	2.39	2.36	2.31

Note: Feed temperature, 733°K; Feed pressure, 2.25 MPa; CH_4 feed rate, 710 kmol/h;
Feed composition, $H_2O/CH_4 = 4.6$, $H_2/CH_4 = 0.25$, $CO_2/CH_4 = 0.06$,
$CO/CH_4 = 0.0$
[Adris et al., 1991]

is given in Table 10.5 where the reactor diameter, bed height and bed volume, catalyst
quantity and heat pipe capacity for configuration I are held constant. The thermodynamic
limit on reaction conversion is lifted by the equilibrium displacement due to the
preferential removal of hydrogen by the membrane. This results in considerable
improvements of the following performances. The total methane and steam conversions
increase from 67% and 23% to 89% and 32%, respectively. The hydrogen yield jumps
from 2.7 to 3.6. The product gas temperature, however, increases slightly from 743°C to
764°C due to the presence of the membrane.

The beneficial effects of a fluidized-bed membrane reactor on methane conversion and
hydrogen yield can be optimized by systematically varying the operating parameters
such as catalyst amount, membrane area for permeation, rate of heat supply, and reactor
bed height and bed volume. A hydrogen yield of 3.70 and an accompanying methane
conversion of 92% can be attained at a relatively moderate temperature of 733°C which

is very close to the operating temperature of an industrial reformer. The maximum theoretical hydrogen yield has a value of 4.0 which is imposed by the stoichiometry of the steam reforming of methane. The maximum value corresponds to the extreme case where all the methane is reformed to carbon dioxide.

TABLE 10.5
Effect of permselective membrane tubes for configuration I

	With membrane	Without membrane
Reactor diameter (m)	3.14	3.14
Reactor bed height (m)	1.0	1.0
Reactor bed volume (m^3)	7.78	7.78
Catalyst mass (kg)	4470	4470
Heat pipe capacity (GJ/h)	301	301
Area of permeation (m^2)	112.8	--
Methane conversion	0.889	0.666
Steam conversion	0.321	0.232
Product gas temperature (°K)	1036.7	1016.4
Hydrogen yield	3.57	2.66

[Adris et al., 1991]

As mentioned earlier, configuration II involves a feed stream split into two streams: one going directly into the membrane tubes and the other moves through the fluidized-bed reaction zone where some of the hydrogen product permeates into the Pd tubes. The part of the feed which enters the reactor reaches a very high conversion. When this reacted stream is mixed with the part of the feed that flows through the membrane tubes and picks up the permeated hydrogen, the final product will have a net conversion lower than that at the reactor outlet. Thus, the overall conversion can be optimized through the adjustment of the feed split ratio.

Adris et al. [1991] also determined that the reactor performance is weakly sensitive to the bubble size, bed porosity at minimum fluidization and flow distribution between bubble and dense phases. Furthermore, the bubbles which remove the products from the reaction mixture in the dense phase enhance the forward reaction and consequently breaks the barrier of the reaction equilibrium.

While most of the membrane reactor studies on ethylbenzene dehydrogenation employ fixed-bed membrane reactors, Abdalla and Elnashaie [1995] evaluated the concept of a fluidized-bed membrane reactor through modeling. Since hydrogen is released from the reaction, a palladium-based membrane can be used for this application.

Abdalla and Elnashaie [1995] found that a fluidized-bed reactor without the use of a selective membrane does not perform as well as a typical industrial fixed-bed reactor under most conditions. Although the conversion of ethylbenzene is higher in some cases, the styrene yield is for the most part lower than that in a fixed-bed reactor. This is mainly attributed to the potential disadvantages of a fluidized-bed reactor: reactant by-pass by bubbles and a high degree of mixing.

However, when membrane tubes are inserted in the fluidized-bed reactor, hydrogen is continuously removed from the reaction mixture; thus, the main reaction of ethylbenzene dehydrogenation continues to move in the direction of forward reaction. The ethylbenzene conversion and the yield of styrene increase as a result of the selective permeation of hydrogen through the membrane. Both the conversion and the yield exceed those of the industrial fixed-bed reactors and fluidized-bed reactors without membranes. When 16 membrane tubes are used, the selectivity to styrene is expected to be almost 100% due to suppression of by-products such as toluene [Abdalla and Elnashaie, 1995]. A high ethylbenzene conversion (96.5%) along with a high styrene yield (92.4%) is possible under properly selected realistic conditions.

The above discussions pertain to models assuming three regions: the dense phase, bubble phase and separation side of the membrane. The membrane is assumed to be inert to the reactions. There are, however, cases where the membrane is also catalytic. In these situations, a fourth region, the membrane matrix, needs to be considered. The mass and heat balance equations for the catalytic membrane region will both contain reaction-related terms.

10.6 CATALYTIC NON-PERMSELECTIVE MEMBRANE REACTORS WITH AN OPPOSING REACTANTS GEOMETRY (CNMR/ORG)

The membrane reactors and their models discussed so far utilize the permselective properties of the membranes. The membranes which can be catalytic or inert with respect to the reactions of interest benefit the reactor performances mostly by selectively removing a product or products to effect the equilibrium displacement.

10.6.1 Concept of a Catalytic Non-Permselective Membrane Reactor with an Opposing Reactants Geometry

There are, however, some studies demonstrating another concept of using porous membranes. In this concept, the permselective property of a membrane is immaterial and not utilized. Instead, the well structured porous matrix of the membrane serves to provide a well controlled reaction zone. Specifically, two reactants are fed separately to the opposing sides of a catalytic membrane. For those reactions the rate of which is faster than the diffusion rate of the reactants in the membrane, the reaction can take place inside the catalytically active membrane. This type of membrane reactor, where the membranes are catalytic but not selective, are called catalytic non-permselective membrane reactors (CNMR).

In a CNMR/ORG as shown in Figure 10.15 [Sloot et al., 1990], the reactants A and B introduced to a catalytic membrane from its opposite sides react inside a small reaction zone in the membrane. If the reaction is instantaneous and irreversible, the reaction zone shrinks to a reaction plane theoretically. At the reaction zone or plane, the molar fractions of both reactants will be very low. In principle, it is possible to control the location of the reaction zone/plane so that slip or penetration of one reactant to the opposing side of the membrane is avoided. The molar fluxes of the two reactants are then always in stoichiometric ratio. Thus, the CNMR's are particularly attractive to those chemical processes which normally require strict stoichiometric feed rates of reactants. An example is the Claus reaction which involves hydrogen sulfide and sulfur dioxide.

GA 35294.1

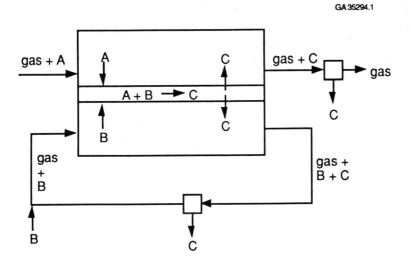

Figure 10.15 Schematic of a catalytic non-permselective membrane reactor with an opposing reactants geometry [Sloot et al., 1990]

10.6.2 CNMR/ORG Models

Some mathematical models describing a CNMR have been developed by various investigators [Harold et al., 1989; Sloot et al., 1990; Sloot et al., 1992; Veldsink et al., 1992; Harold et al., 1992].

Sloot et al. [1990] presented a simplified isothermal CNMR/ORG model which assumes that the two chambers divided by the membrane are well mixed. In practical applications, the model needs to be incorporated into a more complex model which, for example, considers the effect of flow configuration (cocurrent or countercurrent mode). In their model, mass transfer in the direction perpendicular to a flat membrane (i.e., y-direction) is considered for a general instantaneous, reversible reaction

$$\sum_{j}^{J} v_j A_j = 0 \qquad\qquad (10\text{-}95)$$

where J is the total number of chemical species including the inert carrier gas (A_J) and the stoichiometric coefficients, v_j, are characterized by

$$v_j \begin{cases} >0 \ \ if \ A_j \ is \ a \ product \\ <0 \ \ if \ A_j \ is \ a \ reactant \\ =0 \ \ if \ A_j \ is \ an \ inert \ gas \end{cases} \qquad\qquad (10\text{-}95a)$$

Mass balance for each species A_j states that the change of the molar flux of species A_j in the y direction must be equal to the reaction rate at that position. Thus the molar flux J_j and reaction rate r_j for species A_j are related by

$$\frac{dJ_j}{dy} = r_j \qquad (j=1,....J) \qquad\qquad (10\text{-}96)$$

The reaction stoichiometry results in (except the inert gas)

$$\frac{r_1}{v_1} = \frac{r_2}{v_2} = \cdots = \frac{r_j}{v_j} = \cdots \qquad (j=1,...J\text{-}1) \qquad\qquad (10\text{-}97)$$

Combining Equations (10-96) and (10-97) yields

$$\frac{dJ_j}{dy} = \frac{v_j}{v_1} r_1 = \frac{v_j}{v_1} \frac{dJ_1}{dy} \qquad (j=2,....J) \qquad\qquad (10\text{-}98)$$

As a special case, the inert gas ($j=J$) calls for

$$\frac{dJ_J}{dy} = 0 \qquad (inert \ gas) \qquad\qquad (10\text{-}98a)$$

Furthermore, reaction equilibrium is assumed to be valid through the entire membrane and hence

$$K_e = \left(\prod_{j}^{J} X_j^{v_j} \right) \Big/ \left(\prod_{j}^{J} p_T^{v_j} \right) = \left(\prod_{j}^{J} X_j^{v_j} \right) \Big/ \left(p_T^{\sum_{j}^{J} v_j} \right) \qquad\qquad (10\text{-}99)$$

where K_e is the equilibrium constant, p_T the total pressure and X_j the mole fraction of the species A_j.

By definition

$$\sum_{j}^{J} X_j = 1 \tag{10-100}$$

Instead of using a lumped diffusivity D_j^m to account for transport through the membrane as in Equation (10-5), Sloot et al. [1990] assumed two mechanisms of transport: viscous flow and molecular diffusion according to Fick's law. These two mechanisms are assumed to be independent of each other and thus their fluxes are additive:

$$J_j = X_j J_{visc} + (J_j)_{diff} \tag{10-101}$$

where

$$J_{visc} = -\frac{1}{RT}\left[(D_j)_{Kn} + \frac{B_o p_T}{\mu}\right]\frac{dp_T}{dy} \tag{10-101a}$$

and

$$(J_j)_{diff} = -\frac{(D_j)_{eff}}{RT}\frac{d(X_j p_T)}{dy} \tag{10-101b}$$

$(D_j)_{Kn}$ is the Knudsen diffusion coefficient of the inert gas and μ the gas viscosity. Equation (10-101a) represents the flow of a compressible gas through a porous medium.

The parameter B_o is related to the porosity (ε_m), tortuosity (k_t) and mean pore diameter (d_{pore}) of the membrane by the following relationship

$$B_o = \frac{\varepsilon_m}{k_t} \cdot \frac{d_{pore}^2}{32} \tag{10-101c}$$

and the effective diffusion coefficients in the membrane can be determined according to resistances in series as

$$(D_j)_{eff} = \frac{1}{\frac{1}{\frac{\varepsilon_m}{k_t}(D_j^o)_{Kn}} + \frac{1}{\frac{\varepsilon_m}{k_t}(D_{jj}^o)}} \tag{10-101d}$$

where $(D_j)_{Kn}$ and D_{jj}^o are the Knudsen diffusion coefficient of species j and binary diffusion coefficient of species j in the inert gas, respectively, and can be estimated from the kinetic theory of gases [Reid et al., 1977].

Equations (10-98) through (10-100) constitute $J+1$ governing equations for $J+1$ variables: X_j ($j=1,....J$) and p_T. They can be solved numerically, for example, by a discretization technique where a set of N coupled differential equations is replaced by a set of $N \times M$ finite difference equations on a grid consisting of M mesh points. The necessary boundary conditions can be established by requiring the reaction equilibrium (i.e., Equation (10-99)) and the sum of the mole fractions equal to one (i.e., Equation (10-100) at the membrane interface and equality of the pressure at the membrane interface and the pressure in the adjacent gas phase. Additional boundary conditions can be obtained from mass balances coupling the molar fluxes from the gas phase to the membrane interface with those at the interface. Details can be found elsewhere [Sloot et al., 1990].

The above model has been refined based on the dusty-gas model [Mason and Malinauskas, 1983] for transport through the gas phase in the pores and the surface diffusion model [Sloot, 1991] for transport due to surface flow. Instead of Equation (10-101), the following equation gives the total molar flux through the membrane pores which are assumed to be cylindrically shaped

$$J_j = \left(J_j\right)_g + \frac{4}{d_{pore}}\left(J_j\right)_s \tag{10-102}$$

where the molar flux of species j through the gas phase in the membrane pores, $\left(J_j\right)_g$, and the molar flux due to surface diffusion, $\left(J_j\right)_s$, will be given as follows.

By assuming the gas phase in the pores of the membrane to behave as an ideal gas, the gas phase molar flux follows

$$\sum_i^J \left\{ \frac{X_i\left(J_j\right)_g - X_j\left(J_i\right)_g}{p_T D_{ji}} \right\} + \frac{\left(J_j\right)_g}{p_T \left(D_j\right)_{Kn}}$$

$$= -\frac{X_j}{RT}\left\{ \frac{1}{p_T} + \frac{B_o}{\mu\left(D_j\right)_{Kn}} \right\} \frac{dp_T}{dy} - \frac{1}{RT}\frac{dX_j}{dy} \qquad (j=1,....J) \tag{10-102a}$$

where the two terms on the left-hand side account for the continuum diffusion (with an effective continuum diffusion coefficient D_{ji}) and Knudsen diffusion (with an effective Knudsen diffusion coefficient $(D_j)_{Kn}$) and the two terms on the right-hand side represent transport due to a pressure gradient (pressure diffusion and viscous flow) and transport owing to a concentration gradient. D_{ji} and $(D_j)_{Kn}$ are related to the continuum binary diffusion coefficients D_{ji}^o and $\left(D_j^o\right)_{Kn}$ by

$$D_{ji} = \frac{\varepsilon_m}{k_t} D_{ji}^o$$

(10-102b)

and

$$\left(D_j\right)_{Kn} = \frac{\varepsilon_m}{k_t} \left(D_j^o\right)_{Kn}$$

(10-102c)

The continuum binary diffusion coefficients D_{ji}^o can be obtained from the gas kinetic theory as stated earlier while the Knudsen diffusion coefficient $\left(D_j^o\right)_{Kn}$ is calculated according to

$$\left(D_j^o\right)_{Kn} = \frac{d_{pore}}{3} \sqrt{\frac{8RT}{\pi M_j}}$$

(10-102d)

where M_j is the molecular weight of species j.

The molar flux due to surface diffusion is described by Fick's law

$$\left(J_j\right)_s = -\left(D_j\right)_s C_s \frac{d\theta_j}{dy} \quad (j=1,....J)$$

(10-102e)

where $\left(D_j\right)_s$ is the effective surface diffusion coefficient of species j, C_s the total concentration of adsorption sites and θ_j the fraction of the adsorption sites covered by species j. It is assumed that an instantaneous adsorption-desorption equilibrium is established. It is further assumed that θ_j obeys a Langmuir isotherm

$$\theta_j = \frac{b_j X_j p_T}{1 + p_T \sum_i b_i X_i} \quad (j=1,....J)$$

(10-102f)

where b_j is the adsorption constant of species j. The boundary conditions required and the solution methodology are the same as those previously mentioned and described in details elsewhere [Sloot et al., 1992]. Mass transfer resistances outside the membrane are considered to specify the molar fluxes from the gas phase to the membrane interface. To account for the combined flow and diffusion outside the membrane, a stagnant gas film model is used for each side of the membrane. The Stefan-Maxwell diffusion equation is used instead of the Fick's law. The former is preferred for multicomponent diffusion, especially for the diffusion of the inert gas as the major component of the gas streams.

The operating temperature and the pressure difference across the membrane can impact the location and the size of the reaction zone inside the membrane and any slip of the reactants. Figures 10.16a and 10.16b show the mole fraction profiles of a Claus reaction in a catalytic nonpermselective membrane at 200 and 300°C, respectively, both without a transmembrane pressure difference and surface diffusion. At 200°C, the reaction is restricted to a very small zone and the reaction can be considered essentially irreversible. The associated slip of the reactants to the opposing side is less than 1% of the incoming reactants to the membrane. Although not shown in Figure 10.16a, the local pressure inside the membrane exhibits a minimum in the reaction zone despite no overall pressure difference across the membrane. This phenomenon is caused by the reduction of the number of gas molecules as a result of the reaction. The minimum pressure developed inside a membrane of small pore diameters (≤ 0.01 μm) where Knudsen diffusion dominates is independent of the pore diameter [Sloot et al., 1992]. For membranes of relatively large pore diameters (say, >1 μm), bulk molecular diffusion is operative and is independent of the pore diameter. And so is the molar flux of hydrogen sulfide. It is interesting to mention that the pressure minimum also occurs in membrane pores where surface diffusion, but not a chemical reaction, is present. Surface diffusion can be significant even in large membrane pores under the condition of high surface concentrations relative to those in the gas phase.

GA 35274.3

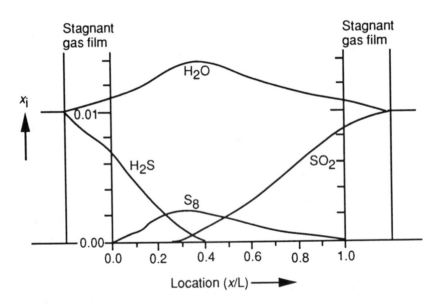

Figure 10.16a Mole fraction profile of a Claus reaction in a catalytic nonpermselective membrane at 200°C in the absence of a transmembrane pressure difference [Sloot et al., 1990]

GA 35274.4

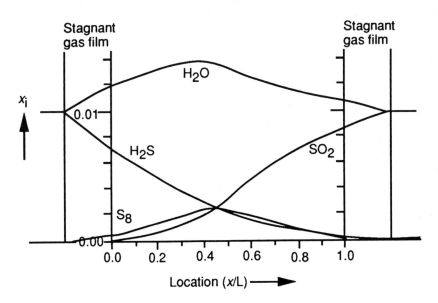

Figure 10.16b Mole fraction profile of a Claus reaction in a catalytic nonpermselective membrane at 300°C in the absence of a transmembrane pressure difference [Sloot et al., 1990]

A pressure maximum, instead of minimum, inside the membrane could result from cases where both chemical reaction and surface diffusion are present [Sloot et al., 1992]. Thus the occurrence of a maximum or minimum local pressure inside the membrane depends on the reaction stoichiometry as well as the mobilities of the reaction species. It is assumed that only hydrogen sulfide adsorbs on the pore surface. Due to a higher transport rate of H_2S enhanced by surface diffusion, the reaction zone is shifted toward the SO_2 side of the membrane. In the reaction zone, larger amounts of the products are formed and higher molar fluxes of the products out of the membrane are expected so that the maxima of the mole fraction profiles of the products at the reaction zone can be sustained.

In contrast, the reaction is definitely reversible at 300°C due to a lower equilibrium constant of the Claus reaction at a higher temperature. The reaction zone is rather broad at 300°C. The slip of hydrogen sulfide to the opposing side is about 8% of its molar flux at the membrane surface. The conversion rate of hydrogen sulfide is calculated by subtracting its slip from its incoming flux at the membrane surface. Although the mole fraction profile and the reactants slip are appreciably different at 200 and at 300°C, the conversion rate at 300°C is only about 6.5% smaller than that at 200°C. It is evident that the mole fractions of the reactants, H_2S and SO_2, decrease substantially and those of the products, S_8 and H_2O, increase significantly from the bulk of the gas to the membrane interface. Thus, mass transfer resistances in the gas phase are important and cannot be ignored.

If a transmembrane pressure difference is imposed at a given constant temperature, the reaction zone will be shifted toward the lower pressure side. The mole fraction of the reactant entering the lower pressure side of the membrane surface drops to a level lower than that in the absence of a pressure difference. It has been shown [Sloot et al., 1990] that the molar fluxes of, say, hydrogen sulfide increases as the pressure on its side increases, thus potentially reducing the membrane area required. A serious drawback with this mode of operation, however, is the amount of inert gas introduced.

As shown in Figure 10.16a and 10.16b, the reaction product, elemental sulfur, exits both sides of the membrane in the absence of a pressure difference. This is due to the molecular diffusion of the reaction components. If it is desirable to force the sulfur to exit only on the side of sulfur dioxide, a significant pressure difference has to be applied. However, as just pointed out, the large quantity of inert gas involved will require further separation downstream of the reactor.

Using the Claus reaction as a model reaction, Sloot et al. [1990] has experimentally measured reaction conversions and verified the above concept of confining and shifting a reaction plane or zone for two opposing reactant streams inside a porous catalytic membrane. The agreement between the experimentally observed and calculated (based on the dusty-gas model) molar flux of H_2S is reasonably good [Sloot et al., 1992]. The simplified model based on Equation (10-101) is a good approximation for dilute systems where the mole fractions of the reactants and products are lower than that of the inert gas.

10.7 CATALYTIC NON-PERMSELECTIVE MEMBRANE MULTIPHASE REACTORS (CNMMR)

Another type of catalytic and yet non-permselective membrane reactors uses one side of the membrane as the phase boundary between gas and liquid reaction streams. An example is the reaction between a gaseous reactant (A) flowing in the tube core while a liquid containing the solution of the other reactant (B) in a suitable solvent flowing in the annular region (Figure 10.17):

$$\upsilon_A A_{(g)} + \upsilon_B B_{(\ell)} \rightarrow Products(s)_{(\ell)} \tag{10-103}$$

The reaction rate is assumed to be first-order with respect to A and zeroth-order with respect to B. This type of membrane reactors is suitable for such reactions as hydrogenation of unsaturated fatty acids in vegetable oils on Pd and Ni catalysts, hydrogenation of 2-ethylhexanol on Ni, and hydrogenation of p-nitrobenzoic acid in an aqueous solution on Pd [Akyurtlu et al., 1988].

GA 35294.2

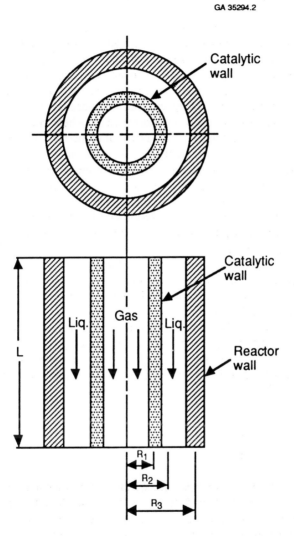

Figure 10.17 Cross-sectional views of the catalytic porous wall gas-liquid reactor. [Akyurtlu et al., 1988]

It is assumed that the gas and liquid streams flow co-currently on the opposite sides of the membrane. It is further assumed that the catalyst pores are completely filled with the liquid and that the gas-liquid interface stays at the pore mouths on the tube side. Consider a steady state, isothermal case where surface tension effects can be neglected, the membrane porosity and catalytic activity are uniform and constant and axial convection and diffusion inside the membrane are negligible.

Harold et al. [1989] have extended a CNMMR model developed by Akyurtlu et al. [1988] to account for transport of gas phase components and for the variation of the reaction rate with the non-volatile reactant B bulk concentration in the liquid phase in the shell (annular) region. Thus three regions are considered: the tube side where the flowing gas contains the volatile reactant A and an inert gas I, the membrane matrix filled with the liquid, and the shell side where the liquid is flowing. The liquid contains A, B, the products and possibly a solvent. Consider those cases where the supply of B from the liquid is sufficient so that B is not depleted anywhere in the membrane. The gas in the tube core is plug flow. Two flow configurations for the CNMMR have been modeled. In the first configuration, the liquid flow in the annular region is assumed to be laminar flow. In the second configuration, the liquid is assumed to be well mixed.

10.7.1 Catalytic Non-permselective Membrane Multiphase Reactor (CNMMR) Model – Laminar Flow Liquid Stream

The governing equations in dimensionless form for the three regions have been derived as follows [Harold et al., 1989].

On tube (gas) side. A momentum balance yields

$$a_I\, \theta_I^t\, U \frac{dU}{dZ} + a_A\, \theta_A^t\, U \frac{dU}{dZ} + b_I \frac{d\theta_I^t}{dZ} + b_A \frac{d\theta_A^t}{dZ} + U = 0 \tag{10-104}$$

The mass balance of species I is described by

$$\frac{d}{dZ}\left(U\theta_I^t\right) = 0 \tag{10-105}$$

and the balance equation for species A contains a plug-flow convective term and a term reflects adsorption of A into the membrane:

$$P_{e_A}^t \frac{d}{dZ}\left(U\theta_A^t\right) - \theta_A^t \left(\frac{d\theta_A^m}{d\xi}\right)_{\xi=\xi_1} = 0 \tag{10-106}$$

Inside membrane. Reaction and diffusion balances for species A and B within the membrane are given by

$$\frac{1}{\xi}\frac{d}{d\xi}\left(\xi\frac{d\theta_A^m}{d\xi}\right)-\phi_0^2\,\theta_B^s(\xi_2,Z)\theta_A^m=0 \qquad (10\text{-}107)$$

$$\frac{1}{\xi}\frac{d}{d\xi}\left(\xi\frac{d\theta_B^m}{d\xi}\right)-m_0\,\phi_0^2\,\theta_A^t(Z)\theta_A^m=0 \qquad (10\text{-}108)$$

where the reaction rate is

$$r=k'\,C_B^s(z)\,C_A^m=k\theta_A^m \qquad (10\text{-}109)$$

and the pseudo rate constant, k, depends on $\theta_B^s(Z)$.

On shell (liquid) side. Mass balances of species A and B contains axial convection terms that are assumed to be uncoupled from the mass transport, and radial and axial diffusion terms:

$$P_{e_A}^s\,\Gamma\left[1-\left(\frac{\xi}{\xi_3}\right)^2+\frac{1-\lambda^2}{ln(1/\Gamma)}ln(\xi/\xi_3)\right]\frac{\partial\theta_A^s}{\partial Z}$$

$$=\frac{1}{\xi}\frac{\partial}{\partial\xi}\left(\xi\frac{\partial\theta_A^s}{\partial\xi}\right)+\left(\frac{R_2-R_1}{2L}\right)^2\frac{\partial^2\left(\theta_A^s\right)}{\partial Z^2} \qquad (10\text{-}110)$$

$$P_{e_B}^s\,\Gamma\left[1-\left(\frac{\xi}{\xi_3}\right)^2+\frac{1-\lambda^2}{ln(1/\Gamma)}ln(\xi/\xi_3)\right]\frac{\partial\theta_B^s}{\partial Z}$$

$$=\frac{1}{\xi}\frac{\partial}{\partial\xi}\left(\xi\frac{\partial\theta_B^s}{\partial\xi}\right)+\left(\frac{R_2-R_1}{2L}\right)^2\frac{\partial^2\left(\theta_B^s\right)}{\partial Z^2} \qquad (10\text{-}111)$$

The above equations are subject to the following boundary conditions. For uniform concentrations and velocity of the gas at the inlet,

$$\theta_I^t(\xi,Z=0)=1 \qquad (10\text{-}112a)$$

$$\theta_A^t(\xi,Z=0)=1 \qquad (10\text{-}112b)$$

$$U(\xi,Z=0)=1 \qquad (10\text{-}112c)$$

For uniform concentrations of the liquid components at the inlet,

$$\theta_A^s\left(\xi, Z = 0\right) = 1 \tag{10-113a}$$

$$\theta_B^s\left(\xi, Z = 0\right) = 1 \tag{10-113b}$$

At the interface of the bulk gas and the catalytic membrane,

$$\theta_A^m\left(\xi = \xi_1, Z\right) = 1 \tag{10-114a}$$

$$\left(\frac{d\theta_B^m}{d\xi}\right)_{\xi=\xi_1} = 0 \tag{10-114b}$$

At the interface of the bulk liquid and the membrane,

$$\theta_A^l(Z)\Delta_A\left(\frac{d\theta_A^m}{d\xi}\right)_{\xi=\xi_2} = \theta_A^o\left(\frac{d\theta_A^s}{d\xi}\right)_{\xi=\xi_2} \tag{10-115a}$$

$$\theta_B^s\left(\xi = \xi_2, Z\right)\Delta_B\left(\frac{d\theta_B^m}{d\xi}\right)_{\xi=\xi_2} = \left(\frac{d\theta_B^s}{d\xi}\right)_{\xi=\xi_2} \tag{10-115b}$$

and finally due to no transport of species across the boundary between the bulk liquid and the shell wall,

$$\left(\frac{d\theta_A^s}{d\xi}\right)_{\xi=\xi_3} = 0 \tag{10-116a}$$

$$\left(\frac{d\theta_B^s}{d\xi}\right)_{\xi=\xi_3} = 0 \tag{10-116b}$$

The dimensionless coordinates and axial velocity of gas are defined by

$$Z \equiv z / L, \xi \equiv \frac{2r}{(R_2 - R_1)}, U \equiv \frac{u_{av}^l(z)}{u_{av}^l(z=0)} \tag{10-117}$$

and the dimensionless concentrations of A, B and I (inert) by

$$\theta_A^l \equiv \frac{C_A^l(z)}{C_A^l(z=0)}, \theta_I^l \equiv \frac{C_I^l(z)}{C_I^l(z=0)}, \theta_A^m \equiv \frac{C_A^m(r,z)}{C_A^l(z=0)/H_C}, \theta_B^m \equiv \frac{C_B^m(r,z)}{C_B^s(r=R_2,z)}$$

$$\theta_A^s \equiv \frac{C_A^s(r,z)}{C_A^s(r,z=0)}, \theta_B^s \equiv \frac{C_B^s(r,z)}{C_B^s(r,z=0)} \tag{10-118}$$

where H_C is the Henry's law constant.

Various dimensionless parameters in the above governing equations are:

$$a_I \equiv \frac{C_I^t(z=0)M_I U_{av}^t(z=0)}{\gamma L}, a_A \equiv \frac{C_A^t(z=0)M_A u_{av}^t(z=0)}{\gamma L},$$

$$b_I \equiv \frac{RT C_I^t(z=0)}{\gamma L u_{av}^t(z=0)}, b_A \equiv \frac{RT C_A^t(z=0)}{\gamma L},$$

$$\gamma \equiv \frac{8\mu_g}{R_1^2}, \Delta_A \equiv \frac{D_{Ae}}{D_A}, \Delta_B \equiv \frac{D_{Be}}{D_B}$$

$$P_{e_A}^t \equiv \frac{R_1(R_2-R_1)u_{av}^t(z=0)H_C}{4 D_{Ae} L}, P_{e_A}^s \equiv \frac{(R_2-R_1)^2 u_{av}^s}{2 D_A L},$$

$$P_{e_B}^s \equiv \frac{(R_2-R_1)^2 u_{av}^s}{2 D_B L}, m_o \equiv \frac{(\upsilon_B/\upsilon_A)D_{Ae}C_A^t(z=0)/H_c}{D_{Be} C_B^s(z=0)},$$

$$\phi_0^2 \equiv \left(\frac{R_2-R_1}{2}\right)^2 \left[\frac{k' C_B^s(z=0)}{D_{Ae}}\right], \xi_i \equiv \frac{2R_i}{R_2-R_1} \quad (i=1,2,3),$$

$$\lambda \equiv \frac{R_2}{R_3}, \theta_A^0 \equiv \frac{C_A^s(r,z=0)}{C_A^t(z=0)/H_C}, \Gamma \equiv \left[1+\lambda^2 - \frac{1-\lambda^2}{ln(1/\lambda)}\right]^{-1} \tag{10-119}$$

where M_A and M_I are the molecular weights of species A and I, respectively, μ_g the gas viscosity, D_{Ae} and D_{Be} the effective diffusivities of species A and B, respectively, in the membrane, and u_{av}^t and u_{av}^s the average velocities of the gas (tube side) and the liquid (shell side), respectively.

The CNMMR model with laminar flow liquid stream in the annular region consists of three ordinary differential equations for the gas in the tube core and two partial differential equations for the liquid in the annular region. These equations are coupled through the diffusion-reaction equations inside the membrane and boundary conditions. The model can be solved by first discretizing the liquid-phase mass balance equations in the radial direction by the orthogonal collocation technique. The resulting equations are then solved by a semi-implicit integration procedure [Harold et al., 1989].

Akyurtlu et al. [1988] have also presented a similar but simpler model. In their model, the gas phase concentrations are assumed to be constant. Thus, Equations (10-104) through (10-106) and Equations (10-112a) and (10-112b) are not needed and θ_A^t in Equation (10-115a) becomes constant. In addition, they assumed that the reaction rate is independent of the bulk concentration of B in the liquid phase (annular region). Furthermore, axial diffusion in the liquid phase is neglected. Akyurtlu et al. [1988] did take into account situations where the reactant may be depleted inside the membrane.

The membrane region is then split into two sub-regions: one region where the concentration of B is zero and the other region where B is present as the case in the model presented above. The more general model of Harold et al. [1989] can be extended to account for the depletion of B by considering the aforementioned two sub-regions.

A similar type of two-phase cross-flow membrane reactor has also been modeled [De Vos and Hamrin, 1982]. In these membrane reactors, liquid flows through channels in one direction and fills the catalytic porous membrane while gas flows at right angles in alternate parallel channels (Figure 10.18). The liquid stream is assumed to be laminar flow. The gas stream is assumed to be plug flow with a uniform concentration. Therefore, only transport in the liquid channel and through the membrane is considered. Mass transport and reaction through the membrane is modeled by focusing on the pore region. Coupling of the pore and channel equations occurs at the membrane walls. Operation of the membrane reactor at total liquid recycle is considered and the concentration of the reactant as a function of time. Hydrogenation of nitrobenzoic acid, which is a first-order reaction with respect to hydrogen and a zero-order reaction with respect to nitrobenzoic acid, is used as an example.

GA 35294.3

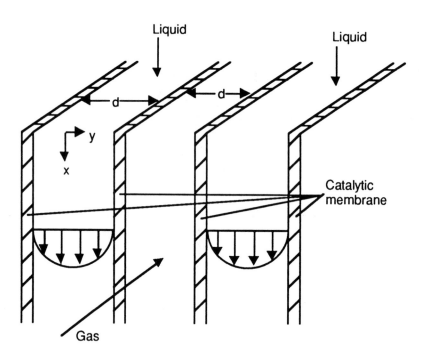

Figure 10.18 Gas-liquid flow in section of cross-flow reactor [De Vos and Hamrin, 1982]

10.7.2 Catalytic Non-permselective Membrane Multiphase Reactor (CNMMR) Model: Well-Mixed Liquid Stream

Another flow configuration of a CNMMR is gas flow through the tube core, but, unlike the preceding case, the liquid in the annular region is well mixed. This may occur when the bulk liquid is given sufficient agitation. In this case, the membrane tube surface is exposed to a uniform concentration.

This model is identical to the CNMMR model with a laminar-flow liquid except the mass balance equations for the annular (liquid) region and the associated boundary conditions. Thus, Equations (10-104) through (10-109) still apply to the tube core and the membrane region but Equations (10-110), (10-111), (10-115a) and (10-115b) need to be replaced. Moreover, Equations (10-113a), (10-113b), (10-116a) and (10-116b) are no longer needed.

On shell (liquid) side. Mass balances for species A and B in the bulk liquid now relate the net rate of material loss to the flux integrated over the external surface of the membrane tube:

$$Q_A\left(\theta_A^0 - \theta_{Ab}^s\right) - Sh_A \int_{Z=0}^{Z=1}\left[\theta_{Ab}^s - \theta_A^t(Z)\theta_A^m(\xi_2,Z)\right]dZ = 0 \qquad (10\text{-}120)$$

$$Q_B\left(1 - \theta_{Bb}^s\right) - Sh_B\,\theta_{Bb}^s \int_{Z=0}^{Z=1}\left[1 - \theta_B^m(\xi_2,Z)\right]dZ = 0 \qquad (10\text{-}121)$$

and at the interface of the bulk liquid and the membrane

$$Sh_A\left[\theta_{Ab}^s / \theta_A^t(Z) - \theta_A^m(\xi_2,Z)\right] - \left(\frac{d\theta_A^m}{d\xi}\right)_{\xi=\xi_2} = 0 \qquad (10\text{-}122a)$$

$$Sh_B\left[1 - \theta_B^m(\xi_2,Z)\right] - \left(\frac{d\theta_B^m}{d\xi}\right)_{\xi=\xi_2} = 0 \qquad (10\text{-}122b)$$

where

$$Sh_A = Sc_A^{0.3}\left[0.35 + 0.34\left(R_e^s\right)^{0.5} + 0.15\left(R_e^s\right)^{0.58}\right]/\xi_2 \qquad (10\text{-}122c)$$

and

$$Sh_B = Sh_A\left(\frac{Sc_B}{Sc_A}\right)^{0.3} \qquad (10\text{-}122d)$$

The above boundary conditions balance the external transport flux with the intraparticle flux at the external membrane tube surface. The Sherwood number correlations given in Equations (10-122c) and (10-122d) apply to situations where liquid is injected into the tube at multiple points along its length without agitation. Therefore, Sh_A and Sh_B are expected to be higher than those calculated from Equations (10-122c) and (10-122d). The new dimensionless concentrations are

$$\theta_B^m \equiv \frac{C_B^m(r,z)}{C_{Bb}^s}, \theta_{Ab}^s \equiv \frac{C_{Ab}^s}{C_A^l(z=0)/H_c}, \theta_{Bb}^s \equiv \frac{C_{Bb}^s}{\left(C_{Bb}^s\right)_0} \tag{10-122e}$$

and the new dimensionless parameters are given by

$$R_e^s \equiv \frac{\rho_\ell q_\ell}{\pi L \mu_\ell}, Sh_A \equiv \frac{(R_2-R_1)k_{cA}\varepsilon_p}{2D_{Ae}}, Sh_B \equiv \frac{(R_2-R_1)k_{cB}\varepsilon_p}{2D_{Be}},$$

$$S_{cA} \equiv \frac{\mu_\ell}{\rho_\ell D_A}, S_{cB} \equiv \frac{\mu_\ell}{\rho_\ell D_B}, Q_A \equiv \frac{q_\ell(R_2-R_1)}{4\pi D_{Ae}R_2 L}, Q_B \equiv \frac{q_\ell(R_2-R_1)}{4\pi D_{Be}R_2 L},$$

$$\phi_0^2 \equiv \left(\frac{R_2-R_1}{2}\right)^2 \left[\frac{k'\left(C_{Bb}^s\right)_0}{D_{Ae}}\right], m_0 \equiv \frac{(\upsilon_B/\upsilon_A)D_{Ae}C_A^l(z=0)/H_c}{D_{Be}\left(C_{Bb}^s\right)_0} \tag{10-122f}$$

The bulk concentrations of the well-mixed liquid in the annular region, C_{Ab}^s and C_{Bb}^s, at the inlet of the reactor are designated as $\left(C_{Ab}^s\right)_0$ and $\left(C_{Bb}^s\right)_0$, respectively. The liquid density, viscosity and volumetric flow rate are ρ_ℓ, μ_ℓ and q_ℓ, respectively, and k_{cA} and k_{cB} represent the external mass transfer coefficients for species A and B, respectively, for the CNMMR with a well mixed liquid stream.

10.7.3 Catalytic Non-permselective Membrane Multiphase Reactor (CNMMR) Model Predictions

Harold et al. [1989] have compared the predicted CNMMR performances based on the above two models with that calculated from a string of pellets reactor model. The string of pellets reactor is considered a prototype trickle-bed reactor.

Shown in Figure 10.19 is a comparison of the overall reaction rate as a function of the liquid flow rate for the three models just mentioned [Harold et al., 1989]. Three different values of the catalytic activity (k') are used as a parameter. The solid line, dashed line (--) and dotted line (-·-) represent the results predicted by the CNMMR model with a well-mixed liquid stream, the CNMMR model with a laminar-flow liquid stream and the string-of-pellets reactor model, respectively. The membrane is assumed to be 0.1 cm thick for the two membrane reactors. The liquid feed is saturated with A and contains a large supply of B $\left(C_B^s = 10^{-3} \text{ gmol/cc}\right)$.

It is evident that the CNMMR with a well-agitated liquid in the annular region out-performs the other two reactors and the CNMMR with a laminar-flow liquid stream on the shell side is superior to the string-of-pellets reactor. The differences among the three models are more pronounced at a higher catalytic activity. The feed liquid film surrounding the wetted pellets contains little volatile reactant *A* and acts as a barrier to species *A* supply along the length of the pellet string. Species *A* can enter the pores where the reaction takes place directly from the gas-filled tube core. Therefore, the catalytic membrane reactors (CNMMR) are not limited by this mass transfer resistance. The advantage of a catalytic membrane reactor becomes more obvious for cases where the liquid contains a small amount (say, 1%) of its equilibrium allotment of the volatile reactant *A*. A reaction rate increase of more than an order of magnitude with the use of a catalytic membrane reactor over a string-of-pellets reactor is possible even at a relatively low catalytic activity [Harold et al., 1989]. It appears that a thinner membrane is more efficient.

GA 35268

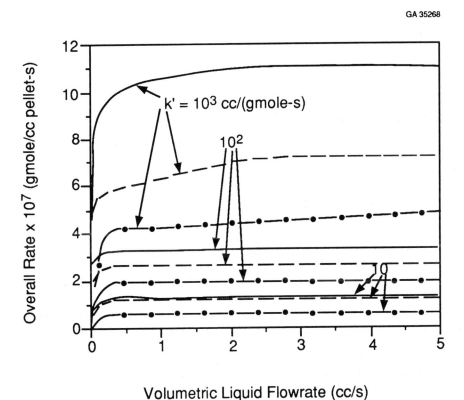

Figure 10.19 Comparison of overall reaction rate as a function of liquid flow rate for string-of-pellets reactor model (–·–), CNMMR model with a well-mixed liquid stream (—) and CNMMR model with a laminar-flow liquid stream (---) [Harold et al., 1989]

The CNMMR, however, is not always advantageous over the pellet-string reactor. When the membrane tube is wetted only on the external surface and the rate of supply of volatile A from the tube core is very high, depletion of the reactant B occurs within the membrane matrix potentially leading to a lower reaction rate than that of a pellet-string reactor.

Akyurtlu et al. [1988] have found that when B becomes depleted inside the membrane matrix, the reaction conversion is very sensitive to the annular space in a CNMMR (see Figure 10.20). The average conversion of B at the reactor outlet increases as the radius ratio $\lambda(\equiv R_2 / R_3)$ increases. The rate of increase is particularly rapid as λ approaches 1. This means that when the annular liquid region becomes small, the reaction conversion is high for a constant throughput and a constant membrane thickness. In the same study, a parameter γ was defined as

$$\gamma \equiv \frac{\phi_B^2}{\phi_A^2} \qquad\qquad\qquad (10\text{-}123)$$

where ϕ_A and ϕ_B are the Thiele modulus for species A and B, respectively. A large Thiele modulus corresponds to a fast reaction rate of that species compared to the diffusion rate inside the catalytic membrane. Akyurtlu et al. [1988] discovered that the conversion of the nonvolatile reactant B increases as γ increases. Thus a higher γ value provides a relatively low Thiele modulus for A which implies a relatively fast diffusion rate and hence a large volatile A concentration. This results in a high conversion of B. The size of the reaction zone becomes small at a high value of γ.

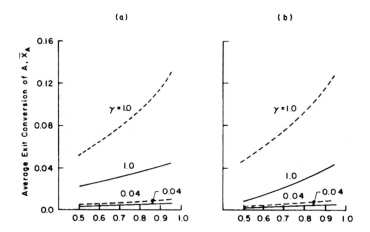

Figure 10.20 Effect of radius ratio on conversion of reactant A at reactor exit with (a) constant throughput and constant catalyst thickness and (b) constant throughput and constant catalyst volume. [Akyurtlu et al., 1988]

10.8 SUMMARY

As a building block for simulating more complex and practical membrane reactors, various membrane reactor models with simple geometries available from the literature have been reviewed. Four types of shell-and-tube membrane reactor models are presented: packed-bed catalytic membrane reactors (a special case of which is catalytic membrane reactors), fluidized-bed catalytic membrane reactors, catalytic non-permselective membrane reactors with an opposing reactants geometry and catalytic non-permselective membrane multiphase reactors. Both dense and porous inorganic membranes have been considered.

The packed-bed catalytic membrane reactors have been modeled most extensively. In many cases the model predictions and the experimental data agree quite well and both show the advantages of the permselective membranes for improving the reaction conversions and selectivities due to the equilibrium displacement caused by the membranes. A range of dehydrogenation reactions are used as the model reactions. Relative advantages of various membrane reactors among themselves and over conventional plug-flow or continuous stirred tank reactors are determined by the flow configurations and conditions (including co-current, counter-current and well-mixed conditions), catalytic zones (packed beds of catalysts vs. catalytic membranes) and other operating parameters such as space velocity, net volume change after reaction and transmembrane pressure drop. Only limited work on fluidized-bed catalytic membrane reactor models has been published.

Membranes have also been used in reactors where their permselective properties are not important. Instead their well-engineered porous matrix provides a well-controlled catalytic zone for those reactions requiring strict stoichiometric feed rates of reactants or a clear interface for multiphase reactions (e.g., a gas and a liquid reactant fed from opposing sides of the membrane). Functional models for these types of membrane reactors have also been developed. The conditions under which these reactors provide performance advantages have been identified.

In real production situations where geometric complexity and flow configurations warrant three-dimensional numerical simulations, computational fluid dynamic codes may be required to capture the complicated physicochemical hydrodynamics. This approach may begin to become feasible with the availability of powerful computers and efficient numerical algorithms.

REFERENCES

Abdalla, B.K., and S.S.E.H. Elnashaie, 1995, J. Membr. Sci. **101**, 31.
Adris, A.M., 1994, A Fluidized Bed Membrane Reactor for Steam Methane Reforming: Experimental Verification and Model Validation, Ph.D. dissertation, Univ. of British Columbia, Vancouver, Canada.

Adris, A.M., S.S.E.H. Elnashaie and R. Hughes, 1991, Can. J. Chem. Eng. **69**, 1061.
Agarwalla, S., and C.R.F. Lund, 1992, J. Membr. Sci. **70**, 129.
Akyurtlu, J.F., A, Akyurtlu and C.E. Hamrin, Jr., 1988, Chem. Eng. Comm. **66**, 169.
Becker, Y.L., A.G. Dixon, W.R. Moser and Y.H. Ma, 1993, J. Membr. Sci. **77**, 233.
Bernstein, L.A., and C.R.F. Lund, 1993, J. Membr. Sci. **77**, 155.
Bird, R. B., W.E. Stewart and E.N. Lightfoot, 1960, Transport Phenomena, Wiley, New York, U.S.A.
Cannon, K.C., and J.J. Hacskaylo, 1992, J. Membr. Sci. **65**, 259.
Champagnie, A.M., T.T. Tsotsis, R.G. Minet and E. Wagner, 1992, J. Catal. **134**, 713.
Collins, J.P., J.D. Way and N. Kraisuwansarn, 1993, J. Membr. Sci. **77**, 265.
De Vos, R., and C.E. Hamrin, Jr., 1982, Chem. Eng. Sci. **37**, 1711.
Fang, S.M., S.A. Stern and H.L. Frisch, 1975, Chem. Eng. Sci. **30**, 773.
Gerald, C.F., and P.O. Wheatley, 1984, Applied Numerical Analysis, Addison-Wesley, Reading, MA, U.S.A.
Harold, M.P., P. Cini, B. Patenaude and K. Venkataraman, 1989, AIChE Symp. Ser. **85** (No. 268), 26.
Harold, M.P., V.T. Zaspalis, K. Keizer and A.J. Burggraaf, 1992, Paper presented at 5th North Am. Membr. Soc. Meeting, Lexington, KY, USA.
Itoh, N., 1987, AIChE J. **33**, 1576.
Itoh, N., 1990, J. Chem. Eng. Japan **23**, 81.
Itoh, N., and R. Govind, 1989a, AIChE Symp. Ser. **85**, 10.
Itoh, N., and R. Govind, 1989b, Ind. Eng. Chem. Res. **28**, 1554.
Itoh, N., and W.C. Xu, 1991, Sekiyu Gakkaishi **34**, 464.
Itoh, N., Y. Shindo, T. Hakuta and H. Yoshitome, 1984, Int. J. Hydrogen Energy **10**, 835.
Itoh, N., Y. Shindo, K. Haraya, K. Obata, T. Hakuta and H. Yoshitome, 1985, Int. Chem. Eng. **25**, 138.
Itoh, N., Y. Shindo, K. Haraya and T. Hakuta, 1988, J. Chem. Eng. Japan **21**, 399.
Itoh, N., Y. Shindo and K. Haraya, 1990, J. Chem. Eng. Japan **23**, 420.
Itoh, N., M.A. Sanchez C., W.C. Xu, K. Haraya and M. Hongo, 1993, J. Membr. Sci. **77**, 245.
Levenspiel, O., 1962, Chemical Reaction Engineering, John Wiley and Sons, New York, USA.
Lund, C.R.F., 1992, Catal. Lett. **12**, 395.
Mason, E.A., and A.P. Malinauskas, 1983, Gas Transport in Porous Media: the Dusty-Gas Model, Chem. Eng. Monographs 17, Elsevier, Amsterdam, Netherlands.
Mohan, K., and R. Govind, 1986, AIChE J. **32**, 2083.
Mohan, K., and R. Govind, 1988a, AIChE J. **34**, 1493.
Mohan, K., and R. Govind, 1988b, Ind. Eng. Chem. Res. **27**, 2064.
Mohan, K., and R. Govind, 1988c, Sep. Sci. Technol. **23**, 1715.
Oertel, M., J. Schmitz, W. Welrich, D. Jendryssek-Neumann and R. Schulten, 1987, Chem. Eng. Technol. **10**, 248.
Reid, R.C., J.M. Prausnitz and T.K. Sherwood, 1977, The Properties of Gases and Liquids (3rd ed.) McGraw-Hill, New York, USA.
Shinji, O., M. Misono and Y. Yoneda, 1982, Bull. Chem. Soc. Japan **55**, 2760.
Sloot, H.J., 1991, A Nonpermselective Membrane Reactor for Catalytic Gas-Phase Reactions, PhD. dissertation, Twente Univ. of Technol., Netherlands.
Sloot, H.J., G.F. Versteeg and W.P.M. van Swaaij, 1990, Chem. Eng. Sci. **45**, 2415.

Sloot, H.J., C.A. Smolders, W.P.M. van Swaaij and G.F. Versteeg, 1992, AIChE J. **38**, 887.

Soliman, M.A., S.S.E.H. Elnashaie, A.S. Al-Ubaid and A.M. Adris, 1988, Chem. Eng. Sci. **43**, 1803.

Song, J.Y., and S.T. Hwang, 1991, J. Membr. Sci. **57**, 95.

Sun, Y.M., and S.J. Khang, 1988, Ind. Eng. Chem. Res. **27**, 1136.

Sun, Y.M., and S.J. Khang, 1990, Ind. Eng. Chem. Res. **29**, 232.

Tsotsis, T.T., A.M. Champagnie, S.P. Vasileiadis, Z.D. Ziaka and R.G. Minet, 1992, Chem. Eng. Sci. **47**, 2903.

Tsotsis, T.T., A.M. Champagnie, S.P. Vasileiadis, Z.D. Ziaka and R.G. Minet, 1993a, Sep. Sci. Technol. **28**, 397.

Tsotsis, T.T., R.G. Minet, A.M. Champagnie and P.K.T. Liu, 1993b, Catalytic membrane reactors, in: Computer-aided Design of Catalysts (Ed. E.R. Becker and C.J. Pereira), p. 471.

Uemiya, S., N. Sato, H. Ando and E. Kikuchi, 1991, Ind. Eng. Chem. Res. **30**, 589.

Veldsink, J.W., R.M.J. van Damme, G.F. Versteeg and W.P.M. van Swaaij, 1992, Chem. Eng. Sci. **47**, 2939.

Wang, W., and Y.S. Lin, 1995, J. Membr. Sci. **103**, 219.

Wu, J.C.S., and P.K.T. Liu, 1992, Ind. Eng. Chem. Res. **31**, 322.

Wu, J.C.S., H. Sabol, G.W. Smith, D.L. Flowers and P.K.T. Liu, 1994, J. Membr. Sci. **96**, 275.

CHAPTER 11

INORGANIC MEMBRANE REACTORS -- ENGINEERING CONSIDERATIONS

As in the case of any other types of reactors, the performance of a membrane reactor depends on the kinetics of the reaction involved and the transport of fluids (and particulates in some cases), heat and species. Thus, material, design and operational parameters related to the membrane and catalyst(s) and their relative placements all can have significant impacts on the reaction conversion, yield and selectivity. Given an accurate kinetics expression, a good approximation to the reactor performance can be obtained by applying sound knowledge of the underlying transport phenomena. Therefore, systematic process understanding, optimal design or scale-up of a membrane reactor can be facilitated by some validated mathematical models such as those reviewed in Chapter 10.

Furthermore, various membrane reactor parameters and configurations result in different performance levels. All the above factors and other engineering aspects will be reviewed in this chapter with both modeling predictions and experimental data.

11.1 CONSIDERATIONS OF FLUID TRANSPORT

Under an applied driving force (pressure, voltage or concentration difference), the fluid streams in a membrane reactor are split or combined possibly at various locations of the reactor. The hydrodynamics of the fluid streams, its interactions with the process and reactor parameters and the fluid management method largely determine the reactor behavior.

11.1.1 Detailed Hydrodynamics in Membrane Reactor

The permeate is continuously withdrawn through the membrane from the feed stream. The fluid velocity, pressure and species concentrations on both sides of the membrane and permeate flux are made complex by the reaction and the suction of the permeate stream and all of them depend on the position, design configurations and operating conditions in the membrane reactor. In other words, the Navier-Stokes equations, the convective diffusion equations of species and the reaction kinetics equations are coupled. The transport equations are usually coupled through the concentration-dependent membrane flux and species concentration gradients at the membrane wall. As shown in Chapter 10, for all the available membrane reactor models, the hydrodynamics is assumed to follow prescribed velocity and sometimes pressure drop equations. This makes the species transport and kinetics equations decoupled and renders the solution of the resulting membrane reactor models manageable. This decoupling assumption has also been used invariably even for non-reactive membrane separators. In the past, by necessity, the complex hydrodynamics inside a membrane separator has been greatly simplified by making crude assumptions and approximations.

Thus, expectedly no rigorous mathematical models are available that can accurately describe the detailed flow behavior of the fluid streams in a membrane separation process or membrane reactor process. Recent advances in computational fluid dynamics (CFD), however, have made this type of problem amenable to detailed simulation studies which will assist in efficient design of optimal membrane filtration equipment and membrane reactors.

Only a limited number of studies has begun to emerge in this area. The available studies encompass various levels of details. Some use CFD to focus the simulation on the flow of liquid and particles through membrane pores while others model fluid flow through a single membrane tube. CFD simulation of fluid flow through a multi-channel monolithic membrane element has also been done.

Some detailed transport behavior of a fluid containing the rejected species through a porous membrane has been simulated by considering multiple moving particles in a moving fluid inside a capillary pore [Bowen and Sharif, 1994]. This is particularly relevant for microfiltration. The finite element method is used to provide space resolution required to solve the equations of momentum and mass conservation inside the pore. Specifically creeping flow of an incompressible Newtonian fluid through a cylinder containing a system of axisymmetric particles is numerically simulated. Direct calculations have been made of flow fields, drag correction factors and pressure drops. The maximum flux achievable in the absence of membrane fouling has been calculated and found to be substantially less than the pure solvent flux through the same membrane under otherwise identical conditions. This rejects the usual assumption that the flux decline of a particle-containing fluid through a porous membrane is largely or entirely due to specific interactions between the particulate species and the membrane (e.g., adsorption and shear-induced deposition inside a pore). It is the additional pressure drop due to the presence of the particles in the pore that causes a lower permeate flux than the pure solvent flux. This pure hydrodynamic effect on the permeate flux, shown in Figure 11.1, is an important factor to consider when designing a membrane or membrane reactor system. In Figure 11.1, the symbols are the results of numerical simulations and the solid line is derived from analytical equations.

Ilias and Govind [1993] also used the CFD approach to solve coupled transport equations of momentum and species describing the dynamics of a tubular ultrafiltration or reverse osmosis unit. An implicit finite-difference method was used as the solution scheme. Local variations of solute concentration, transmembrane flux and axial pressure drop can be obtained from the simulation which, when compared to published experimental data, shows that the common practice of using a constant membrane permeability (usually obtained from the data of pure water flux) can grossly overestimate the membrane flux. If, however, gel polarization is considered, the simulation matches the experiment closely. A similar CFD-type approach has been taken to study the effects of concentration-dependent solute viscosity on the transmembrane flux and concentration polarization of a thin-channel ultrafiltration unit [Ilias et al., 1995]. It is found that a constant-viscosity assumption may grossly overestimate the membrane permeation velocity and underestimate the solute concentration at the membrane wall.

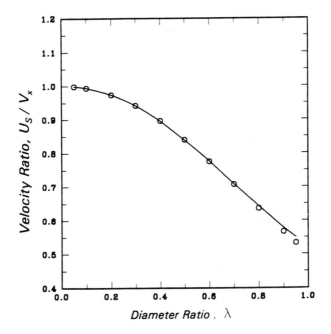

Figure 11.1 Ratio of particle velocity (U_s) to maximum flow velocity (V_x) as a function of the ratio of the particle to pore diameter for the zero-drag case [Bowen and Sharif, 1994]

Wai and Furneaux [1990] applied CFD to crossflow membrane filtration to provide an array of data such as local pressures and fluid velocities on both sides of a membrane, shear stresses on the membrane surface and local concentrations of retentate species. This type of information is useful for designing the membrane unit as a separator or a reactor. With a commercial CFD code, the authors simulated the fluid flow, on both the feed and permeate sides, along the membrane channel and through the membrane. Frictional effects between the fluid and membrane surfaces depend on the nature of the fluid flow. For flow parallel to the essentially flat membrane surface, standard wall friction expressions based on logarithmic velocity profiles adjacent to the wall are used.

For fluid flow in a porous medium such as a porous membrane, the frictional loss of laminar flow in a uniform channel is used.

In addition to the Navier-Stokes equations, the convective diffusion or mass balance equations need to be considered. Filtration is included in the simulation by preventing convection or diffusion of the retained species. The porosity of the membrane is assumed to decrease exponentially with time as a result of fouling. Wai and Furneaux [1990] modeled the filtration of a 0.2 μm membrane with a central transverse filtrate outlet across the membrane support. They performed transient calculations to predict the flux reduction as a function of time due to fouling. Different membrane or membrane reactor designs can be evaluated by CFD with an ever decreasing amount of computational time.

Finally, CFD simulations have been performed on a more practical membrane separation system, a multi-channel ceramic membrane element described in details in Chapter 5. Dolecek [1995] studied permeate flow through the porous support structure of a rectangular honeycomb ceramic membrane element (Figure 11.2) by solving a Darcy's law-based mathematical model using a finite-element method. The porous structure of the membrane element body is considered an isotropic homogeneous porous medium. The permeate conduits are plugged at the feed end. Most of the permeate flows in permeate conduits towards slots where it exits the element. The slots are plugged at the retentate end. By assuming that the pressure drop in the flow channels is small compared to the average pressure difference between the flow channels and the permeate conduits, the problem reduces to a two-dimensional one. When the ratio of the membrane layer permeability to the specific permeability of the porous body is small, the permeate flux per unit volume of the element is proportional to the membrane packing density and the membrane element configuration is not important here. However, when the ratio increases, the membrane element configuration becomes increasingly influential over the permeate flux. An example of such a geometric effect is given in Figure 11.3 which shows isobars and streamlines of permeate flowing inside the porous body of a multi-channel membrane element for three different values of the number of flow channels between two adjacent rows of permeate conduits (N). As N increases, the overall permeate flux decreases due to the growing "dead zone" with flat pressure profile around the inner channels.

None of the above studies, however, deals with the detailed hydrodynamics in a membrane reactor. It can be appreciated that detailed information on the hydrodynamics in a membrane enhances the understanding and prediction of the separation as well as reaction performances in a membrane reactor. All the reactor models presented in Chapter 10 assume very simple flow patterns in both the tube and annular regions. In almost all cases either plug flow or perfect mixing is used to represent the hydrodynamics in each reactor zone. No studies have yet been published linking detailed hydrodynamics inside a membrane reactor to reactor models. With the advent of CFD, this more complete rigorous description of a membrane reactor should become feasible in the near future.

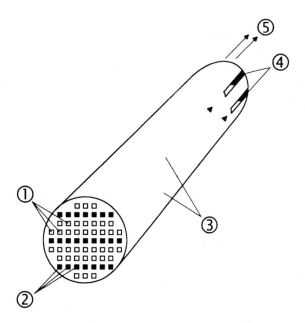

Figure 11.2 Schematic diagram of a honeycomb ceramic membrane element with rectangular channels [Dolecek, 1995]

11.1.2 Flow Configurations of Feed, Permeate and Retentate Streams

The flow patterns of the feed, permeate and retentate streams can greatly influence the membrane reactor performance. First of all, the crossflow configuration distinctly differs from the flow-through membrane reactor. In addition, among the commonly employed crossflow arrangements, the relative flow direction and mixing technique of the feed and the permeate have significant impacts on the reactor behavior as well. Some of these effects are the results of the contact time of the reactant(s) or product(s) with the membrane pore surface.

As in the case of gas separation discussed in Chapter 7, which reaction component(s) in a membrane reactor permeates through the membrane determines if any gas recompression is required. If the permeate(s) is one of the desirable products and needs to be further processed downstream at a pressure comparable to that before the membrane separation, recompression of the permeate will be required. On the other hand, if the retentate(s) continues to be processed, essentially no recompression will be necessary. Recompressing a gas can be rather expensive and its associated costs can be pivotal in deciding whether a process is economical.

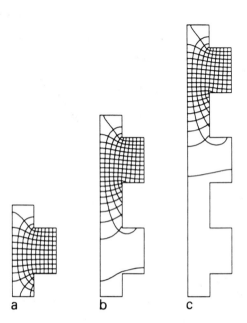

a   b   c

Figure 11.3 Isobars and streamlines of permeate flowing inside porous body of multi-channel membrane element for three values of the number of flow channels between two adjacent rows of permeate conduits: (a) $N=1$; (b) $N=3$; (c) $N=5$ [Dolecek, 1995]

Sweep gas for permeate stream. An inert carrier or sweep gas is often used in gas/vapor phase processes. As applied to membrane reactors, the use of an inert gas may serve multiple purposes. On the reaction side, the inert may be added to the reactant(s) stream to decrease the partial pressure of the product(s) in the case of a reaction involving a volume increase. On the permeate side, employing an inert sweep gas has been a proven method of increasing the permeate production rate by decreasing the partial pressure of the permeate (thus increasing the driving force for membrane permeation). Often the higher the sweep rate of the carrier gas, the greater the permeation rate. This effect is shown in Figure 11.4 for the decomposition of methanol in a gamma-alumina membrane reactor using helium as the carrier gas [Zaspalis et al., 1991a]. It can be seen that at 500°C the conversion of methanol exceeds 70% with a purge rate of helium almost double the feed rate of methanol in comparison to a conversion of about 15% without the carrier gas. Collins et al. [1993] modeled the decomposition of ammonia in a co-current or counter-current catalytic membrane reactor and analyzed the effects of the sweep gas rate relative to the reactant feed rate. The co-current or counter-current configuration refers to the direction of the feed stream relative to the permeate stream. Similarly it is found that as the ratio of the sweep rate to the feed rate increases, the reactor conversion monotonically increases. At 600°C, the conversion jumps from essentially 0 to greater than 45% when the rate ratio is raised from 0 to 2.

GA 35297.4

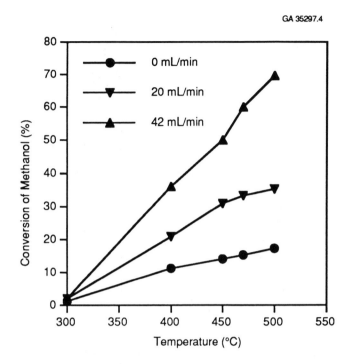

Figure 11.4 Effect of sweeping rate of carrier gas on conversion of methanol decomposition [Zaspalis et al., 1991a]

Apparently when the permeated products are removed by the sweep gas, their build-up and back-diffusion into the membrane is reduced leaving more chance for the reactant(s) to undergo the intended reaction. Purging by the carrier gas leads to an effective concentration (and thus partial pressure) gradient through the membrane and consequently a greater driving force for permeation. Rapid sweeping of the permeate minimizes build-up of the reaction products in the membrane and promotes more reaction. This effect has been observed in many reactions occurring in inorganic membrane reactors [e.g., Kameyama et al., 1981; Itoh, 1987].

Introducing a significant quantity of the carrier (or sweep) gas to the permeate side of the membrane, however, has two major implications. One is that it may necessitate the need to separate the permeate from the carrier gas downstream of the membrane reactor operation if the permeate is a valuable product. The other is the cost associated with the sweep gas. If chemically compatible, air can be used as the least expensive sweep gas available.

Flow directions on feed and permeate sides. There are numerous possible combinations of the flow directions of the feed and permeate streams in relation to the membrane (and the support layer in the case of a composite membrane).

An interesting comparison of the effects of three feed/permeate (purge) flow directions (see Figure 11.5) on reaction conversion and selectivity can be seen through the example of the decomposition reaction of methanol to form formaldehyde at 450°C. This is shown in Table 11.1 [Zaspalis et al., 1991a]. The reactor consists of an inherently catalytic alumina membrane on a support layer. In flow patterns A and C in Figure 11.5, all the feed and permeate/purge streams are in the crossflow mode. The flow pattern A involves the reactant methanol flowing on the side of the selective membrane layer while the carrier gas, helium, and the permeate on the side of the porous support layer. The arrangement for flow pattern C is just the opposite. In flow pattern B, the methanol enters the membrane side in a flow-through mode while the purge/permeate stream in a crossflow mode on the side of the support. In all cases, a slightly higher pressure is maintained on the feed (methanol) side than the permeate side.

Table 11.1 shows that the flow directions of the reactant(s) and the permeate(s)/purge gas strongly influence the conversion and selectivity. The differences have a lot to do with the contact time of the reactant with the catalytic zone in the membrane. In flow pattern B, all the methanol is forced to contact with the catalytic membrane, thus resulting in the highest conversion among the three configurations. The accompanying selectivity to formaldehyde, however, is the lowest due to the greatest chance of consecutive reactions leading to products other than formaldehyde. In flow pattern C, both the conversion and the selectivity are less than those of flow pattern A. The support layer in flow pattern C poses as additional mass transfer barrier which lowers the concentration profile of methanol in the membrane and thus the effectiveness factor of the catalyst (membrane). The effect is a lower conversion than that of flow pattern A.

A special case arises when the "skin" (membrane) layer of a normal composite membrane element is immobilized with a catalyst and not intended for separating reaction species. Consider the example of an enzyme, invertase, for the reaction of sucrose inversion. Enzyme is immobilized within a two-layer alumina membrane element by filtering an invertase solution from the porous support side. After enzyme immobilization, the sucrose solution is pumped to the skin or the support side of the membrane element in a crossflow fashion. By the action of an applied pressure difference across the element, the sucrose solution is forced to flow through the composite porous structure. Nakajima et al. [1988] found that the permeate direction of the sucrose solution has pronounced effects on the reaction rate and the degree of conversion. Higher reaction rates and conversions occur when the sucrose solution is supplied from the skin side. The effect on the reaction rate is consistently shown in Figure 11.6 for two different membrane elements: membrane A is immobilized by filtering the enzyme solution from the support layer side while membrane B from the skin layer side.

The two most common flow directions of the permeate/sweep stream relative to the feed stream are co-current and counter-current. As expected, these two flow configurations produce different reactor performances. They will be treated in the next subsection.

GA 34568.5

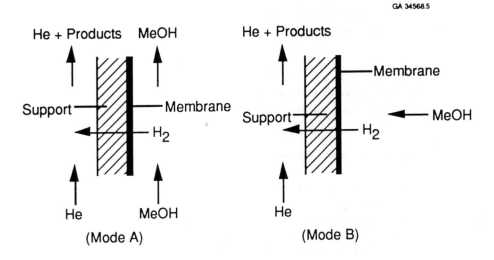

(Mode A)   (Mode B)

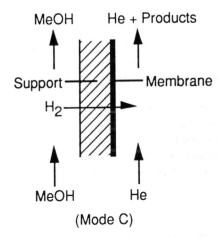

(Mode C)

Figure 11.5 Schematic diagram of three different flow patterns of feed and permeate/purge streams used for comparison in Table 11.1 [Zaspalis et al., 1991a]

TABLE 11.1
Conversion of methanol and selectivity to formaldehyde for three different flow patterns
of feed and permeate streams at 450°C

Flow pattern	Conversion of methanol	Selectivity to formaldehyde
A (Figure 10.5)	70	17
B (Figure 10.5)	90	4
C (Figure 10.5)	40	12-14

[Zaspalis et al., 1991a]

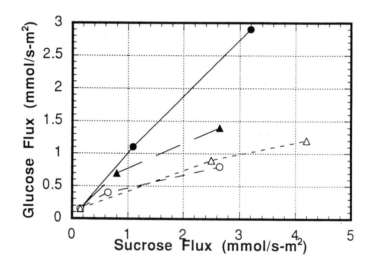

Figure 11.6 Effect of permeate direction on reaction rate of sucrose inversion (circles
refer to membrane A and triangles to membrane B; open symbols are associated with
permeate flow from skin layer side and filled symbols from support layer side)
[Nakajima et al., 1988]

Co-current versus counter-current flows. It is noted that in the operation of a
separation system, a counter-current flow has always given a larger average
concentration gradient than a co-current flow. Thus, it is expected that counter-current
flow configuration is preferred between the two in a membrane unit. In a membrane
reactor, however, an additional factor needs to be considered. To obtain a high
conversion of a reversible reaction, it is necessary to maintain a high forward reaction
rate.

As will become evident later, the counter-current flow configuration appears to provide a clear-cut advantage over the co-current flow configuration with respect to the reaction conversion in a dense membrane reactor.

The choice between the two flow modes for an imperfect membrane with a finite permselectivity (e.g., microporous membrane), however, is not as clear-cut as for the case of an essentially perfect membrane such as a dense membrane which only allows the passage of a product. Here, the membrane is permeable to the various reaction components to varying degrees. Thus an increase in the permeation rate is accompanied by an increase of reactant loss from the reaction side through the membrane. This loss of reactant(s) can negate the conversion enhancement effects of a membrane reactor due to equilibrium displacement and prevent the reaction from reaching complete conversion. The extent of the reactant loss determines the maximum conversions attainable in such reactors. As will become evident later, reactant loss and its associated permeation rate in comparison to the reaction rate play a vital role in reactor performance and explain a lot of seemingly perplexing phenomena pertaining to membrane reactors.

The counter-current flow membrane reactor has a relatively high product concentration on the permeate side towards the feed end; therefore, the driving force for product permeation is relatively low which tends to hinder the conversion enhancement effect of equilibrium displacement. On the other hand, the driving forces for membrane permeation are likely to be high for all reaction species including the reactant(s). This loss of reactant through an imperfect membrane also suffers from the potential of reduced conversion. The following examples illustrate the critical parameters for selecting either of the two configurations.

Mohan and Govind [1988a] modeled co-current and counter-current plug-flow microporous membrane reactors for the following reaction

$$2A \leftrightarrow B + C \tag{11-1}$$

and found that the choice between the two flow configurations depend on a number of factors such as the permeation flux, permselectivity, equilibrium constant, ratio of the reaction to permeation rate and inert gas flow rate relative to reactant flow rate. The conversion comparison is made in Table 11.2 for a dimensionless inert flow rate on the reaction (feed) side of 3.0 (i.e., the molar flow rate of the inert gas equal to three times the molar flow rate of the reactant). The product B is assumed to be the most selective gas and the permselectivities of the other two gas species, namely A and C, are defined as α_A and α_B, respectively. The permselectivity of gas A is $\alpha_A=0.274$ in the figure. For those cases where species C and B have the same permeability which is greater than that of A (that is, $\alpha_C=1.0$), co-current flow is preferred over the counter-current flow because equilibrium displacement is the controlling mechanism. On the other hand, when the permselectivity of C ($\alpha_C=0.1$) is smaller than that of A ($\alpha_A=0.274$) which, in turn, is smaller than that of B, counter-current flow performs better at low equilibrium constants and high permeation to reaction rate ratios due to significant reactant loss.

The trend is quite different when no inert gas is employed on the reaction (feed) side. At α_C=0.1, the reactant loss becomes a problem and the counter-current flow gives higher conversions regardless of the values of the equilibrium constant and rate ratio. At the other extreme of α_C=1.0, one flow configuration performs better than the other depending on the equilibrium constant and rate ratio [Mohan and Govind, 1988a].

TABLE 11.2

Comparison of conversion of co-current and counter-current membrane reactors

Rate Ratio	$\alpha_c = 1.0$			$\alpha_c = 0.1$		
δ	$K_E = 0.06$	$K_E = 0.15$	$K_E = 0.5$	$K_E = 0.06$	$K_E = 0.15$	$K_E = 0.5$
0.0	0.0	0.0	0.0	0.0	0.0	0.0
0.1	0.1	0.21	0.21	0.1	0.09	0.06
0.5	1.2	1.62	1.8	0.3	0.626	0.6
1.0	1.3	2.28	3.1	-0.4	0.505	1.0
1.5	1.2	2.42	3.9	-2.0	-0.16	1.4
2.0	0.8	2.52	4.7	-2.7	-0.6	1.7

Note: 1. Listed are conversion difference (co-current - counter-current), $\Delta X \times 100\%$; 2. $\alpha_A = 0.274$; $2A = B + C$; Da No. = 10.0; $P_r = 0.3$; $F_{inert} = 3.0$; $Q_{inert} = 1.0$ [Mohan and Govind, 1988a]

Another case is related to catalytic ammonia decomposition in a packed-bed shell-and-tube porous membrane reactor. Through modeling, Collins et al. [1993] found that the co-current configuration is generally preferred for membrane reactors with permselectivities governed by Knudsen diffusion (i.e., microporous membranes). See Figure 11.7. This can be explained by a higher initial reaction rate near the reactor entrance and decomposition of back permeated ammonia near the reactor exit. For other membranes with higher hydrogen permselectivities, the choice between the two flow configurations depends on the sweep gas flow rate and operating pressure ratio. The counter-current flow is better when the pressure ratio (ratio of the total pressure on the tube side to that on the shell side) is 1 and a sweep gas is used (Figure 11.8). When no sweep gas is used and the pressure ratio is low, ammonia conversion is essentially the same for both flow configurations.

Idealized membrane reactors. Mohand and Govind [1986] have suggested that, to achieve a maximum conversion due to equilibrium displacement, an idealized simple co-current or counter-current membrane reactor should have the following attributes: infinitely long reactor to consume all reactant(s), essentially zero pressure on the permeate side, an infinitesimally small (but not zero) ratio of permeation to reaction rate to reduce reactant loss. For a membrane reactor with finite values of the reactor length and permeability of the reactant(s), the theoretically attainable maximum conversion will not be as high as the idealized membrane reactor and is dictated by the loss of reactant(s) from the feed (reaction) to the permeate side. Examples of the values for the theoretical maximum conversions for some reactions with given membrane permselectivities are provided in Table 11.3. Increasing reactant loss due to permeation of reactant(s) causes the membrane reactor to deviate further from the above idealized condition. There are some measures to lower the reactant loss from the membrane reactor system. The use of recycle streams and feeding reactant(s) at intermediate locations, which will be discussed later, can enhance the conversions in simple membrane reactors.

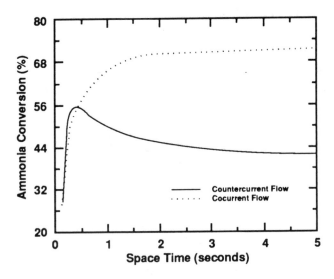

Figure 11.7 Comparison of co-current and counter-current porous membrane reactors for decomposition of ammonia with a hydrogen permselectivity determined by Knudsen diffusion at 627°C [Collins et al., 1993]

Flow patterns on feed and permeate sides. Not only the directions of the process streams carrying the feed and the permeate affect the membrane reactor performance as demonstrated above, the flow (mixing) patterns on both sides of the membrane also can have substantial effects on the reactor behavior. The well known mixing patterns vary from plug flow, perfect mixing (continuous stirred tank reactor), packed bed to fluidized bed. Many real systems lie between the plug flow and the perfect mixing modes. In a packed-bed, plug flow is often a good approximation. If the reaction involved is very fast and no catalyst is needed, it is essential to have a uniform temperature and rapidly supply the necessary heat for the reaction. In this case, a fluidized bed for the reaction side should be considered [Itoh et al., 1990].

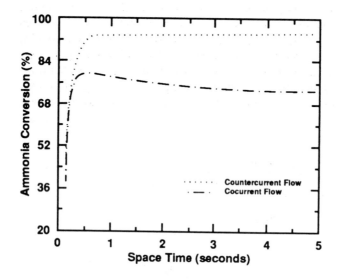

Figure 11.8 Comparison of co-current and counter-current porous membrane reactors for decomposition of ammonia with a high hydrogen/nitrogen separation factor (= 50) at 627°C [Collins et al., 1993]

TABLE 11.3
Maximum conversion in membrane reactors

| | Conversion | | |
Reaction	X_E	X_{max}	X_{Rmax}
$A = B$	0.33	0.70	2.1
($\alpha_A = 0.25$)			
$A = B$	0.33	0.50	1.5
($\alpha_A = 0.5$)			
$A = B$	0.33	0.40	1.2
($\alpha_A = 0.75$)			
Cyclohexane	0.25	0.60	2.4
Hydrogen iodide	0.21	0.45	2.2
Propylene (Vycor glass)	0.34	0.42	1.2
Propylene (hollow fiber)	0.34	0.40	1.1

[Mohan and Govind, 1988a]

All the possible combinations of two ideal flow (mixing) patterns, plug flow and perfect mixing, for packed beds on both sides of the membrane in a membrane reactor have been analyzed and compared by Itoh et al. [1990]. Illustrated in Figure 11.9 are five possible flow or mixing patterns. It is noted that two variations of the plug flow-plug flow combination are co-current and counter-current flows. Other combinations are plug flow-perfect mixing, perfect mixing-plug flow and perfect mixing-perfect mixing for the feed and permeate side, respectively. By assuming isothermal and steady-state conditions and

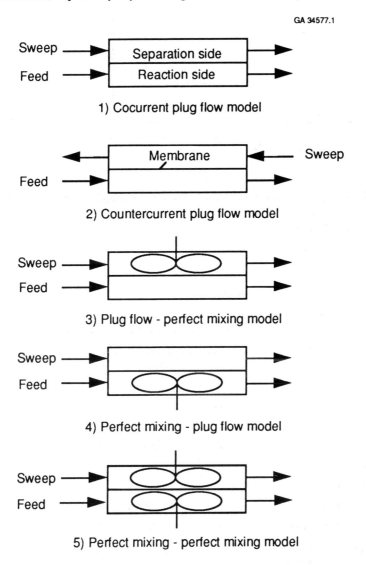

Figure 11.9. Five different flow (or mixing) patterns for two sides of a dense palladium membrane [Itoh et al., 1990]

no pressure drop along the length of a dense Pd membrane reactor, Itoh et al. [1990] developed mathematical models similar to those presented in Chapter 10 for the five flow patterns. Representative of the comparison among the above five flow patterns is Figure 11.10 for a case with a moderate reaction rate (indicated by Da in the figure) and hydrogen permeation rate (determined by Tu in the figure). The dehydrogenation reaction conversion is plotted against the ratio of the flow rate of the sweep stream to that of the feed, V_i^0. The larger the value of the rate ratio is, the more hydrogen permeates. The counter-current plug flow configuration shows a distinctly higher conversion than any other flow patterns except when the rate ratio is below 10 which results in a slightly lower conversion than the co-current plug flow mode. Generally speaking, the conversion performance decreases in the following order: counter-current plug flow, co-current plug flow, plug flow-perfect mixing, perfect mixing-plug flow and perfect mixing-perfect mixing.

For cases where both reaction and hydrogen permeation rates are higher, the same trend prevails with the counter-current flow clearly being the preferred mode. The corresponding differences among the other four patterns are even smaller. On the other hand, when both the reaction and hydrogen permeation rates are small, the counter-current flow is still advantageous over the other flow patterns the differences of which become larger. It is noted that in all cases the conversions exceed the equilibrium value for a given temperature.

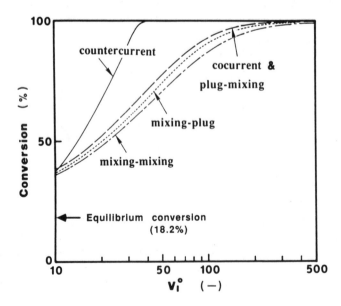

Figure 11.10. Comparison of conversions in a dense Pd membrane reactor with five different ideal flow patterns [Itoh et al., 1990]

The above discussion pertains to dense membrane reactors. For the case of a semipermeable membrane reactor which has finite permselectivities for the various reaction components, isothermal operations favor the plug flow membrane reactor over the perfect mixing membrane reactor for both endothermic and exothermic reactions. In the case of exothermic reactions, the difference between the two flow patterns is rather small for low feed temperatures [Mohan and Govind, 1988b].

A drastically different flow pattern in the reaction zone of a membrane reactor is the fluidized bed configuration which is a strong contrast to the fixed bed configuration discussed above. By combining the advantages of a fluidized bed (such as reduced intraparticle resistance and improved heat transfer) and a permselective Pd membrane (equilibrium displacement and in-situ product separation), the concept of a fluidized bed membrane reactor has been mentioned in Chapter 10. The concept has been evaluated by modeling [Adris et al., 1991; Adris, 1994; Abdalla and Elnashaie, 1995] and experimentally [Adris et al., 1994a; Adris et al., 1994b]. It has been consistently demonstrated that a fluidized bed membrane reactor outperforms a conventional fluidized bed reactor or a conventional perfect mixing reactor for steam reforming of natural gas. An example of comparison is given in Table 11.4. No comparison, however, has been made between a fluidized and a fixed bed configuration under a comparable set of conditions.

TABLE 11.4
Comparison of performance of fluidized bed membrane reactor (FBMR), fluidized bed reactor (FBR) and continuous stirred tank reactor (CSTR)

	FBMR	FBR	CSTR	FBMR	FBR	CSTR
Steam-to-carbon ratio	4.1	4.1	4.1	2.3	2.3	2.3
Methane conversion	0.703	0.690	0.678	0.546	0.539	0.493
% change (wrt CSTR)	+4	+2	0.00	+11	+9	0.00
Steam conversion	0.309	0.293	0.281	0.396	0.370	0.352
% change (wrt CSTR)	+10	+4	0.00	+12	+5	0.00

Note: $T = 925$ K; $P = 0.69$ MPa; methane feed = 74.2 mol/h
(Adapted from Adris et al. [1994a])

Preferred flow patterns in nonisothermal membrane reactors. The discussions so far focus on flow patterns in an isothermal membrane reactor. In many situations, however, the membrane reactor is not operated under a uniform temperature. The choice between a plug flow (PFMR) and a perfect mixing membrane reactor (PMMR) depends on a number of factors. First of all, it depends on whether the reaction is endothermic or exothermic.

In the case of an endothermic reaction, the PFMR always performs better than the PMMR. The operating temperature and reactant concentration in a PMMR requiring supply of heat for the endothermic reaction are usually lower than those of the feed at the reactor inlet. This results in a lower forward reaction rate than the average rate attainable in a PFMR.

Shown in Figure 11.11 is the difference in conversion between a PFMR and PMMR as a function of the permeation to reaction rate ratio for the case of exothermic reactions [Mohan and Govind, 1988b]. Included in the figure are two sensitive parameters: the feed temperature and the heat generation index which is proportional to the heat of reaction and a measure of the maximum temperature drop in the reactor. Under a given set of conditions, the advantage of a PFMR over a PMMR decreases as the feed temperature decreases or as the permeation to reaction rate ratio increases. Beyond a critical rate ratio, a PMMR may actually outperform a PFMR. Both effects are related to severe reactant loss.

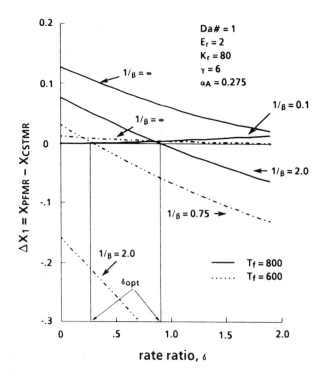

Figure 11.11 Difference in conversion between plug flow membrane reactor and perfect mixing membrane reactor as a function of permeation to reaction rate ratio for an exothermic reaction [Mohan and Govind, 1988b]

The conditions near the feed end are critical in determining the performance of a PFMR. In a PFMR, the reactant concentration is high at the feed end. Therefore, at low feed temperatures, the PFMR is unable to take advantage of the high reactant concentration and the conversion is low. In contrast, the temperature in the PMMR is likely to be higher than the feed temperature for an exothermic reaction. With this favorable temperature effect, the forward reaction rate in the PMMR may exceed the average rate in the PFMR, thus resulting in higher conversions in a PMMR.

As the rate ratio becomes large, the reactant loss due to a high permeation rate is significant. The loss is particularly severe in the case of a PFMR in view of the fact that the reactant concentration on the feed side near the feed end is higher in a PFMR. In a PMMR, the composition in the reactor is uniform and the reactant loss due to membrane permeation is not as notable when compared to a PFMR. Consequently, as the rate ratio increases beyond a critical value for a fixed set of conditions, the PMMR may be preferred.

Recycle streams. In many large-scale chemical processing systems, recycle or tear streams are often used and feed streams introduced at intermediate locations of reactor units are practiced. As mentioned earlier, a membrane reactor with a finite reactor length and reactant permeability offers a reaction conversion lower than that by the idealized membrane reactor due to the loss of reactant through the membrane. Thus, when the reactant permeability is high, some measures to lower the reactant loss from the membrane reactor system are required to maintain a high conversion. Both properly using recycle streams and shifting the feed location toward the middle of the membrane reactor can enhance the conversions in simple membrane reactors.

Two types of useful recycles as given in Figure 11.12 have been analyzed by Mohan and Govind [1988a].

In the bottom product recycle mode, the retentate stream containing unconverted reactant(s) is recycled and this helps drive the equilibrium displacement towards a maximum limit determined by the permeation to reaction rate ratio, permselectivity of the reactant relative to that of the product(s), pressure ratio, among other factors. The maximum conversion corresponds to the limiting case of a total recycle with no bottom or retentate product(s). This maximum limit decreases with increasing reactant loss. This occurs, for example, when the permeation to reaction rate ratio increases, as shown in Table 11.5 [Mohan and Govind, 1988a]. Total recycle is equivalent to having a sufficiently long membrane reactor in which all the reactant(s) on the feed side is consumed. To realize the benefits of this type of recycle, the feed rate needs to be decreased as the recycle rate is increased. Recycling of the retentate stream without decreasing the feed amount does not result in increase in conversion. What happens is that, as the recycle rate increases, the reaction components on the reaction side begins to be well mixed which can lead to decreased net reaction rate.

GA 34577.2

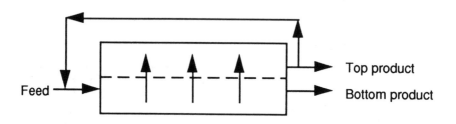

Type 1: Bottom product recycle

Type 2: Top product recycle

Figure 11.12. Two types of useful recycle in a membrane reactor [Mohan and Govind, 1988a]

TABLE 11.5
Maximum conversion ratio$^{(*)}$ of reaction $A \leftrightarrow B$ in a porous membrane reactor at total (bottom product) recycle

Rate ratio$^{(\S)}$	$\alpha_A = 0.274$	$\alpha_A = 0.5$
0.1	1.895	1.462
0.5	1.776	1.330
1.0	1.641	1.200
1.5	1.530	1.100
2.0	1.477	

Note: $^{(*)}$ conversion ratio = reactor conversion/equilibrium conversion; $^{(\S)}$ rate ratio = reactant permeation rate/reaction rate
[Mohan and Govind, 1988a]

In the top product (permeate) recycle, additional energy is required to recompress the recycled portion of the permeate stream. As expected, the ratio of the recycle flow rate to that of the feed rate (called the compressor load) affects the conversion. There is an optimal compressor load (q_{opt}^s) beyond which the conversion begins to suffer. This is due to the negating effect on equilibrium displacement of backreaction resulting from the significant presence of product(s) in the reaction zone. In Table 11.6, the conversion and the optimal compressor load are given for different values of the membrane permselectivity for the following two types of reaction: $aA \leftrightarrow bB$ and $aA \leftrightarrow bB + cC$ with B being the most selective gas species [Mohan and Govind, 1988a]. It is evident from the table that as the product or both products have higher permeabilities than that of the reactant, the conversion increase is limited by the maximum conversion, X_{max}, given by the idealized membrane reactor discussed earlier. On the other hand, if the permeability of the reactant is in between those of the two products, the top product recycle can reach a maximum conversion greater than that achievable with the bottom product recycle. This further enhancement effect can be attributed to the fact that equilibrium displacement also depends on separation of products.

TABLE 11.6
Performance of membrane reactor as a function of top product recycle

Da Number	$A = B$ $\alpha_A = 0.274$ $X_{max} = 0.63$ $\delta = 1.0$		$A = B + C$ $\alpha_A = 0.274$ $\alpha_C = 0.451$ $X_{max} = 0.78$ $\delta = 3.0$		$A = B + C$ $\alpha_A = 0.274$ $\alpha_C = 0.100$ $X_{max} = 0.70$ $\delta = 3.0$	
	X	q_{opt}^s	X	q_{opt}^s	X	q_{opt}^s
0.0	0.0	0.0	0.0	0.0	0.0	0.0
1.0	0.37	0.0	0.46	0.0	0.43	1.0
3.0	0.55	0.45	0.65	2.9	0.62	3.5
6.0	0.59	2.1	0.72	7.6	0.74	8.5
12.0	0.62	5.1	0.75	17.3	0.82	21.0
25.0	0.63	11.8	0.78	38.5	0.89	52.0
∞	0.63	∞	0.78	∞	1.0	∞

[Mohan and Govind, 1988a]

In addition to the co-current flow mode shown in Figure 11.12 are other situations where the sweep/permeate stream flows in the opposite direction to the feed stream (counter-current flow mode). The effect of recycle on the preferred flow mode has been studied for those reactions $aA \leftrightarrow bB + cC$ where the permeability of the reactant lies between those of the two products [Mohan and Govind, 1988a]. It has been found that at low values of the recycle ratio (i.e., the fraction of the product stream recycled), the counter-current flow gives higher conversions than the co-current flow as a result of lower reactant losses and greater separation of the products. As the recycle ratio increases, the difference between the two flow modes diminishes to the extent that both flow modes offer the same conversion at the extreme of a total recycle.

Intermediate feed locations. Another way of improving the conversion of a simple membrane reactor is to introduce the feed stream at one or multiple intermediate locations along the reactor length. In the following type of reactions involving one reactant and two products

$$aA \leftrightarrow bB + cC \tag{11-2}$$

conditions exist that make intermediate feed locations attractive. Shown in Figure 11.13, a membrane reactor operated in this mode resembles a continuous membrane column with an enriching and stripping section [Hwang and Thorman, 1980].

When the permeability of the reactant lies between those of the two products, there is an optimal feed location to the membrane reactor as demonstrated in Table 11.7. This occurs because, beyond a certain size or length of the enriching section (for separation of products and reactants) any gains in reaction rate by separation of products cannot offset the loss of unconverted reactant in the stream leaving the stripping section (for introduction of the feed). The feed location is determined by how much reaction conversion can be gained by separating the two products. As the permeation rate increases relative to the reaction rate, a smaller stripping section is needed to reduce backreaction and, therefore, the feed should be introduced at a location closer toward the reactor exit. As the difference in permeabilities of the two products becomes smaller, the optimal feed location also moves toward the reactor exit because a longer enriching section is required to achieve the increasingly difficult separation.

GA 35297.1

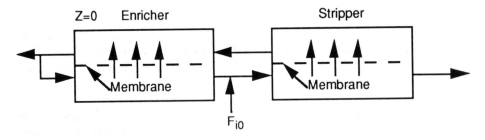

Figure 11.13. Schematic of a membrane reactor with an intermediate feed location and recycle [Mohan and Govind, 1988a]

TABLE 11.7
Effect of feed location on equilibrium displacement for a membrane
reactor when permeability of reactant is in between those of two products

	Conversion			
	$\delta = 1.0$		$\delta = 3.0$	
Da Number	$F_L = 0.0$	$F_L = $ opt	$F_L = 0.0$	$F_L = $ opt
4.0	0.37	0.38	0.46	0.48
10.0	0.46	0.48	0.58	0.60
20.0	0.55	0.57	0.67	0.69
40.0	0.64	0.66	0.75	0.76

Note: $Pr = 0.0$; $A = B + C$; $\alpha_A = 0.4$; $\alpha_C = 0.1$
[Mohan and Govind, 1988a]

In contrast, for those cases where the permeability of the reactant is either smaller or larger than those of both products or product (i.e., $P_A < P_B$ or $P_A < P_B$ plus $P_A < P_C$ being one case and $P_A > P_B$ or $P_A > P_B$ plus $P_A > P_C$ being the other), the optimal feed location is at the reactor entrance ($z=0$).

Mohan and Govind [1988a] have summarized various factors which favor equilibrium displacement and, therefore, optimal reactor conversions. The factors include the relative permeabilities of the reactant and products, separation of products, fraction of recycle, the Damköhler number, dimensionless diffusivity ratio and the ratio of the total pressure on the sweep side to that on the reaction side. This is shown in Figure 11.14 for those reactions of the following type: $A \leftrightarrow B$ or $A \leftrightarrow bB + cC$. The strategy given in the summary may be deployed when designing and operating isothermal packed-bed membrane reactors.

11.1.3 Reactant Composition

The flow rate and the concentration (dilution) of the feed stream entering the membrane reactor both have substantial influence over the reactor performance [Itoh et al., 1993; Collins et al., 1993]. The effect of flow rate of the reactant(s) will be treated later in this chapter.

Whether and how much a component in the entering reactant stream has any effects depend on its role in the reaction. In a study of ammonia decomposition in a counter-current microporous packed-bed membrane reactor, the inlet concentration of hydrogen greatly influences the decomposition rate. As expected from Figure 11.15, ammonia conversion increases as the hydrogen concentration in the feed stream decreases at a given temperature [Collins et al., 1993]. On the contrary, the inlet nitrogen concentration does not appreciably affect ammonia conversion. This can be attributed to the relatively minor role nitrogen plays in the decomposition rate compared to hydrogen.

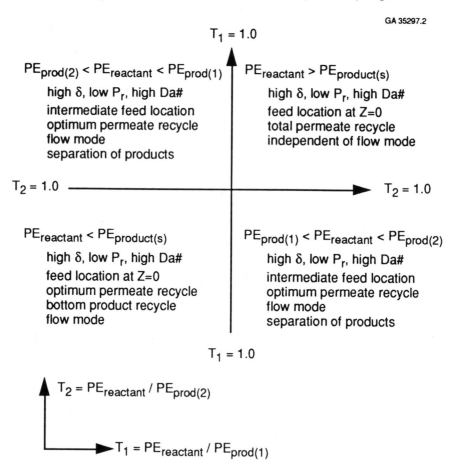

Figure 11.14 Summary of various factors which optimize maximum reactor conversions [Mohan and Govind, 1988a]

In series-parallel reactions such as

$$A + B \rightarrow R + T \tag{11.3}$$

$$R + B \rightarrow S + T \tag{11.4}$$

the feed composition can have significant impacts on the performance of the membrane reactors. For example, the molar feed ratio of the reactant B to that of A, as shown in

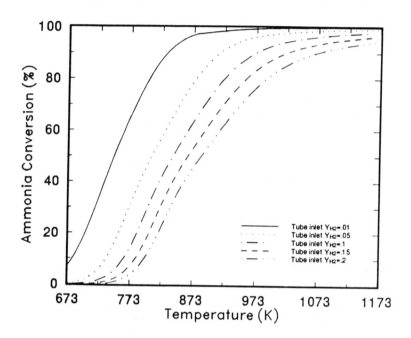

Figure 11.15 Ammonia conversion as a function of hydrogen concentration in the feed [Collins et al., 1993]

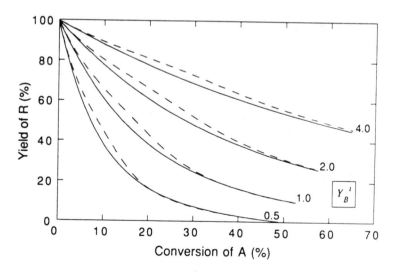

Figure 11.16 Yield of the desired product as a function of reaction conversion with the molar ratio of two rectants as the parameter [Bernstein and Lund, 1993]

Figure 11.16, is key to the selectivity of the desired product R. Here the reaction stoichiometry controls the yield of R. If the reactant A is consumed before B (molar feed ratio of B to A, Y_B^t, greater than 1), the only reaction for the remaining B is the undesirable reaction according to Eq. (11-4) which further converts R. The end result is a reduced yield of R. If, on the other hand, Y_B^t is less than 1, B runs out before A and the conversion of A is limited to low values.

11.1.4 Permeation Rate and Permeate Recovery

The permeation rate and its profile along the membrane reactor length can substantially determine the reactor performance. Moreover, to increase the reaction conversion, a lower permeate concentration on the permeate side is often adopted. This has implications on energy consumption and downstream separation costs. These issues will be addressed here.

Effects of average permeation rate. The effects of average permeation rate of oxygen through a dense oxide membrane on the reactor performance have been investigated for the case of oxidative coupling of methane where air is supplied to one side of the membrane and pure oxygen permeates to the other side where oxygen reacts with methane. This controlled addition of one reactant (oxygen) to the reaction zone helps improve the yield and selectivity of C_2 products [Wang and Lin, 1995]. The oxygen permeation flux can be varied in principle by selecting a proper membrane material or modifying a given membrane material (by doping or forming a composite membrane), changing the membrane thickness, or adjusting the transmembrane pressure difference.

Shown in Figure 11.17 are the responses of the C_2 product yield and selectivity of oxidative coupling of methane at the exit of a 0.5-m long plug-flow membrane reactor to the change in the relative oxygen permeation rate, j_0 (i.e., ratio of oxygen permeation rate to methane flow rate at the reactor entrance). A larger j_0 in Figure 11.17 represents a greater oxygen permeation rate. The selectivity is essentially constant for small and large relative oxygen permeation rates and decreases with increasing j_0 in the 5×10^{-5} and 5×10^{-4} range. At small j_0, the selectivity is very close to 100% even though the conversion is low. The yield first increases, then reaches a maximum before decreasing with increasing j_0 and finally approaches an asymptotic level. At a large j_0, the yield and selectivity are essentially the same and both approach a constant value indicating that methane and oxygen are completely consumed under those conditions. Wang and Lin [1995] have further concluded that a dense oxide membrane with a lower oxygen permeance can provide a higher C_2 selectivity and yield for oxidative coupling of methane in a plug-flow membrane reactor but at the expense of the greater reactor length required.

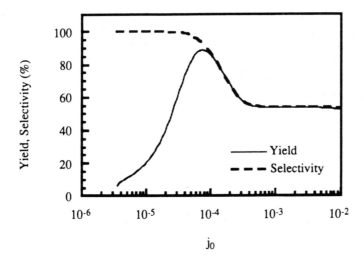

Figure 11.17 C_2 product yield and selectivity of oxidative coupling of methane of a 0.5-m long plug-flow membrane reactor as a function of j_0 (ratio of oxygen permeation rate to methane flowrate at the reactor entrance) [Wang and Lin, 1995]

They have also modeled a perfectly mixed membrane reactor for the same reaction. For this membrane configuration, the C_2 selectivity is essentially 100% for a j_0 value of less than 0.2 and drops off rapidly with increasing j_0.. The yield becomes greater as the relative oxygen permeation rate increases. The yield reaches a maximum and then decreases with the permeation rate. There is only a limited range of j_0 in which the yield can reach beyond 20%.

Effects of variations in local permeability. Pressure drop along the membrane length can be significant and different for the two sides of the membrane. This is particularly true if only the feed or permeate side incurs appreciable longitudinal pressure drop. In many cases where a membrane is utilized to selectively remove one or more products to effect conversion increase, the longitudinal pressure drop due to the packed bed of catalysts on the feed (reaction) side is substantially more than the corresponding pressure drop on the permeate side where no packed bed is involved. Furthermore, in some other cases where a membrane is used to distribute one of the key reactants to the reaction zone to improve the yield and selectivity of a desired intermediate product, the longitudinal pressure drop on the high pressure (supply) side is small compared to that on the low pressure (reaction) side which is packed with a bed of catalyst particles. In both situations above, the transmembrane pressure difference varies along the reactor length. Therefore, given a membrane with a uniform permeability lengthwise, the resulting local permeation rate of the permeating product or reactant varies from the reactor entrance to exit.

This longitudinal variation of the permeation rate has some undesirable implications. Take the case of a controlled addition of oxygen into a catalyst-packed reaction zone on the opposite side of an oxygen selective membrane. Using a membrane with a uniform porosity and permeability gives rise to a higher oxygen permeation rate near the outlet of the packed-bed membrane reactor because of a higher transmembrane pressure difference in that zone. This can cause two undesirable effects. First, the oxygen having permeated near the reactor exit does not have enough contact time to react, thus resulting in a low oxygen conversion. Second, the higher oxygen permeation rate near the reactor exit increases the local oxygen concentration. It so happens that the hydrocarbon concentrations in this region are also very high, potentially leading to a reduced selectivity.

Thus, the profile of the permeability along the membrane reactor length should be and can be tailored to provide the maximum yield. This concept has been demonstrated experimentally by Coronas et al. [1994] using porous ceramic membranes for oxidative coupling of methane. Starting with a porous alumina membrane, they immersed the membrane in a silica sol followed by drying and calcination. Then 75% of the treated membrane is again subject to the immersion-drying-calcination cycle. Similarly 50 and 25% of the membrane is sequentially deposited, dried and calcined. The end result is that the four quarters of the membrane tube have received different degrees of thermochemical treatment and have correspondingly different permeabilities. Two different membranes are compared in Figures 11.18 and 11.19 in terms of their reactor performance at a nominal temperature of 750°C. The membrane with a graded reduction in the permeability outperforms that with a uniform permeability profile. The membrane with a decreasing permeability consistently gives higher methane conversions for a given oxygen concentration in the feed (Figure 11.18). Although not shown, the same trend is true for oxygen conversion. These improvements are the results of correcting the preferential permeation of oxygen near the end of the membrane when a uniform-permeability membrane is employed. In Figure 11.19, a decreasing permeability along the membrane length also provides higher selectivities of hydrocarbons containing C_{2+} for a given methane conversion because excessive oxygen concentration near the reactor exit is avoided. This effect becomes more pronounced at a higher methane conversion.

Permeate recovery. To achieve high conversions, it is often desirable to maintain a very low permeate partial pressure which leads to an increase in the permeation rate. Vacuum or a sweep gas is usually employed to attain a low permeate pressure. Vacuum adds some energy cost while the use of a sweep gas may require further downstream processing for the recovery of the permeate (if it contains the desired species) or the separation of the permeate from the sweep gas. The use of a condensable gas or vapor as the sweep gas will facilitate the recovery of the permeate. For example, steam can be condensed at a relatively lower temperature and easily separated from many other gases. However, the issue of hydrothermal stability of the membrane, discussed in Chapter 9, can be critical. Air, on the other hand, is a convenient sweep or carrier gas to use because of its availability. In contrast, the use of vacuum eliminates the need for any further separation of the sweep gas and the permeate.

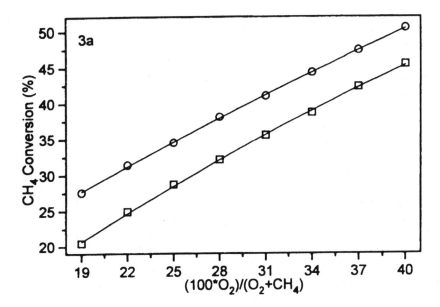

Figure 11.18 Comparison of methane conversion as a function of the molar fraction of oxygen in the total feed in two reactors using porous membranes with a uniform permeability and a decreasing permeability profile at 750°C and a total flowrate of 172 standard cm^3/min (open circles represent the membrane with a decreasing permeability along its length and open squares represent the membrane with a uniform permeability) [Coronas et al., 1994]

In cases where the primary purpose of the membrane reactor system is not to produce and separate or purify the permeate, say, hydrogen, the heating value of hydrogen can be recovered or utilized by burning it with air as the sweep gas to provide the required heat of reaction if the reaction involved is an endothermic one. Thermal decomposition of hydrogen sulfide and dehydrogenation of ethylbenzene are just two examples. Consumption of the permeate in these cases can significantly lower the permeate partial pressure. For example, Edlund and Pledger [1993] have estimated that a permeate-side hydrogen partial pressure as low as 10^{-9} bar can ultimately be achieved by catalytically burning the permeate hydrogen with air as a sweep gas.

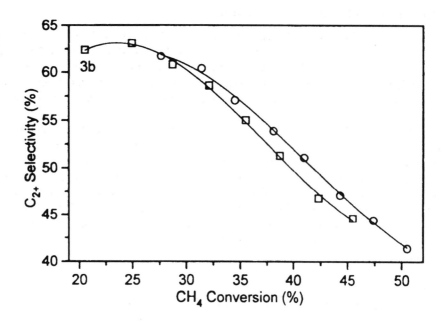

Figure 11.19 Comparison of hydrocarbon selectivity as a function of the methane conversion in two reactors using porous membranes with a uniform permeability and a decreasing permeability profile at 750°C and a total flowrate of 172 standard cm^3/min (open circles represent the membrane with a decreasing permeability along its length and open squares represent the membrane with a uniform permeability) [Coronas et al., 1994]

11.2 CONSIDERATIONS OF ENERGY TRANSFER

Energy management of a conventional reactor is at times quite a challenge and often constitutes a major operating cost component. It can become an even greater issue for membrane reactors as other process streams such as sweep gases and/or diluents are potentially involved. For a plant containing a membrane reactor system to become energy efficient, various streams (feed, permeate and retentate streams) must exchange heat in an overall optimal manner. Many hydrogenation and dehydrogenation reactions are prime candidates to benefit from the use of membrane reactors for carrying out the required reactions with higher conversions and reduced or eliminated downstream separation or purification requirements. As a general rule, dehydrogenation reactions are endothermic while hydrogenation reactions, once heated up to the required temperatures, are exothermic.

11.2.1 Supply and Removal of Heat of Reaction

Given in Table 11.8 are the heats of reactions for selected hydrogenation and dehydrogenation reactions. As can be seen, many hydrogenation reactions are strongly exothermic. To maintain the reaction temperature in the membrane reactor at an acceptable level, heat evolved would need to be removed at an adequate rate. On the contrary, heat needs to be supplied to a dehydrogenation reaction to sustain the reaction.

It is often desirable to operate the reactor and the catalyst under isothermal conditions to achieve high reactor performance. Heat requirement of an endothermic reaction in a membrane reactor to maintain an isothermal condition can be challenging as in most of the dehydrogenation reactions such as conversions of ethylbenzene to styrene and propane to propylene. Maintaining an isothermal condition implies that some means must be provided to make the adequate heat input (e.g., from a burner) that is longitudinally dependent. It is not trivial to make the temperature independent of the longitudinal position because the permeate flow also varies with the location in the axial direction.

TABLE 11.8
Heats of selected hydrogenation and dehydrogenation reactions

Reaction	Energy (kJ/mol)
Nitro group to amine	-550
C≡C to C-C	-300
Aromatic ring saturation	-210
C=C to C-C	-125
Carbonyl group to alcohol	-65
Dehalogenation	-65
Hydrogenation of toluene	-50
Hydrogenation of methylnaphthalene	-50
Dehydrogenation of ethylbenzene	121
Dehydrogenation of hexane	130
Dehydrogenation of butane	134
Dehydrogenation of butene	134
Dehydrogenation of naphthenes to aromatics	221

This is particularly true for a ceramic membrane reactor system where arrays of tubular or multichannel membranes are used. The underlying issue is how to provide uniform and adequate heating to such a class of materials and configurations. Ceramic membranes can be heated by applying external fired heaters or condensing steam. A possible arrangement for heating a membrane reactor is given in Figure 11.20 which displays a membrane reactor heated by flame generated by the combustion of a fuel gas [Minet and Tsotsis, 1991]. The membrane is enclosed in a metallic housing which is directly exposed to the flame. On a production scale, banks of membrane reactor

housings can be lined up and heated by multiple flames inside a furnace equipped with
refractory brick walls.

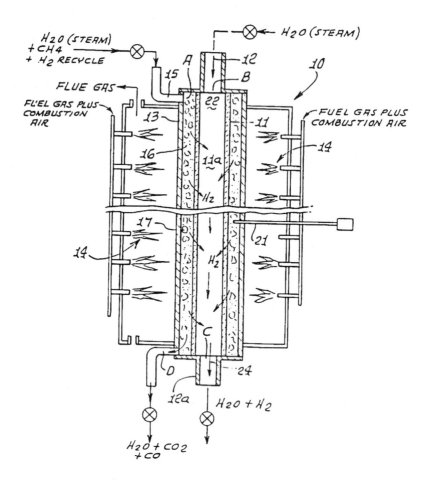

Figure 11.20 A membrane reactor module heated by a burner [Minet and Tsotsis, 1991]

It has been suggested that the design and operational practices for cracking furnace
reactors can be put to use for heat supply to inorganic membrane reactors [Tsotsis et al.,
1993]. Having sufficient electrical conductivity at typical reaction temperatures for
dehydrogenation reactions, ceramic membranes allow the use of electric current through

the permeable wall of the membrane to supply the heat required. Cylindrical band heaters can be employed.

Alternatively steam can be introduced to the feed stream to supply the heat of reaction required. For example, for dehydrogenation of organic compounds, a steam to organic compound molar ratio of 5 to 13 at a temperature 600 to 635°C has been suggested [Bitter, 1988]. The direct addition of steam has the side benefit of depressing potential formation of carbonaceous deposits for those reactions which have the tendency to generate cokes.

Another possible layout is to place the membrane reactor systems (e.g., membrane elements) in a furnace. This can be relatively easy to install and operate in a laboratory setting. To scale up the underlying process to a production level, however, large furnaces will be required. Multi-zone furnaces are available which are capable of controlling the reactor temperature profile within a few degrees. Heat can be continuously supplied to an array of membrane reactor elements in series or in parallel inside a furnace. As in many other chemical processes, the heat demand should be balanced and integrated with the rest of the plant to optimize energy utilization and improve the economical viability of the membrane reactor operation. It is noted that a higher reaction conversion can reduce not only the reactor size (and therefore the size of the membrane system) but also the size of the heat load and any other associated downstream equipment required. For a given conversion, a membrane reactor possibly can be operated at a lower temperature, thus reducing the heat input required.

In addition to the above heating methods, there is another way of securing heat sources for an endothermic reaction such as a dehydrogenation reaction. This involves carrying out two reactions: one endothermic reaction on one side of the membrane and an exothermic reaction on the other side of the membrane. It is called reaction coupling as mentioned in Chapter 8. An example of two coupled reactions partitioned by a membrane is to conduct the endothermic thermal decomposition of hydrogen sulfide on the feed side of a metal composite membrane while catalytically burning the permeate hydrogen using an air sweep stream [Edlund and Pledger, 1993]. This arrangement can provide all or most of the energy required to drive the thermolysis of hydrogen sulfide.

Another possible method of providing or removing heat of reaction is through the use of an inert gas stream on either or both sides of the membrane. This helps maintain the desirable temperature level for the entire reactor.

The above discussion primarily applies to membrane reactors where heat needs to be supplied to or removed from the catalytic zone which consists of a packed bed of catalysts in the tube core or a catalytic membrane itself. Transfer of the heat of reactions can be easier when the catalytic zone lies outside the membrane tubes. This is the case when, for example, a packed bed of catalysts is placed on the shell side of a shell-and-tube type membrane reactor. In a fluidized-bed membrane reactor [Adris et al., 1991; Aldris et al., 1994], the catalytic zone is in the emulsion phase (solids-containing) of the fluidized bed inside a reactor vessel. This also makes it easier to supply or remove heat from the reaction zone. Heat pipes can be used in these cases, especially for those highly endothermic or highly exothermic reactions.

Heat pipes are known for their usage as heat flux transformers. Their rapid heat transfer rate makes them attractive for situations where a small heat transfer area is desired because of limited space or for a compact reactor design. Refer to Figures 10.14a and 10.14b for an example where a fluidized-bed membrane reactor can utilize sodium heat pipes. Heat pipes inserted into a conventional fluidized-bed reactor [McCalliste, 1984] or fixed-bed reactor [Richardson et al., 1988] without the use of membranes have been previously proposed for supplying heat to hydrocarbon steam reforming reactions. The heat pipe will form at least one loop for circulating a thermal or heating fluid.

Heat can be supplied into the reactor bed to maintain the reaction with a very small temperature drop by using the heat pipes. Each heat pipe has its evaporator section connected to the heat source and its condenser section connected to the reactor bed. The heat source may comprise a furnace containing a burner. To further improve heat transfer, the heat pipes can have heat transfer fins on their outer surfaces within the reactor.

Other means of supplying heat to a fluidized bed are available. An example is to withdraw particles from the bed, heating them and then returning them to the bed [Guerrieri, 1970].

11.2.2 Thermal Management of Nonisothermal Membrane Reactors

While isothermal condition is preferred to prevent undesirable reactions and their products from occurring and makes process control of the membrane reactors easier, it is sometimes difficult, if not impossible, to control the temperature of the reactor containing the reaction components uniformly across the reactor at a fixed level. Moreover, as will be pointed out later, certain nonisothermal conditions actually give better reactor performance than isothermal conditions.

Plug flow membrane reactor. For this type of reactor, the preferred strategy of thermal management very much depends on whether the reaction involved is endothermic or exothermic.

Modeling the performance of an adiabatic plug-flow membrane reactor (PFMR) for a generic reversible reaction of the form $A \leftrightarrow B$, Mohan and Govind [1988b] presented the results as shown in Figure 11.21. For endothermic reactions, the conversion increases monotonically with increasing feed temperature. Therefore, in this case, the feed temperature should be as high as economically and technically feasible since both reaction rate and thermodynamic equilibrium are favored at high temperatures. For a given feed temperature, the conversion is maximum at an optimal ratio of the reactant permeation rate to the reaction rate. This, as discussed previously, can be explained by the reactant loss by permeation through the membrane.

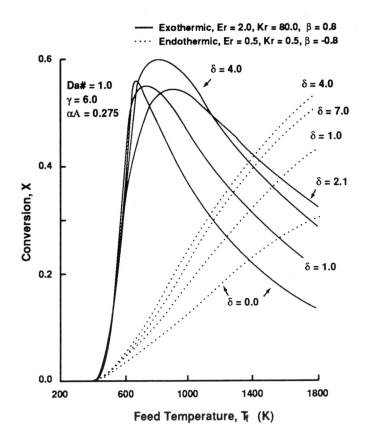

Figure 11.21. Conversion of endothermic and exothermic reaction $A = B$ in a plug flow membrane reactor with permeation to reaction rate ratio as parameter [Mohan and Govind, 1988b]

Itoh and Govind [1989] have compared plug-flow isothermal and adiabatic Pd membrane reactors with catalysts packed in the tube core for dehydrogenation of 1-butene as shown in Figure 11.22. Hydrogen permeating through the membrane reacts with oxygen on the permeate side. The parameter γ in the figure represents dimensionless heat transfer coefficient. As the coefficient increases, the tube and shell side temperatures become closer. When compared with an isothermal condition, the adiabatic condition drives the conversion to completion in a very short reactor length.

Over a wide range of conditions, especially those at moderate to high γ, the adiabatic membrane reactor performs better than the isothermal membrane reactor.

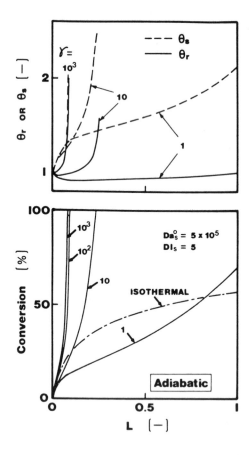

Figure 11.22 Comparison of plug-flow isothermal and adiabatic palladium membrane reactors for dehydrogenation of 1-butene [Itoh and Govind, 1989]

Unlike endothermic reactions, exothermic reactions under a fixed set of conditions in a membrane reactor generate a conversion curve which peaks at an optimal feed temperature. The conversion decrease at high feed temperatures is the result of backreaction favored at high temperatures. With heat release from the reaction, a trade-off between a high conversion and a high reaction rate (for a small reactor volume) has to be balanced. For an exothermic reaction in a PFMR, an optimal temperature profile along the reactor length exists that maximizes the conversion. Such a profile and its corresponding maximum conversion are given in Figure 11.23. As the ratio of the reactant permeation rate to reaction rate increases, the optimal temperature profile as a whole increases. So does the maximum conversion profile.

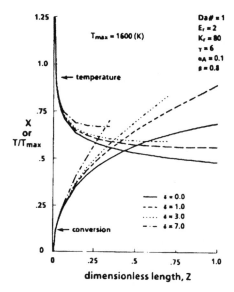

Figure 11.23. Optimal temperature and corresponding maximum conversion profiles in a plug-flow membrane reactor for an exothermic reaction [Mohan and Govind, 1988b]

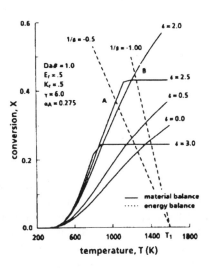

Figure 11.24. Conversion of a reversible reaction $A \leftrightarrow B$ in a perfect-mixing membrane reactor for an endothermic reaction [Mohan and Govind, 1988b]

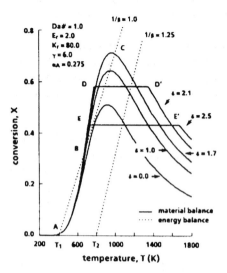

Figure 11.25. Conversion of a reversible reaction $A \leftrightarrow B$ in a perfect mixing membrane reactor for an exothermic reaction [Mohan and Govind, 1988b]

Perfect-mixing membrane reactor. Mohan and Govind [1988b] have also analyzed the behavior of perfect-mixing membrane reactors (PMMR). Similar to the case of plug-flow membrane reactors, the reaction conversion in a PMMR is determined by the nature of the heat of reaction.

The conversions in a PMMR for an endothermic and an exothermic reaction as a function of the reactor temperature are given in Figure 11.24 and Figure 11.25, respectively. The intersection of a material balance curve with a given ratio of the permeation to reaction rate and an energy balance line with a known heat generation index provides the conversion under the operating conditions.

For an endothermic reaction, as displayed in Figure 11.24, the conversion increases continuously with the operating temperature except at high permeation rates the conversion stays constant beyond a certain temperature and the composition in the reactor is constant. The level-off represents those conditions in which there is no retentate stream. That is, the permeation rate is so high that all the feed contents go to the permeate. In this case, the maximum conversion extends over a range of operating temperatures. At a fixed temperature, the conversion first increases and then decreases with increasing rate ratio. Thus an optimal rate ratio which maximizes the conversion exists due to the significant reactant loss through the membrane when the reactant permeation rate is high. If the membrane does not allow the passage of the reactant, there is no optimal permeation rate and the conversion can be driven to completion.

In contrast to the endothermic case, an exothermic reaction in a PMMR leads to a maximum conversion at an optimal temperature or over a temperature range for a given set of conditions (Figure 11.25). The rising portion of a given curve is a result of equilibrium displacement and the declining portion is due to the back reaction which is favored at high temperatures. For a given temperature, there is an optimal rate ratio above which the conversion decreases instead of increasing because of reactant loss through permeation. The plateaus correspond to the conditions in which the permeate rate equals feed rate and the reactor content has a constant composition over a range of temperatures. From the standpoint of achieving a maximum conversion, it is desirable to operate the energy balance line such that it intersects with the material balance curve at or near the maximum conversion point. To that end, external heat removal may be necessary.

Membrane reactor stability. Multiple steady states have been found in continuous stirred tank reactors (perfect-mixing reactors) or other reactors where mixing of process streams take place. This phenomenon is also evident in membrane reactors. The thermal management of a membrane reactor should be such that the reactor temperatures provide a stable range of operation.

As shown in Figure 11.25 for an exothermic reaction, the energy balance line T_1C crosses the material balance curve with a rate ratio (reactant permeation to reaction rate) of 1.7 three times, generating three steady states represented by three points: A, B and C. The middle state B is unstable as any small perturbation in temperature will move the reactor condition to either of the two stable steady states. Under certain nonadiabatic conditions, the reactor may show oscillatory instability at an intersection such as A and C. Mohan and Govind [1988b] have developed two necessary and sufficient conditions for stability in a perfect-mixing membrane reactor. The procedure involves rewriting the steady state material and energy balance equations in their transient forms. This is followed by linearizing the resulting equations about their respective steady state positions by applying first order Taylor series. The membrane reactor may be expected to be stable when the characteristic equation associated with the linearized equations has roots whose real parts are negative.

Plug-flow membrane reactors are not faced with potentially unstable states since no back mixing of mass or heat is involved. However, when the product is recycled, heat is exchanged between the product and the feed streams or dispersive backmixing exists, multiple steady states can occur and membrane reactor stability needs to be considered.

Heat recovery. Heat contained in the reaction gas streams can be recovered to help supply the compression or recompression energy required within the membrane reactor system. For example, to perform a dehydrogenation reaction, a membrane reactor module is heated to supply the heat of reaction. The hydrogen-containing gas stream from the membrane reactor system can be directed to a heat exchanger where the waste heat is used to heat water to produce high-pressure steam. The steam can be introduced to the feed or permeate stream or to a steam turbine which drives a generator for a compressor. The compressor provides the high pressure required of the feed stream to the reactor.

The heat recovered from the exiting process streams can also be directed to a steam generator for general plant usage or a process gas preheater. Heat can also be exchanged with other sections of the entire plant.

11.2.3 Heat Transfer between Coupled Membrane Reactors

When two reactions on the two opposite sides of a membrane are coupled or conjugated through the membrane, not only species balances including reaction and permeation need to be maintained but also heat balance needs to be established. From an engineering standpoint, it is desirable to have one of the coupled reactions endothermic while the other exothermic and the heat transfers through the shared membrane. If the supply and demand of heat is not properly balanced as it often occurs, net heat input or removal from the reactor will be required. This poses a technical challenge.

An interesting effect of heat transfer coefficient on the conversion of a coupled membrane reactor has been presented by Itoh [1990]. Shown in Figure 11.26 are the temperature profiles on the reaction and permeate sides and the conversion of the endothermic cyclohexane dehydrogenation in a packed-bed dense Pd membrane reactor. The exothermic reaction, oxidation of hydrogen, takes place on the permeate side. For a small dimensionless heat transfer coefficient (γ in the figure), both feed and permeate side temperatures are higher than those corresponding to a large heat transfer coefficient. Similarly, the reactor length required to achieve 100% conversion decreases as the dimensionless heat transfer coefficient decreases. The above observations apply as long as the dimensionless heat transfer coefficient is not smaller than a threshold value which depends on the operating conditions. In other words, the dehydrogenation reaction proceeds faster at a smaller heat transfer coefficient.

These results first appear to be unusual but can be explained as follows. As the heat transfer coefficient decreases, it becomes more difficult for heat to transfer from the permeate side (where oxidation of hydrogen takes place) to the feed side (where cyclohexane is dehydrogenated). The resulting excess heat goes to increasing the temperature of the permeate side which makes the oxidation reaction faster. This, in turn, leads to higher temperatures on both the feed and permeate sides. Therefore, at a given

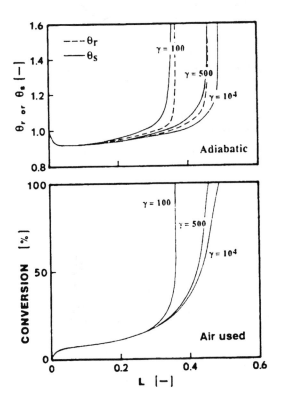

Figure 11.26 Temperature profiles on the reaction and permeate sides and conversion of the endothermic cyclohexane dehydrogenation in a packed-bed dense Pd membrane reactor [Itoh, 1990]

longitudinal position, the dehydrogenation conversion is higher for a lower heat transfer coefficient. The above trend continues with decreasing dimensionless heat transfer coefficient until it reaches a threshold value. Below the threshold heat transfer coefficient, heat generated by the oxidation reaction on the permeate side is no longer sufficient to maintain the endothermic dehydrogenation reaction at a fast pace on the feed side. Thus a longer membrane reactor is required for dimensionless heat transfer coefficients smaller than the threshold values as demonstrated in Figure 11.27 which displays a minimum of the reactor length for complete dehydrogenation at a given Damköhler number (Da_s^0) of the permeate side. That minimum is equivalent to the optimal operating condition of the membrane reactor. It is also observed that as the

Damköhler number increases at a fixed heat transfer coefficient the complete conversion reactor length decreases due to the increased rate of heat evolution from the oxidation reaction.

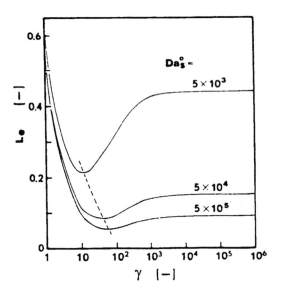

Figure 11.27 Length of packed-bed dense Pd membrane reactor to attain 100% conversion [Itoh, 1990]

11.2.4 Pressure Recompression of Recycled Streams

The top product recycle mode in Figure 11.12 brings part of the permeate stream at a lower pressure to join the feed stream at a higher pressure. Thus, additional energy external to the membrane reactor will be required to recompress the recycled permeate. On the contrary, in the bottom product recycle, also shown in Figure 11.12, only the transmembrane pressure difference and the longitudinal pressure drop need to be overcome between the recycled portion of the bottom product (or retentate) and the feed. Therefore, the required pressure recompression is expected to be small compared to the top product recycle mode.

As indicated in Chapter 7, compression and recompression costs can be very significant in determining if the underlying membrane process, including membrane reactors, is economically competitive. The compression cost can be considered to be proportional to the compression ratio which is the ratio of the outlet to inlet pressures.

11.2.5 Thermal Treatments and Disturbances

As will be pointed out later, catalysts and membranes may become gradually deactivated and, when this occurs, they need to be regenerated often by some combustion process using dilute oxygen. In cases where coke deposit becomes significant on the catalyst or the membrane and needs to be removed by combustion, the membrane reactors are potentially subject to thermochemical cycling as part of the normal operation.

Thermal treatment with an inert gas can be an effective means of improving the catalytic activity of inherently catalytic membranes. Palladium and palladium alloy membranes benefit from some special heat treatment. In this treatment, the membranes are heated to 700°C for 20 minutes or so in an inert atmosphere containing nitrogen, argon, helium or hydrogen and then cooled down rapidly to the reaction temperature (about 400°C) in 1 to 5 minutes. After the above treatment, a Pd-Ni (5.5% Ni) membrane increases the yield of benzene by hydrodealkylation of toluene to 33% from 7% prior to treatment [Gryaznov et al., 1973b]. The same treatment, when applied to Pd-Rh (10% Rh) membranes, doubles the yields of benzene and toluene in the dehydrocyclization of n-heptane [Gryaznov, 1986]. Slow cooling does not impart the improvement in catalytic activity. The difference was attributed to the possibility that some mobile surface Pd atoms at the high temperature may remain isolated on the crystal surface after quenching and serve as catalytically active sites [Gryaznov et al., 1973b]. This type of procedure may be implemented during normal operation.

Other potential temperature variations or cycling may also occur as a result of feed or control disturbances. Thus it is important to ensure that the temperature of the system is in control so that no damage is done to the membrane, catalyst or other system components. It is well known, for example, that palladium can become embrittled under hydrogen-rich conditions particularly at a temperature near 300°C or so. Some remedy in the case of palladium is to introduce certain metal such as silver as the alloying element.

11.3 CONSIDERATIONS OF SPECIES TRANSPORT

The primary function of a membrane is to selectively allow the transport of certain species while rejecting the flow of other species. Except a perfect membrane (with a dense membrane approaching to being a perfect membrane), the majority of membranes permit the passage of all species to varying degrees. This imperfection of the membrane selectivity has significant implications which will be discussed here.

11.3.1 Transport of Species through Imperfect Membranes

Most of the commercial inorganic membranes are not perfect membranes and have finite values of the permselectivity. Therefore, most of the reaction components permeate through the membranes to various degrees. When a high fraction of the less selective reaction component permeates through a membrane, the reaction conversion is likely to deteriorate for an equilibrium-displaceable reaction.

It is then expected that the permselectivity of the membrane in a membrane reactor has a significant impact on the reactor performance. An example is shown in Figure 11.28 for the decomposition of ammonia using a porous membrane with a hydrogen/nitrogen separation factor of 50 [Collins et al., 1993]. It is obvious that the conversions using a membrane reactor are all higher than the equilibrium values. As the hydrogen selectivity improves (or as the membrane becomes more permselective), the conversion consistently increases. The rate of conversion improvement with the hydrogen selectivity is particularly more pronounced at a lower selectivity. Figures 11.28 represents the results for a counter-current flow permeate stream. Similar results have been obtained for a co-current flow configuration although the conversion is lower than that for the counter-current flow at a given temperature. This is due to the opposing flows of the streams on the tube and shell sides which allows more hydrogen to permeate and a lower hydrogen concentration in the tube (reaction zone). A lower hydrogen concentration will enhance equilibrium displacement and improve the conversion.

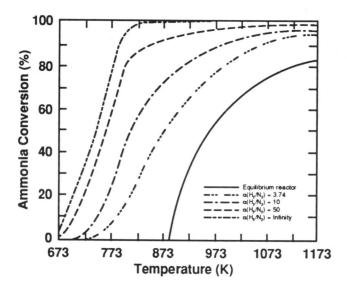

Figure 11.28 Effect of hydrogen selectivity on conversion of ammonia decomposition in a counter-current porous membrane reactor [Collins et al., 1993]

11.3.2 Separation of Reaction Components in Membrane Reactors

Besides improving conversion or enhancing selectivity of a reaction, a membrane reactor is often involved in separating components of a reaction mixture. In a simple but quite generic reaction system containing a reactant A and two products, B and C, an important goal of membrane separation could be separation of the reactant from either of the two products (i.e., A from B or A from C) or separation of the two products (B from C).

To measure the separation efficiency of a membrane reactor involving multiple reaction components, the extent of separation, briefly introduced in Chapter 7, was used to replace the more commonly used separation factor by Mohan and Govind [1988a]. This alternative index of separation performance is based on the flow quantities of the process streams involved while the separation factor is calculated from the compositions instead. The goals of a high conversion and a high separation sometimes contradict each other. The choice or, more often than not, compromise of the two goals depends, on one hand, on the downstream separation costs and, on the other, on the process parameters such as the ratio of the reactant permeation to reaction rate and the relative permeabilities of the reaction components.

For those cases where the permeability of reactant A is in between those of the two products, B and C, both the conversion and extent of separation increase with increasing permeation rate or permeation to reaction rate ratio (Table 11.9). The corresponding optimal compressor load (recycle flow rate to feed flow rate) also increases with the rate ratio. The top (permeate) stream is enriched with the most permeable product (i.e., B) while the bottom (retentate) stream is enriched with the least permeable product (i.e., C). It is noted from Table 11.9 that the optimal compressor loads for achieving the highest conversion and extents of separation can be quite different and a decision needs to be made for the overall objective.

TABLE 11.9
Separation efficiency of membrane reactor when permeability of reactant
is in between those of two products

δ	X / q_{opt}^{s}	ξ_{BC} / q_{opt}^{s}	ξ_{AC} / q_{opt}^{s}	ξ_{AB} / q_{opt}^{s}
$\alpha_A = 0.5, \alpha_C = 0.1$				
0.5	0.35/1.2	0.68/0.9	0.40/0.8	0.35/1.3
1.0	0.40/2.4	0.81/2.3	0.64/1.7	0.42/3.2
2.0	0.46/5.5	0.88/5.7	0.75/4.5	0.50/7.5
4.0	0.53/12.8	0.92/13.5	0.83/11.0	0.55/16.6
$\alpha_A = 0.7, \alpha_C = 0.5$				
0.5	0.31/1.6	0.33/1.5	0.19/1.4	0.16/1.7
1.0	0.32/3.5	0.40/3.8	0.24/3.5	0.20/3.8
2.0	0.33/7.7	0.47/8.2	0.28/7.7	0.24/8.4
4.0	0.34/16.7	0.54/17.2	0.32/16.5	0.29/17.8

Note: $Pr = 0.0$; $A = B + C$; Da No. $= -6.0$; $K_e = 0.1$
[Mohan and Govind, 1988a]

In the case where the permeabilities of both products are greater than that of the reactant, the situation is more complicated. As shown in Figure 11.29, while the conversion increases with increasing rate ratio, the extent of separation for each pair of reaction components exhibits a maximum at a different rate ratio. Rational compromises need to be made for achieving high conversion and high separation, not necessarily optimal for both at the same time. For example, a logical decision may be one that designs and operates the membrane reactor at the maximum separation of *A* and *B* with the corresponding conversion.

11.4 REACTION ENGINEERING CONSIDERATIONS

The kinetic and catalytic parameters of the reaction involved as well as the design and operating variables of the membrane reactor all play important roles in the reactor performance. They will be treated in some detail in this chapter.

11.4.1 Reactions Amenable to Inorganic Membrane Reactors

Much of the research and development activity on inorganic membrane reactors in recent years has been directed toward hydrogenation or dehydrogenation of hydrocarbons, especially the latter, and other reactions involving production or depletion of hydrogen. This can be attributed to a number of factors.

First of all, hydrocarbons constitute a major force in the very important petroleum and petrochemical industry. Generation or consumption of hydrogen, a very valuable commodity chemical, is one of the key steps in many chemical processes. Even a modest success in the improvement of reaction conversion, yield or selectivity by the use of a membrane reactor can represent a substantial economic benefit due to the volume of streams involved.

Secondly, many if not most of these high temperature reactions are equilibrium limited. Therefore, the removal of one of the reaction products (often hydrogen) by the use of an inorganic membrane displaces the reaction equilibrium toward more product formation, thus increasing the reaction conversion as has been demonstrated extensively. This favorable effect of equilibrium displacement or shift is particularly pronounced for those reactions where the stoichiometric coefficient of the product to be removed is high.

Finally, current technology commercializes porous inorganic membranes that offer high permeabilities which are directly related to the throughput and are operated predominantly based on Knudsen diffusion. In the Knudsen diffusion regime, as indicated previously, acceptable separation factors or performances can be attained relatively easily when hydrogen is separated from hydrocarbons or other compounds that have relatively high molecular weights.

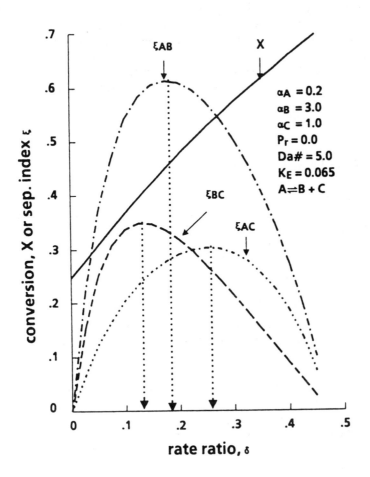

Figure 11.29 Conversion and separation index of a membrane reactor as a function of permeation to reaction rate ratio when reactant permeability is smaller than product permeabilities [Mohand and Govind, 1988a]

The organic polymer membranes have an upper operating limit of approximately 200°C due to their thermal stability. Below this temperature, there exist organic polymer membranes suitable for homogeneous catalytic reactions in some compatible solvents. This temperature also represents the lower limit for most heterogeneously catalyzed reactions [Armor, 1992].

As mentioned in Chapter 8, the understanding of how the material composition of a dense membrane interplays with the operating conditions and the reaction components is far from being sufficient for reliable design of a dense membrane reactor, particularly with hydrogenation reactions. It is expected that the exploitable opportunities for dense metal membranes lie in the gas/vapor phase dehydrogenation reactions which will be developed at a faster pace to the implementation stage than hydrogenations.

Based on the above considerations, the types of reactions that are amenable to inorganic membrane reactors in the first wave of industrial implementation will probably be as follows: (1) The reactions are heterogeneous catalytic reactions, particularly dehydrogenation processes; (2) The reaction temperature exceeds approximately 200°C; (3) When the reactions call for high-purity reactant(s) or product(s) and the volume demand is relatively small, dense membrane reactors (e.g., Pd-based) can be used. On the other hand, if high productivity is critical for the process involved, porous membrane reactors are necessary to make the process economically viable.

As the inorganic membrane technology evolves, membranes with higher selectivities than what current generation of products can offer will become practical. The types of the reactions amenable to membrane reactors will correspondingly broaden.

11.4.2 Space Time and Velocity

The extent of contact between the reactant(s) and the catalyst is often believed to be very important in the reactor performance but is difficult to determine. This problem is complicated by several factors. In gas/vapor reactions, the volumetric flow rate may change as the reaction proceeds. Furthermore, the true void volume of the catalyst bed or membrane pores (and therefore, the void fraction of the reactor) is not easy to measure.

Space time and space velocity. In practice, a superficial contact time called space time is generally used. Its reciprocal is the space velocity.

The space time can be defined in two ways. It is often calculated as

$$\tau = V / F_f \qquad\qquad (11\text{-}5)$$

where V and F_f are the (unpacked) reactor volume and the volumetric flow rate of the feed at a specified condition (often at the entry condition or some standard condition), respectively. Alternatively, the space time is defined as

$$\tau = W / F_f \qquad\qquad (11\text{-}6)$$

where W is the catalyst weight in the reactor. In essence, space time or space velocity is a convenient capacity measure for the chemical reactor and is frequently used to relate to the reactor performance such as conversion, yield or selectivity.

Effects of space time under isothermal conditions. Some indication of the effects of space time on the performance of a co-current membrane reactor compared to a conventional reactor can be seen in Figures 11.30 and 11.31 where experimental data are given for dehydrogenation of isobutane to produce isobutene at 500°C [Ioannides and Gavalas, 1993]. Plotted in Figure 11.30 is the isobutene yield against the space time which in this case is calculated according to Eq. (11-6). At a low space time, the difference in the yield between the two reactor configurations is small because the rate of the reverse reaction for the main equilibrium reaction:

$$i\text{-}C_4H_{10} \leftrightarrow i\text{-}C_4H_8 + H_2 \qquad\qquad (11\text{-}7)$$

is low and the continuous removal of hydrogen from the reaction zone by the membrane does not affect the reaction conversion (and therefore the yield) to any significant degree. As the space time increases, the yield increases for both reactor configurations. At a higher space time, however, the yield for the conventional reactor approaches the equilibrium value before beginning to decline while the removal of hydrogen continues to improve the yield which becomes increasingly greater than that for the conventional reactor. The slight decline of the yield at very high space times is very likely due to the occurrence of side reactions. In addition to the main reaction described by Eq. (11-7), there are other side reactions associated with the dehydrogenation of isobutane. The major side reaction is stoichiometrically described by

$$i\text{-}C_4H_{10} \leftrightarrow CH_4 + C_3H_6 \qquad\qquad (11\text{-}8)$$

The side reactions continue even as the main reaction slows down as the equilibrium is approached.

The conversion of an equilibrium-displaceable reaction does not always increase with increasing space time. The trend depends on the flow configuration of the permeate stream, the permselectivity of the membrane and whether the reaction proceeds isothermally, among other factors. In Figure 11.7 presented earlier, the calculated conversion data of ammonia decomposition in a porous membrane reactor at 627°C are plotted against space time for two sweep stream configurations: co-current vs. counter-current. Knudsen diffusion is assumed to dominate the separation mechanism with a hydrogen/nitrogen separation factor of only 3.7. With a co-current permeate flow configuration, the ammonia conversion increases with increasing space time consistently until the conversion curve approaches a plateau representing the equilibrium dictated by the permeation rate of ammonia. Contrary to the case of a co-current permeate flow, the counter-current curve reaches a peak in the ammonia conversion and then slowly declines with further increase of space time which has been similarly observed during dehydrogenation of isobutane. This maximum conversion is significantly lower than the limiting conversion for the co-current case.

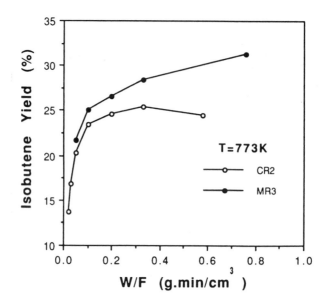

Figure 11.30 Effects of space time on the yield of isobutane from its dehydrogenation reaction in a membrane reactor versus a conventional reactor [Ioannides and Gavalas, 1993]

The higher conversion associated with the co-current flow configuration at high space times is attributable to back permeation of ammonia from the sweep side to the reaction side. In a co-current configuration, the driving force for ammonia permeation to the sweep side is greatest near the reactor entrance. Therefore, a significant amount of ammonia is lost to the sweep side before it decomposes. The situation is just the opposite downstream close to the reactor exit. Sufficient ammonia on the tube side has decomposed to the extent that the ammonia partial pressure on the tube side is lower than that on the sweep side. This creates back permeation of unconverted ammonia into the tube side where it is finally converted and contributes to the higher total conversion of ammonia.

As mentioned briefly earlier, the conversion-space time behavior can be quite different for a membrane with a higher hydrogen permselectivity (e.g., Figure 11.8 where the hydrogen/nitrogen separation factor is 50 and the total pressures on both sides of the membrane are equal). In this case, the conversion for a counter-current flow configuration increases rapidly with the space time and approaches an equilibrium limit determined by the ammonia permeation rate through the membrane. In contrast, the co-current flow configuration results in a lower conversion than the counter-current flow case and exhibits a maximum conversion followed by a slowly declining tail end. This maximum can be explained by a number of factors [Collins et al., 1993].

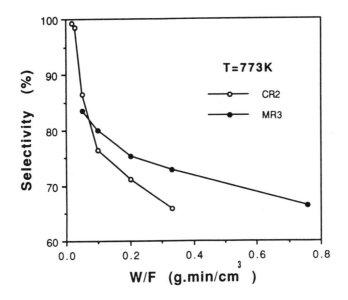

Figure 11.31 Isobutene selectivity versus space time at 773°K for a conventional and a membrane reactor [Ioannides and Gavalas, 1993]

Another important reactor performance is the selectivity, Figure 11.31 [Ioannides and Gavalas, 1993]. Figure 11.31 shows the relative extent of the side reactions and the main reaction at 3 hours time on stream. The selectivity declines sharply with the space time. This is especially true for the conventional reactor because the side reactions in the conventional reactor continue despite the slow-down of the main reaction by the equilibrium. In contrast, both the main and the side reactions continue in the membrane reactor. Consequently, the decrease of selectivity with space time is slower in the membrane reactor than a conventional reactor.

Effects of space time under nonisothermal conditions. The above discussions around the effects of space time on a membrane reactor performance are limited to isothermal conditions. The behavior of the reaction conversion in response to space time can be further complicated under nonisothermal conditions.

First of all, the space time defined in Eq. (11-5) or (11-6) depends on the volume of the reactor and the total volumetric feed rate. Thus, for a given reactor volume, space time is inversely proportional to the total feed rate. Itoh et al. [1993] studied the use of a dense yttria-stabilized zirconia membrane reactor for thermal decomposition of carbon dioxide. The reactor temperature was not kept constant everywhere in the reactor but varying with the reactor length instead. The resulting temperature profile is parabolic with the maximum temperature at the midpoint of the reactor length. This nonisothermal characteristic of the reactor introduces an unusual effect on the reaction conversion as a function of space time. Figure 11.32 presents their conversion data as a function of the total feed rate with the concentration of Ar (as a carrier gas) in the feed stream as the parameter. The dilution ratio α is defined as the ratio of the carrier gas (Ar) flow rate in the reactant feed stream to the carbon dioxide feed rate. Since the reactor volume is fixed, the total flow rate of the feed stream is inversely proportional to space time. That is, a high feed rate is equivalent to a low space time and vice versa. The conversion curves with the dilution ratio smaller than 14 all show a local minimum and maximum. The authors attributed the behavior between the minimum and the maximum points to the parabolic temperature profile effect just mentioned. Had this parabolic temperature effect not been present, the conversion would have been a continuously decreasing function of the feed rate or increasing function of the space time, as discussed for the isothermal case.

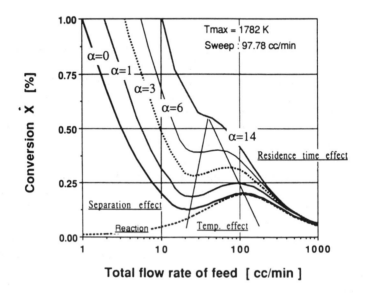

Figure 11.32 Nonisothermal decomposition of carbon dioxide in a dense zirconia membrane with the dilution ratio of carbon dioxide as a parameter [Itoh et al., 1993]

11.4.3 Contact Time of Reactant(s) through Catalytic Membrane

In addition to space time which is indicative of the residence time of the feed (reactant) stream through the reactor length, the contact time of the reactant(s) through the membrane pores can be very important. This is particularly true with inherently catalytic or catalyst-impregnated membranes. Here the residence time of the reactant(s) inside the membrane zone is expected to exert influence on the reaction involved. One obvious way of controlling the residence time, for a given pore size and tortuosity of the membrane, is to vary the membrane thickness or the applied pressure gradient (except in the Knudsen diffusion regime). In those cases where Knudsen diffusion predominates, a higher pressure gradient does not result in a shorter residence time.

Contrary to a crossflow configuration, a flow-through catalytic membrane generally has better control of the residence time of the permeating species. This is particularly suitable for fast reactions. A porous flow-through catalytic combustor using a ceramic foam as the membrane has been applied to the combustion of natural gas containing low NO_x [Philipse et al., 1989]. The homogeneous distributions of pores and residence time reduce the number of hot spots which produce NO_x. Similarly, Itaya et al. [1992] found that a flow-through porous methane burner for boiler applications could produce high specific thermal energy.

11.4.4 Reaction Rate versus Reactant Permeation Rate

In many situations the conversion of a membrane reactor increases as the total permeate rate increases. This is to be expected if the membrane has a perfect or very high permselectivity. In many commercially available porous inorganic membranes, however, the permselectivity is moderate and some reactants as well as products other than the most selective species "leak" through the membrane. This leakage rate often increases with the total permeate rate, for example, as the feed side pressure increases. This has an important consequence on the reactor performance.

In Figure 11.33, the reaction conversion is plotted as a function of the ratio of the total permeation rate to reaction rate for a packed-bed microporous membrane reactor with the permeate side under vacuum. Three generic types of reactions are considered, each assuming a constant permselectivity. Thus the "leakage" rate of the reactant is proportional to the permeation rate of the most selective gas component B. The rate ratio shown in the figure is then directly related to the ratio of the leakage rate to the reaction rate of the reactant. The conversion in all three cases exhibit a maximum at a certain optimal rate ratio. The initial conversion increase with the rate ratio is attributable to the selective permeation of the product(s) which leads to equilibrium displacement and conversion enhancement. As the rate ratio increases to a point where the amount of unconverted reactant permeating to the permeate side is significant, the reaction conversion starts to decrease with further increase of the rate ratio.

The maximum conversion as a function of the ratio of the permeation rate to reaction rate does not always occur. Shown in Figure 11.34 is the reaction conversion of cyclohexane dehydrogenation, again, as a function of the permeation to reaction rate ratio for two general cases: one with the permeate side under vacuum ($P_r=0$) and the other with the total permeate side pressure equal to the total pressure on the feed side ($P_r=1$). Under the latter situations, the conversion may, instead of reaching a maximum, arrive at a plateau particularly at a high dimensionless reactor length (Z in the figure).

This is due to backpermeation of the reactant and product from the permeate to the feed side, causing the reaction to remain proceeding in the forward direction slowly.

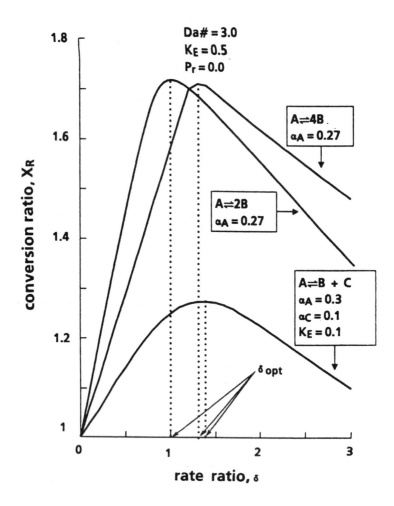

Figure 11.33, Reaction conversion as a function of the ratio of total permeation rate to reaction rate for a packed-bed microporous membrane reactor with the permeate side under vacuum [Mohan and Govind, 1988a]

11.4.5 Coupling of Reactions

A proven approach to enhancing chemical conversion in the reaction zone is to carry out another reaction in the permeate zone to consume the permeate, thus creating a greater driving force for the permeate through the membrane. Examples have been given in

Chapter 8 to demonstrate the benefits of coupling a second reaction on the permeate side to the first on the retentate side.

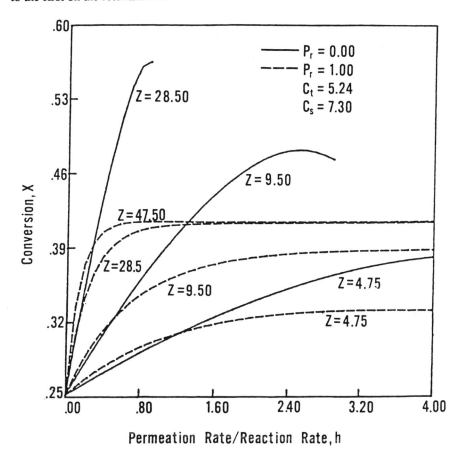

Figure 11.34 Effect of permeation to reaction rate ratio on membrane reactor conversion [Mohan and Govind, 1986]

Conjugation of two reactions in a single reactor compartmentalized by a membrane reduces the number of degrees of freedom compared to having two independent reactors. For example, the reaction stoichiometry may require the addition or removal of the permeating species which potentially can lower the maximum total conversion. Likewise, heat may need to be added to or taken away from one of the reaction zones.

In dehydrogenation reactions, hydrogen is generated and transported through a hydrogen-selective membrane. If the permeating hydrogen reacts with a gaseous mixture containing oxygen (e.g., air, mixtures of air or oxygen with an inert diluent such as nitrogen) in the permeate zone, the resulting conversion on the dehydrogenation reaction side is higher than that obtained by direct catalytic dehydrogenation where the hydrogen is swept away by an inert carrier gas. Therefore, oxidative dehydrogenation by the use of a permselective membrane is a case of coupled reactions.

Oxidation of hydrogen using oxygen is known to be a fast reaction releasing a large amount of heat. Thus, the heat generated can be used to support the dehydrogenation reaction. The oxygen concentration should be adjusted to effect the combustion of hydrogen at a rate which will provide the desirable reaction temperature and avoid overheating of the reactor. The oxygen-hydrogen reaction can be effected at atmospheric pressure or as high as 14 bar [Pfefferle, 1966]. Palladium is a well known catalyst for this reaction.

Itoh [1990] simulated a Pd membrane reactor coupling the cyclohexane dehydrogenation reaction on the feed side with oxidation of hydrogen on the permeate side. Given in Figure 11.35 is the predicted conversion of the dehydrogenation reaction as a function of the total flow rate of the sweep gas with the Damköhler number for the permeate side as a parameter:

$$Da_s{}^\circ \equiv k_s A_m p_s{}^3 / u_c{}^\circ \qquad\qquad (11\text{-}9)$$

where k_s is the rate constant for the oxygen-hydrogen reaction, A_m the membrane separation area, p_s the pressure on the permeate side and $u_c{}^\circ$ the cyclohexane feed rate at the reactor inlet. Thus the Damköhler number $Da_s{}^\circ$ is a measure of the maximum forward reaction rate (in this case for the oxygen-hydrogen reaction) that can be achieved in a membrane reactor. As shown in Figure 11.35, even in the case of no coupling reaction on the permeate side the dehydrogenation conversion increases beyond the equilibrium value as the sweep gas flow rate becomes higher as one would expect. As the reaction rate of oxygen-hydrogen accelerates (a higher $Da_s{}^\circ$), the dehydrogenation conversion further increases to reach a complete conversion at a lower sweep gas flow rate. As the permeating hydrogen reacts catalytically with oxygen on the Pd membrane surface, hydrogen concentration (pressure) on the permeate side further decreases. This creates an even larger hydrogen pressure difference across the membrane and promotes more dehydrogenation reaction on the feed side.

11.4.6 Membrane Reactor Configuration

The direction of the feed stream relative to the orientation of the membrane surface and the general membrane reactor configuration also affect the catalytic and reaction aspects of the membrane reactor.

Crossflow membrane reactor. For the same catalytic reaction, various types of reactor configurations discussed in Chapter 8 can result in different performances. For example, for an equilibrium-shift reaction such as dehydrogenation of cyclohexane to produce benzene, a membrane impregnated with catalyst provides a higher conversion than that not attached with the catalyst (e.g., packed catalyst enclosed in a membrane element) provided there is sufficient residence time for most of the reactants to be in contact with the catalyst [Sun and Khang, 1990]. Ioannides and Gavalas [1993] found that, for all the temperatures and space times studied, the membrane reactor gives higher isobutene yield and selectivity than a conventional reactor. The performance difference between the two configurations increases with the space time.

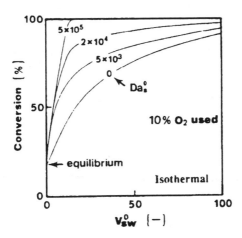

Figure 11.35 Predicted conversion of dehydrogenation of cyclohexane as a function of total flow rate of the sweep gas with the Damköhler number for the permeate side as a parameter [Itoh, 1990]

Membranes that are catalytically active or impregnated with catalyst do not suffer from any potential catalyst loss or attrition as much as other membrane reactor configurations. This and the above advantage have the implication that the former requires a lower catalyst concentration per unit volume than the latter. It should be mentioned that the catalyst concentration per unit volume can be further increased by selecting a high "packing density" (surface area per unit volume) membrane element such as a honeycomb monolith or hollow fiber shape.

There is some indication that a high "packing density" membrane shape may result in a greater conversion than, for example, membrane tubes. In a study of using an α-alumina hollow fiber (1.6 mm in diameter) coated with an γ–alumina membrane as a membrane reactor for dehydrogenation of cyclohexane, Okubo et al. [1991] found that a given overall conversion can be achieved in a much shorter space time in a hollow fiber reactor than in typical tubular reactors used by most of the studies.

Flow-through multistage membrane reactors. For most of the industrially important processes, crossflow membrane reactors are the choice. There are, however, some special situations in which a flow-through multistage membrane reactor is worth considering. Such a type of membrane reactor has been suggested [Michaels, 1989; Saracco and Specchia, 1994]. In this reactor, several layers of different porous materials are produced sequentially and placed upon one another. A specific catalyst is impregnated into each layer for a specific reaction. This type of membrane reactor is particularly suitable for a sequence of consecutive reactions (see Figure 11.36) where the catalytic membranes A, B, and C are permselective only to species A, B, and C, respectively, and the catalysts impregnated in the respective membranes are catalytic only to the reactions $A{\rightarrow}B$, $B{\rightarrow}C$, $C{\rightarrow}D$, respectively. Each layer can be tailored for the desirable transport and catalytic characteristics.

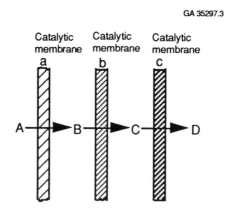

Figure 11.36 Schematic diagram of a flow-through multistage membrane reactor [Michaels, 1989; Saracco and Specchia, 1994]

A membrane reactor confined by two different membranes. When two (or more) products are generated in a membrane reactor, it is desirable to perform in-situ separation of the products from each other. This can be accomplished effectively in a two-membrane reactor under certain conditions [Tekić et al., 1994]. In this concept, the reactor is confined by two separate membranes having different separation properties (Figure 11.37). The two-membrane reactor has the advantage over the single-membrane reactor in that both high reactant conversion and product-product separation are possible. The concept has been demonstrated for the reversible reaction: $A \rightleftharpoons bB + cC$. For the same reaction side volume, overall membrane area and membrane thickness, the overall process efficiency (high conversion and high separation of reaction components) can be improved by the introduction of the second membrane provided that the reactant is the slowest permeating component. As shown in Figure 11.38, the conversion ratio ($\equiv$ reactor conversion / equilibrium conversion) is higher for the two-membrane reactor than the single-membrane reactor. The difference is more pronounced at a large ratio of the reactant permeation rate to the reaction rate.

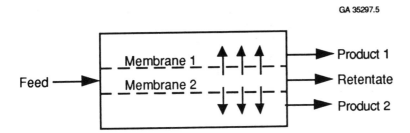

GA 35297.5

Figure 11.37 The two-membrane reactor concept [Tekić et al., 1994]

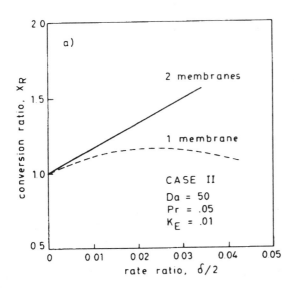

Figure 11.38 Comparison of single- and two-membrane reactors as a function of the ratio of the reactant permeation rate to the reaction rate [Tekić et al., 1994]

11.4.7 Hybrid Reactors Combining Membrane and Conventional Reactors

While inorganic membrane reactors perform more efficiently than conventional reactors in most cases, there are situations calling for the combined usage of these two types of reactors for reasons to be discussed. The conventional reactors in these special cases serve as either the pre-processing or post-processing step for the inorganic membrane reactor system to derive a maximum overall reaction conversion. These hybrid types of reactors consist of conventional reactors at the front end or tail end or both of the membrane reactor.

A membrane reactor preceded or followed by a conventional reactor. Consider a typical commercial porous membrane currently available that exhibits a moderate to low gas separation factor and a high gas permeance even for the gas intended to be the retentate. Much of this less permeable gas in the feed stream is lost to the permeate (low pressure) side in the entrance section of the reactor due to its high partial pressure difference across the membrane layer. This leads to the undesirable effect of a low reactant conversion in that section. An effective way of reducing this reactant loss is to have a membrane enclosed section preceded by an impermeable reaction zone. To achieve a maximum total conversion, the impermeable length relative to the membrane length needs to be optimized.

The above problem has been encountered in different reaction systems. To solve the reactant loss problem in a membrane reactor for the reaction of dehydrogenation of ethylbenzene, Wu and Liu [1992] proposed a hybrid reactor system which consists of a traditional packed-bed reactor followed by a membrane reactor (Figure 11.39). The yields of the two side products, toluene and benzene, monotonically increase with the length of the second packed-bed reactor. With the add-on membrane reactor as the second stage instead, the yields of both side products also increase with the reactor length but at a significantly slower pace. The yield of the main reaction product, styrene, continuously increases with the length of the membrane reactor. In contrast, when the second stage is another conventional packed-bed reactor, the styrene yield begins to decline after reaching a maximum at some reactor length. The reason for the styrene yield to continue to increase in the add-on membrane reactor is that styrene generation is enhanced while the side reactions are inhibited or slowed down by the selective removal of hydrogen by the membrane. The styrene yield can be maximized by searching for the optimal operating conditions.

Gokhale et al. [1993] further simplified the hybrid reactor system by combining the two separate reactors into a single reactor with an impervious section at the reactor entrance followed by a permeable membrane section. The total reaction conversion has been calculated to be a function of the ratio of the impermeable section length to the total reactor length, as shown in Figure 11.40 for butane dehydrogenation with a total pressure of 151.5 kPa on both sides of the membrane. An optimal length ratio for the maximum conversion exists for a given total length of the reactor. It appears that the optimal length ratio decreases with increasing reactor length. A large reactor length is equivalent to a large space time for the reactant.

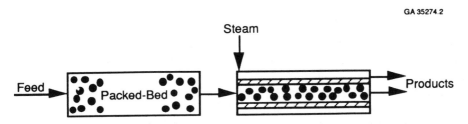

Figure 11.39 A membrane reactor preceded by a conventional packed-bed reactor [Wu and Liu, 1992]

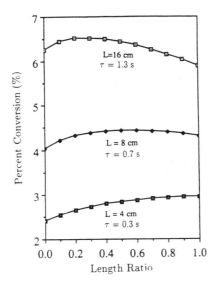

Figure 11.40 Conversion of butane dehydrogenation as a function of the ratio of the impermeable section length to the total reactor length [Gokhale et al., 1993]

Similarly when a reactant is distributed through a membrane to the reaction zone on the other side to mix and react with other reactant(s), a hybrid of a membrane reactor and a conventional fixed-bed reactor may be advantageous over either type of reactor under certain conditions. By placing a conventional fixed-bed reactor after a membrane reactor for the oxidative coupling of methane, Coronas et al. [1994] discovered that the resulting methane and oxygen conversions and C_{2+} selectivity can be higher than the corresponding values for a membrane reactor with a uniform permeability. However, as the permeation area for a given length of the hybrid reactor is reduced, the local concentration of the permeating reactant increases which may lead to a lower selectivity. There appears to be an optimal length of the permeable section for the hybrid reactor.

A membrane reactor preceded and followed by conventional reactors. In equilibrium-limited reactions, membrane reactors enhance reaction conversions by continuously removing one or more of the products, say, hydrogen, thus displacing the equilibrium toward more product formation. Furthermore, the permeate stream through a porous membrane will inevitably contain some unconverted reactant(s). An idea of maximizing the total reaction yield has been proposed where the membrane reactor is preceded by a conventional reactor and followed by another conventional reactor [Bitter, 1988]. The first conventional reactor is designed to carry out the reaction close to the equilibrium for the reason just discussed earlier (to reduce the reactant loss due to permeation) and the membrane reactor is then used to displace the equilibrium to elevate the conversion. The same reaction proceeds again with the permeate stream which still contains some unconverted reactant(s) in the second conventional reactor, thereby further increasing the total yield.

The above reactor configuration has been used for the dehydrogenation of dehydrogenatable organic compounds such as propane and 2-methylbutenes, among many others. In Figure 11.41 the pre-dehydrogenation zone or reactor is positioned below the dehydrogenation membrane reactor. A permeate (after) dehydrogenation zone or reactor is physically arranged below the pre-dehydrogenation zone. The pre-dehydrogenation fixed-bed reactor zone is freely connected to the membrane reactor zone which in turn is connected to the permeate dehydrogenation fixed-bed reactor zone. In all three zones, the same catalyst is used. Improvements in the conversion and selectivity were observed.

11.4.8 Requirements of Catalytic and Membrane Surface Areas

The requirement for catalytic surface area may determine in what form the catalyst should be incorporated in the membrane reactor. If the underlying reaction calls for a very high catalytic surface area, the catalyst may need to be packed as pellets and contained inside the membrane tubes or channels rather than impregnated on the membrane surface or inside the membrane pores due to the limited available area or volume in the latter case.

11.4.9 Deactivation of Catalytic Membrane Reactors

The nature of many high-temperature hydrocarbon reactions which potentially can benefit from inorganic membrane reactors (particularly catalytic membrane reactors) is such that the catalysts or the catalytic membranes are subject to poisoning over time. Deactivation and regeneration of many catalysts in the form of pellets are well known, but the same issues related to either catalyst-impregnated membranes or inherently catalytic membranes are new to industrial practitioners. They are addressed in this section.

GA34561.2

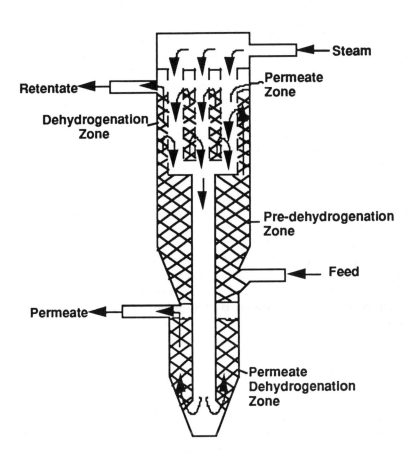

Figure 11.41 Schematic diagram of a compact dehydrogenation membrane reactor preceded and followed by conventional reactors [adapted from Bitter, 1988]

Poisoning. Both the membrane and the catalyst in a membrane reactor may become deactivated over time in the application environment. This poisoning arises from some species present in the feed stream or from some product(s) of the reaction. When the poison is present in the feed stream at a relatively high concentration and is weakly adsorbed onto the catalyst or membrane surface or when the poison is formed by reaction, it is uniformly distributed throughout the catalyst. On the other hand, if the poison is present in the feed stream in a relatively low concentration and is strongly adsorbed, the outer pore surfaces can completely lose catalytic activity before the inside pore surfaces do. When significant deactivation of either or both occurs, effective prevention measures and treatments will be required to make the membrane reactor technically as well as economically viable.

A catalyst can lose its activity or selectivity in many ways: poisoning, fouling, loss of some active area or species [Satterfield, 1980]. For metal catalysts, poisoning and reduction of the active area are predominant. Most of poisoning takes place when some impurity in the process stream adsorbs on the active sites of the catalyst and causes the catalyst activity to be lowered. Different compounds, sometimes even in trace quantity, can become poisons to different catalysts.

Coke deposits. A special type of poisoning in many hydrocarbon reactions (e.g., catalytic cracking, dehydrogenations and reforming) is the formation of coke or carbonaceous deposits as a result of reaction on catalyst particles or membranes or both, depending on the reactor configuration. Carbonaceous or coke deposits block active sites of a catalyst or catalytic membrane and ultimately leads to deactivation of the catalyst or the membrane.

The nature of the these deposits depends on the reactions and conditions under which they are formed. When formed at relatively low temperatures, say below 300 or 400°C, the deposits appear to be some form of high molecular weight polymers while a substantial amount of those formed above 400 - 500°C show a pseudographite form and may be represented empirically as CH_x where x lies between 0.5 and 1.0 [Satterfield, 1980]. In several hydrogenation reactions, hydrogen used comes from carbonaceous sources and the resultant carbon monoxide may be decomposed to form coke. Coke or carbonaceous deposits can also form on catalysts or membranes in a reducing chemical environment as a result of the decomposition of methane or other reactions.

Carbon deposits on metallic catalysts, particularly those consisting of heavy metals, in general have little or no hydrogen while those on nonmetallic catalysts contain a significant amount of hydrogen. Coke can attack and cause disintegration of the metal and can have deleterious effects on the catalytic reactions. Typical hydroprocessing catalysts are metal sulfides supported on high surface area alumina substrates. Carbon and metal deposition causes catalyst deactivation during hydrotreating of heavy oils. Two possible mechanisms of catalyst deactivation by coking are believed to be blocking of the catalyst pores and chemical modifications of the pore surface by carbon deposits. On the other hand, metal deposition leads to poisoning of catalytically active sites and, more severely, plugging of pore mouths [Nourbaksh et al., 1989b].

It is interesting to note that, due to its uniform and nearly straight pore structure, planar anodic alumina membranes have been used as the probes for monitoring catalyst deactivation at the single pore level [Nourbaksh et al., 1989a]. The nearly idealized structure is suitable for various surface analysis techniques such as scanning and transmission electron microscopy and associated EDX and XPS to be applied to fresh as well as spent catalysts. In the case of hydrotreating of crude-derived heavy oils, catalyst experiences the heaviest coking at the reactor inlet and pores as large as 100 nm in diameter can be blocked completely. In another case where catalyst is deactivated during metaloporphyrin hydrodemetalation, the use of anodic alumina membranes has shown that the catalyst is deactivated by the deposition of metal crystallites inside individual pores.

Enhancement of reaction conversion by employing a permselective membrane often has the implication that, for a given conversion, it is possible to run the reaction at a lower temperature in the membrane reactor than in a conventional reactor. Catalyst deactivation due to coke formation generally becomes more severe as the reaction temperature increases. Therefore, the use of a membrane reactor to replace a conventional one should, in principle, reduce the propensity for coke formation because for the same conversion the membrane reactor configuration may be operated at a lower temperature than a conventional reactor. This is particularly true for such reactions as dehydrogenation.

Coking in membrane reactors. Coking can affect a membrane reactor on three major aspects. The first is that coking deactivates the catalyst. The second is that it can potentially reduce the permeability of the membrane. Finally, coking occurs on some metals frequently used in processing vessels, pipings and connections such as stainless steel and can influence the overall reactor performance. Accumulation of cokes to render the catalysts inactivated can occur in less than an hour as in dehydrogenation of butane to a few hours as in dehydrogenation of isobutane or over a period of months as in catalytic reforming. Once deactivated, the membrane reactor needs to be regenerated.

Hydrogen is believed to delay or suppress carbon deposits on the catalyst surface, thus inhibiting catalyst deactivation. Therefore, in dehydrogenation reactions where hydrogen is continuously removed from the reaction zone, it could be expected that carbon deposition may accelerate. The problem, however, depends on the specific dehydrogenation reaction. For the reactions of ethane or cyclohexane dehydrogenation, for example, the catalysts do not become seriously deactivated due to coke formation. In contrast, catalytic butane or isobutane dehydrogenation suffers from rapid carbon deposition leading to catalyst deactivation. Consider the case of dehydrogenation of butane using an inorganic membrane reactor. Zaspalis et al. [1991b] studied the reaction in a porous alumina membrane reactor containing a bed of the standard catalyst for the reaction: silica-supported platinum catalyst (6.2% of Pt by weight) precipitated from $Pt(NH_3)_4Cl_2$. It is found that a major problem is carbon deposition leading to deactivation of the catalyst. Shown in Figure 11.42 is the total conversion of butane as a function of time for the membrane reactor operated at 480°C. The effect of coking is serious. After only half an hour of operation, the conversion decreases to about 50% of the initial level and after one hour the catalyst is practically inactive. It has been suggested that, for this reaction, coke is formed by the following sequence: n-butane → n-butene → coke [Toei et al., 1975; Uchida et al., 1975].

Dehydrogenation of isobutane similarly leads to serious catalyst deactivation due to coking. Extensive loss of catalyst activity can occur within the first two or three hours on stream [Ioannides and Gavalas, 1993]. Hydrogen is indeed found to inhibit coking and the presence of isobutane alone (without introducing isobutene in the feed stream) leads to catalyst deactivation. This seems to further support the proposed general mechanism of coke formation for butane dehydrogenation just mentioned.

Although hydrogen is believed to suppress the formation of carbon deposits, coking may still occur in hydrogenation reactions as well. During ethylene hydrogenation, extensive coke deposits are noticed on the feed side of Pd-Y membranes and are believed to eventually lead to the embrittlement and rupture of the dense membranes due to carbon diffusion inside the membrane [Al-Shammary et al., 1991]. A modest deposit of carbon could actually increase the selective formation of ethane which may be indicative of some reaction taking place on the carbon deposit.

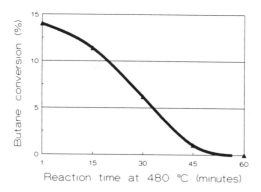

Figure 11.42 Decrease of butane conversion with time at 480°C as a result of catalyst deactivation due to coke deposits [Zaspalis et al., 1991b]

At high temperatures catalysts or catalytic membranes may suffer from irreversible deactivation to the extent that after a few cycles of reaction and regeneration the catalyst or membrane will need to be replaced. Ioannides and Gavalas [1993] estimates from differential reaction experiments that for dehydrogenation of isobutane 35% of the chromia/alumina catalytic activity is irreversibly lost after four cycles of reaction and regeneration.

Coke deposits are not limited to the catalysts. The membranes, in the case of packed bed catalysts, are also subject to the adverse effects of carbonaceous deposits leading to possible reduction in the membrane permeability. Catalytic membranes such as Pd-based membranes can become deactivated as a result of poisoning due to strong adsorption of sulfur compounds on the membranes. Palladium membranes are also subject to coking due to the presence of carbon monoxide which, like sulfur compounds, are often present in gas streams in petroleum or petrochemical operations. Small quantities of these impurities can even profoundly affect the catalytic activity of palladium-based membranes. They are believed to preferentially block the catalytically active sites. The activation energy for adsorption on those sites is lower than that on the remaining surface of the membrane [Al Shammary et al., 1991].

Porous membranes, catalytically active or passive, are prone to the blockage of their pores due to coking. By comparing an untreated (porous) Vycor glass membrane tube to a similar one coated with a dense silica layer, Ioannides and Gavalas [1993] observed that coke deposition is obvious within the pores of the untreated glass tubes during the reaction of isobutane dehydrogenation while no change in the membrane permeability occurs over an extended period of time in the case of silica-coated glass tubes. Presumably, the dense silica membrane layer prevents access of the hydrocarbon molecules to the internal pore surface area.

Finally, coking is known to take place on certain metals including stainless steel due to its nickel content. When stainless steel is used as the housing material for a membrane reactor, the contact of hydrocarbons with the metal promotes coke formation. This problem is more pronounced when the membrane is not very selective, potentially leading to a reduced total conversion and selectivity when considering reaction products on both the feed and permeate sides. Stainless steel is often used in membrane reactor experiments on the bench but rarely utilized in production vessels and equipment. Therefore, care should be exercised when considering scale-up of the bench results. To better simulate the results on a production scale, bench-scale membrane reactor systems can employ some linings such as glass on the contact surface of the stainless steel housing to reduce the effects of coking [Liu, 1995].

Prevention of poisoning. One of the potential methods for preventing the catalyst poisoning on a catalytic membrane is to form a layer of a material, such as a polymer, which permits the passage of the reactant(s) but not the poison [Gryaznov et al., 1983]. This method, however, places a restriction on the temperature range in which the membrane reactor can be utilized.

11.4.10 Regeneration of Catalytic Membrane Reactors

The commonly practiced method of regenerating coked catalysts or catalytic membranes is by combustion with air, air/steam mixtures, air/nitrogen mixtures, or dilute oxygen streams with another carrier gas.

Coke generally contains a significant amount of hydrogen. During the regeneration reaction, the hydrogen is first preferentially removed leaving behind a carbon skeleton on the catalyst or catalytic membrane. So the critical step in the regeneration process is the gasification of carbon. At relatively high temperatures, the rate-determining step of the gasification reaction is the mass transfer of oxygen to the carbon. The intrinsic rate of regeneration in general has been found to be proportional to oxygen concentration and, in cases where the coke concentration is low, carbon concentration. The presence of chromia, other transition metals and some potassium compounds tends to catalyze combustion of carbon.

The use of water or steam also seems to be effective in minimizing the coking problem in membrane reactors where dehydrogenation of several organic compounds take place [Setzer and Eggen, 1969; Bitter, 1988; Wu et al., 1990]. In the well known reaction of dehydrogenation of ethylbenzene, superheated steam is added for the de-coking reaction to regenerate the catalyst. The same approach can be taken for dehydrogenation of propane to produce propene and dehydrogenation of 2-methylbutenes to form isoprene. The use of steam or water, however, introduces some impurities such as carbon monoxide, carbon dioxide and methane in this case.

Sometimes, however, despite in-process regeneration of the catalyst, coke deposit continues to pose a serious operational problem for membrane reactors. An oxidant such as oxygen can be introduced to the reaction zone to continuously regenerate the catalyst and thus prolonging the service life of the catalyst. This beneficial effect can be reflected in the total conversion of methanol as a function of the operation time (Figure 11.43). In Figure 11.43 [Zaspalis et al., 1991a], two cases are compared. The first case involves the use of oxygen (4% by volume relative to methanol feed) introduced from the permeate side and the second case does not use oxygen. In each case, the membrane reactor is operated at 450°C for three hours a day and held at 80-100°C for the remainder of the day with helium continuously flowing on both sides of the membrane. It is seen that higher conversions are maintained when oxygen is supplied to the reaction zone from the permeate side during the decomposition reaction of methanol.

However, the addition of an oxidant such as oxygen is not without some trade-off. To help solve the problem of catalyst deactivation due to carbon deposit in an alumina membrane reactor for dehydrogenation of butane, oxygen is introduced to the sweep gas, helium, on the permeate side at a concentration of 8% by volume. The catalyst service life increases from one to four or five hours, but the selectivity to butene decreases from 60 to 40% at 480°C [Zaspalis et al., 1991b]. If oxygen is added to the feed stream entering the membrane reactor in order to inhibit coke formation, the butene selectivity decreases even more down to 5%.

For those reactions such as dehydrogenation of butane or isobutane which require frequent regeneration of the catalyst or catalytic membrane, a cyclic process involving reaction and coke-burnout steps in sequence may need to be considered. It is essential, however, that a purge step be installed between the reaction and the burnout steps to prevent any potential explosion due to some dangerous gas mixtures. The worse scenario, of course, would be to shut down the operation and replace the deactivated packed catalyst or catalytic membrane with fresh material.

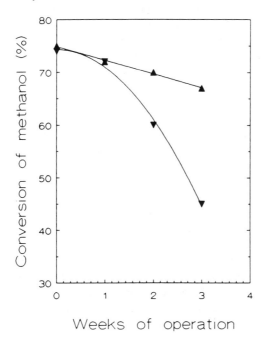

Figure 11.43 Conversion of methanol as a function of operation time in the presence (▲) and in the absence (▼) of oxygen [Zaspalis et al., 1991a]

11.5 DESIGN CONSIDERATIONS OF MEMBRANE REACTOR

In the crossflow configuration widely used in the membrane separation industry today, the direction of the feed flow is perpendicular to that of the permeate flow. To increase the driving force for the permeate, a carrier stream to "sweep" away the permeate is often used. In most of the membrane configurations such as tubes and multichannel monoliths described in Chapter 5, the carrier stream and the feed stream are generally co-current or counter-current in direction.

The geometry of the membrane reactor and the relative locations and flow directions of the feed, permeate and retentate streams all play important roles in the reactor performance. The simplest, but not efficient, membrane reactors consist of disk or foil membranes with a flow-through configuration [Mischenko et al., 1979; Furneaux et al., 1987]. The same type of membrane reactor can also be constructed and operated in the more common crossflow mode.

Single tubular membrane reactors are often used in experimental and feasibility studies. Its justification for use in production environments can sometimes be made in small volume applications. As mentioned in Chapters 4 and 5, inorganic composite membranes consist of multiple layers. The inner most layer in a tubular composite membrane reactor does not necessarily possess the finest pores. For example, a two-layered tubular ceramic membrane reactor used for enzymatic reactions has an inner layer containing pores larger than those in the outer layer [Lillo, 1986]. The pores of the inner layer are immobilized with enzymes. Under the influence of an applied pressure difference across the membrane matrix, a solution entering the hollow central core of the tube flows into the inner layer where the solution reacts with the enzyme. The product which is smaller than the enzyme passes through the permselective outer layer membrane which retains the enzyme. Thus the product is removed from the reaction mixture.

Multiple-tube or tube-bundle geometry provides some advantages of a higher packing density and cost savings of reactor shells. The shell-and-tube type of membrane reactor design similar to that of floating head heat exchangers has also been used (for example [Gryaznov et al., 1973a]). General process economics and ease of operation and maintenance, however, can be further improved by the use of multichannel monolithic membrane reactors. The latter have inherently higher surface areas for the catalysts and higher permeation rates per unit volume than individual tubular membrane reactors.

A still higher membrane packing density can be accomplished with some geometries available to certain commercial organic membranes. One such attempt is hollow fibers [Baker et al., 1978; Beaver, 1986]. Sintered monolithic hollow fibers of metals or metal oxides have been developed [Dobo and Graham, 1979] but their burst strengths appear to be critical. Further developments are needed before commercialization.

In addition to the above conventional configurations, there have been some novel designs proposed for various types of inorganic membrane reactors. They will be discussed here.

11.5.1 High Surface Area Membrane Reactors

Conventional membrane surfaces are either flat (as a foil, plate or square channel) or regularly curved (as in a tube or circular channel). Potentially there may be stagnation zones in which carbonaceous materials are deposited. Moreover, not every site on the membrane surface participates in the catalytic reaction.

Membranes with non-traditional surfaces have been suggested [Gryaznov et al., 1975]. Pd-based membranes in the form of cellular foils were fabricated and used to carry out a hydrogen-consuming and a hydrogen-generating reaction on the opposite sides of the membrane. The foils have oppositely directed, alternate projections of hemispherical or half-ellipsoidal shapes. This type of configuration, in principle, provides uniform washing of the membrane surface, thus reducing the formation of coke deposits. Furthermore, uniform contact of the reaction components with the membrane surface should ensure the participation of the entire membrane surface in the catalytic conversion of hydrocarbons.

11.5.2 Double Spiral-Plate or Spiral-Tube Membrane Reactors

As a result of many years of development, Gryaznov and his co-workers have fabricated Pd-based membrane reactors with two unusual configurations aiming at high packing density [Gryaznov, 1986].

In the first configuration (Figure 11.44), a Pd or Pd-alloy membrane is made of a spiral plate and forms two compartments on the opposite sides of the membrane with double spirals. The edges of the double spiral coiled plate are fixed to the walls of the reactor shell. The number of spirals can be as high as 200.

In the second configuration (Figure 11.45), a membrane is fabricated into a bank of seamless thin-walled tubes to form flat double spiral partitions. One spiral is placed on top of the other such that each succeeding spiral is in a mirror symmetry with the adjacent one. This type of membrane reactor has been utilized for hydrogenation of liquid compounds [Gryaznov et al., 1981].

11.5.3 Fuel Cells

Some sophisticated thin-film fuel cell configurations have been designed with industrial implementation in mind. Sverdrup et al. [1973] have shown a fuel cell with a tubular geometry developed at Westinghouse for a coal reactor. Its cross section is schematically shown in Figure 11.46. The fuel cell is constructed by sequential deposition of layers of various materials each about 20 to 50 μm thick on a porous tube of stabilized zirconia. The cell consists of a fuel electrode and an air electrode made from cobalt- or nickel-zirconia cermets and tin-doped indium oxide, respectively. An interconnection made from manganese-doped cobalt chromite links the air electrode to the fuel electrode of an adjacent cell. The choice of material for the interconnect must meet the requirements of high conductivity and thermodynamic stability in reducing and oxidizing atmospheres, prevailing in the fuel and air electrodes, respectively, at high temperatures.

Similar tubular cell designs based on stabilized zirconia and β-alumina solid electrolyte membranes have been developed for hydrogen production [Dönitz and Erdle, 1984] and load-leveling applications [Sudworth and Tilley, 1985].

The configuration of the fuel cells is not limited to the tubular geometry. Michaels et al. [1986] fabricated monolithic fuel cells which basically have a crossflow configuration and all unit cells connected in series (Figure 11.47) or in parallel (Figure 11.48) for the oxidation of hydrogen. The cells consist of ribbed sheets of stabilized zirconia of about 300 μm thick. The sheets are prepared first by blending zirconia powder, polyethylene resin and a mineral plasticizer into a mixture which is then hot pressed onto a grooved metal mold to form plasticized sheets that are smooth on one side and ribbed on the other. Upon removing the plasticizer by a solvent, the sheets are sintered at 1,500°C followed by deposition of platinum electrodes on both sides. The fuel cells are finally assembled by stacking the sintered sheets vertically which are cemented with a high-temperature ceramic adhesive.

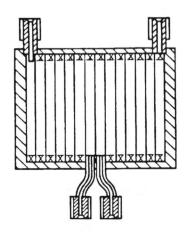

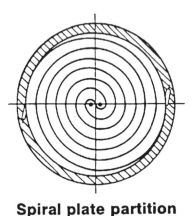

Spiral plate partition

Figure 11.44 Double spiral coiled plate membrane reactor [Gryaznov, 1986]

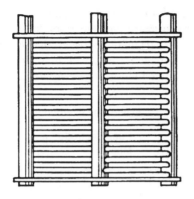

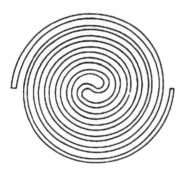

Spiral tube partition

Figure 11.45 Flat double spiral tube membrane reactor [Gryaznov, 1986]

11.5.4 Crossflow Dual-compartment Membrane Reactors.

De Vos et al. [1982] have designed and tested a simple and yet innovative membrane reactor configuration using porous Pd-based membranes [Figure 11.49]. Corrugated spacer plates are used to provide a lower flow resistance on both sides of the membrane. Two coupled reactions (e.g., a dehydrogenation and a hydrogenation reaction) take place on the opposite sides of the membrane. The flow directions in the two compartments form a crossflow pattern. The design has two unique basic features. First, it is modular such that a large-sized membrane reactor can be scaled up, constructed and maintained more easily than many other designs. Second, the crossflow pattern of the two process streams on the two sides of the membrane allows separate lengths of the two reactor compartments.

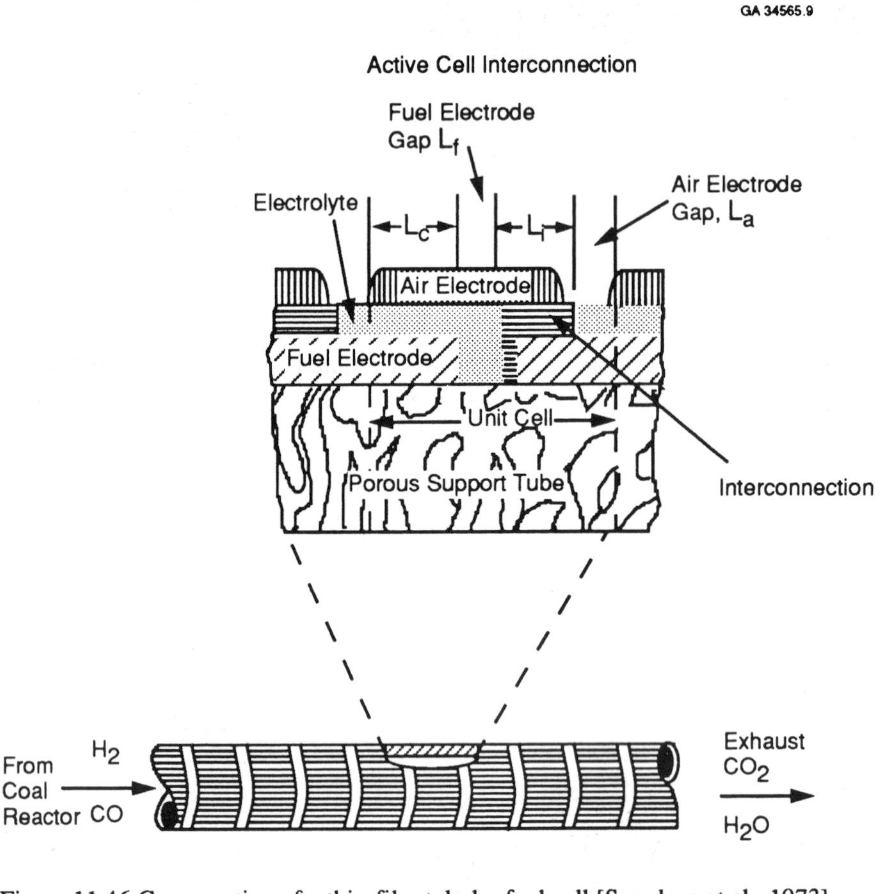

GA 34565.9

Figure 11.46 Cross section of a thin-film tubular fuel cell [Sverdrup et al., 1973]

11.5.5 Hollow-Fibers Membrane Reactors

As mentioned earlier, one of the membrane element shapes with the highest packing density is hollow fibers. Typically several fibers are bundled to provide higher strength. In a packed-bed membrane reactor of this type, catalyst particles are packed around the bundles.

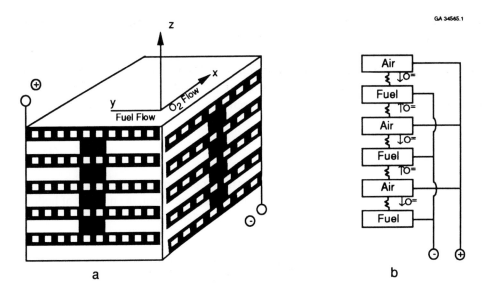

Figure 11.47 Crossflow monolithic fuel cells with all unit cells connected in series: (a) schematic showing external connections and (b) equivalent circuit [Michaels et al., 1986]

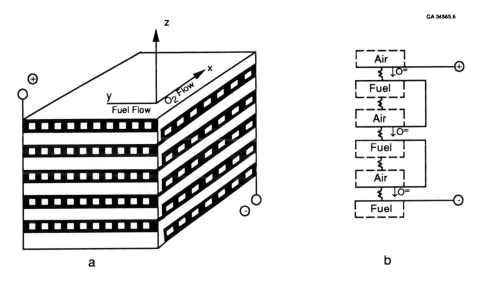

Figure 11.48 Crossflow monolithic fuel cells with all unit cells connected in parallel: (a) schematic showing external connections and (b) equivalent circuit [Michaels et al., 1986]

Itoh et al. [1984] modeled a 1.5-m long packed-bed membrane reactor for catalytic decomposition of HI. The reactor consists of a large number of microporous hollow fiber membranes with an outer diameter and a thickness of 1 mm and 0.1 mm, respectively. The reaction conversion at the exit of the membrane reactor (with 1 mm outer diameter fibers) at 427°C was calculated. Three different membrane thicknesses were considered. The maximum conversion for those three cases is identical at 41%; however, the number of hollow fibers achieving the maximum conversion is 1370 for a thickness of 0.1 mm, 3150 for 0.2 mm and 5630 for 0.3 mm, respectively. It has been found that the ratio of membrane area to membrane thickness

$$A_p / t_m = 2n\pi / \ln(R_o / R_i) \tag{11-7}$$

is a constant at the maximum conversion under fixed operating conditions. It implies that once the outer and inner diameters of the hollow fibers are given, the number of hollow fibers needed to achieve the maximum conversion is determined.

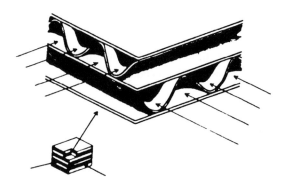

Figure 11.49 Schematic diagram of a crossflow membrane reactor with two reaction compartments involving coupled reactions [De Vos et al., 1982]

11.5.6 Fluidized-Bed Membrane Reactors

To overcome heat transfer and diffusional limitations associated with fixed bed reactors, fluidized-bed membrane reactors (FBMR) using dense Pd membranes have been proposed for the steam reforming of methane [Adris et al., 1991; Adris et al., 1994a; Adris et al., 1994b]. In FBMR, bundles of membrane elements or modules are employed to increase the packing density for more hydrogen removal which in turn leads to more conversion of the forward reaction due to equilibrium displacement.

The choice between membrane elements and modules for use in a FBMR is primarily determined by balancing the considerations of having a higher packing of bare membrane elements on one hand, and protecting the membrane elements from severe abrasion on the other. To reduce abrasion of a bundle of membrane tubes, an outer protective sheath or tube is used to surround the tubes to form a module. The protective sheath material should be selected to be abrasion resistant and permeable to most or all gases but not to particulates such as catalysts. Porous ceramic or certain metallic materials are good candidates. It is feasible that, in practical applications, several modules each containing several membrane elements are placed inside a FBMR.

The gases not permeating the membranes move to the bottom of a long tube and pass through a cyclone to remove entrained particles which are returned to the fluidized-bed region. The particle-lean, hydrogen-lean gases are then directed to downstream processing steps. If desired, some portion of these gases can be recycled.

Horizontal versus vertical membrane tubes or modules. Two general types of fluidized-bed membrane reactors have been tested. The first type places the membrane elements or modules perpendicular to the general direction of the fluidizing reaction gases (see Figures 10.14a and 10.14b). In the second type of FBMR, the membrane elements or modules are essentially parallel to the fluid flow direction inside the reactor, as schematically shown in Figure 11.50. It appears that the vertical type exhibits more advantages for practical implementation.

In the vertical design, membrane tubes or modules extend through both the fluidized-bed and the enlarged freeboard zones to maximize conversion and hydrogen yield. Hydrogen removal within the freeboard also avoids the reverse reaction (i.e., methanation) which is favored by the lower temperature in the freeboard. A major advantage of the vertical design over the horizontal type is minimizing interference of fluid flow in the fluidized bed and potential abrasion of the membrane material (and consequently shorter service life of the membrane) due to high impact of the particle-laden gases directly on horizontal membrane elements in the reactor. For the same reasons, it is preferred to align any heat pipes or heat exchangers vertically.

The presence of vertical membrane elements or modules also helps to prevent bubble coalescence, thus favoring heat and mass transfer in the reactor. Their spacing should be sufficiently small so that the maximum number of the membrane tubes or modules may be provided in the reaction zone and large enough that no blockage or bridging of the fluidized bed occurs.

Fluidized-bed membrane reactors with various fluid routings. Two particular designs have been shown in Figures 10.14a and 10.14b. The major difference between the two designs is the nature of the process stream that passes through the membrane tubes to remove the permeating hydrogen. In one design, the reaction product stream is combined with the permeating hydrogen while passing through the membrane tubes. The driving force for hydrogen permeation comes from a pressure reduction of the product stream before entering the membrane tubes. Near complete conversion of methane can be achieved by optimization at temperatures within the range used industrially. As mentioned in Chapter 10, this design generates products that have high hydrogen contents and is quite suitable for hydrogen plants in petroleum refineries and for hydrotreating and hydrocracking, for methanol plants and for iron ore reduction processes [Adris et al., 1991].

In the other design, the feed reactant stream is split into two: one passing through the membrane tubes and the other moving through the fluidized bed in the reactor vessel. The converted stream, the hydrogen permeating from the fluidized bed side to the tube side and the unreacted, split feed stream are all combined to form the final product stream. The driving force for hydrogen permeation in this case is the difference in hydrogen concentrations between the product and feed streams. This second design is appropriate when it is desirable to maintain some unconverted hydrocarbons in the final product stream for further processing such as ammonia production plants where the product stream from the primary reformer is further processed in a secondary reformer. Hydrocarbon conversion from this second design, as expected, will not be as high as that from the first design.

11.6 SUMMARY

Transport phenomena of process streams in an inorganic membrane reactor is an integral part of the reactor design and operation. Their interplay with kinetic and catalytic parameters largely determines the reactor performance. Detailed hydrodynamics based on computational fluid dynamics (CFD) provides more accurate reactor design. With the progresses in computer hardware and computational algorithms, the use of CFD to model membrane reactors is becoming increasingly feasible.

The flow directions (e.g., co-current, counter-current and flow-through) and flow patterns (e.g., plug flow, perfect mixing and fluidized bed) of feed, permeate and retentate streams in a membrane reactor can significantly affect the reaction conversion, yield and selectivity of the reaction involved in different ways. These variables have been widely investigated for both dense and porous membranes used to carry out various isothermal and non-isothermal catalytic reactions, particularly dehydrogenation and hydrogenation reactions.

Recycling some portion of the permeate or retentate stream and introducing feed at intermediate locations are effective methods for improving the reaction conversion. The relative permeabilities of the reactant(s) and product(s) and whether the permeate or retentate is recycled all affect the effectiveness of these measures for conversion enhancement. To compensate for the variations in the transmembrane pressure difference and consequently in the permeation rate, the concept of a location-dependent membrane permeability has been proposed. The effects of this approach and the average permeation rate are discussed.

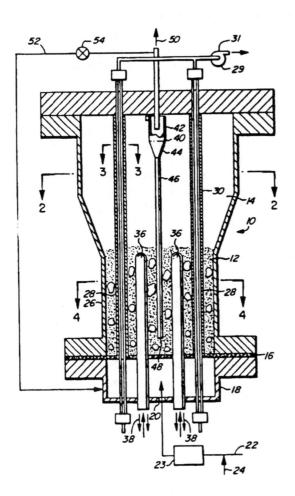

Figure 11.50 Schematic diagram of a fluidized bed membrane reactor with vertical membrane elements [Adris et al., 1994b]

Various techniques of supplying or removing heat from an inorganic membrane reactor are presented and their implications discussed. Thermal management of plug-flow, perfect-mixing and reaction-coupling membrane reactors is surveyed together with a brief discussion on the thermal stability of a membrane reactor.

Some portion of the reactant inevitably "leaks" through a porous membrane. The rate of reactant loss through membrane permeation in relation to the reaction rate has profound impact on the reactor performance and can be used to explain some unusual reactor behavior.

In view of the state-of-the-art developments in inorganic membranes, those reactions amenable to inorganic membrane reactors are characterized. The effects of space time in isothermal and non-isothermal membrane reactors are reviewed.

Finally, possible causes for deactivation of catalytic membranes are described and several aspects of regenerating catalytic membrane reactors are discussed. A variety of membrane reactor configurations are mentioned and some unique membrane reactor designs such as double spiral-plate or spiral-tube reactor, fuel cell unit, crossflow dual-compartment reactor, hollow-fiber reactor and fluidized-bed membrane reactor are reviewed.

REFERENCES

Abdalla, B.K., and S.S.E.H. Elnashaie, 1995, J. Membr. Sci. **101**, 31.

Adris, A.M., 1994, A Fluidized Bed Membrane Reactor for Steam Methane Reforming: Experimental Verification and Model Validation, Ph.D. dissertation, Univ. of British Columbia, Vancouver, Canada.

Adris, A.M., S.S.E.H. Elnashaie and R. Hughes, 1991, Can. J. Chem. Eng. **69**, 1061.

Adris, A.M., C.J. Lim and J.R. Grace, 1994a, Chem. Eng. Sci. **49**, 5833.

Adris, A-E M., J.R. Grace, C.J. Lim and S.S. Elnashaie, 1994b, U.S. Patent 5,326,550.

Al-Shammary, A.F.Y., I.T. Caga, J.M. Winterbottom, A.Y. Tata and I.R. Harris, 1991, J. Chem. Tech. Biotechnol. **52**, 571.

Armor, J.N., 1992, Chemtech **22**, 557.

Baker, R.A., G.D. Forsythe, K.K. Likhyani, R.E. Roberts and D.C. Robertson, 1978, U.S. Patent 4,105,548.

Beaver, R.P., 1986, European Pat. Appl. 186,129.

Bernstein, L.A., and C.R.F. Lund, 1993, J. Membr. Sci. **77**, 155.

Bitter, J.G.A., 1988, U.K. Pat. Appl. 2,201,159A.

Bowen, W.R., and A.O. Sharif, 1994, J. Colloid Interf. Sci. **168**, 414.

Collins, J.P., J.D. Way and N. Kraisuwansarn, 1993, J. Membr. Sci. **77**, 265.

Coronas, J., M. Menendez and J. Santamaria, 1994, Chem. Eng. Sci. **49**, 4749.

De Vos, R., V. Hatziantoniou and N.H. Schoon, 1982, Chem. Eng. Sci. **37**, 1719.

Dobo, E.J., and T.E. Graham, 1979, U.S. Patent 4,175,153.

Dolecek, P., 1995, J. Membr. Sci. **100**, 111.

Dönitz, W., and E. Erdle, 1984, High-temperature electrolysis of water vapor; status of development and perspectives for application, in: Proc. 5th World Hydrogen Ener. Conf., Toronto, Canada (T.N. Veziroglu and J.B. Taylor, Eds.), Pergamon Press, New York, p. 767.

Edlund, D.J., and W.A. Pledger, 1993, J. Membr. Sci. **77**, 255.

Furneaux, R.C., A.P. Davidson and M.D. Ball, 1987, European Pat. Appl. 244,970.

Gokhale, Y.V., R.D. Noble and J.L. Falconer, 1993, J. Membr. Sci. **77**, 197.

Gryaznov, V.M., 1986, Plat. Met. Rev. **30**, 68.

Gryaznov, V.M., A.P. Mishchenko, V.S. Smirnov and S.I. Aladiyshev, 1973a, U.K. Patent 1,358,297.

Gryaznov, V.M., V.S. Smirnow and M.C. Slinko, 1973b, Heterogeneous catalysis with reagent transfer through the selectively permeable catalysts, in: Proc. 5th Int. Congr. Catal., North-Holland Publishing Co., Amsterdam, Netherlands, p. 1139.

Gryaznov, V.M., V.S. Smirnov and A.P. Mischenko, 1975, U.S. Patent 3,876,555.

Gryaznov, V.M., A.P. Mishchenko, A.P. Maganjuk, V.M. Kulakov, N.D. Fomin, V.P. Polyakova, N.R. Roshan, E.M. Savitsky, J.V. Saxonov, V.M. Popov, A.A. Pavlov, P.V. Golovanov and A.A. Kuranov, 1981, U.K. Pat. Appl. 2,056,043A.

Gryaznov, V.M., V.S. Smirnow, V.M. Vdovin, M.M. Ermilova, I.D. Gogua, N.A. Pritula and G.K. Fedorova, 1983, U.S. Patent 4,394,294.

Guerrieri, S.A., 1970, U.S. Patent 3,524,819.

Hwang, S.T., and J.M. Thorman, 1980, AIChE J. **26**, 558.

Ilias, S., and R. Govind, 1993, Sep. Sci. Technol. **28**, 361.

Ilias, S., K.A. Schimmel and G.E.J.M. Assey, 1995, Sep. Sci. Technol. **30**, 1669.

Ioannides, T., and G.R. Gavalas, 1993, J. Membr. Sci. **77**, 207.

Itaya, Y., K. Miyoshi, S. Maeda and M. Hasarani, 1992, Int. Chem. Eng. **32**, 123.

Itoh, N., 1987, AIChE J. **33**, 1576.

Itoh, N., 1990, J. Chem. Eng. Japan **23**, 81.

Itoh, N., and R. Govind, 1989, Ind. Eng. Chem. Res. **28**, 1554.

Itoh, N., Y. Shindo, T. Hakuta and H. Yoshitome, 1984, Int. J. Hydrogen Energy **10**, 835.

Itoh, N., Y. Shindo and K. Haraya, 1990, J. Chem. Eng. Japan **23**, 420.

Itoh, N., M.A. Sanchez C., W.C. Xu, K. Haraya and M. Hongo, 1993, J. Membr. Sci. **77**, 245.

Kameyama, T., M. Dokiya, M. Fujishige, H. Yokohawa and K. Fukuda, 1981, Ind. Chem. Eng. Fundam. **20**, 97.

Lillo, E., 1986, U.S. Patent 4,603,109.

Liu, P.K.T., 1995, private communication.

McCalliste, R.A., 1984, German Patent 3,331,202.

Michaels, A.S., 1989, presented at 7th ESMST Summer School, Twente University.

Michaels, J.N., C.G. Vayenas and L.L. Hegedus, 1986, J. Electrochem. Soc. **133**, 522.

Minet, R.G., and T.T. Tsotsis, 1991, U.S. Patent 4,981,676.

Mishchenko, A.P., V.M. Gryaznov, V.S. Smirnov, E.D. Senina, I.L. Parbuzina, N.R. Roshan, V.P. Polyakova and E.M. Savitsky, 1979, U.S. Patent 4,179,470.

Mohan, K., and R. Govind, 1986, AIChE J. **32**, 2083.

Mohan, K., and R. Govind, 1988a, AIChE J. **34**, 1493.

Mohan, K., and R. Govind, 1988b, Ind. Eng. Chem. Res. **27**, 2064.

Nakajima, M., N. Jimbo, K. Nishizawa, H. Nabetani and A. Watanabe, 1988, Process Biochem. **23**, 32.

Nourbakhsh, N., I.A. Webster and T.T. Tsotsis, 1989a, Appl. Catal. **50**, 65.

Nourbakhsh, N., A. Champagnie, T.T. Tsotsis .and I.A. Webster , 1989b, Appl. Catal. **50**, 65.

Okubo,, T., K. Haruta, K. Kusakabe, S. Morooka, H. Anzai and S. Akiyama, 1991, Ind. Eng. Chem. Res. **30**, 614.

Pfefferle, W.C., 1966, U.S. Patent 3,290,406.

Philipse, A.P., F.P. Moet, P.J. van Tilborg and D. Bootsma, 1989, EUROCERAMICS **3**, 600.

Richardson, J.T., S.A. Paritpatyadar and J.C. Chen, 1988, AIChE J. **34**, 743.

Rony, P.R., 1968, Sep. Sci. **3**, 239.

Saracco, G., and V. Specchia, 1994, Catal. Rev.-Sci. Eng. **36**, 305.

Satterfield, C.N., 1980, Mass Transfer in Heterogeneous Catalysis, Robert E. Krieger Pub., New York (USA).

Setzer, H.J., and A.C.W. Eggen, 1969, U.S. Patent 3,450,500.

Sudworth, J.D., and A.R. Tilley, 1985, The Sodium-Sulfur Battery, Chapman and Hall Pub., New York (USA).

Sun, Y.M., and S.J. Khang, 1990, Ind. Eng. Chem. Res. **29**, 232.

Sverdrup, E.F., C.J. Warde and R.L. Eback, 1973, Ener. Conv. **13**, 129.

Tekić, M.N., R.N. Paunović and G.M. Ćirić, 1994, J. Membr. Sci. **96**, 213.

Toei, R., K. Nakanishi, K. Yamada and M. Okazaki, 1975, J. Chem. Eng. Japan **8**, 131.

Tsotsis, T.T., A.M. Champagnie, R.G. Minet and P.K.T. Liu, 1993, Catalytic membrane reactors, in: Computer-Aided Design of Catalysts (E.R. Becker and C.J. Pereira, Eds), Marcel Dekker (USA).

Uchida, S., S. Osuda and M. Shindo, 1975, Can. J. Chem. Eng. **53**, 666.

Wai, P.C., and R.C. Furneaux, 1990, Applying mathematical modeling to cross flow filtration process design, in: Proc. Int. Congr. Membr. Membr. Processes, Chicago, IL, USA, p.744.

Wang, W., and Y.S. Lin, 1995, J. Membr. Sci. **103**, 219.

Wu, J.C.S., and P.K.T. Liu, 1992, Ind. Eng. Chem. Res. **31**, 322.

Wu, J.C.S., T.E. Gerdes, J.L. Pszczolkowski, R.R. Bhave, P.K.T. Liu and E.S. Martin, 1990, Sep. Sci. Technol. **25**, 1489.

Zaspalis, V.T., W. van Praag, K. Keizer, J.G. van Ommen, J.R.H. Ross and A.J. Burggraaf, 1991a, Appl. Catal. **74**, 205.

Zaspalis, V.T., W. van Praag, K. Keizer, J.G. van Ommen, J.R.H. Ross and A.J. Burggraaf, 1991b, Appl. Catal. **74**, 223.

CHAPTER 12

FUTURE TRENDS OF INORGANIC MEMBRANES AND MEMBRANE REACTORS

Inorganic membranes and their uses as membrane reactors have just grown from an infant stage, particularly the latter. They are currently limited in the choice of membrane material, pore size distribution, element shape, end-seal material and design and module configuration. But these limitations mostly reflect the degree of maturity in the technology development. As more research and development efforts are directed to these areas as witnessed worldwide today, inorganic membranes are expected to be more technically feasible and economically viable as separators and reactors in selected applications.

12.1 ECONOMIC ASPECTS

While technical data of inorganic membranes and inorganic membrane reactors are generated at an accelerated pace, very little information on their economics has been published. This can be attributed to two major factors: (1) Inorganic membrane technology is an emerging field such that new developments in its preparation or process development as separators or reactors can drastically change the capital investments and operating costs required; (2) Even in the more established applications, the process economics is in general fairly competitive and well guarded as a trade secret and has not been much discussed in the open literature.

Thus, it is not easy to obtain reliable cost data for separation processes, let alone catalytic reaction processes, using inorganic membranes. Some general guidelines, however, have been provided for separation processes in isolated cases and will be summarized in this chapter. Understandably no definitive economics related to inorganic membrane reactors has been presented in the literature due to the evolving nature of the technology.

12.1.1 Cost Components

Like all other chemical processes, the separation processes by inorganic membranes have two major cost issues: capital investments and operating costs. Capital costs are affected by the membrane area required (which in turn are determined by the permeability and permselectivity), compression or recompression energy requirements as dictated by the operating pressure, piping and vessels, instrumentation and control, and any pretreatment requirements (depending on the nature of the feed material and the membrane). The operating costs are determined by the required membrane replacement frequency and costs, membrane regeneration (e.g., backwashing, cleaning, etc.), system maintenance, pumping energy, labor, depreciation and any feed pretreatment.

There are two seemingly not so obvious cost elements that should be considered. One is capital and operating costs related to any compression or recompression requirements. The other element is product losses [Spillman, 1989].

Recompression may be an issue, for example, in the case of hydrogen separation. In the majority of cases, most of the hydrogen from a hydrogen-containing feed gas mixture appears in the permeate which is on the lower pressure side of the membrane. Most applications, however, require the hydrogen so obtained to be at high pressures for subsequent processing. Thus, in a case like this, recompression-related costs must be included for an equitable comparison and these costs sometimes are critical to the economic feasibility of the gas separation process.

In a more complete membrane process optimization or cost comparison of alternative membrane processes, the cost associated with any loss of the valuable gas component(s) needs to be considered as another operating expense just like any other utility expense. For example, in the case of the removal of carbon dioxide from natural gas (methane), there will be some amount of methane in the permeate (albeit in a small quantity) due to imperfect rejection of methane by the membrane. Since methane is, in this case, the desirable product component, its presence in the permeate represents a loss or an expense. The value of the loss should be included in any cost comparison.

The basic guiding principle for the economics of membrane or membrane reactor processes is the operating cost per unit of the desired product. If the unit cost is higher than those of the competitive separation or separation/reaction processes, economic justification will be difficult. The operating cost of a membrane unit strongly depends on its permeate flux and permselectivity. The permeate flux is, in turn, essentially proportional to the permeability and the available permeation area and inversely proportional to the membrane thickness. Among these factors, the available permeation area is the easiest one to effect the economy of scale in membrane processes, using either organic or inorganic membranes. Chan and Brownstein [1991] demonstrated that as the membrane filtration area increases from 100 to 1000 m^2 the average operating cost per unit filtration area is decreased by 1/3.

Even available cost information can not be easily generalized. However, for rough estimates in scaling up an inorganic membrane system, cost data from other similar applications using the same type of membrane and equipment may be applied with caution. For example, some cost data from the applications of ceramic membranes in the dairy industry has been utilized as a basis for cost estimating fish processing applications also using ceramic membranes [Quemeneur and Jaouen, 1991].

12.1.2 General Guidelines for Various Cost Components

Based on limited available literature information [e.g., Spillman, 1989; Chan and Brownstein, 1991; Muralidhara, 1991; Quemeneur and Jaouen, 1991; Johnson and Schulman, 1993], some guidelines or assumptions that are general in nature for ceramic membranes may be developed:

(1) The average service life of a ceramic membrane is approximately 3 to 5 years compared to 1 to 3 years for an organic polymer membrane although the service life can greatly depend on the module design and the nature of the feed streams (e.g., fouling propensity and corrosion tendency). The replacement cost of an inorganic membrane may be close to twice that of an organic membrane currently but is expected to be lower as the demand increases in the future. The unit cost of an inorganic membrane is highly sensitive to the production volume.

(2) Maintenance cost is 5 to 10% of capital investments or approximately $0.05 per 1,000 liter liquid processed.

(3) Straightline depreciation is usually assumed over a 5 to 10 year period.

(4) The total system cost is significantly larger than the membrane module cost, typically 3 to 5 times as much.

(5) For scale-up, the general power law applies. This can be expressed in terms of the available membrane permeation area:

$$C_2/C_1 = (A_2/A_1)^m \qquad (12\text{-}1)$$

where C and A are the membrane or system cost and the available membrane permeation area, respectively, and the exponent m is 0.9 for the membranes and 0.67 for the systems. An illustration of this rule of thumb has been given by Chan and Brownstein [1991]. They made a cost estimate for the case of increasing the filtration area from 100 to 1000 m^2. This tenfold increase in the throughput results in an increase of capital cost by only approximately 5 times. Part of the reason for this economy of scale is that at a low capacity the quantities of required instrumentation and control devices are not proportionally reduced and thus may constitute a significant portion of the capital costs.

An alternative approach to costing membrane systems for scale-up is based on the volumetric flow rate of the permeate:

$$C_2/C_1 = (V_2/V_1)^n \qquad (12\text{-}2)$$

where C and V are the membrane element or system cost and the maximum volumetric flow rate of the permeate, respectively, and the exponent n is 0.32 [Lahiere and Goodboy, 1993]. An example of the general trend described by Eq.

(12-2) is given in Figure 12.1 which shows the decreasing ceramic membrane
system cost with increasing module size for microfiltration applications. An
assumption implied in the above equation is that the density of the process streams
is constant.

(6) General breakdown of the operating costs (excluding return on investment) is
     given in Table 12.1. The depreciation cost is a strong function of the system
     capacity and the accounting practice and therefore can vary significantly from one
     case to another.

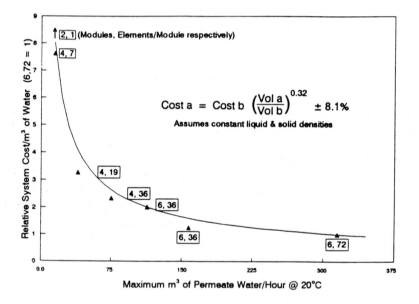

$$Cost\ a\ =\ Cost\ b\ \left(\frac{Vol\ a}{Vol\ b}\right)^{0.32}\ \pm\ 8.1\%$$

Assumes constant liquid & solid densities

Figure 12.1 General trend of ceramic membrane element or system cost as a function of
the volumetric permeation rate [Lahiere and Goodboy, 1993]

Not explicitly included in Table 12.1 is the cost associated with any pretreatment
required. For the extreme cases where the feed stream needs to be treated prior to
entering the membrane system to facilitate membrane separation and prevent frequent
shutdown, the pretreatment costs may be quite significant. Flux enhancement schemes
are critical to continuous and consistent operation of membrane separation systems.
They can add significant expenses to the total operating costs. For example, it has been
reported that backflushing can reduce fouling and result in a cost savings of $0.41 per
1,000 liter of processed liquid stream, mostly through savings in the membrane
replacement costs [Muralidhara, 1991]. In the absence of any prior cost knowledge for

extrapolation, the above cost-related guidelines may be used with caution for other applications employing ceramic membranes.

TABLE 12.1
Distribution of operating costs of inorganic membrane systems

Cost component	% of total cost
Cleaning products	4-8
Depreciation	10-40
Power	10-15
Labor	10-20
Maintenance	3-10
Membrane replacement	10-20

12.1.3 High-Temperature Gas Separations and Membrane Reactors

When the application temperature increases as in many gas separation and membrane reactor applications, additional capital as well as operating costs associated with fluid containment can be expected. For many such high-temperature applications, inorganic membranes have the distinct advantages over other separation processes including organic membrane separation in the savings of recompression costs. Quite likely for those cases where recompression of the permeate or the product stream is needed, the recompression costs may be close to or more than the operating costs of the membrane. As a preliminary estimate for the cost of gas separation in the petroleum refining industry (e.g., hydrogen recovery) using inorganic membranes, the unit cost of 13¢ per million standard cubic feet (Mscf) may be used [Johnson and Schulman, 1993].

When combining the membrane and the reactor into an integrated system, cost savings may be significant. This difference can be attributed to the convection and other transport modes in the separation steps by membranes in contrast to the traditional diffusion mode by other separation techniques. For example, the use of membrane reactors instead of the traditional reactors for bioengineered products can reduce the operating costs by as much as 25% [Chan and Brownstein, 1991].

For high-temperature processes where inorganic membranes may be employed, the potential energy savings are considerable. Humphrey et al. [1991] made an estimate on the energy savings that may be realized if inorganic membrane reactors are used in a number of large-volume dehydrogenation and dehydration reactions. Potential industrially important dehydrogenation applications include ethylene from ethane, propylene from propane, styrene from ethylbenzene, hydrogen from hydrogen sulfide, butadiene from butene and finally benzene, toluene and xylene (BTX) from cycloalkanes with their possible energy savings in the descending order. The energy savings in the above cases primarily come from reduced reactor preheat and elimination of distillation recovery of unconverted reactants. Target dehydration reactions are ketene from acetic acid and diethylether from ethanol. The combined energy savings of the above dehydrogenation and dehydration reactions total about 0.309 quad (10^{15} Btu) per year for the U.S. alone (see Table 12.2). The authors assumed that the energy values are

TABLE 12.2
Potential energy savings via membrane reactor technology

Application	Reasons for energy reductions	Energy savings (Btu/lb product)	U.S. production (billion lb/yr)	Total national energy savings (quad/yr)
Dehydrogenation:				
Styrene from ethylbenzene	Reduced reaction temperature, elimination of ethylbenzene recovery	4,095	8.13	0.033
Ethylene from ethane	Reduced reaction temperature, elimination of ethane recovery	4,655	34.95	0.163
Propylene from propane	Reduced reaction temperature, elimination of propane recovery	2,794	20.23	0.057
Butadiene from butene	Reduced reaction temperature, elimination of the butene recovery	4,375	3.09	0.014
BTX from cycloalkanes	Eliminating need to recycle (via compressor) hydrogen, reduced reaction temperature	1,128	9.32	0.011
Hydrogen from hydrogen sulfide	Producing hydrogen as a by-product	60,000	0.49	0.029
Dehydration:				
Ketene from acetic acid	Reduced reaction temperature, elimination of recovery of ketene	1,781	0.64	0.001
Diethylether from ethanol	Elimination of an azeotropic distillation tower	65,000	0.012	0.001
	Total energy savings			0.309

(Adapted from Humphrey et al. [1991])

1,000 Btu/scf for natural gas, 3,412 Btu/kwh for electricity, 1,000 Btu/lb for steam and 60,000 Btu/lb for hydrogen and 100% of the reactants are converted to products. The last assumption may not be validated in the foreseeable future. Nevertheless, the magnitude of the potential energy savings is reasonable.

Humphrey et al. [1991] further analyzed the amount of energy that can be saved by partially or completely replacing distillation operations with membrane technologies which encompass both gas separation and membrane reactor applications. A significant portion of the estimated savings will be derived from the usage of inorganic membranes. Given in Table 12.3 are the potential energy savings progression due to the use of membrane technologies as they evolve and also the estimated energy savings per installation. The potential energy savings are substantial and the savings per installation become much larger if distillation for difficult separations is completely replaced by the membranes. In their estimates, assumptions were made that: (1) distillation-membrane hybrid systems will become materialized within fifteen years (some are within five or ten years); (2) complete replacement of distillation for difficult separations (where low relative volatility is involved) with membranes will take fifteen years (applicable to new plants); and (3) one-half of the targeted membrane reactor applications for dehydrogenation and dehydration reactions can be realized within fifteen years (mostly applicable to new plants).

TABLE 12.3
Energy savings of membrane technologies

Technology	Quads/year saved at end of			Applications	Estimated energy savings per installation
	5 yrs	10 yrs	15 yrs		
Distillation-membrane hybrid systems	0.02	0.03	0.07	Ethanol and isopropanol separations - all applications realized within five years. Olefins, miscellaneous hydrocarbons, water-separations - one-half of applications realized within ten years and all within fifteen years	34%

TABLE 12.3 – Continued

Technology	Quads/year saved at end of			Applications	Estimated energy savings per installation
	5 yrs	10 yrs	15 yrs		
Membranes to replace distillation	0	0	0.18	To replace distillation for difficult separations. Olefins, miscellaneous hydrocarbons, water-oxygenated hydrocarbons, and aromatic separations. All applications realized within fifteen years. Applicable to new plants.	84%
Membrane reactors	0	0	0.16	Dehydrogenation and dehydration reactions. One-half applications realized within fifteen years. Mostly applicable to new plants.	83%

(Adapted from Humphrey et al. [1991])

12.2 POTENTIAL MARKETS

As stated earlier, the field of inorganic membranes for separation and reaction is evolving and both process and product R&D can significantly change the production costs, process economics or pricing. Therefore, any projection of the market sizes in the years ahead can be speculative.

However, based on certain assumptions of some technological breakthroughs, the market sizes of inorganic membranes have been estimated [Business Communications Co., 1994] and an example is shown in Table 12.4. It can be seen that ceramic membranes dominate the market of inorganic membranes currently and in the next decade. At the present, many inorganic membranes are largely used in the food and beverage sector. But greater growth is expected in other applications. It is estimated that the average annual growth rates in the U.S. alone within the next ten years of inorganic membrane usage in the food/beverage, biotechnology/pharmaceutical and environmental sectors are 13, 19 and 30%, respectively [Business Communications Co., 1994]. By the year 2003, the expected market size in the environmental sector will almost match that in the industrial processing sector.

The above market projection hinges on the assumption that critical technical hurdles will be removed over time. Some of the breakthroughs required before the market of inorganic membranes can reach the estimated size are discussed below.

TABLE 12.4
Market sizes of inorganic membranes in the U.S. and worldwide

Year	U.S. market ($million)		World market ($million)	
	Inorganic	Ceramic	Inorganic	Ceramic
1989			32	19
1993	39	27		
1994			108	
1998	112			
1999			546	~440
2003	288	249		

12.3 MAJOR TECHNICAL HURDLES

It is obvious that some major technical issues are related to the preparation of more permselective and robust inorganic membranes. In addition, system packaging and installation and operating knowledge also await to be advanced to realize full potentials of inorganic membranes for both separation and reaction applications. While breakthroughs in one of the above developments can accelerate those in the others, only co-current advancements in all fronts can facilitate the acceptance of inorganic membranes by potential users. History of the membrane technology has shown that the process of gaining acceptance to replace other separation technologies with membranes is often accompanied by persistent demonstration of values and continuous technology evolution.

12.3.1 Synthesis of Inorganic Membranes with High Permselectivities and Permeabilities

Dense palladium and palladium alloy membranes have been repeatedly demonstrated to show extremely high selectivities of hydrogen and certain solid electrolyte membranes made of metal oxides such as stabilized zirconia exhibit very high permselectivities of oxygen. However, they all suffer from low permeabilities which are directly related to throughput of a separation or reaction/separation process. In fact, their permeabilities are often at least an order of magnitude lower than those of porous membranes. Some developmental efforts are being contemplated to improve the permeation fluxes of dense inorganic membranes. One pursuit is to deposit a very thin dense inorganic membrane layer on a porous support. Similar to the concept of a porous composite membrane, this approach segregates two of the most important functions of a membrane into two separate layers: an integral, defect-free permselective layer (membrane) and a non-permselective porous support layer for mechanical strength. The thin membrane layer assumes most of the permeate flow resistance, but, due to its very small thickness, a relatively high permeability is still possible for a given transmembrane pressure difference. The challenge, however, remains in the manufacturing of a very thin and yet defect-free layer on a porous support. The quality of the porous support surface becomes an even more critical issue for dense membranes that deposit on the support.

To make an inorganic membrane an economically attractive separation medium or reactor, a high permselectivity imparted by the membrane to the feed components is not only desirable but necessary in many cases. Size-selective separation is generally expected to be favored for achieving a high permselectivity. To this end, it has been suggested that inorganic ultramicroporous membranes with mean pore diameters smaller than approximately 0.8 nm are needed [Armor, 1992].

Inorganic membranes with pore sizes in that range are still a great challenge particularly from the standpoint of defect-free membranes which are essential in many separation and membrane reactor applications. The synthesis and assembly of these membranes will continue to be a major hurdle before successful commercial implementation of inorganic membrane gas separation and reactors can be realized. Some proof-of-concept studies have been conducted successfully on the preparation of these ultramicroporous inorganic membranes. For example, as mentioned in Chapter 3, zeolite membranes have been prepared by precipitation of zeolite inside the pores of mesoporous inorganic membranes. Key questions, however, still remain as to how the membranes can be produced and assembled consistently on a large scale.

12.3.2 High Packing-Density Inorganic Membranes

Current commercial inorganic membranes come in a limited number of shapes: disk, tube and monolithic honeycomb. Compared to other shapes such as spiral-wound and hollow-fiber that are available to commercial organic membranes, these types of membrane elements have lower packing densities and, therefore, lower throughput per unit volume of membrane element or system.

There have been some attempts to address this issue. However, the size of the inorganic fibers or spiral wound unit is still quite larger than that of the organic counterparts. To be more competitive, the packing density of inorganic membranes preferably exceeds 1,000 m^2/m^3.

12.3.3 Prevention or Delay of Flux Decline

Permeation flux of a membrane declines with run time as a result of concentration polarization or fouling in separation processes or coking in catalytic reaction processes. Fouling is frequently observed to decrease the flux of porous organic or inorganic membranes. This problem occurs most often in liquid-phase systems across various industries from chemical to petroleum to pharmaceutical applications. Dehydrogenation reactions are the most promising application area for inorganic membrane reactors; however, the removal of hydrogen also tends to promote coking in the membrane reactors. The resulting coking not only can deactivate the catalyst but also block membrane pores.

In addition to slowing down the fouling rate on a membrane, timely regeneration of a slightly fouled porous membrane is also very important and necessary in many cases. Flow dynamics of the permeate as well as the foulant passing across the membrane surface and through the membrane layer plays a key role. Furthermore, experimental studies and theoretical analyses on backflushing (or backpulsing) improve the design and efficiency of membrane regeneration systems. The operating conditions and the feed compositions in a membrane reactor have great influences over the coking propensity. Thermal and/or chemical cleaning are also vital in preserving the service life of an inorganic membrane. Care needs to be exercised to address the issue of chemical compatibility between the membrane material and the cleaning chemical and conditions.

Despite the aforementioned efforts, membrane flux decline due to fouling continues to be a major operational issue. Attempts have been made to modify inorganic membranes, mostly their surfaces, to render them less prone to foulant adsorption. One of the frequently encountered fouling problems in biotechnology and food applications is protein adsorption. In membrane reactor applications which are largely associated with hydrocarbons, carbonaceous deposits pose as one of the operational problems.

Novel designs may be required to further prolong the run time cycle between the thermal and chemical cleaning cycles.

12.3.4 Fluid Containment at High Temperatures

Critical to both gas separation and membrane reactor applications, fluid leakage and any potential re-mixing of the separated species have to be avoided. The problems could arise if pin-holes or structural defects exist or if the ends of the membrane elements or the connections between the membrane elements and assembly housings or pipings are not properly sealed.

Many methods have been proposed to address this issue (see Chapter 9). Beside thermal and chemical resistances of the sealing materials, other issues need to be considered as well. One such important issue is the mismatch of the thermal expansion coefficients between the membrane element and the sealing material or joining material. While similar material design and engineering problems exist in ceramic, metal and ceramic-metal joining, developmental work in this area is much needed to scale up gas separation units or membrane reactors for production. The efforts are primarily performed by the industry and some national laboratories.

12.4 OUTLOOK OF INORGANIC MEMBRANE TECHNOLOGY

Twenty some years after the massive installations of porous inorganic membranes for production of enriched uranium started, commercial porous alumina membranes were introduced into the marketplace and became the first major commercial inorganic membrane. It has been about a decade since the market entry that inorganic membranes are starting to become a contender in selected liquid-phase separation applications. Early applications were focused on the food and beverage industries with encouraging successes. Later application developments spread into other uses such as produced water, waste or recycle water treatment and clean air filtration. Inorganic membranes are also being seriously considered for future petroleum and petrochemical processing. This is evident by a number of major chemical and ceramic companies currently involved in marketing as well as developing new porous inorganic membranes.

An anticipated major application of inorganic membranes is high-temperature gas separation which requires, in addition to acceptable membrane materials and geometries, compatible sealing materials and methods for the ends of the membrane elements and the connection between the elements and assembly housings or pipings. Moreover, within the range of commercial and developmental inorganic membranes available today, Knudsen diffusion appears to be the dominant separation mechanism and offers only limited permselectivities in many gas separation requirements. To take advantage of other potentially useful transport mechanisms for gas separation purposes, membranes with pore sizes finer than 3 to 4 nm range or with controlled pore modifications are needed. Promises have begun to appear in bench developments of these ultramicroporous membranes.

No commercial installations of inorganic membrane reactors are in operation today. This technology has the most potential in high-temperature reversible dehydrogenation, hydrogenation and dehydration reactions. Their potential payoffs are widely viewed as very high and are considered high-priority research and development areas by government agencies. However, the high risks associated with the current status of the technology keep the industry from major R&D investments at this time. On the contrary, the breadth and depth of academic research in this area have increased immensely in recent years.

Traditionally, acceptance of organic membranes as a substitute for other separation processes has been slow. A major reason for this hesitation is the general uncertainty associated with the scale-up of the membrane separation process relative to that of the well established bulk separation processes such as distillation, adsorption and others. This mindset is likely to prevail for some time before critical mass of the acceptance of inorganic membranes occurs, particularly for the inorganic membrane reactor technology. The ongoing gradual acceptance of inorganic membranes in several liquid-phase applications will be followed by that in gas- and vapor-phase separations which may be five to ten years away from key breakthroughs. The inorganic membrane reactors will probably not be a significant unit operation until one or two decades from now. This, of course, is contingent upon the development of inorganic membranes capable of making precise gas separations at a cost effective permeation flux level and the resolution of the technical hurdles previously described.

12.5 SUMMARY

Like many other large-scale chemical processes, the economics of inorganic membrane separation and reaction processes is very sensitive to capital investments and operating costs. While little accurate cost data is available for extrapolation, some guidelines on various cost components have been extracted from the literature for rough cost estimates. Caution should be exercised to use the data for comparison purposes.

While it is generally believed that high-temperature gas separation and catalytic reactors using inorganic membranes can benefit from significant energy savings, reliable cost information in these areas is practically not available, probably due to the emerging nature of the technology.

Based on certain assumptions of some technological breakthroughs, the market size of inorganic membranes have been projected to reach about six million dollars (US) worldwide by the year of 2000. Ceramic membranes are expected to continue to dominate the inorganic membrane market within the next decade. While the food and beverage applications currently are the largest usage of inorganic membranes, the environmental applications will grow to become a major force within the next decade.

Breakthroughs are needed to overcome three major technical hurdles: synthesis of inorganic membranes with high permselectivities and permeabilities, prevention and delaying of permeate flux decline due to fouling or coking, and fluid containment at high temperatures. Some key efforts are identified.

Finally, the current status of the inorganic membrane technology is summarized for an overall perspective. The future is speculated based on that perspective to provide a framework for future developments in the synthesis, fabrication and assembly of inorganic membranes and their uses for traditional liquid-phase separation, high-temperature gas separation and membrane reactor applications.

REFERENCES

Armor, J.N., 1992, Chemtech **22**, 557.

Business Communications Co., 1994, "Inorganic membranes: markets, technologies, and players" Report No. A6GB-112R.

Chan, K.K., and A.M. Brownstein, 1991, Ceram. Bull. **70**, 703.

Humphrey, J.L., A.F. Selbert and R.A. Koort, 1991, Separation technologies - advances and priorities, Final report (Report No. DOE/ID/12920-1) to U.S. Dept. of Energy.

Johnson, H.E., and B.L. Schulman, 1993, Assessment of the Potential for Refinery Applications of Inorganic Membrane Technology -- An Identification and Screening Analysis, Report DOE/FE-61680-H3 for U.S. Department of Energy.

Lahiere, R.J., and K.P. Goodboy, 1993, Env. Progr. **12**, 86.

Muralidhara, H.S., 1991, Key Eng. Mat. **61&62**, 301.

Quemeneur, F., and P. Jaouen, 1991, Key Eng. Mat. **61&62**, 585.

Spillman, R.W., 1989, Chem. Eng. Progr. **85**, 41.

APPENDIX

INORGANIC MEMBRANE MANUFACTURERS

ORGANIZATION	LOCATION	PRODUCTS
Anotec/Whatman	Anotec Operations & Development Unit Whatman Scientific Ltd. Beaumont Road Banbury Oxon, OX16 7RH UK Whatman Inc. 9 Bridewell Place Clifton, NJ 07014	Ceramic disk filters MF and UF
Asahi Glass	2-1-2 Marunouchi Chiyoda-ku, Tokyo 100 Japan 1185 Avenue of the Americas 20th Floor New York, NY 10036 U.S.A.	Tubular SiO_2 based microfilters
Carbone-Lorraine	Le Carbone Lorraine Tour Manhattan Cedex 21, F-92095 Paris La Defense 2 France Carbone-Lorraine Industries Corporation 400 Myrtle Avenue Boonton, NJ 07005 U.S.A.	Tubular MF and UF carbon membranes
Carre/Du Pont	Du Pont Separation Systems; Glasgow Wilmington, DE 19898 U.S.A.	Dynamic ceramic and composite membranes on tubular support

Ceramem	Ceramem Corporation 12 Clematis Avenue Waltham, MA 02154 U.S.A.	Honeycomb multi-channel ceramic membranes on micro-porous cordierite support
Ceramesh	Ceramesh Ltd. Unit 1 Edward Street Business Centre Thorpe Way Banbury Oxon OX16 2SA UK	Flat sheet Zr-Inconel membranes
Ceram Filtre	FITAMM B.P. 60 34402, Lunel Cedex, France	Tubular SiC membranes for MF and UF
Du Pont	Du Pont Separation Systems; Glasgow Wilmington, DE 19898 U.S.A.	PRD-86 spiral-wound ceramic microfilters
Fairey	Fairey Industrial Ceramics, Ltd. Filleybrooks Stone Staffs ST15 OPU UK	Porous tubular and disk ceramic microfilters
Fuji Filters	Fuji Filter Co., Ltd. 2-4 Nihonbashi-Muromachi Chuo-ku Tokyo 103 Japan	Porous tubular glass glass filters for MF and UF
Gaston County	Gaston County Filtration Systems P.O. Box 308 Stanley, NC 28164 U.S.A.	Dynamic ZrO_2 membranes on carbon tubes for UF
Hoogovans	Hoogovans Industrial Ceramics BV Postbus 10000 1970 CA Ijmuiden The Netherlands	Tubular ceramic membranes

Imeca/CTI	Imeca B.P. 94 34800 Clermont L'Herault France	Tubular ceramic membranes for MF
	CTI B.P. 12 36500 Buzancais France	
Kubota	2-47 Shikitsu-higashi Osaka 0556 Japan	Tubular composite Al_2O_3 membranes for MF and UF
Mott	Mott Metallurgical Corporation Farmington Industrial Park Farmington, CT 06032 U.S.A.	Disk and tubular metallic microfilters
NGK	NGK Insulators, Ltd. Shin Maru Building 1-5-1, Marunouchi Chiyodo-ku, Tokyo 100 Japan	Tubular and multichannel ceramic MF and UF membranes
Nihon Cement	Nihon Cement Co. Ltd. 6-1 Otemachi 1-Chome Tokyo, 0100 Japan	Tubular and disk alumina membranes
NOK	NOK Corp. Seiwa Building, 12-15 Shiba Tokyo, 0105 Japan	Tubular alumina membranes
Osmonics	Osmonics, Inc. 5951 Clearwater Drive Minnetonka, MN 55343 U.S.A.	Ceratrex™ ceramic microfilters (disk and tubes)
Pall	Pall Porous Metal Filters Courtland, NY 13045 U.S.A.	Disk and tubular metallic microfilters

Poral/Pechiney	Voie des collines 38800 Le Pont de Claix France	Metallic and composite ceramic tubes for MF and UF
Schott	Schott Glaswerke Postfach 2480 D-6500 Maniz 1 Germany	Bioran® tubular SiO2- based ultrafilters
SCT/U.S. Filter	Societe Ceramiques Techniques Usine de Bazet B.P. 113 650001, Tarbes France U.S. Filter 181 Thorn Hill Road Warrendale, PA 15086 U.S.A.	Membralox® tubular and multichannel ceramic MF and UF membranes
Tech Sep- Rhone Poulenc (formerly SFEC)	Boite Postale No. 201 84500 Bollene France	Carbosep® ceramic MF and UF tubular membranes
TOTO	Toto Co., Ltd. 1-1 Nakajima 2-Chome Kita Kyashu 0802 Japan	Tubular composite ceramic filters for MF and UF

KEY WORD INDEX